Algorithms and Computation in Mathematics · Volume 9

Editors

Arjeh M. Cohen   Henri Cohen
David Eisenbud   Michael F. Singer   Bernd Sturmfels

T0207236

Sergei Matveev

Algorithmic Topology and Classification of 3-Manifolds

Second Edition

With 264 Figures and 36 Tables

 Springer

Author

Sergei Matveev

Chelyabinsk State University
Kashirin Brothers Street, 129
Chelyabinsk 454021
Russia
E-mail: matveev@csu.ru

Mathematics Subject Classification (2000): 57M

ISSN 1431-1550
ISBN 978-3-642-07960-3   e-ISBN 978-3-540-45899-9

Springer is a part of Springer Science+Business Media
springer.com
© Springer-Verlag Berlin Heidelberg 2007
Softcover reprint of the hardcover 2nd edition 2007

Cover design: *design & production* GmbH, Heidelberg

Preface to the First Edition

The book is devoted to algorithmic low-dimensional topology. This branch of mathematics has recently been undergoing an intense development. On the one hand, the exponential advancement of computer technologies has made it possible to conduct sophisticated computer experiments and to implement algorithmic solutions, which have in turn provided a motivation to search for new and better algorithms. On the other hand, low-dimensional topology has received an additional boost because of the discovery of numerous connections with theoretical physics.

There is also another deep reason why algorithmic topology has received a lot of attention. It is that a search for algorithmic solutions generally proves to be a rich source of well-stated mathematical problems. Speaking out of my experience, it seems that an orientation towards "how to" rather than just "how is" serves as a probing stone for choosing among possible directions of research – much like problems in mechanics led once to the development of calculus.

It seemed to me, when planning this book, that I had an opportunity to offer a coherent and reasonably complete account of the subject that nevertheless would be mainly accessible to graduate students. Almost all parts of the book are based upon courses that I gave at different universities. I hope that the book inherits the style of a live lecture. Elementary knowledge of topology and algebra is required, but understanding the concepts of "topological space" and "group" is quite sufficient for most of the book. On the other hand, the book contains a lot of new results and covers material not found elsewhere, in particular, the first proof of the algorithmic classification theorem of Haken manifolds. It should be therefore useful to all mathematicians interested in low-dimensional topology as well as to specialists.

Most of the time, I consider 3-manifolds without geometric structures. There are two reasons for that. On the one hand, I prefer to keep the exposition within the limits of elementary combinatorial approach, which is a natural environment for considering algorithmic questions. On the other hand, geometric approach, which become incorporated into 3-manifold topology after works of Thurston, is presented in mathematical literature comprehensively [62, 100, 120, 121]. Also, I touch upon computer investigation of hyperbolic manifolds only briefly, since this subject is completely covered by the outstanding program SNAPPEA of Weeks [44].

When I embarked on this project I had not intended the writing of this book to take so long. But once the ground rules were set up, I had no choice. The rules are:

1. The book should be maximally self-contained. Proofs must be complete and divided into "easy-to-swallow" pieces. Results borrowed from other sources must be formulated exactly in the same form as they appear there, and no vague references to technique or proofs are allowed. The sources must be published in books or journals with world-wide circulation.
2. The text should be written so that at each moment the reader could know what is going on. In particular, statements must always precede proofs, not vice versa. Also, proofs must be straightforward, whenever possible.
3. Different parts of the book (sections, subsections, and even individual statements) must be as independent as possible, so that one could start reading the book from any point (after taking a few steps backwards or looking at the index, if necessary, but without having to read everything else before).
4. Each mathematical text is a cipher to encode information. The main difficulty for the reader often lies not in understanding the essence of a statement, but in decoding what the authors really want to say. Therefore, each paragraph should contain a redundant amount of information to avoid misunderstanding. In particular, rephrasing is welcome.

Consistently with the algorithmic viewpoint, the book begins with a comprehensive overview of the theory of special spines. The latter encode 3-manifolds in a rather comfortable, user-friendly, way, which also is easily turned into a computer presentation. This chapter also contains an existence result giving a criterion for two special polyhedra to be spines of the same manifold.

While special spines allow us to work with a single manifold, the theory of complexity (Chap. 2) attempts to overview the whole set of 3-manifolds introducing an order into their chaos. Specifically, the set of 3-manifolds is supplied with a filtration by finite subsets (of 3-manifolds of a bounded complexity), and this allows us to break up the classification problem for all 3-manifolds into an infinite number of classification problems for some finite subsets. This approach is implemented in Chap. 7, where we describe a way to enumerate manifolds of a given complexity. To be precise, we describe a computer program that enumerates manifolds and conducts a partial recognition. The final recognition is done by computing first homology groups and Turaev–Viro invariants. The latter are described in Chap. 8; the exposition was intentionally made as elementary and prerequisite-free as possible. The resulting tables of manifolds up to complexity 6, of their minimal special spines, and of values of their Turaev–Viro invariants are given in Appendix.

Chapter 3 contains the first ever complete exposition of Haken's normal surface theory, which is the cornerstone of algorithmic 3-dimensional topology. Almost all known nontrivial algorithms in low-dimensional topology use it or at least are derived from it. Several key algorithms are included into Chap. 4. In Chap. 5 the Rubinstein–Thompson algorithm for recognition of S^3 is presented. My approach is in a sense dual to the original proof of Thompson and seems to me more transparent.

Chapter 6 is the central part of the book. There, I prove the algorithmic classification theorem for Haken manifolds. Surprisingly, although already in 1976 it was broadly announced that the theorem is true [58, 59, 131], no proof appeared until 1995. Moreover, it turned out that the ideas described in the above-mentioned survey papers, in Hemion's book [42], and in other sources are insufficient. I prove the theorem using some facts from the algorithmic version of the Thurston theory of surface homeomorphisms [9].

For closed Haken manifolds other proofs of the algorithmic classification theorem can be composed now. They are based on Thurston's hyperbolization theorem for Haken manifolds containing no essential tori and annuli and on Sela results on algorithmic recognition of hyperbolic manifolds [114, 115]. A brief survey is contained in [73], where the author explains how a hyperbolic structure on a 3-manifold can be constructed algorithmically once one is known to exist.

This book began with several lectures that I gave first at Tel-Aviv University in October–December 1990 and then at the Hebrew University of Jerusalem in October 1991–January 1992. The lectures were extended to a lecture course that I read in 1993 at the University of British Columbia as a part of the Noted Scholar Summer School Program. Chapters 1, 2, and 8 are based on the notes taken by Djun Kim and Mark MacLean; my thanks to them. Later on I returned regularly to these notes and extended them by including new parts, once I had found what I hoped was the right way to expose them. Lectures on the subject which I gave at Pisa University in 1998, 2002 and in the J.-W. Goethe Univesität Frankfurt in April–May 2000 were especially valuable for me; you always profit when an attentive audience wishes to understand all details and forces the lecturer to find the most precise arguments. I hope I have managed to capture the spirit of those lectures in this text.

Many mathematicians have helped me during my work on the book. Among them are M. Boileau, A. Cavicchioli, N. A'Campo, I. Dynnikov, M. Farber, C. Hog-Angeloni, A. Kozlowski, W. Metzler, M. Ovchinnikov, E. Fominykh, E. Pervova, C. Petronio, M. Polyak, D. Rolfsen, A. Shumakovich, M. Sokolov, A. Sossinsky, H. Zieschang.

There are two more persons I wish to mention separately, my teacher A. Chernavskii, who introduced me to the low-dimensional topology, and my colleague A. Fomenko, whose outstanding personality and mathematical books influenced significantly the style of my thinking and writing.

A great part of this book was written during my stay at Max-Planck-Institut für Mathematik in Bonn. I am grateful to the administration of the institute for hospitality and for a friendly and creative atmosphere. I also thank the Russian Fund of Basic Research and INTAS for the financial support of my research.

Of course, the book could not even have been started without the encouraging support of Chelyabinsk State University, which is my home university. I am profoundly grateful to all my colleagues for their help. Special thanks to my beloved wife L. Matveeva, who is also a mathematician, for her patience and help.

Chelyabinsk,   *Sergei Matveev*
                                                                              March 2003

Preface to the Second Edition

The book has been revised, and some improvements and additions have been made. In particular, in Chap. 7 several new sections concerning applications of the computer program "3-Manifold Recognizer" have been included.

March 2007   *Sergei Matveev*

Contents

1

Simple and Special Polyhedra

1.1 Spines of 3-Manifolds

We wish to study the geometry and topology of 3-manifolds. To this end we will need the central notion of *spine* of a 3-manifold. Indeed, we will be able to refine the notion of spine to get a class of spines that give us a natural presentation of 3-manifolds.

1.1.1 Collapsing

In order to discuss spines, we need to define precisely *collapsing*. We start with the definition of an *elementary simplicial collapse*.

Let K be a simplicial complex, and let $\sigma^n, \delta^{n-1} \in K$ be two open simplices such that σ is *principal*, i.e., σ is not a proper face of any simplex in K, and δ is a *free* face of it, i.e., δ is not a proper face of any simplex in K other than σ.

Definition 1.1.1. *The transition from K to $K \backslash (\sigma \cup \delta)$ is called an* elementary simplicial collapse, *see Fig. 1.1.*

Definition 1.1.2. *A polyhedron P collapses to a subpolyhedron Q (notation: $P \searrow Q$) if for some triangulation (K, L) of the pair (P, Q) the complex K collapses onto L by a sequence of elementary simplicial collapses.*

In general, there is no need to triangulate P to construct a collapse $P \searrow Q$; for this purpose one can use larger blocks instead of simplexes. It is clear that any n-dimensional cell B^n collapses to any $(n-1)$-dimensional face $B^{n-1} \subset \partial B^n$. It follows that the collapse of P to Q can be performed at once by removing pairs of cells. Let $P = Q \cup B^n$, $P \cap B^n = B^{n-1}$, where B^n is an n-cell and B^{n-1} is an $(n-1)$-dimensional face of B^n.

Definition 1.1.3. *The transition from P to Q is called an elementary poly-hedral collapse, see Fig. 1.2.*

Fig. 1.1. Elementary simplicial collapse

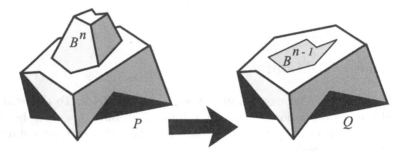

Fig. 1.2. Elementary polyhedral collapse

It is easy to see that an elementary simplicial collapse is a special case of an elementary polyhedral collapse. Likewise, it is possible to choose a triangulation of the ball B^n such that the collapse of B^n onto its face B^{n-1} can be expressed as a sequence of elementary simplicial collapses. It follows that the same is true for any elementary polyhedral collapse.

By a *simplicial collapse* of a simplicial complex K onto its subcomplex L we mean any sequence of elementary simplicial collapses transforming K into L. Similarly, a *polyhedral collapse* is a sequence of elementary polyhedral collapses.

1.1.2 Spines

Definition 1.1.4. *Let M be a compact connected 3-dimensional manifold with boundary. A subpolyhedron $P \subset M$ is called a* spine *of M if $M \searrow P$, that is, M collapses to P. By a spine of a closed connected 3-manifold M we mean a spine of $M \setminus Int B^3$ where B^3 is a 3-ball in M. By a spine of a disconnected 3-manifold we mean the union of spines of its connected components.*

Remark 1.1.5. A simple argument shows that any compact triangulated 3-manifold M always possesses a spine of dimension ≤ 2. Indeed, let M collapse to a subcomplex K. If K contains a 3-simplex, then K contains a 3-simplex with a free face, so the collapsing can be continued.

Fig. 1.3. The mapping cylinder and the cone

It is often convenient to view 3-manifolds as *mapping cylinders* over their spines and as regular neighborhoods of the spines. Theorem 1.1.7 justifies these points of view. We first recall the definition of a mapping cylinder.

Definition 1.1.6. *Let* $f : X \to Y$ *be a map between topological spaces. The mapping cylinder* C_f *is defined as* $Y \cup (X \times [0,1])/ \sim$, *where the equivalence relation is generated by identifications* $(x,1) = f(x)$. *If* Y *is a point, then* C_f *is called the* cone *over* X. *See Fig. 1.3.*

Theorem 1.1.7. *The following conditions on a compact subpolyhedron* $P \subset$ Int M *of a compact 3-manifold* M *with boundary are equivalent:*

(a)   *P is a spine of* M
(b) M *is homeomorphic to a regular neighborhood of* P *in* M
(c) M *is homeomorphic to the mapping cylinder of a map* $f: \partial M \to P$
(d)   *The manifold* $M \setminus P$ *is homeomorphic to* $\partial M \times [0,1)$

Proof. (a)$\Rightarrow$(b). This implication is valid in view of the following property of a regular neighborhood of P in M: it is a submanifold of M that can be collapsed onto P, see Corollary 3.30 of [110].

(b)$\Rightarrow$(c) Let a pair (K,L) of simplicial complexes triangulate the pair (M,P). Denote by $St(L,K'')$ the star of L in the second barycentric subdivision K'' of K. According to the theorem on regular neighborhoods [110], M can be identified with the underlying space $N = | St(L,K'') |$ of the star. The possibility of representing the manifold N in the form of the cylinder of a map $f : \partial N \to P$ is one of the properties of the star.

(c)$\Rightarrow$(d). This implication is obvious.

(d)$\Rightarrow$(a). Suppose the manifold $M \setminus P$ is homeomorphic to $\partial M \times [0,1)$. Denote by N a small regular neighborhood of P in M. Since we have proved the implications (b)$\Rightarrow$(c)$\Rightarrow$(d), we can apply them to N. Therefore the manifold $N \setminus P$ is homeomorphic to $\partial N \times [0,1)$. Note that the embedding of $N \setminus P$ into $\partial M \times [0,1)$ is proper in the following sense: the intersection of any compact set $C \subset \partial M \times [0,1)$ with $N \setminus P$ is compact. In this case the manifold $Cl(M \setminus N)$ is homeomorphic to $\partial N \times I$. Since $\partial N \times I \searrow \partial N \times \{0\}$ and $N \searrow P$, it follows that $M \searrow P$. $\square$

1.1.3 Simple and Special Polyhedra

A spine of a 3-manifold M carries much information about M. In particular, if $\partial M \neq \emptyset$ then any spine P of M is homotopy equivalent to M and
hence determines the homotopy type of M. Nevertheless, it is possible for
two nonhomeomorphic manifolds to have homeomorphic spines. The simplest
way to see this is to think about the 2-dimensional annulus and the Möbius
strip. Both of these 2-manifolds collapse to a circle, yet they are clearly not
homeomorphic. To get a 3-dimensional example, it is sufficient to multiply the
annulus and the Möbius strip by the segment I. We obtain a solid torus and
a solid Klein bottle, which have circles as spines.

In order to eliminate this difficulty, we will restrict our class of spines to
those called *special spines*. We will give a precise definition shortly afterward.
First we must define the notion of *simple polyhedron*.

Definition 1.1.8. *A compact polyhedron P is called* simple *if the link of each
point $x \in P$ is homeomorphic to one of the following 1-dimensional polyhedra:*

(a) *A circle (such a point x is called* nonsingular*)*
(b) *A circle with a diameter (such an x is a* triple point*)*
(c) *A circle with three radii (such an x is a* true vertex*)*

Typical neighborhoods of points of a simple polyhedron are shown in
Fig. 1.4. The polyhedron used here to illustrate the true vertex singularity
will be denoted by E. Since E will be used very often, it deserves a name: We
will call it *a butterfly*. It is a very strange butterfly indeed. Its body consists
of four segments having a common endpoint, and it has six *wings*. Each wing
spans two segments, and each pair of the segments is spanned by exactly one
wing. Perhaps it may be illuminating to look at some different forms of E, see
Fig. 1.5. The third model is placed inside the regular tetrahedron Δ to emphasize that the singularity is totally symmetric. It can be viewed as the union
$\cup_i \mid \mathrm{lk}(v_i, \Delta') \mid$ of the links of all four vertices of Δ in the first barycentric
subdivision Δ'.

Definition 1.1.9. *The set of singular points of a simple polyhedron (that is,
the union of its true vertices and triple lines) is called its* singular graph *and
is denoted by SP.*

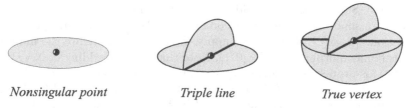

Nonsingular point   *Triple line*   *True vertex*

Fig. 1.4. Allowable neighborhoods in a simple polyhedron

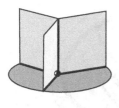 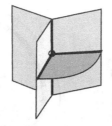 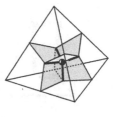

Fig. 1.5. Equivalent ways of looking at the butterfly

In general, SP is not a graph whose vertices are the true vertices of P, since it can contain closed triple lines without true vertices. If there are no closed triple lines, then SP is a regular graph of degree 4, i.e., every true vertex of SP is incident to exactly four edges.

Let us describe the structure of simple polyhedra in detail. Each simple polyhedron is naturally stratified. In this stratification each stratum of dimension 2 (*a 2-component*) is a connected component of the set of nonsingular points. Strata of dimension 1 consist of open or closed triple lines, and dimension 0 strata are true vertices. Sometimes it is convenient to imagine true vertices as transverse intersection points of triple lines.

It is natural to demand that each such stratum be a cell that is, we would like P to be cellular. We will make this demand in our future considerations, as can be seen in the following definition.

Definition 1.1.10. *A simple polyhedron P is called* special *if:*

1. *Each 1-stratum of P is an open 1-cell.*
2. *Each 2-component of P is an open 2-cell.*

Remark 1.1.11. If P is connected and contains at least one true vertex, then condition 1 in the above definition follows from condition 2.

1.1.4 Special Spines

Definition 1.1.12. *A spine of a 3-manifold is called* simple *or* special *if it is a simple or special polyhedron, respectively.*

Two examples of special spines of the 3-ball are shown in Fig. 1.6: Bing's House with two rooms and the Abalone (a marine mollusk with an oval, somewhat spiral shell). Bing's House is a cube B decomposed by the middle section into two rooms. Each room has a vertical tube entrance joined to the walls by a quadrilateral membrane. The Abalone consists of a tube spanned by an artful membrane with a triple line. The tube is divided by a meridional disc.

Let us describe a collapse of the 3-ball onto Bing's House. First we collapse the 3-ball onto a cube B which is contained in it. Next we penetrate through

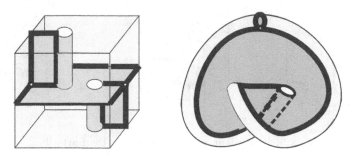

Fig. 1.6. Bing's House and Abalone

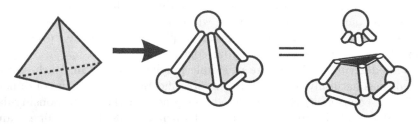

Fig. 1.7. Going from a triangulation to a handle decomposition

the upper tube into the lower room and exhaust the interior of the room keeping the quadrilateral membrane fixed. Finally, we do the same with the upper room.

To collapse the 3-ball onto the Abalone, one may collapse it onto the Abalone with a filled tube and then, starting from the ends of the tube, push in the 3-dimensional material of the tube until we get the meridional disc.

Theorem 1.1.13 ([20]). *Any compact 3-manifold possesses a special spine.*

Proof. Let M be a 3-manifold with boundary and let T be a triangulation of M. Consider the handle decomposition generated by T. This means the following: We replace each vertex with a ball B_i (a handle of index 0), each edge with a *beam* C_j (a handle of index 1), and each triangle with a *plate* P_k (a handle of index 2), see Fig. 1.7. The rest of M consists of index 3 handles. Let P be the union of the boundaries of all handles: $P = \bigcup_{i,j,k} \partial B_i \cup \partial C_j \cup \partial P_k$ (the boundaries of index 3 handles do not contribute to the union). Then P is a special polyhedron and is indeed a special spine of M with an open ball removed from each handle. Alternatively, one can construct a special spine of multipunctured M by taking the union of ∂M and the 2-dimensional skeleton of the cell decomposition dual to T.

It remains to show that if M with $m > 1$ balls removed has a special spine, then M with $m - 1$ balls removed also has a special spine. We do that in two steps. First, we show that as long as the number of removed balls is greater than one, there exist two distinct balls separated by a 2-component of P. This can be achieved by considering a general position arc connecting two

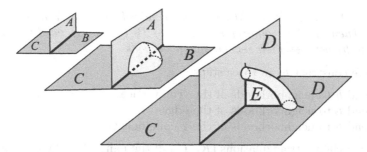

Fig. 1.8. The arch construction

distinct balls and observing that it must pass transversally through at least one separating 2-component.

The second step consists in puncturing the spine to fuse these two balls into one so that the remaining spine is also special. If we just made a hole to cut our way through the 2-component, the boundary of the hole would contain points of forbidden types. One can try to collapse the punctured spine as long as possible with hoping to get a special polyhedron, but sometimes we would end up with a polyhedron which is not even simple. So we must find a way to avoid this. The *arch construction* illustrated in Fig. 1.8 gives us a solution.

The arch connects two different balls separated by a 2-cell C in such a way as to form a special polyhedron. To see this, consider how we get such an arch: first add a "blister" to the spine as illustrated in Fig. 1.8. This is done by considering a neighborhood of the spine and then collapsing most of it (except the blister) back down to the spine. Squeeze in the blister until what remains is a filled tube attached by a membrane F to the spine. From each end of the tube, push in its contents until all that remains is a disk in the middle of the tube. Now remove this disk.

The claim is that we get a special spine for M with the number of removed balls decreased by one. The crux of the matter is that each of the 2-components of the new spine is a 2-cell. Actually the only suspicious 2-component is D, that appeared after joining 2-components A and B by the arch. Clearly, D is a 2-cell provided $A \neq B$ (if $A = B$, we get either an annulus or a Möbius band). To see that the proviso always holds, one should use the fact that we have started with two *distinct* balls separated by the 2-component C: A differs from B, since they separate different pairs of balls.

After a few such steps we get a special spine P' of once punctured M. If M is closed, then we are done. If not, we slightly push P' into the interior of M and use the arch construction again to unite the ball and a component of $M \setminus P'$ homeomorphic to $\partial M \times [0, 1)$.   □

Remark 1.1.14. Bing's House with two Rooms is a good illustration for the above proof. This polyhedron is obtained from a 2-sphere with a disk in the middle by applying the arch construction twice.

Lemma 1.1.15. *Let $P_i \subset M_i$ be special subpolyhedra of 3-manifolds M_i, $i = 1, 2$. Then any homeomorphism $h : P_1 \to P_2$ can be extended to a homeomorphism h_1 between their regular neighborhoods.*

Proof. We construct h_1 in three steps:

1. Extend h to neighborhoods of the true vertices.
2. Extend h to neighborhoods of the edges.
3. Extend h to neighborhoods of the 2-components.

For $i = 1, 2$ choose triangulations (K_i, L_i) of the pairs (M_i, P_i) such that h is simplicial with respect to triangulations L_1, L_2.

To extend the homeomorphism to neighborhoods of the true vertices, observe that for any true vertex v of L_1 the link $\mathrm{lk}(v, K_1'')$ of v in the second barycentric subdivision K_1'' of K_1 is a 2-sphere, and the link $\mathrm{lk}(v, L_1'')$ is a complete graph Γ_4 with four vertices. Certainly, the same holds for the true vertices of L_2. One of the properties of the graph Γ_4 is that it can be imbedded into a 2-sphere in a unique way (up to homeomorphisms of the sphere). It follows that the homeomorphism between $|\mathrm{lk}(v, L_1'')|$ and $|\mathrm{lk}(h(v), L_2'')|$ induced by h can be extended to a homeomorphism between the 2-spheres $|\mathrm{lk}(v, K_1'')|$ and $|\mathrm{lk}(h(v), K_2'')|$. By using the cone construction, we get an extension to the ball neighborhood of v. To extend the homeomorphism to neighborhoods of the edges, we use a similar argument: A polyhedron that consists of two disks and three arcs connecting them can be embedded in S^2 in a unique way.

Finally, we extend the homeomorphism to neighborhoods of 2-components. This extension is possible because every 2-component is a 2-cell, and a neighborhood of a 2-cell is a direct product of that 2-cell by an interval. This completes the proof.   □

Remark 1.1.16. If P_1, P_2 are simple but not special, then the above proof shows that any homeomorphism $h: P_1 \to P_2$ can be extended to homeomorphism between neighborhoods of their singular graphs SP_1, SP_2. However, further extension of h onto neighborhoods of 2-components of P_1 may be impossible.

Theorem 1.1.17. *If two compact connected 3-manifolds have homeomorphic special spines and either both are closed or both have nonempty boundaries, then these 3-manifolds are homeomorphic.*

Proof. If both 3-manifolds have nonempty boundaries, we identify them with regular neighborhoods of special spines and apply Lemma 1.1.15. If both manifolds are closed, we apply Lemma 1.1.15 to the corresponding punctured manifolds and use the cone construction to get a desired homeomorphism between the manifolds.   □

The meaning of Theorem 1.1.17 is that any special spine of a manifold determines it uniquely. It follows that special spines may be viewed as presentations of 3-manifolds.

One should point out that, in contrast to group theory where every presentation determines a group, not every special polyhedron presents a 3-manifold. It is because there exist *unthickenable* special polyhedra which cannot be embedded into 3-manifolds.

Example 1.1.18. We attach the disc D^2 by its boundary to the projective plane RP^2 along the projective line RP^1. All the 2-components of the 2-polyhedron P obtained in this way are 2-cells. However, P cannot be embedded into a 3-manifold M. Indeed, if this were possible, the restriction to RP^1 of the trivial normal bundle of D^2 in M would be isomorphic to the nontrivial normal bundle of RP^1 in RP^2.

Since P has no true vertices, it is not special. Nevertheless, it is easy to attach to P additional 2-cells (bubbles) to get an unthickenable special polyhedron.

It turns out that the "normal bundle obstruction" described above is the only thing that can make a special polyhedron unthickenable. To make this claim precise, let us describe the behavior of a special polyhedron in a neighborhood of the boundary of a 2-component.

Let α be a 2-component of a special polyhedron P. Then there is a *characteristic map* $f : D^2 \to P$, which takes Int D^2 onto α homeomorphically and whose restriction onto $S^1 = \partial D^2$ is a local embedding. We will call the curve $f_{|\partial D^2} : \partial D^2 \to P$ (and its image $f_{|\partial D^2}(\partial D^2)$) the *boundary curve* of α.

Denote by $A \cup D$ the annulus $S^1 \times I$ with a disc D^2 attached along its middle circle, and by $M \cup D$ a Möbius band with a disc D^2 attached along its middle line.

Definition 1.1.19. *We say that the boundary curve of a 2-component α of a special polyhedron P has a* trivial *or* nontrivial normal bundle *if its characteristic map $f : D^2 \to P$ can be extended to a local embedding $f^{(A \cup D)} : A \cup D \to P$ or to a local embedding $f^{(M \cup D)} : M \cup D \to P$, respectively.*

A simple way to determine the type of a normal bundle is to go around the boundary curve and follow the behavior of a normal vector (in the PL sense). Nothing happens when we are moving along a triple line. The events near true vertices are shown in Fig. 1.9. The normal bundle is trivial or nontrivial

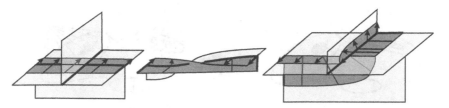

Fig. 1.9. Normal vectors passing through true vertices. Since the typical neighborhood of a true vertex is totally symmetric, both pictures are equivalent

depending on whether we get the same vector we have started with or the opposite one.

Theorem 1.1.20. *A special polyhedron P is thickenable if and only if the boundary curves of all its 2-components have trivial normal bundles.*

Proof. Denote by $N(SP)$ a small regular neighborhood of the singular graph SP in P. Blowing up each vertex of SP to a 3-ball and each edge of SP to an handle of index 1, one can easily construct a not necessarily orientable handlebody H and an embedding $N(SP) \to H$ such that $N(SP) \cap \partial H$ is a union of circles and SP is a spine of H.

Suppose that all the normal bundles are trivial. Then all the circles in $N(SP) \cap \partial H$ have annular neighborhoods in ∂H. Therefore, all the discs in $\mathrm{Cl}(P \setminus N(SP))$ can be expanded to index 2 handles attached to H. we get a 3-manifold M together with an embedding $P \to M$.

If at least one boundary curve of a 2-component of P has a nontrivial normal bundle, we use the same argument as in Example 1.1.18 to prove that P is unthickenable. $\qquad\square$

1.1.5 Special Polyhedra and Singular Triangulations

The aim of this section is to bring together two dual ways of presenting 3-manifolds: special spines and triangulations.

Definition 1.1.21. *A compact polyhedron Q is called a singular 3-manifold if the link $F_x = lk(x, Q)$ of every point $x \in Q$ is a closed connected surface.*

It follows from the definition that every point $x \in Q$ has a conic regular neighborhood $N(x) \approx \mathrm{Con}(F_x)$, where $\mathrm{Con}(F_x)$ is the cone with the vertex x. If F_x is not a 2-sphere, x is called *singular*. Since all other points of $\mathrm{Con}(F_x)$ are nonsingular, Q contains only finitely many singular points. All other points of Q have ball neighborhoods.

Remark 1.1.22. A simple way to construct an example of a singular 3-manifold is to take a genuine 3-manifold M with boundary and add cones over all the boundary components, see Fig. 1.10.

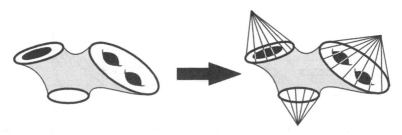

Fig. 1.10. Singular manifold

It is readily seen that any singular 3-manifold Q can be obtained in this way. An easy calculation of the Euler characteristic of Q shows that $\chi(Q) = \chi(M) + \sum_i (1 - \chi(F_i))$, where F_i are the components of ∂M. Since $\chi(M) = (1/2)\chi(\partial M) = (1/2)\sum_i \chi(F_i)$, we have $\chi(Q) = (1/2)\sum_i(2-\chi(F_i))$. It follows that Q is a genuine manifold if and only if $2 - \chi(F_i) = 0$ for all i, i.e., if $\chi(Q) = 0$. Indeed, since $\chi(F_i) \leq 2$ for all i, $(1/2)\sum_i \chi(F_i) = 0$ if and only if all the summands are zeros, i.e., all F_i are 2-spheres.

Suppose we are given a finite set $\mathcal{D} = \{\Delta_1, \Delta_2, \ldots, \Delta_n\}$ of disjoint tetrahedra and a finite set $\Phi = \{\varphi_1, \varphi_2, \ldots, \varphi_{2n}\}$ of affine homeomorphisms between triangular faces of the tetrahedra such that every face has a unique counterpart. In other words, the set of all faces of the tetrahedra should be divided into pairs, and the faces of every pair should be related by an affine isomorphism. We will refer to the pair $(\mathcal{D}, \Phi)$ as to a *face identification scheme*.

Let us identify now all the faces of the tetrahedra via the homeomorphisms $\{\varphi_j, 1 \leq j \leq 2n\}$. The resulting polyhedron will be called the *quotient space* and denoted by $\hat{M}(\mathcal{D}, \Phi)$. The following result is classical (see [112]).

Proposition 1.1.23. *The quotient space $\hat{M} = \hat{M}(\mathcal{D}, \Phi)$ of any identification scheme $(\mathcal{D}, \Phi)$ is a singular manifold. $\hat{M}$ is a genuine 3-manifold if and only if $\chi(\hat{M}) = 0$.*

Proof. If a point $x \in \hat{M}$ corresponds to a point in the interior of a tetrahedron or to a pair of points on the faces, then the existence of a ball neighborhood of the point x in $\hat{M}$ is obvious. Suppose that x comes from identifying points $x_1, x_2, \ldots, x_n$ which either are vertices of tetrahedra or lie inside edges. The link of each point x_i in the corresponding tetrahedron is a polygon: A biangle, if x_i lies inside an edge, and a triangle, if it is a vertex. The link of the corresponding point of $\hat{M}$ is obtained from these polygons by pairwise identifications of their edges. It follows that it is a closed connected surface. The second statement of Proposition 1.1.23 follows from Remark 1.1.22. $\square$

Remark 1.1.24. All singular points of the singular manifold $\hat{M} = \hat{M}(\mathcal{D}, \Phi)$ correspond either to vertices or to barycenters of edges of the tetrahedra of $\mathcal{D}$.

The latter happens when the projection map $p : \cup_i \Delta_i \to \hat{M}$ folds the edges so that symmetric points (with respect to the barycenters) have the same image.

Definition 1.1.25. *A face identification scheme $(\mathcal{D}, \Phi)$ is called* admissible *if all singular points of the quotient manifold $\hat{M} = \hat{M}(\mathcal{D}, \Phi)$ correspond to vertices of the tetrahedra of $\mathcal{D}$, i.e., if the projection map folds no edges.*

Let $(\mathcal{D}, \Phi)$ be an admissible face identification scheme and $p: \cup_i \Delta_i \to \hat{M}$ the projection map. Then the images under p of the tetrahedra and their closed faces of all dimensions of the projection $p : \cup_{i=1}^n \Delta_i \to \hat{M}$ can be considered as *singular simplices*. We will call the decomposition of $\hat{M}$ into singular simplices a *singular triangulation* of $\hat{M}$. Simplices of a singular triangulation are not necessarily embedded (but their interiors are), and the intersection of different simplices may consist of several faces. Nevertheless, singular triangulations possess almost all the properties of usual triangulations: one can take barycentric subdivisions, consider dual decompositions, and so on. It is worth mentioning that any genuine triangulation is a singular one, and taking the second barycentric subdivision of any singular triangulation makes it genuine. The advantage of singular triangulations is that they are more economical in the sense of having a smaller number of simplices. For example, any closed 3-manifold admits a one-vertex triangulation, i.e., a singular triangulation with a single vertex.

Let $(\mathcal{D}, \Phi)$ be a face identification scheme. Recall that any tetrahedron $\Delta \in \mathcal{D}$ contains a copy $E = \cup_i \mid \mathrm{lk}(v_i, \Delta') \mid$ of the butterfly, see Definition 1.1.8 and Fig. 1.5 (right). Since the face identifications Φ are affine, gluing the tetrahedra induces gluing the corresponding butterflies together. We get a special polyhedron $P = P(\mathcal{D}, \Phi)$. Another way of viewing P is to recall that it coincides with $P_T = \cup_j \mid \mathrm{lk}(v_j, T') \mid$, where T' is the barycentric subdivision of the singular triangulation T and $v_j, 1 \leq j \leq m$, are all the vertices of T.

Theorem 1.1.26. *The correspondence* $(\mathcal{D}, \Phi) \rightarrow P(\mathcal{D}, \Phi)$ *induces a bijection between (the equivalence classes of) face identification schemes and (the homeomorphism classes of) special polyhedra. Moreover, a face identification scheme is admissible if and only if the corresponding special polyhedron is thickenable.*

Proof. To prove the bijectivity, it is sufficient to describe the inverse map. Let P be a special polyhedron. Choose a point a_C inside every 2-component C of P and a point b_e inside every edge e of P. Connect each point a_C by arcs in C with all those points b_e that lie on the boundary curve of C. We get a decomposition of P into a collection $\{E_j\}$ of butterflies. In other words, P can be obtained by gluing together several butterflies E_j. Replace each E_j by a copy Δ_j of the standard tetrahedron such that $E_j \subset \Delta_j$. Now gluing the E_j determines the face identification of Δ_j.

To prove the second statement, consider an admissible face identification scheme $(\mathcal{D}, \Phi)$. Let us glue together *truncated tetrahedra* instead of genuine ones. We get a genuine manifold $M \supset P(\mathcal{D}, \Phi)$. Since any truncated tetrahedron Δ_j can be collapsed onto a butterfly E_j, the polyhedron $P(\mathcal{D}, \Phi)$ is a special spine of M.

Conversely, let P be a special spine of a 3-manifold M with nonempty boundary and $(\mathcal{D}, \Phi)$ the face identification scheme corresponding to P. Then M embeds into $\hat{M} = \hat{M}(\mathcal{D}, \Phi)$ so that the complement consists of regular neighborhoods of the vertices of $\hat{M}$. This means that $(\mathcal{D}, \Phi)$ is admissible. $\square$

The correspondence above looks especially simple if we restrict it to the class of one-vertex singular triangulations of genuine closed manifolds. Let T be a singular triangulation of a closed 3-manifold M. Then it determines a face identification scheme and thus a special polyhedron, which we denote by $P(T)$.

Corollary 1.1.27. *The correspondence $T \to P(T)$ induces a bijection between (the equivalence classes of) one-vertex triangulations of a closed manifold M and (the homeomorphism classes of) special spines of M such that the number of tetrahedra in T is equal to the number of true vertices of $P(T)$.*

Proof. Let T be a one-vertex triangulation of M and v the unique vertex of T. Then P_T and $M \setminus P_T$ can be identified with the link $| \operatorname{lk}(v, T') |$ and the open star $M \setminus P_T$ of v in T, respectively. Since M is a manifold, $M \setminus P_T$ is an open ball. It follows that P_T is a special spine of M.

Conversely, let P be a special spine of M and $(\mathcal{D}, \Phi)$ the corresponding admissible face identification scheme. By construction of $(\mathcal{D}, \Phi)$, P embeds into the corresponding singular manifold $\hat{M}(\mathcal{D}, \oplus)$ and determines a singular triangulation T of $\hat{M}(\mathcal{D}, \oplus)$. If we replace coherently every tetrahedron of T by the corresponding truncated one, we get a manifold $M_0 \supset \hat{M}(\mathcal{D}, \oplus)$ such that M_0 collapses to P and $\hat{M}(\mathcal{D}, \oplus)$ is obtained from M_0 by taking cones over all the components of ∂M_0. The vertices of the cones coincide with the vertices of T. Since P is a special spine of the closed manifold M, ∂M_0 is a sphere. It follows that T has only one vertex and $\hat{M}(\mathcal{D}, \oplus)$ is homeomorphic to M. □

Let M be a compact 3-manifold whose boundary consists of tori. By a *topological ideal triangulation* of M we mean a decomposition of Int M into tetrahedra with their vertices removed. Let P be a special spine of M. Then one can easily construct an ideal triangulation $T = T(P)$ of M by taking the corresponding singular triangulation and removing all its vertices.

Corollary 1.1.28. *If the boundary of a 3-manifold M consists of tori, then the correspondence $P \to T(P)$ induces a bijection between (the homeomorphism classes of) special spines of M and (the equivalence classes of) ideal triangulations of M and such that the number of the tetrahedra in $T(P)$ is equal to the number of true vertices of P.*

Proof. Follows from Theorem 1.1.26. □

Corollary 1.1.28 facilitates the construction of ideal triangulations: All that we Have to do is to construct a special spine, which is much easier. This observation is particularly important for constructing hyperbolic 3-manifolds by Thurston's method [120].

1.2 Elementary Moves on Special Spines

Any 3-manifold possesses infinitely many different special spines. How to describe all of them? In this section we solve this problem by showing that any two special spines of the same 3-manifold are related by a sequence of elementary moves.

1.2.1 Moves on Simple Polyhedra

Definition 1.2.1. *A compact polyhedron P is called a* simple polyhedron with boundary *if the link of each of its points is homeomorphic to one of the following 1-dimensional polyhedra:*

(a) *A circle*
(b) *A circle with a diameter*
(c) *A circle with three radii*
(d) *A closed interval*
(e) *A wedge of three segments with a common endpoint*

The set of points of P that have no neighborhoods of types (a), (b), (c) is called the boundary *of P and denoted by ∂P.*

Comparing this definition with the definition of a simple polyhedron (see Definition 1.1.8), we see two new allowed singularities (d) and (e) that correspond to the boundary points. The boundary of a simple polyhedron can be presented as a graph such that any vertex is incident to two or three edges. For example, the boundary of the standard butterfly is the complete graph Γ_4 with four vertices.

Definition 1.2.2. *Let P be a simple polyhedron and Q a simple polyhedron with boundary contained in P. Then Q is called* proper *if $Q \setminus \partial Q$ is open in P. In other words, $Q \subset P$ is proper if after cutting P along ∂Q we get a copy of Q disjoint from the rest of P.*

Let us describe now two special polyhedra E_T, E_T' with boundary. E_T is a typical neighborhood of an edge in a simple polyhedron. It consists of a "cap" and a "cup" joined by a segment, with three attached "wings" (see Fig. 1.11, on the left). E_T' is the union of the lateral surface of a cylinder, a middle disc, and three wings (see Fig. 1.11, to the right). Note that there is a natural identification of ∂E_T with $\partial E_T'$.

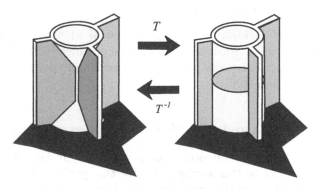

Fig. 1.11. The T-move

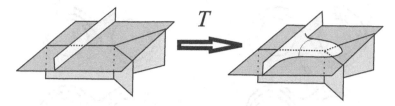

Fig. 1.12. An alternative form of T

Definition 1.2.3. *The elementary move T on a simple polyhedron P consists in removing a proper subpolyhedron $E_T \subset P$ and replacing it by E'_T.*

Notice that T increases the number of true vertices in a polyhedron by one, while the inverse move T^{-1} decreases it. Also note that we start and finish with simple polyhedra that have more than one true vertex each. Another way of viewing T is shown in Fig. 1.12. We present E_T as a butterfly E with one additional wing and then deform the attaching curve of that wing by an isotopy through the vertex of the butterfly.

Remark 1.2.4. We have defined the moves $T^{\pm 1}$ as moves on abstract simple polyhedra, without referring to embeddings in 3-manifolds. On the other hand, we can think of the pairs $(E_T, \partial E_T), (E'_T, \partial E'_T)$ as being embedded into (D^3, S^2) such that $\partial E_T = \partial E'_T$ and D^3 collapses onto E_T and E'_T. It follows that if we apply the moves $T^{\pm 1}$ to a special subpolyhedron P of a 3-manifold M, then they can be realized inside M. In particular, if P is a special spine of M, then we get another special spine of the same manifold. Note that T is dual to one of Pachner's moves, see [104, 105].

Theorem 1.2.5 tells us that any two spines of the same 3-manifold M are related by a sequence of moves $T^{\pm 1}$. It means that we get a complete description of the set of all special spines of M. This set consists of those special polyhedra that can be obtained from a particular special spine of M by successive applications of the moves. This theorem is very important, since it helps us to introduce and investigate new properties of manifolds by introducing and investigating those properties of spines which are invariant under the moves. The only restriction is that we should consider spines with two or more true vertices, since neither T nor T^{-1} can be applied to a spine with one vertex. This restriction is not burdensome, since there are only four special spines with one vertex. See Fig. 1.13, where the spines are presented by regular neighborhoods of their singular graphs. The corresponding 3-manifolds are: lens spaces $L_{4,1}, L_{5,2}$, the Abalone (see Fig. 1.6), and $S^2 \times I$.

Theorem 1.2.5. *Let P and Q be special polyhedra with at least two true vertices each. Then the following holds:*

1. *If P and Q are special spines of the same 3-manifold, then one can transform P into Q by a finite sequence of moves $T^{\pm 1}$.*

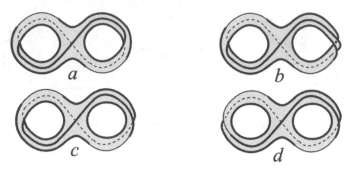

Fig. 1.13. Special spines of (a) $L_{4,1}$, (b) $L_{5,2}$, (c) the Abalone, and (d) $S^2 \times I$

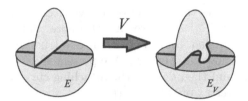

Fig. 1.14. The vertex move

2. *If one can transform P into Q by a finite sequence of moves $T^{\pm 1}$ and one of them is a special spine of a 3-manifold, then the other is a special spine of the same manifold.*

Conclusion 2 of the theorem follows from Remark 1.2.4 and Theorem 1.1.17. For the proof of the difficult first conclusion see the Sect. 1.3.

We now introduce two auxiliary moves related to the T-move above: the *vertex* move V and the *lune* move L. Let A, B be two opposite wings of the butterfly E, see Fig. 1.4. Present E as a disc D^2 with the wings A, B attached along two diameters d_1, d_2 of D^2. Let us cut out A from E and reattach it to the remaining part of E along a simple curve $l \subset D^2$ that has the same ends as d_1 and crosses d_2 transversally at three points. The resulting polyhedron will be denoted by E_V.

Definition 1.2.6. *The* vertex move *replaces a butterfly E in a simple polyhedron P by the fragment E_V, see Fig. 1.14.*

Remark 1.2.7. Note that E and E_V can be embedded into D^3 such that $\partial E = \partial E_V$ and D^3 collapses on each of them. It follows that if we apply V or V^{-1} to a polyhedron embedded into a 3-manifold, than that move can be realized inside the manifold.

The V-move is needed to work with special polyhedra having exactly one true vertex. In the case of ≥ 2 true vertices this move is superfluous.

Proposition 1.2.8. *If a special polyhedron P has more than one true vertex then the V-move is a composition of the moves $T^{\pm 1}$.*

Proof. Figure 1.15 shows how one can express the V-move as a product of three T-moves and one move T^{-1}.   □

Let us describe now a new move called an *ambient lune move*. In contrast to the moves T and V, we define it for simple subpolyhedra of a 3-manifold M, not for *abstract* simple polyhedra. Let E_T, E_T' be the simple polyhedra with boundary which participate in the definition of the move T. We may think of them as being embedded into a 3-ball B^3. Let the pair (B^3, E_L) be obtained from (B^3, E_T) by removing a wing passing through both vertices of E_T. Similarly, (B^3, E_L) is obtained from (B^3, E_T') by removing the corresponding wing.

Definition 1.2.9. *Let P be a simple subpolyhedron of a 3-manifold M and B^3 a ball in M. Suppose that the pair $(B^3, B^3 \cap P)$ is homeomorphic to the pair (B^3, E_L) described above. Then the* lune move *replaces the fragment $E_L = B^3 \cap P$ of P by $E_L' \subset B^3$, see Fig. 1.16.*

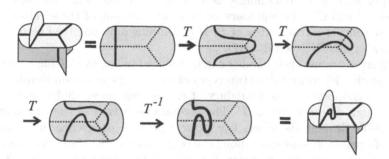

Fig. 1.15. V-move is a composition of $T^{\pm 1}$-moves

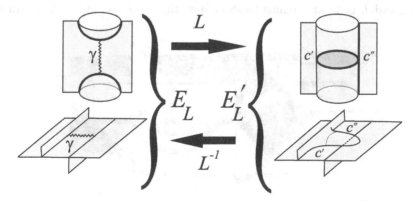

Fig. 1.16. The lune move

Remark 1.2.10. The lune move L differs from $T^{\pm 1}$ and $V^{\pm 1}$ in two respects. First, it is applicable only to simple polyhedra that lie in 3-manifolds. We emphasize this by speaking of an *ambient* lune move. The abstract definition (similar to the ones given for $T^{\pm 1}$ and $V^{\pm 1}$) is inadequate due to the fact that the graph ∂E_L homeomorphic to a circle with two disjoint chords can be embedded into S^2 in two inequivalent ways. Indeed, it suffices to permute a chord spanning an arc of the circle with that arc. In general, abstract lune moves cannot be realized within 3-manifolds.

Second, whereas application of any one of the moves $T^{\pm 1}, V^{\pm 1}, L$ to a special polyhedron automatically yields another special polyhedron, L^{-1} may give rise to an annular or a Möbius 2-component. This happens when the wings c', c'' (see Fig. 1.16) lie in the same 2-component of Q.

Lemma 1.2.11. *If an ambient lune move L transforms a special spine P of a 3-manifold M into another spine P', then it can be represented as a composition of moves $T^{\pm 1}, V^{\pm 1}$. Equivalently, any move L^{-1} transforming a spine of M into a special spine of M is a composition of moves $T^{\pm 1}, V^{\pm 1}$.*

Proof. It is sufficient to prove the first statement. The result of applying L is completely determined by the curve γ, see Fig. 1.16. Since P is special, the 2-component c of P containing γ is a 2-cell. Therefore γ decomposes c into two cells c' and c''. The boundary curve of at least one of these cells (say, of c') must contain a true vertex of P. If it contains only one true vertex, then L coincides with V (see Fig. 1.17).

If there are more, their number can be reduced to one by shifting γ within c. To be precise, let us introduce two types of elementary moves on simple curves in c with endpoints in the boundary of c. The first move shifts an endpoint of c through a true vertex of P. The second one permutes the endpoints by shifting one through the other. The latter is possible only when the boundary curve of c passes at least twice along the edge e containing the endpoints and γ approaches e along different wings. It is evident that these moves are sufficient for transforming γ into the desired position when c' contains only one true vertex and L is V. It remains to show how the moves on curves in c can be

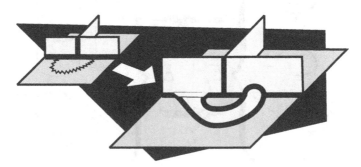

Fig. 1.17. Vertex move as a special case of the lune move

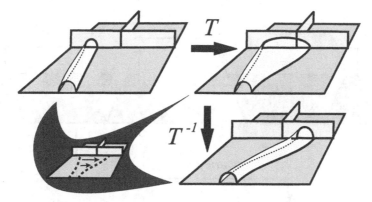

Fig. 1.18. Two steps that realize the endpoint-through-vertex move

realized by moves $T^{\pm 1}$ on P. The realization for the first case is explicitly presented in Fig. 1.18: We bypass the true vertex by taking two steps T and T^{-1}. For a realization of the second case see Fig. 1.19.   □

1.2.2 2-Cell Replacement Lemma

In this section we prove that, under certain conditions, replacement of one 2-cell of a simple subpolyhedron of a 3-manifold by another can be realized by a sequence of moves $T^{\pm 1}, L\pm 1$.

Definition 1.2.12. *Let P be a simple polyhedron in a 3-manifold M. An open ball $V \subset M \setminus P$ is called* proper *(with respect to P), if $Cl(V) \setminus V \subset P$.*

It is worth mentioning that in general $Cl(V)$ may be not a 3-ball. this can happen when V approaches a 2-cell of P from two sides. In particular, if V is the complement to a spine of a closed 3-manifold M, then $Cl(V) = M$.

Lemma 1.2.13. *Let an open 3-ball V in a 3-manifold M be proper with respect to a simple subpolyhedron $P \subset M$. Then closure $Cl(V)$ is a compact submanifold of M whose boundary is contained in P.*

Proof. Let x be a point of $Cl(V) \setminus V$. By Definition 1.2.12, we have $x \in P$. If x is an interior point of a 2-component of P, then a neighborhood of x in $Cl(V)$ is either a half-ball or a ball depending on whether V is adjacent to the 2-component from one side or from both sides. If x is a triple point or a true vertex of P, then we have the same result: x is either an interior point of $Cl(V)$ (if V surrounds x from all sides) or lies in the boundary of $Cl(V)$ (otherwise). It follows that $Cl(V)$ is a 3-manifold.   □

Definition 1.2.14. *If V is a proper ball in a 3-manifold M with a simple subpolyhedron P, then $Free(V) = \partial Cl(V)$ is called the* free boundary *of V.*

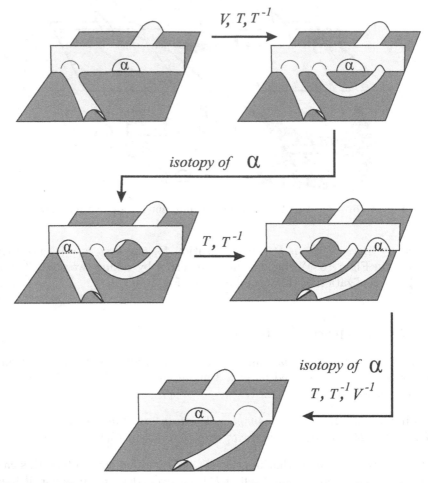

Fig. 1.19. How to realize the endpoint-through-endpoint move of the endpoints

It follows from Lemma 1.2.13 that the definition makes sense and that Free(V) is a closed surface.

Definition 1.2.15. *Let P be a simple polyhedron and l a curve in P. Then l is* in general position *if it possesses the following properties:*

(1) *l is locally simple (by a* locally simple *curve we mean the image of a map $f : S^1 \to P$ such that f is a local embedding).*
(2) *l intersects itself transversally.*
(3) *l contains only double crossing points.*
(4) *l intersects triple lines of P transversally.*
(5) *Crossing points of l with itself do not lie on the triple lines.*
(6) *l contains no true vertices of P.*

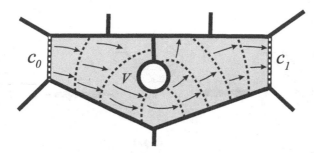

Fig. 1.20. The homotopy of the 2-cell through the proper ball

Lemma 1.2.16. (2-cell replacement) *Let $P \subset M$ be a simple polyhedron, $V \subset M$ a proper ball, and $c_0, c_1 \subset Free(V)$ two 2-components of P such that they are open 2-cells and the polyhedron $P \setminus (c_0 \cup c_1)$ is simple. Then the simple polyhedra $P_0 = P \setminus c_1$ and $P_1 = P \setminus c_0$ can be connected by a sequence consisting of $T^{\pm 1}$ and $L^{\pm 1}$ moves.*

Proof. Since V is a ball, there exists a homotopy $h_t : D^2 \to Cl(V), 0 \leq t \leq 1$, such that the following holds:

1. $h_0(D^2) = Cl(c_0)$ and $h_1(D^2) = Cl(c_1)$.
2. If $0 < t < 1$, then $h_t(D^2) \subset Cl(V) \setminus (c_0 \cup c_1)$ and $h_t(\mathrm{Int}\ D^2) \subset V$.
3. The restriction of h_t onto $\mathrm{Int}\ D^2$ is an isotopy.
4. The restriction of h_t onto ∂D^2 is a local isotopy, i.e., an isotopy near each point $(x, t) \in \partial D^2 \times I$, see Fig. 1.20.

Let us call a moment of time t *singular* if the curve $l_t = h_t(\partial D^2)$ is not in general position. It means that at least one of the following holds:

(a) l_t intersects itself nontransversally.
(b) l_t intersects itself in a triple point of P.
(c) l_t intersects an edge of P nontransversally.
(d) l_t passes through a true vertex of P.

Note that if l_t is self-transversal, then it cannot pass through a point of P three times, since l_t is the boundary curve of the disc $h_t(D^2)$ with the embedded interior.

By the general position argument, we may assume that there are only finitely many singular moments and that the behavior of l_t near each singular moment $(x, t) \in \partial D^2 \times I$ is canonical, as is illustrated in Fig. 1.21. The labels a–d in the figure indicate the types of the corresponding singular moments. We can see that the changes consist in applying one of the moves T, L or one of their inverses. This is naturally enough, since T and L had been invented for describing modifications of a simple polyhedron when the boundary curve of a 2-component is moving with respect to the other triple lines. $\square$

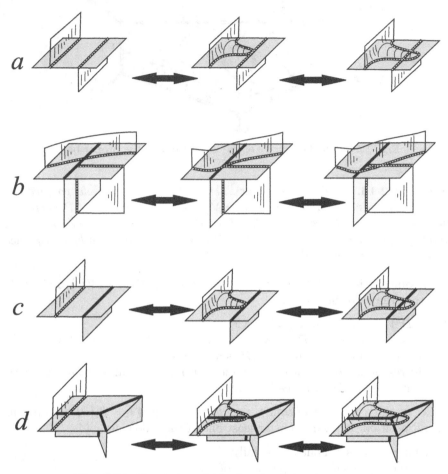

Fig. 1.21. Events in neighborhoods of singular moments

1.2.3 Bubble Move

We introduce now another move B illustrated in Fig. 1.22. It is called the *bubble move* and consists in attaching a 2-disc to a neighborhood of a point $x \in P$. There are three types of the bubble move depending on the type of x. Let us describe this move rigorously.

Let P be a simple subpolyhedron of a 3-manifold M. Choose a point $x \in P$ and its ball neighborhood $N(x) \subset M$ such that for the polyhedron $S(x) = P \cap N(x)$ the following holds:

1. If x is a nonsingular point of P, then $S(x)$ is a disc.
2. If x lies on a triple line of P, then $S(x)$ is homeomorphic to $Y \times I$, where Y is a wedge of three segments with a common endpoint.

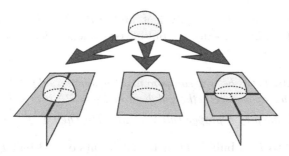

Fig. 1.22. The bubble move

3. If x is a true vertex, then $S(x)$ is a butterfly.

Evidently, $S(x)$ is a simple polyhedron whose boundary $\partial S(x)$ decomposes the sphere $\partial N(x)$ into 2, 3, or 4 discs, depending on the type of x. Choose from them one disc D and add it to P. We get a new simple subpolyhedron $P' \subset M$.

Definition 1.2.17. *The transition from P to P' is called a bubble move at x and denoted by B.*

Let us describe a few properties of the bubble move. First, it is an ambient move; one can apply it only to subpolyhedra of 3-manifolds. Of course, one can define a bubble move for abstract simple polyhedra, but we do not need that.

Second, the resulting polyhedron $P' = P \cup D$ does not depend on the choice of D. If we add to P another disc $D \subset S(x)$ such that $D' \cap P = \partial D'$, then we get an isotopic polyhedron.

Third, in contrast to moves T, V, and L, the bubble move changes the homotopy type of P, in particular, increases the Euler characteristic of P by one. For example, if P is a simple spine of M, then P' is a simple spine of $M \setminus \text{Int } D^3$, where D^3 is a ball in Int M.

Finally, there are three types of bubble moves, which depend on the type of the initial point x. To compare them, let us force x to slide along a 2-component α of P to a point y in a triple line l and then along l to a true vertex v. Denote by $P'(x), P'(y)$, and $P'(v)$ polyhedra, obtained from P by attaching bubbles at x, y, and v, respectively. It is easy to see that they are related as follows: $P'(y)$ is obtained from $P'(x)$ by move L, $P'(v)$ is obtained from $P'(y)$ by move T. We may conclude that all three types of the bubble move are equivalent up to the moves $T^{\pm 1}, L^{\pm 1}$.

In what follows we will use different combinations of the moves T, L, B.

Definition 1.2.18. *Two simple subpolyhedra P_1, P_2 of a 3-manifold M are T-equivalent (notation: $P_1 \overset{T}{\sim} P_2$) if one can pass from P_1 to P_2 by a finite sequence of moves $T^{\pm 1}$. If, in addition, moves $L^{\pm 1}$ are allowed, then we say* *that the polyhedra are (T, L)-equivalent and write $P_1 \overset{T,L}{\sim} P_2$. The moves $B^{\pm 1}$ together with $T^{\pm 1}$, $L^{\pm 1}$ produce the bubble equivalence that will be denoted by $P_1 \overset{T,L,B}{\sim} P_2$.*

Lemma 1.2.19. *Let c be a 2-component of a simple subpolyhedron $P \subset M$ such that c is an open 2-cell, the polyhedron $P_1 = P \setminus c$ is simple, and c lies in the free boundary of a proper ball $V \subset M \setminus P$. Then $P \overset{T,L,B}{\sim} P_1$.*

Proof. We add to P_1 a bubble D such that it cuts off a 3-ball B from V and does not intersect c. Then $V_1 = V \setminus B$ is a proper 3-ball for $P \cup D$ such that the cells c and D are contained in its free boundary. Applying Lemma 1.2.16, we obtain that the polyhedron $P_1 \cup D$ is (T, L)-equivalent to P. It follows that $P \overset{T,L,B}{\sim} P_1$. See Fig. 1.23. □

Theorem 1.2.20. *Any two simple spines of the same manifold M are bubble equivalent.*

Proof. Let $K \subset M$ be a simplicial complex. Assign to K a simple polyhedron $W(K)$ as follows: Replace each vertex by a handle of index zero (ball), each edge by a handle of index one (beam) and each triangle by a handle of index two (plate). Then $W(K)$ is defined as the union of the boundaries of all these handles. We used this construction in the proof of Theorem 1.1.13. We claim that:

(1) If $|K|$ is a simple polyhedron, then $|K| \overset{T,L,B}{\sim} W(K)$.
(2) If $K \searrow L$, then $W(K) \overset{T,L,B}{\sim} W(L)$.
(3) If K_s is a stellar subdivision of K, then $W(K_s) \overset{T,L,B}{\sim} W(K)$.

Let us prove (1). To replace a vertex by the boundary of a ball it is sufficient to perform a bubble move that creates a bubble at this vertex. To create the boundary of a beam, we add a bubble at the corresponding edge and expand it over the edge by two T-moves. Similarly, we create bubbles and expand them to the boundaries of plates.

To prove (2), it is sufficient to analyze the behavior of $W(K)$ under elementary collapses $K \searrow K_1 = K \setminus (\sigma^n \cup \delta^{n-1})$, where σ^n is a principal

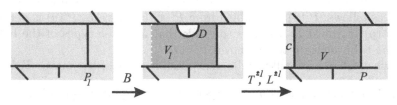

Fig. 1.23. Removing a 2-cell from the free boundary of a proper ball is a bubble equivalence

simplex and δ^{n-1} is a free face of it. In all three cases $(n = 1, 2, 3)$, $W(K_1)$ can be obtained from $W(K)$ by removing two 2-cells. The first cell lies in the boundary of the handle containing δ^{n-1}, the second cell is responsible for the handle corresponding to σ. The transformation $W(K) \to W(K_1)$ is a bubble equivalence by Lemma 1.2.19 (proper balls required for the application of the Lemma are just the interiors of handles).

In order to prove (3), we also analyze the behavior of $W(K)$ under elementary stellar subdivisions and show that its (T, L, B)-class remains the same.

Now let P_1 and P_2 be simple spines of M. There exist triangulations T_1, T_2 of M such that P_1 and P_2 can be presented as subcomplexes $K_1 \subset T_1, K_2 \subset T_2$ and $T_1 \searchow K_1$, $T_2 \searchow K_2$. We have

$$P_1 \overset{T,L,B}{\sim} W(K_1) \overset{T,L,B}{\sim} W(T_1) \overset{T,L,B}{\sim} W(T_2) \overset{T,L,B}{\sim} W(K_2) \overset{T,L,B}{\sim} P_2,$$

where the first and last equivalences come from (1), the second and forth ones come from (2). Then we apply the Alexander Theorem [1] to show that T_1 can be transformed to T_2 by a sequence of stellar subdivisions and their inverses, and use (3).

1.2.4 Marked Polyhedra

In the last section we have proved that any two simple spines P_1, P_2 of the same 3-manifold are bubble equivalent. A stronger statement is true – we may dispense with the B-move entirely. This is a more satisfying situation in that the intermediate polyhedra are also simple spines for the given manifold. Note that in situation of the earlier theorem one can get a polyhedron which is not a spine, already after the first bubble move.

In order to show that the bubble move is unnecessary, we will introduce a new concept of *marked polyhedra*.

Definition 1.2.21. *A mark m on a simple subpolyhedron P of a 3-manifold M is a simple arc in a 2-component c of P such that one endpoint of m is in the interior of c, the other in a triple line and $m \subset Free(V)$ for some proper ball $V \subset M$. A simple polyhedron $P \subset M$ with a mark m is called a* marked polyhedron *and denoted (P, m).*

Definition 1.2.22. *Two marked polyhedra P_1, P_2 are (T, L, m)-equivalent (notation: $P_1 \overset{T,L,m}{\sim} P_2$) if one can pass from one to the other by the following moves:*

(1) *$T^{\pm 1}$- and $L^{\pm 1}$-moves carried out far from the mark. This means that the mark must lie outside the fragments of P that are replaced during the moves.*

(2) *m-move consisting of transferring the mark from one 2-cell of a bubble to another 2-cell of the same bubble.*

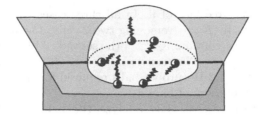

Fig. 1.24. Marks on a bubble

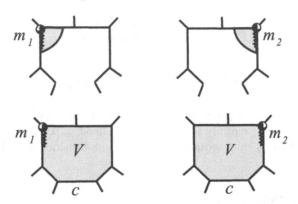

Fig. 1.25. Transplantation of marks

For example, if the bubble is created at a triple line, then six distinct marks on it are possible, see Fig. 1.24. The m-move allows marks to be moved from any one of these positions to any other.

Theorem 1.2.23 shows that the location of a mark on a polyhedron is not so important. The only requirement is that the mark should be placed in the free boundary of a proper ball.

Theorem 1.2.23. *Suppose two marked polyhedra* (P_1, m_1), (P_2, m_2) *are such that* $P_1 \overset{T,L}{\sim} P_2$. *Then* $(P_1, m_1) \overset{T,L,m}{\sim} (P_2, m_2)$.

Proof. We divide the proof into two steps. STEP 1. First we consider the case when $P_1 = P_2$. Suppose that the marks m_1 and m_2 lie on the free boundary of the same proper ball V. Choose a 2-component $c \subset \mathrm{Free}(V)$ whose closure contains no endpoint of m_1 or m_2; if there is no such 2-component, it can always be introduced by a lune move in a neighborhood of an arbitrary triple point in $\mathrm{Free}(V)$. Using the 2-cell replacement (see Lemma 1.2.16), we shift c in two different ways so that the marks lie on two bubbles created at triple points, see Fig. 1.25.

We then carry one bubble to the other. It remains to show how one can do that by means of moves $L^{\pm 1}$ and m. This is illustrated in Fig. 1.26: Each time when the mark prevents us from making a move, we transfer it to another cell of the bubble.

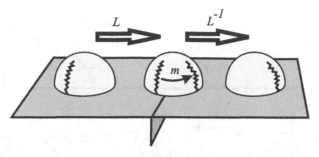

Fig. 1.26. Bubbles may transport marks

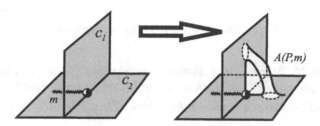

Fig. 1.27. Creating an arch

Suppose now that m_1, m_2 are on the free boundaries of two different proper balls V_1, V_2. As before, we use the 2-cell replacement lemma to get the marks on two bubbles. The bubbles then can be permuted by moves $L^{\pm 1}$ and m.

STEP 2. Suppose that $P_1 \overset{T,L}{\sim} P_2$. We transform (P_1, m_1) to (P_2, m_2) as follows: If a T-move or an L-move on P_1 is far from the mark, we make it. Otherwise we use the arguments of (1) to relocate the mark and then make the move.   □

We were motivated to define marked polyhedra by our wish to show that the bubble move is redundant. A mark is used to indicate where an arch construction (see Sect. 1.1.4 and Fig. 1.8) should be used to kill the corresponding proper ball. Let us describe this in detail.

Consider a simple polyhedron $P \subset M$ with a mark $m \subset P$. Denote by c_1, c_2 those 2-components of P whose closures do not contain the mark, but contain its endpoint. Remove open discs $D_1 \subset c_1, D_2 \subset c_2$ and attach an unknotted tube to the two boundary circles arising in this way. Attach also to the polyhedron P' thus obtained a disc D_3 (a membrane) whose boundary circle passes once along the tube and intersects the singular graph of P at the endpoint of the mark.

Definition 1.2.24. *The transition from a marked polyhedron (P, m) to the simple polyhedron $A(P, m) = P' \cup D_3$ will be called* creating an arch, *see Fig. 1.27.*

As we have indicated above, arches annihilate bubbles.

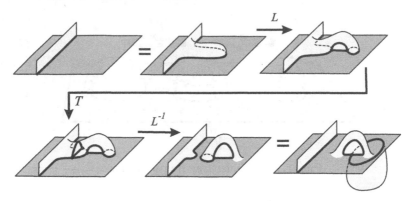

Fig. 1.28. Creating a bubble with an arch

Lemma 1.2.25. *Suppose that a marked polyhedron (Q, m) is obtained from a simple polyhedron $P \subset M$ with a nonempty set of triple points by adding a bubble B and selecting a mark on the bubble. Then $P \overset{T,L}{\sim} A(Q, m)$.*

Proof. We create a bubble with an arch by the moves L, T, L^{-1}, see Fig. 1.28. First steps create an arch with a disc inside it, the last one converts the disc to a bubble.   □

Lemma 1.2.26. *Suppose two marked polyhedra $(Q_1, m_1), (Q_2, m_2)$ in a 3-manifold M are (T, L, m)-equivalent. Then $A(Q_1, m_1) \overset{T,L}{\sim} A(Q_2, m_2)$.*

Proof. Arguing by induction, we can suppose that (Q_2, m_2) is obtained from (Q_1, m_1) by a single move. The case of moves $T^{\pm 1}, L^{\pm 1}$ performed far from the mark is obvious: one needs only perform the moves far from the arch. To demonstrate the equivalence of $A(Q_1, m_1)$ and $A(Q_2, m_2)$ in the case of an m-move (transfer of the mark from one 2-cell of a bubble to another 2-cell of the same bubble), we apply Lemma 1.2.25 twice: First remove the arch and the bubble, then re-create the bubble with the arch in the other position.   □

Theorem 1.2.27. *Any two simple spines of the same manifold are (T, L)-equivalent.*

Proof. Suppose we have two simple spines P_1 and P_2 of a 3-manifold M. By Theorem 1.2.20 they are bubble equivalent. Note that P_1, P_2, being spines of the same manifold, have the same Euler characteristic, and that each bubble move increases the characteristic by 1. Therefore, the numbers of moves B and B^{-1} in the equivalence are equal. Using an induction on the number of $B^{\pm 1}$-moves, we may suppose that P_2 is obtained from P_1 as follows: Add a bubble to get a polyhedron Q_1, apply moves $T^{\pm 1}, L^{\pm 1}$ to get a polyhedron Q_2 with another bubble, and destroy this bubble to get P_2. Choose marks m_1, m_2 on the bubbles.

Thus we have

$$P_1 \overset{T,L}{\sim} A(Q_1, m_1) \overset{T,L}{\sim} A(Q_2, m_2) \overset{T,L}{\sim} P_2,$$

where the first and last equivalences are by Lemma 1.2.25 and the middle equivalence is by Theorem 1.2.23 and Lemma 1.2.26.   □

Our next goal is to eliminate all those moves L^{-1} that create annular or Möbius 2-components. By Lemma 1.2.11, such elimination would be sufficient for converting any (T, L)-equivalence between special spines into a T-equivalence. The idea is that whenever we run into the risk of getting a nonspecial polyhedron by making L^{-1}, we create an arc with a membrane described in Sect. 1.1.4.

Let us recall the construction of the arch. Let D_1, D_2 be open discs in two neighboring 2-components of a special spine P. Remove them and attach an unknotted tube $C = \partial D^2 \times I$ with the middle disc $D = D^2 \times \{*\}$ to the two boundary circles arising in this way. Attach also a disc membrane D_3 whose boundary circle passes once over the tube and intersects the triple line separating the 2-components. We get a new special spine P_1 of the same manifold.

Definition 1.2.28. *The transition from P to P_1 will be called* creation of an arch with a membrane.

Lemma 1.2.29. *An arch with a membrane can be created by means of one move L.*

Proof. See Fig. 1.29.   □

Theorem 1.2.30. *Let P_1 and P_2 be special spines of the same manifold M with at least two true vertices each. Then they are T-equivalent.*

Proof. By Theorem 1.2.27, one can pass from P_1 to P_2 by moves $T^{\pm 1}, L$, which preserve the property of a spine of having only disc 2-components, and by moves L^{-1}, which may create annular and Möbius 2-components. Instead of doing L^{-1}, we erect an arch with a membrane and only then carry out the move. This can be realized by the moves L and T^{-1}, see Fig. 1.30.

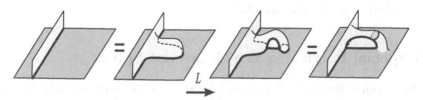

Fig. 1.29. Creating an arch with a membrane

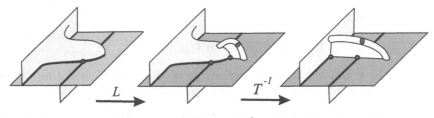

Fig. 1.30. Replacing L^{-1} by L and T

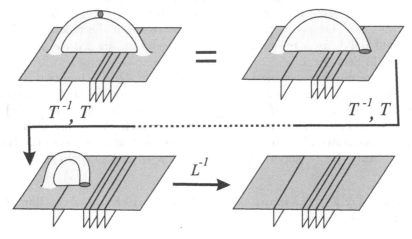

Fig. 1.31. Removing an arch

The arches created by previous steps are very flexible and thus do not prevent us from carrying out further steps. Whenever an arch intersects a fragment we wish to replace to make the next move, we shift the arch away by moves $T^{\pm 1}$ or L. At the end we get P_2 with some extra arches. These extra arches can be removed by moves $T^{\pm 1}$ and $L^{\pm 1}$, see Fig. 1.31 and Lemma 1.2.29. Since P_2 and all intermediate polyhedra that appear during this removal are special, moves $T^{\pm 1}$ and $V^{\pm 1}$ are sufficient for transforming P_1 into P_2, see Lemma 1.2.11. It remains to note that by Proposition 1.2.8 and by the assumption on the number of true vertices of P_1, P_2, every V-move is a composition of T-moves. $\square$

We conclude this section with the remark that Theorem 1.2.30 is exactly the "difficult part" of Theorem 1.2.5.

1.3 Special Polyhedra Which are not Spines

We have seen in Sect. 1.1.4 (Example 1.1.18) that there exist special polyhedra which are not spines of any 3-manifold. Let us describe a systematic way to construct such examples. We start with any special polyhedron P and

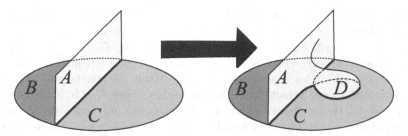

Fig. 1.32. Creating a loop

transform it to produce a new special polyhedron P_1 that does not embed into a 3-manifold. Choose a point on a triple line and modify a neighborhood of this point by reattaching a sheet incident to this line such that there appears a new loop as shown in Fig. 1.32.

To prove that the result of this transformation is still a special polyhedron it suffices to verify that the endpoint of the loop has a neighborhood of the type allowed by Definition 1.1.8. Indeed, we have a butterfly: The sheets B, C, D form a disc while the sheet A passes the point twice and thus produces two wings.

Note that a regular neighborhood of the loop in the modified polyhedron P_1 contains a Möbius band in the union of wings A and C. Assuming that P_1 embeds into a 3-manifold M, we get an embedding of the Möbius band with the disc D attached along its core circle. But this is impossible, since the normal bundle of the disc, hence its restriction to the boundary, is trivial.

1.3.1 Various Notions of Equivalence for Polyhedra

In this section we would like to extend the theory of elementary moves on special spines to a more general category of special polyhedra. We need to replace the property of being spines of the same 3-manifold by another suitable notion of equivalence. Let us discuss briefly different types of equivalence relations for 2-dimensional polyhedra, see Chaps. 1–3, 11, 12 of [47] for a complete account of the subject. Let X, Y be 2-dimensional polyhedra.

(I) *Isomorphism of fundamental groups.* $X \sim Y$ if $\pi_1(X) = \pi_1(Y)$. This relation is fairy rough. For example, taking the one-point union with S^2 preserves the fundamental group but increases the Euler characteristic by 1. It might be used as a substitute for the bubble equivalence, since the bubble move (see Definition 1.22) is essentially an addition of S^2.

(II) *The same fundamental group and Euler characteristic.*

$$X \sim Y \text{ if } \pi_1(X) = \pi_1(Y) \text{ and } \chi(X) = \chi(Y).$$

(III) *Homotopy equivalence.* Recall that two topological spaces X and Y are *homotopy equivalent* if there exist maps $f: X \to Y$ and $g: Y \to X$ such that the maps $fg : X \rightarrow X$ and $gf : Y \rightarrow Y$ are homotopic to the identity. The problem of classification of 2-dimensional polyhedra up to homotopy equivalence is hard. Given a finitely presented group G, one can construct a 2-dimensional polyhedron K_G that realizes G geometrically. To do this, take a wedge of circles that correspond bijectively to generators. Relations show how one should attach 2-cells to the wedge to get K_G. By construction, $\pi_1(K_G) = G$. Hence classification of 2-dimensional polyhedra up to homotopy equivalence is intimately related to the isomorphism problem for finitely presented groups. The latter problem is known to be unsolvable.

Surprisingly, the problem of constructing homotopically distinct 2-dimensional polyhedra with the same π_1 and χ turned out to be very difficult. It was solved only in 1976 by Dunwoody [26] and Metzler [92].

(IV) *Simple homotopy equivalence.* This relation was introduced by Whitehead. A basic reference for this material is Milnor's paper [95]. Originally, Whitehead worked with simplicial complexes, but later found the theory easier to express in terms of CW complexes. If we prefer to stay inside the polyhedral category, we can work with *elementary polyhedral collapses* (see Definition 1.1.1) and inverse transformations called *elementary polyhedral expansions*. Each elementary move (i.e., an elementary collapse or expansion) has a dimension that by definition equals the dimension of the cell that disappears, respectively, appears during the move.

Two polyhedra X, Y are said to be *simple homotopy equivalent* ($X \overset{s}{\sim} Y$) if there is a sequence of elementary expansions and collapses taking X to Y. The dimension of a simple homotopy equivalence is the maximum of $\dim(X)$, $\dim(Y)$, and dimensions of expansions and collapses in the corresponding sequence. It is clear that any simple homotopy equivalence (i.e., the sequence of expansions and collapses) determines a homotopy equivalence. The converse is not true. Whitehead's key result states that there is an obstruction $\tau(f)$ for deforming a homotopy equivalence $f : X \rightarrow Y$ to a simple one. It takes values in an abelian group $Wh(\pi)$ that depends only on the group $\pi = \pi_1(X) = \pi_1(Y)$, and vanishes if and only if f originates from a simple homotopy equivalence. The group $Wh(\pi)$ and the obstruction $\tau(f)$ are now called *Whitehead group* and *Whitehead torsion*.

One should point out that the existence of a nontrivial obstruction does not automatically imply that the simple homotopy type and homotopy type do not coincide. In other words, the existence of a homotopy equivalence $f : X \rightarrow Y$ with $\tau(f) \neq 0$ does not exclude the existence of a simple homotopy equivalence $g : X \rightarrow Y$. Indeed, g does exist if and only if there is a self-equivalence $X \rightarrow X$ (or $Y \rightarrow Y$) with torsion $-\tau$. A solution of this hard problem was suggested in 1990 by Metzler [93]. Let $K_{2,4}$ be the standard geometric realization of the presentation

$$\langle x, y \mid x^2 = [x, y] = y^4 = 1 \rangle.$$

It can be visualized as a union of a 2-dimensional torus T^2, projective plane RP^2, and a "four-fold hat" (the quotient space of the disc D^2 by the free action of Z_4 on the boundary) such that a projective line of RP^2 and the four-fold line of the hat are attached to the meridian x and longitude y of the torus, respectively. Since $\pi_1(K_{2,4}) = Z_2 \times Z_4$, the group $Wh(\pi_1(K_{2,4}))$ is trivial. It follows that for any n and any 2-dimensional polyhedron X we have

$$Wh(\pi_1(X)) = Wh(\pi_1(X \vee_{i=1}^{n} K_{2,4})),$$

where $\vee_{i=1}^{n} K_{2,4}$ is the wedge of n copies of $K_{2,4}$.

Metzler proved that $K_{2,4}$ possesses two properties. The first one is that for any 2-dimensional polyhedron X and any $\tau \in Wh(\pi_1(X))$ there is a 2-dimensional polyhedron $Y = Y(X, \tau)$, an integer number n, and a homotopy equivalence $f : X \vee_{i=1}^{n} K_{2,4} \to Y$ such that $\tau(f) = \tau$. In other words, all elements of $Wh(\pi_1(X))$ are realizable (at the expense of taking one-point unions with several exemplars of $K_{2,4}$). This is not very surprising since many 2-dimensional polyhedra (for instance, S^2) have this property. The second property is more interesting. It turns out that one can choose X so that *not all* elements of $Wh(\pi_1(X))$ can be realized as Whitehead torsions of homotopy equivalences $X \vee_i K_{2,4} \to X \vee_i K_{2,4}$. Therefore, there exist Y and n such that $X\vee_{i=1}^{n} K_{2,4}$ and Y have the same homotopy type but distinct simple homotopy types.

The simplest examples of this kind belong to Lustig [71]. They are the standard geometric realizations for the group presentations

$$\langle x, y, z \mid y^3, yx^{10}y^{-1}x^{-5}, [x^7, z] \rangle,$$

$$\langle x, y, z \mid y^3, yx^{10}y^{-1}x^{-5}, x^{14}zx^{14}z^{-1}x^{-7}zx^{-21}z^{-1} \rangle.$$

Note that for polyhedra the property of being simply homotopy equivalent is fairly close to the property of being spines of the same manifold or, more generally, of the same polyhedron. Indeed, if we can apply an elementary collapse followed by an elementary expansion, nothing prevents us from performing the moves in the reverse order. Hence any simple homotopy equivalence can be replaced by another one such that all elementary expansions are made first and the elementary collapses later. It follows that for any simple homotopy equivalence between X, Y there exists a polyhedron Z such that $Z \searrow X$ and $Z \searrow Y$. The dimension of Z equals the dimension of the equivalence. It would be desirable to be able to reduce it to 3. Theorem 1.3.1 sheds light on this problem.

Theorem 1.3.1. *[135] If $m > 2$ then for any two simple homotopy equivalent polyhedra X, Y of dimension $\leq m$ there is a sequence of elementary moves of dimension at most $m + 1$ that takes X to Y.*

Remark 1.3.2. It follows from Theorem 1.3.1 that any simple homotopy equivalence between 2-dimensional polyhedra can be replaced by an equivalence of dimension 4.

(V) *3-Deformation equivalence*. This relation is obtained by restricting the dimension of simple homotopy equivalence.

Definition 1.3.3. *Two polyhedra* X, Y *of dimension* ≤ 2 *are* 3-deformation equivalent (*we write* 3*d-equivalent or* $X^2 \overset{3d}{\sim} Y^2$), *if there is a sequence of elementary moves of dimension no greater than 3 taking* X–Y.

Clearly, two 2-dimensional polyhedra X, Y which are 3-deformation equivalent are simple homotopy equivalent, and there is a 3-dimensional polyhedron Z such that $Z \searrow X$ and $Z \searrow Y$. In a sense, X and Y may be considered as spines of the same 3-dimensional body, and the only difference between this and the equivalence relation used in Sect. 1.2 is that we do not require that the body is a 3-manifold. A natural question at this point is:

Does simple homotopy equivalence imply 3-deformation equivalence?

It may be recast as a conjecture, which is now known as the *generalized Andrews–Curtis conjecture*:

Conjecture (Generalized AC). Any simple homotopy equivalence between 2-dimensional polyhedra can be deformed to a 3-deformation equivalence.

Remark 1.3.4. The original conjecture in [4] states the following. If F is a free group on generators $x_1, \ldots, x_n$, and a set $\{r_1, \ldots, r_n\}$ of elements of F is such that its normal closure is F, then $r_1, \ldots, r_n$ may be transformed into $x_1, \ldots, x_n$ by a finite sequence of operations of the following types:

(i) Replace r_1 by its inverse r_1^{-1}.
(ii) Interchange r_1 and r_i, leaving the other elements of the set unchanged.
(iii) Replace r_1 by $r_1 r_2$, leaving the other elements fixed.
(iv) Replace r_1 by the conjugate $g r_1 g^{-1}$, where g is any element of F.

The motivation for the conjecture (now known as the *balanced Andrews–Curtis conjecture*) was that if it is true, then any regular neighborhood of any contractible 2-dimensional polyhedron in R^5 is a ball. An anonymous referee remarked that this result would follow from a somewhat weaker conjecture: Suppose $\langle x_1, \ldots, x_n \mid r_1, \ldots, r_n \rangle$ is a balanced presentation of the trivial group. Then it can be reduced to the empty presentation by operations (i)–(iv) on the relators, and two new operations:

(v) Introduce a new generator x_{n+1} and a new relator r_{n+1} that coincides with x_{n+1}.
(vi) The inverse of (v).

The referee pointed out that the weaker conjecture (now known as the Andrews–Curtis conjecture) can be formulated in an equivalent geometric form.

Conjecture (AC). Any contractible 2-dimensional polyhedron 3-deforms to a point.

Remark 1.3.5. One should point out that the difference between equivalence relations (I) (isomorphism of π_1) and (V) (3-deformation) is fairly delicate. For instance, it disappears if one is allowed to add spherical bubbles. Indeed, it follows from the Tietze theorem that any two 2-dimensional polyhedra X and Y with isomorphic fundamental groups satisfy $X \vee S_1^2 \vee \ldots \vee S_k^2 \overset{3d}{\sim} Y \vee S_1^2 \vee \ldots \vee S_m^2$ for suitably chosen finite numbers k, m of 2-spheres.

Henceforth we will be primarily interested in special polyhedra up to 3-deformation equivalence.

1.3.2 Moves on Abstract Simple Polyhedra

Suppose that P_1, P_2 are 3-deformation equivalent special polyhedra. Are they T-equivalent? The following example shows that in general the answer is "no."

Example 1.3.6. Suppose P_1 and P_2 are special spines of a solid torus $S^1 \times D^2$ and a solid Klein bottle $S^1 \tilde{\times} D^2$, respectively. Then they are 3d-equivalent. Indeed, we can expand P_1 to $S^1 \times D^2$ and then collapse $S^1 \times D^2$ to its core circle. The same can be done with P_2: We expand it to $S^1 \tilde{\times} D^2$ and collapse onto a circle. It follows that P_1, P_2 are 3d-equivalent. On the other hand, they are not T-equivalent, since $S^1 \times D^2$ and $S^1 \tilde{\times} D^2$ are not homeomorphic.

In this section we show that only one additional move is needed to relate any two 3d-equivalent special polyhedra. Present a butterfly E (see Fig. 1.4) as a disc D^2 with wings A, B attached along two diameters of D^2. Cut off A and reattach it to the remaining part of E along a simple curve with the same ends such that it climbs on B, makes an U-turn, and returns to the other half of D^2. The resulting polyhedron will be denoted by E_U.

Definition 1.3.7. *The elementary move U on a simple polyhedron P consists in removing a proper butterfly $E \subset P$ and replacing it by E_U, see Fig. 1.33.*

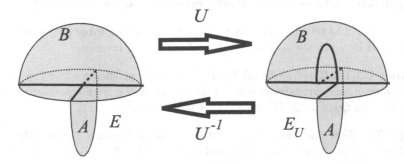

Fig. 1.33. The U-move

Notice that U increases the number of true vertices in a polyhedron by one, and that the E_U does not embed into R^3.

Theorem 1.3.8. *Let P, Q be special polyhedra. Then $P \overset{3d}{\sim} Q$ if and only if one can transform P into Q by a finite sequence of moves $T^{\pm 1}, U^{\pm 1}$.*

The "if" part of the proof follows from Lemma 1.3.10.

Definition 1.3.9. *Let 2-dimensional polyhedra X_1, X_2 be obtained from a polyhedron Y by attaching discs D_1^2, D_2^2 with homotopic attaching maps $f_1, f_2 : S^1 \to Y$. Then the transition from X_1 to X_2 will be called a* transient move.

Lemma 1.3.10. *Any transient move is a 3-deformation.*

Proof. This is a special case (for 2-dimensional polyhedra) of Lemma 13 of [134] and of Lemma 1 of [135]. Attach a ball B^3 to $Z = Y \cup D_1^2 \cup D_2^2$ by a map $S^2 \to Z$ that takes the "polar caps" of S^2 onto D_1^2, D_2^2, and the annular region between them to the trace of a homotopy between f_1 and f_2. By construction, $Z \searrow X_1$ and $Z \searrow X_2$.   □

Now both the T-move and the U-move may be regarded as moves that change the attaching map for a disc by a homotopy. The same is true for the lune move. By Lemma 1.3.10, these moves can be realized by 3-deformations.

Definition 1.3.11. *Two simple polyhedra P_1, P_2 are (T, U)-equivalent (notation: $P_1 \overset{T,U}{\sim} P_2$) if one can pass from P_1 to P_2 by a finite sequence of moves $T^{\pm 1}, U^{\pm 1}$. If in addition moves $L^{\pm 1}$ are allowed, then we say that the polyhedra are (T, U, L)-equivalent.*

Remark 1.3.12. We do not make use of the bubble move. The reason is that we are considering abstract simple polyhedra and have no 3-manifold they are contained in. It means that after performing a few moves $T^{\pm 1}, U^{\pm 1}$ we would lose the control over attached bubbles, and the bubble move would be essentially equivalent to taking the one-point union with a 2-dimensional sphere. This would annihilate any difference between 3d-equivalent polyhedra and polyhedra with isomorphic fundamental groups, see Remark 1.3.5.

Lemma 1.3.13 tells us that any transient move of a simple polyhedron can be realized by a sequence of moves T, U, L and their inverses.

Lemma 1.3.13. *(2-cell shifting) Let P be a simple polyhedron and $f, g : S^1 \to P$ two homotopic curves in general position. Then the simple polyhedra $Q_1 = P \cup_f D^2$ and $Q_2 = P \cup_g D^2$ are (T, U, L)-equivalent.*

Proof. We follow the main lines of the proof of Lemma 1.2.16. The only difference is that there is no 3-manifold, where D_1, D_2, and the trace of the homotopy between the curves could bound a proper 3-ball.

Let $f_t : S^1 \to P$ be a homotopy between f and g. Define a map $F : S^1 \times I \to P \times I$ by the rule $F(x, t) = (f_t(x), t)$. We say that f_t is *regular* at a point $(x, \tau) \in S^1 \times I$, if the restriction of F to a neighborhood of this point is an embedding. We call a moment of time τ *singular*, if F is not regular at some point (x, τ) or the curve $f_\tau(S^1)$ is not in general position. By a general position argument we may assume that f_t has only finitely many singular moments. We may also arrange that in a small neighborhood of each singular moment (that is, when t runs from $\tau - \varepsilon$ to $\tau + \varepsilon$) the curve $l_t = f_t(S^1)$ undergoes local modifications (and their inverses) of the following type:

(A_1) Creation of a new crossing point of l_t, i.e., creation of a loop.
(A_2) Creation of a pair of new crossing points of different arcs of l_t.
(A_3) Taking an arc of l_t through a crossing point of two other arcs. The number of crossing points of l_t with itself should remain unchanged.
(A_4) Creation of a pair of new points where l_t intersects a triple line of P.
(A_5) Replacement of a crossing point with a triple line by a pair of new crossing points of l_t with the same line.
(A_6) Shifting of an arc of l_t through a point where another arc of l_t intersects a triple line. One new crossing point of l_t with itself should appear.
(A_7) Taking an arc of l_t through a true vertex of P.

Remark 1.3.14. Modifications A_1–A_3 should take place in a nonsingular part of P. They may remind you of the Reidemeister moves on knot projections. Modifications A_4, A_5 are responsible for singular moments when l_t intersects a triple line nontransversally. A_6 corresponds to the case when a double point of l_t lies on a triple edge, and A_7 realizes a passage of l_t through a true vertex of P.

To conclude the proof, it suffices to show that modifications $A_i, 1 \le i \le 7$, induce (T, U, L)-equivalences of polyhedra $P \cup_{f_t} D^2$ for $t = \tau \pm \varepsilon$. Cases $i = 2, 4, 6$, and 7 were considered in the proof of Lemma 1.2.16, see Fig. 1.21. Figure 1.32 may serve as an illustration of modification A_1 that creates a loop. The corresponding move on special polyhedra is a composition of the moves U, U and T^{-1} as shown in Fig. 1.34.

A_5 coincides with the U-move. For a realization of A_3 we assume that two of the three arcs of l_t that take part in the modification are fixed, and that the corresponding parts of D^2 are already attached. We use them to express A_3 as a composition of T, U^{-1}, U, T^{-1}, see Fig. 1.35.   □

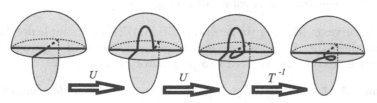

Fig. 1.34. Creating a loop by the moves U, U, T^{-1}

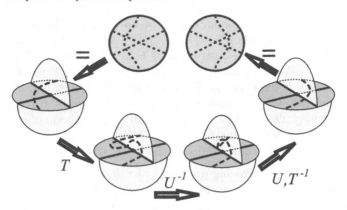

Fig. 1.35. Realization of the "triangle" Reidemeister move

Remark 1.3.15. Lemma 1.3.13 remains true if in its statement one replaces the disc D^2 by an arbitrary simple polyhedron whose boundary consists of a circle. The proof is the same.

Let us assign to a given 2-dimensional complex simplicial complex K a simple polyhedron $W(K)$. The assignment is done by means of a construction. Given a simplicial complex K, let $|K^{(1)}|$ denote its 1-skeleton. The construction is carried out in four steps:

1. Choose an orientable or nonorientable handlebody H^3 such that $\chi(H^3) = \chi(|K^{(1)}|)$. Note that $|K^{(1)}|$ and H are homotopy equivalent.
2. Choose a homotopy equivalence $\varphi: |K^{(1)}| \to H$. One may think of the pair (H, φ) as of a *thickening* of $K^{(1)}$; in a sense, we blow-up $|K^{(1)}|$ to a 3-dimensional handlebody.
3. Replace H^3 by some choice of a simple spine P of H^3. A collapse r of H^3 to P induces a well-defined homotopy equivalence, so from the point of view of homotopy theory we may identify them.
4. For every 2-simplex Δ_i of K, define the map $\varphi_i: \partial\Delta_i \to P$ as the composition of the maps $\partial\Delta_i \subset |K^{(1)}| \xrightarrow{\varphi} H^3 \xrightarrow{r} P$, and adjust φ_i by homotopy to get curves $\varphi_i(\partial\Delta_i)$ in general position. Finally, use φ_i to attach the 2-simplices of K to P.

Definition 1.3.16. *Any 2-dimensional polyhedron $W(K)$ that can be obtained by the above construction is called* a blow-up *of K.*

Since the curves $\varphi_i(\partial\Delta_i)$ are in general position, $W(K)$ is a simple polyhedron. We must check that $W(K)$ is independent of the choices made, up to (T, U, L)-equivalence. To do this, we need two preparatory lemmas.

Lemma 1.3.17. *Any homotopy equivalence of an orientable handlebody H into itself is homotopic to a homeomorphism.*

Proof. Present H as a ball with index 1 handles. Since there are no handles of index > 1, the homotopy class of any homotopy equivalence $H \to H$ is completely determined by the induced automorphism of $\pi_1(H)$. This group is freely generated by the cores of the handles. Recall that any automorphism of a free group can be presented as a composition of the following Nielsen moves on generators:

(i) Replacing a generator by its inverse
(ii) Permuting generators
(iii) Multiplying a generator by another one, all others being kept fixed

All these moves can be easily realized by homeomorphisms. For instance, the third move corresponds to the handle sliding, see Fig. 1.36. It follows that any homotopy equivalence $H \to H$ is homotopic to a homeomorphism.   □

Lemma 1.3.18. *An orientable handlebody H and a nonorientable handlebody $\tilde{H}$ of the same genus $g \geq 1$ admit homeomorphic simple (but not special) spines.*

Proof. Let F be a nonorientable surface with connected boundary such that $\chi(F) = \chi(H)$. Present H and $\tilde{H}$, respectively, as an orientable twisted I-bundle $F \tilde{\times} I$ and a trivial I-bundle $F \times I$ over F. Then both of them collapse to F with a thin solid tube running along ∂F. See Fig. 1.37 for the genus 1

Fig. 1.36. Handle sliding

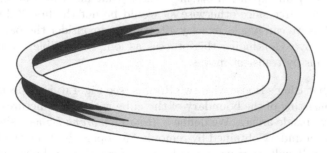

Fig. 1.37. A common spine of the solid torus and solid Klein bottle

case when H is a solid torus and $\tilde{H}$ is a solid Klein bottle. To get a common simple spine of $H, \tilde{H}$, we collapse the tube onto a simple subpolyhedron.   □

Proposition 1.3.19. *If K is a simplicial complex, then $W(K)$ is well defined up to (T, U, L)-equivalence.*

Proof. We will show step by step that arbitrary choices made by constructing $W(K)$ (see items 1–4 above) do not affect its (T, U, L)-type. We begin with the last item:

(1) There are many different ways of taking the curves $\varphi_i(\partial \Delta_i)$ into general position. By Lemma 1.3.13, all of them give equivalent blow-ups.
(2) One may choose another simple spine P_1 of H^3. By Theorem 1.2.27, there is a sequence of moves $T^{\pm 1}, L^{\pm 1}$ transforming P_1 into P. The same sequence can be used to relate the corresponding blow-ups. Attached cells do not prevent us from making the moves. Each time the boundary curve of an attached cell intersects a fragment replaced by the move, we apply Lemma 1.3.13 for shifting the curve away.
(3) Suppose that H^3 is orientable. It follows from Lemma 1.3.17 that any two homotopy equivalences $\varphi, \varphi' : |K^{(1)}| \to H^3$ differ by a homeomorphism $H^3 \to H^3$. This means that the corresponding blow-ups are equivalent.
(4) If we take a nonorientable handlebody $\tilde{H}^3$, then for a specific choice of homeomorphic spines for H^3, $\tilde{H}^3$ (which exist by Lemma 1.3.18) we get equivalent blow-ups. By (1) and (2), the (T, U, L)-type of $W(K)$ does not depend on the choice of a spine. It follows that $W(K)$ does not depend on the choice of the handlebody.

We have shown that any two blow-ups of K are (T, U, L)-equivalent.   □

Lemma 1.3.20. *Let K, L be 2-dimensional simplicial complexes such that $K \overset{3d}{\sim} L$. Then any sequence $s_1, s_2, \ldots, s_n$ of elementary simplicial collapses and expansions of dimension ≤ 3 transforming K into L can be rearranged so that it consists of simplicial transient moves and deformations of dimension ≤ 2.*

Proof. Let s_k be the first 3-dimensional collapse in the sequence. The tetrahedron σ that disappears after s_k (together with its free triangle face δ) has been created by an expansion $s_m, m < k$. At that moment and all the time afterward δ was free since otherwise s_k would be not the first 3-dimensional collapse. Thus we can put s_k exactly after s_m and consider the pair s_m, s_k as a transient move. Continuing this process as long as possible, we replace all 3-deformations by transient moves.   □

Recall that Bing's House with two Rooms (see Fig. 1.6) is a special spine of the cube consisting of the boundary of the cube with two punctures, a middle disc, and two inside arches. We define a *Bing membrane* B^2 as a simple polyhedron with boundary obtained by removing an open disc from the bottom of Bing's House. It will serve us as a substitute for a disc, although in contrast

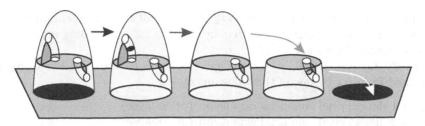

Fig. 1.38. Destroying the Bing membrane

to the disc, the Bing membrane can be 3-deformed into its boundary. Let $f: \partial B^2 \to P$ be a general position contractible curve in a simple polyhedron P. We will say that the simple polyhedron $P \cup_f B^2$ *is obtained from* P *by attaching the Bing membrane.*

Lemma 1.3.21. *Let* Q *be obtained from* P *by attaching a Bing membrane* B^2. *Then* $Q \overset{T,L}{\sim} P$.

Proof. First we shift B^2 into a neighborhood of a nonsingular point of P. This can be done by the moves $T^{\pm 1}, U^{\pm 1}, L^{\pm 1}$, see Remark 1.3.15. Then we move the disc bounded by ∂B^2 into the upper tube to get an arch with the membrane, and destroy the arch. Lemma 1.2.29 shows how it can be done by the move L^{-1}. Similarly, we destroy the second arch, see Fig. 1.38.   □

Proposition 1.3.22. *Blow-ups of 2-dimensional complexes possess the following properties:*

(1) *If the underlying polyhedron* $|K|$ *of a complex* K *is simple, then any blow-up* $W(K)$ *of* $|K|$ *is* (T, L)-*equivalent to* $|K|$.
(2) *If* $|K_1| \overset{3d}{\sim} |K_2|$, *then* $W(K) \overset{T,L}{\sim} W(L)$.

Proof. (1) Suppose that a simplicial complex K triangulates a simple polyhedron $P = |K|$. For every vertex v of K we choose a disc $D_v \subset P$ such that $D_v \cap SP$ either is empty, or is a diameter of D_v, or consists of three radii of D_v, depending on the singularity type of v. Different discs must be disjoint. Then for every vertex v we attach a 3-ball B_v to P along D_v. In other words, we blow-up all the vertices of K to 3-balls. Then we repeat this operation with respect to the edges of K. Namely, in a neighborhood of each edge e of K we attach a beam which runs along e and joins the blow-ups of its vertices.

Denote by U the polyhedron obtained in this way. It consists of the union H of balls and beams (which is a handlebody) and of remnants of the triangles of K (which are 2-cells). Our next step consists in collapsing each ball B_v to a Bing membrane contained in it by penetrating inside B_v through two holes and exhausting the 3-dimensional material of B_v. We certainly can choose the holes in such a way that they intersect neither D_v nor the beams. It follows from the construction that the union Q of P and all the Bing membranes is a blow-up of K. It remains to note that by Lemma 1.3.21 the Bing membranes can be created by moves $T^{\pm 1}, U^{\pm 1}, L^{\pm 1}$.

(2)By Lemma 1.3.20, it is sufficient to consider the case when K_2 is obtained from K_1 by one of the following moves:

(i) A 1-dimensional simplicial expansion (or collapse)
(ii) A 2-dimensional simplicial expansion (or collapse)
(iii) A simplicial transient move

In Cases (i) and (iii) any blow-up of K_2 is simultaneously a blow-up of K_1 since we do not change the homotopy type of $K_1^{(1)}$ and thus we can take the same handlebody H (see the first step of the construction of a blow-up). Consider Case (ii), when there is a simplex $\sigma^2 \subset K_2$ with a free edge δ^1 such that $K^2 = K_1 \cup \sigma^2 \cup delta^1$. We can construct a blow-up $W(K_2)$ by attaching to $W(K_1)$ an arch composed from a tube with a middle disc and a membrane. The tube is obtained from the index 1 handle running along δ^1 by pushing the 3-dimensional material to the middle disc starting from both ends. The membrane corresponds to σ^2. By Lemma 1.2.29, attaching the arch is a (T, L)-equivalence. $\qquad \square$

Remark 1.3.23. The proof of property (1) of blow-ups can be easily modified to show that *any 2-dimensional polyhedron 3-deforms to a special one*, namely, to an appropriate blow-up of its triangulation.

Theorem 1.3.24. *Let* P, Q *be two simple polyhedra such that* $P \overset{3d}{\sim} Q$. *Then* $P \overset{T,L}{\sim} Q$.

Proof. Let simplicial complexes K and L triangulate P and Q, respectively. Then there exists a simplicial 3-deformation of K to L. Thus we have

$$P_1 \overset{T,L}{\sim} W(K_1) \overset{T,L}{\sim} W(K_2) \overset{T,L}{\sim} P_2,$$

where the first and last equivalences are given by part (1) of Proposition 1.3.22, and the middle one is valid by part (2). $\qquad \square$

We are now ready to prove the main result of this section.

Proof of Theorem 1.3.8 (difficult part). Let P, Q be two special polyhedra such that $P \overset{3d}{\sim} Q$. By Theorem 1.3.24, one can pass from P to Q by moves $T^{\pm 1}, U^{\pm 1}, L$ and L^{-1}. Just as in the Proof of Theorem 1.2.30, we use additional arches to dispense with L^{-1}. All the remaining L can be expressed through $T^{\pm 1}$ and $U^{\pm 1}$. $\qquad \square$

1.3.3 How to Hit the Target Without Inverse U-Turns

The main result of this section is Theorem 1.3.25. This fairly deep result will be used in Sect. 1.3.4. Recall that a special polyhedron is unthickenable, if it cannot be embedded into a 3-manifold. By Theorem 1.1.20, a special polyhedron is unthickenable if and only if the boundary curve of at least one 2-component has a nontrivial normal bundle. From Theorem 1.3.8 we know that if two special polyhedra P, Q are $3d$-equivalent, then one can pass from P to Q by a sequence of moves $T^{\pm 1}, U^{\pm 1}$. It turns out that if Q is unthickenable, then one can get rid of move -1.

Theorem 1.3.25. *Let* P, Q *be special polyhedra such that* $P \overset{3d}{\sim} Q$ *and* Q *is unthickenable. Then one can transform* P *into* Q *by a finite sequence of moves* $T^{\pm 1}, U$, *i.e., without using* U^{-1}.

We postpone the proof to the end of this section. Let us introduce now two auxiliary moves. The first one we met earlier: It is the creation of a loop, see Fig. 1.32. It would be natural to call it the *loop move* and denote it by L, but L is occupied by the lune move. So we adopt the above name and denote the loop move by α considering the shape of the letter as a motivation.

Remark 1.3.26. The loop move transforms a regular neighborhood of a triple point that consists of three half-discs with a common diameter. To specify the move, one should choose a moving half-disc that changes its position, and a fixed half-disc that will contain the created loop. The third half-disc is neutral. It turns out that only the choice of the neutral disc is important. For instance, moving A and fixed C (see Fig. 1.32) and moving C and fixed A produce the same result. Therefore, in a neighborhood of a triple point the move α can be made in three different ways.

Remark 1.3.27. Just to fix terminology, let us say that a special polyhedron P *contains a loop* if the singular graph SP of P contains a loop ℓ such that ℓ is the boundary curve of a 2-component of P and has a nontrivial normal bundle. It is evident that P contains a loop if and only if it can be obtained from another special polyhedron by the loop move.

The second auxiliary move β is defined by Fig. 1.39. It will serve as a substitute for the move U^{-1} we would like to avoid.

Lemma 1.3.28. *The moves* α *and* $\beta^{\pm 1}$ *can be presented as compositions of* $T^{\pm 1}, U$.

Proof. Figure 1.34 shows how one can present α and β as compositions of moves U, U, T^{-1} and U, T^{-1}, respectively. For a presentation of β^{-1} see Fig. 1.40. The last move in the Fig. 1.40 is the inverse of the vertex move V, which by Proposition 1.2.8 is a composition of T, T^{-1}. $\square$

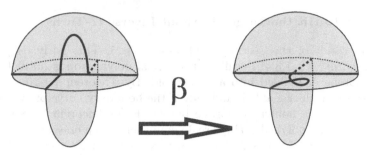

Fig. 1.39. A substitute for U^{-1}: the move β

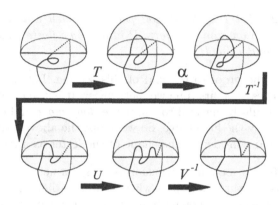

Fig. 1.40. A presentation of the move β^{-1}

The meaning of Lemma 1.3.29 is that loops are very movable: One may transfer them from one place to another using $T^{\pm 1}$ and U, but not U^{-1}.

Lemma 1.3.29. *Let a, b be two triple points of a special polyhedron P. Suppose that special polyhedra P_a and P_b are obtained from P by creating loops at a and b, respectively. Then one can transform P_a into P_b by moves $T^{\pm 1}, U$.*

Proof. We first consider the case when a and b lie in the same edge of P. Then we may assume that they coincide. There are three ways to insert a loop at a given point, see Remark 1.3.26. Suppose that P_a, P_b are obtained by two of them. Then one can transform P_a into P_b by a composition of moves $\beta^{-1}, \alpha, T^{-1}$, and β as shown in Fig. 1.41. We use Lemma 1.3.28 to split $\beta^{\pm 1}$ and α into compositions of moves $T^{\pm 1}, U$.

If a and b lie in edges of P having a common vertex, the transformation of P_a to P_b is carried out by β^{-1} and β: we create a U-turn instead of the loop at a, and then replace it by a loop at b. It follows that we can move loops along the singular graph SP wherever we like.   □

Corollary 1.3.30. (Cancellation of loops). *Applying moves $T^{\pm 1}$ and U, one can replace any two loops by one loop.*

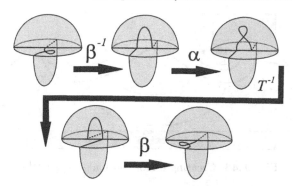

Fig. 1.41. Turning a loop inside out

Fig. 1.42. Canceling loops

Proof. Using α, we add a third loop. Then we use Lemma 1.3.29 to stack them one over another, and cancel a pair of loops by the move V^{-1}, see Fig. 1.42. It remains to recall that α is a composition of $T^{\pm 1}, U$ by Lemma 1.3.28 and V^{-1} is a composition of $T^{\pm 1}$ by Proposition 1.2.8.   □

Proposition 1.3.31. *Any unthickenable special polyhedron P can be transformed by the moves $T^{\pm 1}$ (without $U^{\pm 1}$!) to a special polyhedron with a single loop.*

Proof. By Theorem 1.1.20, there is a 2-component c of P such that the boundary curve of c has a nontrivial normal bundle. We apply the vertex move V to decompose c into two 2-components c', c''. The normal bundle of the boundary curve of the smaller 2-component c' is trivial while the one of c'' is nontrivial. Just as in the proof of Lemma 1.2.11, we apply moves $T^{\pm 1}$ to enlarge c' and diminish c'' until we get a loop, see Fig. 1.43.   □

Proof of Theorem 1.3.25. Let P, Q be special polyhedra such that $P \overset{3d}{\sim} Q$ and Q is unthickenable. By Proposition 1.3.31, we may assume that Q has a loop. By Theorem 1.3.8, one can find a sequence of moves $T^{\pm 1}, U^{\pm 1}$ transforming P into Q. We replace each move U^{-1} that occurs in the sequence by the move β,

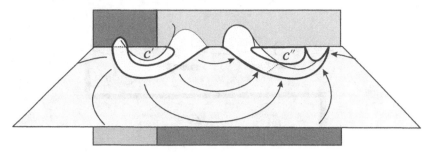

Fig. 1.43. Creating a loop without using $U^{\pm 1}$

which by Lemma 1.3.28 is a composition of moves $T^{\pm 1}, U$. The additional loops that arise in this way do not interfere with carrying out each next move, since by Lemma 1.3.29 we may assume that they are situated far from the place where the move is performed. Finally, we get Q with several superfluous loops. Since Q already has one loop, we can cancel all the other by Corollary 1.3.30.

$\square$

1.3.4 Zeeman's Collapsing Conjecture

The following innocuous-looking statement is known as the Zeeman conjecture [138].

Conjecture (ZC). Let X be a contractible polyhedron of dimension ≤ 2. Then $X \times I$ is collapsible.

Recall that X is called *collapsible* if $X \searrow \{*\}$, i.e., X can be collapsed to a point. Every collapsible polyhedron is *contractible* (it means that the identity map $X \to X$ is homotopic to a constant map $p \colon X \to x_0 \subset X$). The converse is not true. The Bing House and Abalone (see Fig. 1.6) are spines of the 3-ball and thus contractible. Nevertheless, they are not collapsible since there are no free edges to start the collapse. Taking a direct product with an interval produces free 2-dimensional faces on the top and bottom, so one can start. The problem is whether it is always possible to continue the collapse so that we do not become stuck before we reach a point.

Definition 1.3.32. *A polyhedron X is called 1-collapsible, if $X \times I \searrow \{*\}$.*

In these terms Zeeman's Conjecture can be reformulated as follows: *Any contractible 2-dimensional polyhedron is 1-collapsible.* One can easily show that the Bing House and Abalone are 1-collapsible. For a recent detailed account of the contemporary status of Zeeman's Conjecture, Andrews–Curtis Conjecture and other conjectures in low-dimensional topology and combinatorial group theory we refer the reader to the fundamental book [47]. Let us recall now the famous *Poincaré Conjecture.*

Conjecture (PC). Any simply connected closed 3-manifold is homeomorphic to S^3.

Since a closed 3-manifold is simply connected if and only if it is homotopy equivalent to S^3, one can reformulate PC as follows: *Any homotopy 3-sphere is S^3*. Removing from a homotopy sphere an open ball, we get another equivalent statement: *Any compact contractible 3-manifold is homeomorphic to the standard 3-ball*.

Proposition 1.3.33 shows that the Zeeman Conjecture is very strong.

Proposition 1.3.33. *The Zeeman Conjecture implies both the Poincaré and Andrews–Curtis Conjectures.*

Proof. ZC $\Rightarrow$ PC. Let M be a compact contractible 3-manifold. Construct a spine $X \subset M$ of dimension ≤ 2. Assuming ZC, we have $M \times I \searrow X \times I \searrow \{*\}$, i.e., $M \times I$ collapses to a point. Therefore, $M \times I$ is homeomorphic to a regular neighborhood of a point, that is, to a 4-dimensional ball. Taking into account that $M = M \times \{1\}$ lies in $\partial(M \times I) = S^3$ and $\partial M = S^2$, we get M is a 3-ball by the Alexander Theorem (every 2-sphere in S^3 bounds a 3-ball).

ZC $\Rightarrow$ AC. Note that any 2-dimensional polyhedron X 3-expands to $X \times I$. Taking a contractible polyhedron X and assuming ZC, we have $X \overset{3d}{\sim} X \times I \overset{3d}{\sim} \{*\}$. $\quad\square$

Remark 1.3.34. Proposition 1.3.33 remains true if we restrict ZC to the class of special polyhedra. Only minor modifications in the proof are needed. For proving ZC $\Rightarrow$ PC one should take not an arbitrary but a special spine X of M, and for proving ZC $\Rightarrow$ AC one should first 3-deform X to a special polyhedron and only then take the product with I. We may conclude that:

(1) ZC for special spines implies PC.
(2) ZC for unthickenable special polyhedra implies AC.

The aim of this section is to show that the converse is also true:

(1) PC implies ZC for special spines.
(2) AC implies ZC for unthickenable special polyhedra.

This close relation between the three classical conjectures seems to be remarkable. The equivalence PC $\Longleftrightarrow$ ZC for special spines was proved by Gillman and Rolfsen in [36]. The following very important statement played the main role in the proof.

Theorem 1.3.35. *[36] If P is a special spine of a homology 3-ball M, then $P \times I$ collapses to a subset homeomorphic to M.*

It is easy to show that any 3-manifold M with nonempty boundary contains a special spine satisfying the conclusion of Theorem 1.3.35. As we know, any two special spines of M are related by the moves $T^{\pm 1}$ (Theorem 1.2.5).

It would be natural to use this to give another proof of the above mentioned Gillman–Rolfsen result. The idea turns out to be fruitful and allows us to get an even stronger result: The statement is true for any M with $\partial M \neq \emptyset$.

Let P_1, P_2 be simple polyhedra having the same boundary, by which we mean that ∂P_2 is a copy of ∂P_1, and an identification homeomorphism $id \colon \partial P_1 \to \partial P_2$ is fixed.

Definition 1.3.36. *We say that P_2 dominates P_1 (notation: $P_1 \preceq P_2$) if there exists an embedding $i \colon P_1 \times I \to P_2 \times I$ such that the following holds:*

1. *The restriction of i onto $\partial P_1 \times I$ is fiber-preserving and projects to the identification homeomorphism id. This means that for any $(x, t) \in \partial P_1 \times I$ we have $i(x, t) \in \{id(x)\} \times I$.*
2. $P_2 \times I \searrow i(P_1 \times I)$.

Example 1.3.37. The following may help to visualize Definition 1.3.36. For simplicity, we consider 1-dimensional special polyhedra, i.e., graphs with vertices of valence 1 and 3. Let P_1 and P_2 be two trees such that each of them consists of 5 segments and spans the union $\partial P_1 = \partial P_2 = A \cup B \cup C \cup D$ of four points in two different ways. Then P_1 dominates P_2. One of the possible embeddings $i \colon P_1 \times I \to P_2 \times I$ is shown in Fig. 1.44. The property $P_2 \times I \searrow i(P_1 \times I)$ is evident.

Lemma 1.3.38. *Let W be a half-disc bounded by a diameter d and a half-circle c. Denote by C_1, C_2 its corner points. Then any embedding $i \colon d \times I \to W \times I$ that takes each interval $C_j \times I, j = 1, 2$ into itself can be extended to an embedding $i_1 \colon W \times I \to W \times I$ such that:*

1. *The restriction of i_1 onto $c \times I$ is fiber-preserving and projects to the identity map $c \to c$.*
2. $W \times I \searrow i(W \times I)$.

Proof. We extend i to an embedding $i' \colon \partial W \times I \to \partial W \times I$ by selecting a fiber-preserving embedding $c \times I \to c \times I$. Since the closed curve $i'(\partial W \times \{*\})$ is isotopic to the core circle of the annulus $\partial W \times I$, it bounds a half-disc

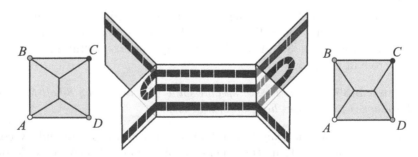

Fig. 1.44. An example of 1-dimensional domination

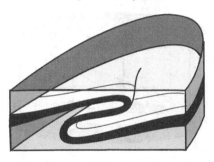

Fig. 1.45. A crafty half-disc (wing) inside the standard one

embedded into $W \times I$. The desired embedding i_1 can be now easily constructed by considering an appropriate product structure (collar) on a close regular neighborhood of this disc, see Fig. 1.45.   □

Let E be a simple polyhedron with boundary. We say that an embedded general position curve $b \subset E$ is *proper* if $b \cap \partial E = \partial b$. If b is a proper curve in E, one can attach to E a half-disc W by identifying the diameter $d \subset \partial W$ with b, and get a new simple polyhedron $E \cup_b W$ with boundary.

Definition 1.3.39. *We will say that the polyhedron $E \cup_b W$ is obtained from E by* attaching a wing *along b.*

Lemma 1.3.40 shows that quite often the domination relation is preserved under attaching of wings.

Lemma 1.3.40. *Let E_1, E_2 be simple polyhedra with identical boundaries and $b \subset E_1, c \subset E_2$ proper curves having the same endpoints in $\partial E_1 = \partial E_2$. Suppose that E_1 is dominated by E_2 such that the embedding $i : E_1 \times I \to E_2 \times I$ that realizes the domination takes $b \times I$ to $c \times I$. Then $E_1 \cup_b W$ is dominated by $E_2 \cup_c W$.*

Proof. Applying Lemma 1.3.38, we embed $W \times I$ into $W \times I$ and thus extend the given embedding $i : E_1 \times I \to E_2 \times I$ to an embedding $i_1 : (E_1 \cup_b W) \times I \to (E_2 \cup_c W) \times I$. To construct a collapse $(E_2 \cup_c W) \times I \searrow i_1((E_1 \cup_b W) \times I)$, we collapse $W \times I$ to $i_1(W \times I)$, and then apply the given collapse $E_2 \times I \searrow i(E_1 \times I)$.   □

Lemma 1.3.40 is a powerful tool for constructing new pairs of special polyhedra such that one is dominated by the other. Given a pair $P_1 \preceq P_2$, one may attach, step by step, additional wings to P_1 and P_2, each time getting a new pair. For a start, let us describe a nontrivial domination of the standard round disc D^2 by itself.

It is convenient to think of D^2 as embedded in $D^2 \times I$ as a horizontal section $D^2 \times \{1/4\}$. Let D_0 be a smaller round disc inside D^2. We modify this embedding by taking the center x_0 of D_0 to a point $(x_1, 3/4)$ and using

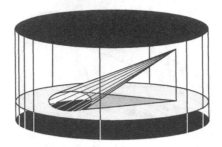

Fig. 1.46. A crafty embedding of $D^2 \times I$ into itself

the cone construction to get a new embedding $\varphi \colon D^2 \to D^2 \times I$. The point x_1 should lie outside D_0 in a diameter of D^2 containing x_0, see Fig. 1.46.

Since the disc $\varphi(D^2)$ lies in $D^2 \times I$ with a collar, one can extend φ to an embedding $i \colon D^2 \times I \to D^2 \times I$ such that the restriction of i onto $\partial D^2 \times I$ is fiber-preserving. Clearly, i is a domination embedding, i.e., $D^2 \times I \searrow i(D^2 \times I)$.

Remark 1.3.41. The embedding φ constructed above is piecewise-smooth, since we have used round discs and cone construction. If one prefers PL category, one may replace the round discs with polygons. A smooth version of the embedding φ can be obtained by gluing together two copies of the half-disc W embedded with a collar into $W \times I$, see Fig. 1.45.

Remark 1.3.42. Let $a_0, a_1 \in \partial D_0$ be the two points where straight lines passing through x_1 are tangent to ∂D_0. It is easy to see that the map $p\varphi \colon D^2 \to D^2$ is a local homeomorphism everywhere except on the union U of the two radii $[x_0, a_0]$, $[x_0, a_1]$, and the smaller arc of ∂D_0 between a_0, a_1. It follows that the embedding i can be assumed to be fiber-preserving everywhere except in a small neighborhood of U.

Denote by d the diameter of D^2 containing x_0 and x_1. We will consider chords and half-chords of D^2 that are orthogonal to d. Note that the embedding $i \colon D^2 \times I \to D^2 \times I$ constructed above takes $d \times I$ into itself. Other useful properties of i are described in Lemma 1.3.43.

Lemma 1.3.43. *There exist a half-chord c_0, a chord c_1, and a simple curve c_2 in D^2 such that the following holds:*

(1) $c_1 \cap c_0 = \emptyset$, $\partial c_2 = \partial c_1$, *and* $c_2 \cap c_0$ *is a point*
(2) i *embeds each fiber* $\{*\} \times I$ *of* $c_0 \times I$ *into itself*
(3) *The restriction of i onto $c_1 \times I$ is a fiber-preserving embedding into $c_2 \times I$*

Proof. Choose a half-chord c_0 such that the point $c_0 \cap d$ lies strictly between D_0 and x_1. Since $c_0 \cap D_0 = \emptyset$, we have (2). We take a chord c_1 such that the point $c_1 \cap d$ lies in D_0 very close to x_0, but outside the segment $[x_0, x_1]$ of d. Since c_1 does not intersect the singular set U (see Remark 1.3.42), the

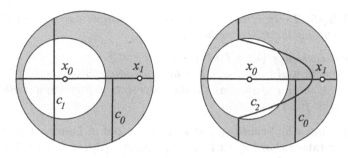

Fig. 1.47. Chord trading

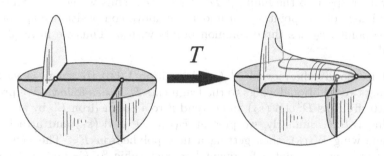

Fig. 1.48. A convenient form of the T-move

restriction of i onto $c_1 \times I$ is a fiber-preserving map. Denoting by c_2 the image of $i(c_1 \times I)$ under the direct product projection $p: D^2 \times I \to D^2$, we get (3) automatically. Since c_1 lies very close to x_0, c_2 passes close to x_1 and thus intersect c_0. See Fig. 1.47. $\square$

Recall that the move T consists in replacing a fragment E_T of a special polyhedron by a fragment E_T' with the same boundary, see Definition 1.2.3.

Proposition 1.3.44. $E_T \preceq E_T'$.

Proof. Let c_0, c_1, c_2, and d be as in Lemma 1.3.43. Present E_T as the disc D^2 with three wings $W(d), W(c_1), W(c_0)$, where $W(d)$ and $W(c_1)$ are attached along d and c_1, respectively, and $W(c_0)$ is attached along the union of c_0 and a half-chord in $W(d)$ (see Fig. 1.48). E_T' has a similar structure. The only difference is that instead $W(c_1)$ we attach another wing $W(c_2)$ along c_2.

Now let us prove that $E_T \preceq E_T'$. Consider the embedding $i: D^2 \times I \to D^2 \times I$ which satisfy conditions 2, 3 of Lemma 1.3.43. In order to convert i to an embedding $i_1: E_T \times I \to E_T' \times I$ that realizes the domination $E_T \preceq E_T'$, we apply Lemma 1.3.40 three times. First, we extend i to an embedding $(E_T \cup_d W(d)) \times I$ into $(E_T' \cup_d W(d)) \times I$. Then we do the same with respect to the wings $W(c_1)$ and $W(c_0)$. Each time we use Lemma 1.3.43 to verify that the assumptions of Lemma 1.3.40 are satisfied.

Lemma 1.3.45. *There exist a half-chord c_0, a simple curve s_1 in D^2 and a chord s_2 such that the following holds:*

(1) *i embeds each fiber $\{*\} \times I$ of $c_0 \times I$ into itself.*
(2) *The restriction of i onto $s_1 \times I$ is an embedding into $s_2 \times I$.*
(3) *s_1 has the same endpoint as s_2 and intersects d three times. Both s_1, s_2 are disjoint from c_o.*

Proof. Choosing a half-chord c_0 just as in the proof of Lemma 1.3.43, we get (1). Then we take a chord s_2 such that the point $s_2 \cap d$ is strictly between x_1 and $c_0 \cap d$, $s_2 \cap c_0 = \emptyset$, and $s_2 \cap D_0$ is an interval. Let s_1 be the inverse image of s_2 with respect to the map $p\varphi \colon D^2 \times I \to D^2$. Thus we get (2). Since the inverse image of the point $s_2 \cap d$ under the above composition map consists of three points, s_2 has three common points with d. Thus we have (3). See Fig. 1.49. □

Let us modify the move V^{-1} (see Definition 1.2.6) to a new move S by attaching two additional wings to the fragments E, E_V as follows. We present the butterfly E as $D^2 \cup W(s_2)$ and extend it to a polyhedron E'_S by attaching the wing $W(c_0)$. Similarly, we present E_V as $D^2 \cup W(s_1)$ and attach to it the same wing $W(c_0)$, thus getting a new polyhedron E_S. The new move S consists in cutting out a fragment homeomorphic to E_S and replacing it by the fragment E'_S. See Fig. 1.50, where the new fragments E_S and E'_S are presented as D^2 with wings. Note that E'_S is homeomorphic with E_T.

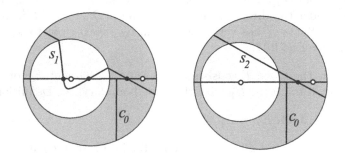

Fig. 1.49. Another chord trading

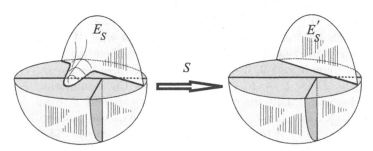

Fig. 1.50. The modified V^{-1}

Proposition 1.3.46. $E_S \preceq E'_S$

Proof. Just as in the proof of Proposition 1.3.44, we apply Lemma 1.3.40 three times, each time using Lemma 1.3.45 to verify the assumptions.   □

Lemma 1.3.47. *Let P_1, P_2 be simple polyhedra such that P_2 can be obtained from P_1 by removing a proper fragment $E_1 \subset P_1$ and replacing it by another fragment E_2 having the same boundary. Then $E_1 \preceq E_2$ implies $P_1 \preceq P_2$.*

Proof. The embedding $i: E_1 \times I \to E_2 \times I$ that realizes the domination $E_1 \preceq E_2$ takes each fiber $\{x\} \times I, x \in \partial E_1$, into itself. Improving it by a fiber-preserving isotopy, we may assume that the restriction of i onto $\partial E_1 \times I$ is the identity map $\partial E_1 \times I \to \partial E_2 \times I$. Now we can extend it by the identity onto the rest of $P_1 \times I$ and get an embedding $P_1 \times I \to P_2 \times I$ that determines the domination $P_1 \preceq P_2$.   □

Theorem 1.3.48. *Let P_1, P_2 be T-equivalent special polyhedra. Then $P_1 \preceq P_2$ and $P_2 \preceq P_1$. In other words, there exist embeddings $i_1: P_1 \times I \to P_2 \times I$ and $i_2: P_2 \times I \to P_1 \times I$ such that $P_2 \times I \searrow i_1(P_1 \times I)$ and $P_1 \times I \searrow i_2(P_2 \times I)$.*

Proof. Since the relation $\preceq$ is transitive, it is sufficient to prove the theorem for the case when P_2 is obtained from P_1 by a single move T. It follows from Proposition 1.3.44 and Lemma 1.3.47 that $P_1 \preceq P_2$. To get $P_2 \preceq P_1$, we present the move T^{-1} from P_2 to P_1 as a composition $P_2 \to Q \to P_1$ of moves T and S, see Fig. 1.51. We get $P_2 \preceq Q \preceq P_1$, where the first domination follows from Preposition 1.3.44 and the second one from Preposition 1.3.46.   □

Corollary 1.3.49. *The Zeeman Conjecture for special spines is equivalent to the Poincaré Conjecture.*

Proof. Theorems 1.2.5 and 1.3.48 together imply that if some special spine P_1 of a 3-manifold M is 1-collapsible, then so is any other spine P_2: We use the domination $P_1 \preceq P_2$ to collapse $P_2 \times I$ to $i(P_1 \times I)$ and then to a point. Recalling that the Bing House with two Rooms (see Fig. 1.6) is 1-collapsible, we get the 1-collapsibility of all special spines of the 3-ball and, assuming

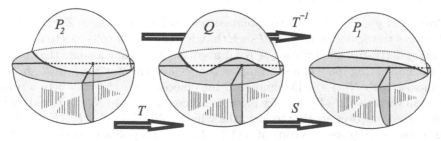

Fig. 1.51. T^{-1} as a composition of T and S

PC, of all contractible special spines. Thus PC implies ZC for special spines. The inverse implication was proved earlier (see Proposition 1.3.33 and Remark 1.3.34).   □

Remark 1.3.50. The original proof of the equivalence PC $\Longleftrightarrow$ ZC for special spines was based on a nontrivial observation that if P is a special spine of a homology 3-ball M, then $P \times I$ collapses to a subset homeomorphic with M, see Theorem 1.3.35. The restriction on the homology groups of M was used only in one place: the 2-components of any special spine of a homology ball can be oriented so that on any edge *two of the three induced orientations are opposite to the third*. One can easily show that any 3-manifold M possesses a special spine P having this property. Therefore $P \times I$ collapses to a subpolyhedron homeomorphic with M. It follows from Theorem 1.3.48 that this remains true for all special spines of M. Thus, Theorem 1.3.35 is true without any restriction on M at all.

Let us turn our attention to the general case of simple polyhedra that are not necessarily spines of 3-manifolds. We know from Theorem 1.3.8 that every 3-deformation of special polyhedra can be replaced by a sequence of moves $T^{\pm 1}, U^{\pm 1}$. The moves $T^{\pm 1}$ preserve the 1-collapsibility. Our aim now is to clarify how U affects it. Recall that the U-move consists in replacing a fragment E of a special polyhedron by another fragment E_U, see Definition 1.3.7. It is not true anymore that $E \preceq E_U$ or $E_U \preceq E$, so one should try other ways.

At first we prove a statement about arbitrary (not necessarily 2-dimensional) polyhedra. Let Y be a subpolyhedron of a polyhedron X. Recall that a regular neighborhood of Y in X is defined as $N(Y) = |\mathrm{St}(L'', K'')|$, where (K, L) is a triangulation of the pair (X, Y). Admitting an abuse of notation (in general, $N(Y)$ is not a manifold), we will write $\partial N(Y) = N(Y) \setminus \mathrm{Int} N(Y)$, and say that $\partial N(Y)$ is the boundary of the regular neighborhood. Consider the quotient space X/Y. It has a natural polyhedral structure that can be described as follows: if we remove Int $N(Y)$ from X and add the cone over $\partial N(Y)$, we get a polyhedron homeomorphic with X/Y. Note that the complements of $N(Y)$ in X and of the cone neighborhood $N(y)$ of the cone vertex y in X/Y do coincide.

Theorem 1.3.51. *Let Y be a collapsible subpolyhedron of a polyhedron X. If $(X/Y) \times I \searrow \{*\}$, then $X \times I \searrow \{*\}$.*

Proof. Let $y \in Z = X/Y$ be the point corresponding to Y. Since $Z \times I \searrow \{*\}$, there is a triangulation K of $Z \times I$ such that K simplicially collapses to a point. Taking derivative subdivisions of K (if necessary), we may assume that $y \times I$ is the underlying space of a subcomplex $J \subset K$ and the star $S = |\mathrm{St}(J'', K'')|$ coincides with $N(y) \times I$. Decompose S into pieces $|\mathrm{St}(v, K'')|$, where v runs over all vertices of J'. Each of the pieces is homeomorphic to $N(y) \times I$, and they are placed one over another to form the whole "pillar" $N(y) \times I \subset Z \times I$. Let us replace each piece by a copy of $N(Y) \times I$. We get a polyhedron $W = W(K)$ homeomorphic with $X \times I$, see Fig. 1.52.

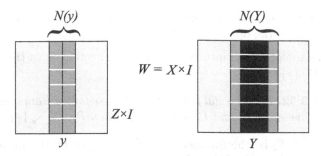

Fig. 1.52. $Z \times I$ and a homeomorphic copy W of $X \times I$

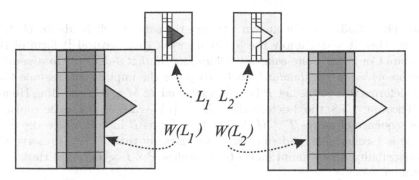

Fig. 1.53. How collapses of $W(L)$ follow the ones of L

To prove that W is collapsible, for every subcomplex $L \subset K$ define a subpolyhedron $W(L) \subset W$ as follows:

1. Add to L all the pieces $N(y) \times I$ that correspond to those vertices of L' that lie in J'.
2. Replace each of the above pieces by a copy of $N(Y) \times I$.

Using the collapsibility of Y, one can easily verify that the given simplicial collapse $K \searrow^s \{*\}$ induces a polyhedral collapse $W(K) \searrow W(\{*\})$. To be more precise, consider an elementary simplicial collapse $L_1 \searrow L_2$ consisting in removing a principal open simplex σ of L_1, together with its free open face δ. We have four possibilities:

1. If σ has no vertices in J, we can apply to $W(L_1)$ actually the same collapse and get $W(L_2)$.
2. If σ has a vertex in J but δ is not in J, then instead of removing σ and δ we collapse the closed simplex $\bar{\sigma}$ to $\mathrm{St}((\partial\bar{\sigma} \setminus \delta)'', \bar{\sigma}'')$. This collapse also produces a collapse $W(L_1) \searrow W(L_2)$.
3. If δ is a 1-dimensional simplex of J, we use the collapsibility of Y to remove the piece $N(Y) \times I$ corresponding to the barycenter of δ. Then we collapse the rest of σ as shown in Fig. 1.53 for the case $L_1 = J \cup \sigma$.

4. If δ is a vertex of J, we collapse the corresponding piece $N(Y) \times I$ and then the rest of the open segment σ.

It remains to note that since $K \searrow \{*\}$, $W(K)$ collapses to the collapsible polyhedron $W(\{*\})$. $\square$

Corollary 1.3.52. *Let a special polyhedron P_2 be obtained from a special polyhedron P_1 by one move U and $P_1 \times I \searrow \{*\}$. Then $P_2 \times I \searrow \{*\}$.*

Proof. Let $E_U \subset P_2$ be the fragment which is inserted by the move U. Since $P_2/E_U = P_1$ and $P_1 \times I \searrow \{*\}$, it follows from Theorem 1.3.51 that $P_2 \times I \searrow \{*\}$. $\square$

Remark 1.3.53. It is not known whether the same result holds for U^{-1}. If it does, then it would imply the following weak (and natural !) form of the Poincaré Conjecture: *any compact 3-manifold M that 3-deforms to a point is homeomorphic to the standard 3-ball.* To prove the implication, assume that M 3-deforms to a point. Let P be a special spine of M and B the Bing House. By Theorem 1.3.8, the 3-deformation $B \to \{*\} \to M \to P$ can be replaced by a sequence of moves $T^{\pm 1}, U^{\pm 1}$ that transforms B into P. Since the Bing House is 1-collapsible, and since we know already that $T^{\pm 1}, U$ preserve the 1-collapsibility, the assumption on U^{-1} implies $P \times I \searrow \{*\}$, and thus M is a genuine 3-ball.

The remark above explains our efforts to get rid of U^{-1}, see Theorem 1.3.25, which shows that quite often it is possible.

Theorem 1.3.54. *Let P_1, P_2 be special polyhedra such that $P_1 \overset{3d}{\sim} P_2$ and P_2 is unthickenable. If P_1 is 1-collapsible, so is P_2.*

Proof. By Theorem 1.3.25, there exists a sequence of moves $T^{\pm 1}$ and U transforming P_1 into P_2. By Theorem 1.3.48 and Corollary 1.3.52, these moves preserve the 1-collapsibility. $\square$

Remark 1.3.55. One can prove (see [78]) that under the same assumptions as in Theorem 1.3.54 a more general conclusion is true: if $P_1 \times I$ collapses to a subpolyhedron X of dimension ≤ 2, then $P_2 \times I$ collapses to a subpolyhedron homeomorphic with X.

Corollary 1.3.56 may serve as a powerful tool for producing a great variety of 2-dimensional polyhedra that satisfy the Zeeman Conjecture.

Corollary 1.3.56. *ZC is true for all unthickenable special polyhedra that can be 3-deformed to a point.*

Proof. Suppose an unthickenable special polyhedron Q can be 3-deformed to a point. Since the Bing House is 1-collapsible, so is Q by Theorem 1.3.54. $\square$

Corollary 1.3.57. ZC *restricted to unthickenable special polyhedra is equivalent to* AC.

Proof. By Corollary 11.3.56, AC implies ZC for unthickenable special polyhedra. Since any 2-dimensional polyhedron 3-deforms to an unthickenable special one, the inverse implication is also true.   □

Combining Corollaries 1.3.49 and 1.3.57, we have

Theorem 1.3.58. *[77]* ZC *restricted to special polyhedra is equivalent to the union of* PC *and* AC.

Theorem 1.3.58 may cast a doubt on the widespread belief that ZC is false. If a counterexample indeed exists, then either it has a "bad" local structure (is not a special polyhedron) or it is a counterexample to either AC or PC.

Complexity Theory of 3-Manifolds

Denote by $\mathcal{M}$ the set of all compact 3-manifolds. We wish to study it systematically and comprehensively. The crucial question is the choice of filtration in $\mathcal{M}$. It would be desirable to have a finite number of 3-manifolds in each term of the filtration, all of them being in some sense simpler than those in the subsequent terms. A useful tool here would be a measure of "complexity" of a 3-manifold. Given such a measure, we might hope to enumerate all "simple" manifolds before moving on to more complicated ones. There are several well-known candidates for such a complexity function. For example, take the Heegaard genus $g(M)$, defined to be the minimal genus over all Heegaard decompositions of M. Other examples include the minimal number of simplices in a triangulation of M and the minimal crossing number in a surgery presentation for M.

Each of these measures has its shortcomings. The Heegaard genus is additive with respect to connected sums of 3-manifolds, but for $g \geq 1$ there are infinitely many distinct manifolds of Heegaard genus g, and already for $g = 2$ one can hardly expect a simple classification. The surgery complexity has the same defect (because of framing). The minimal number of simplices in a triangulation is not a "natural" measure of complexity because the simplest possible closed manifold, S^3, already would have nonzero complexity, and we would have no chances to get the additivity.

In this chapter, an integral non-negative function $c \colon \mathcal{M} \to \mathcal{Z}$ is constructed, which has the following properties:

1. c is additive, that is, $c(M_1 \# M_2) = c(M_1) + c(M_2)$.
2. For any $k \in Z$, there are only finitely many closed irreducible manifolds $M \in \mathcal{M}$ with complexity $c(M) = k$.
3. $c(M)$ is relatively easy to estimate.

2.1 What is the Complexity of a 3-Manifold?

2.1.1 Almost Simple Polyhedra

As we know from Sect. 1.1.4, any homeomorphism between special spines can be extended to a homeomorphism between the corresponding manifolds (Theorem 1.1.17). This means that a special spine P of a 3-manifold M may serve as a presentation of M. Moreover, M can be reconstructed from a regular neighborhood $N(SP)$ in P of the singular graph SP of P: Starting from $N(SP)$, one can easily reconstruct P by attaching 2-cells to all the circles in $\partial N(SP)$, and then reconstruct M. If M is orientable, then $N(SP)$ can be embedded into R^3. This gives us a very convenient way for presenting 3-manifolds: we simply draw a picture, see Fig. 2.1.

Theorem 2.1.1. *For any integer k there exists only a finite number of special spines with k true vertices. All of them can be constructed algorithmically.*

Proof. We will construct a finite set of special polyhedra that a fortiori contains all special spines with k true vertices. First, one should enumerate all regular graphs of degree 4 with k true vertices. Clearly, there is only a finite number of them. Given a regular graph, we replace each true vertex v by a copy of the butterfly E that presents a typical neighborhood of a true vertex in a simple polyhedron, see Definition 1.1.8. Neighborhoods in ∂E of triple points of ∂E (we will call them *triodes*) correspond to edges having an endpoint at v. In Fig. 2.2 the triodes are shown by fat lines. For each edge e, we glue together the triodes that correspond to endpoints of e via a homeomorphism between them. It can be done in six different ways (up to isotopy). We get a simple polyhedron P with boundary. Attaching 2-discs to the circles in

Fig. 2.1. Bing's House with two Rooms and its mutant (another special spine of the cube) presented as regular neighborhoods of their singular graphs

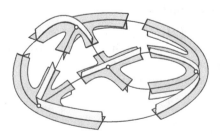

Fig. 2.2. A decomposition of $N(SP)$ into copies of E

∂P, we get a special polyhedron. Since at each step we have had only a finite number of choices, this method produces a finite set of special polyhedra. Not all of them are thickenable. Nevertheless, the set contains all special spines with k true vertices.   □

It would be a natural idea to measure the complexity of a 3-manifold by the number of true vertices of its special spine. This characteristic is convenient in that there exists only a finite number of 3-manifolds having special spines with a given number of vertices. But it has two shortcomings. First, it is not additive with respect to connected sums. Second, restricting ourselves to special spines, we lose the possibility to consider very natural spines such as a point for the ball (and S^3), a circle for the solid torus, and a projective plane for the projective space RP^3. Also, working only with special spines, we are sometimes compelled to make artificial tricks to preserve the special polyhedra structure. For example, in the proof of Theorem 1.1.13 we used a delicate arch construction instead of simply making a hole in a 2-cell.

All these shortcomings have the same root: the property of being special is not hereditary. In other words, a subpolyhedron of a special polyhedron may not be special, even if it cannot be collapsed onto a smaller subpolyhedron. This is why we shall widen the class of special polyhedra by considering a class of what we call almost simple polyhedra. Roughly speaking, the class of almost simple polyhedra is the minimal class which contains special polyhedra and is closed with respect to the passage to subpolyhedra.

Definition 2.1.2. *A compact polyhedron P is said to be* almost simple *if the link of any of its points can be embedded into Γ_4, a complete graph with four vertices. A spine P of a 3-manifold M is* almost simple, *if it is an almost simple polyhedron.*

It is convenient to present Γ_4 as a circle with three radii or as the boundary of the standard butterfly. One usually considers only almost simple polyhedra that cannot be collapsed onto smaller subpolyhedra. It is easy to see that any proper subpolyhedron of the circle with three radii can be collapsed onto a polyhedron L having one of the following types:

(a) L is either empty or a finite set of $n \geq 2$ points.
(b) L is the union of a finite (possibly empty) set and a circle.
(c) L is the union of a finite (possibly empty) set and a circle with a diameter.
(d) L is a Γ_4.

The "cannot start" property assures us that an almost simple polyhedron P cannot be collapsed onto a smaller subpolyhedron if and only if the link L of any point of P is contained in the above list.

For example, a wedge of any simple polyhedron and any graph without free vertices satisfies this condition and hence cannot be collapsed onto a smaller subpolyhedron. This example is very typical, since any almost simple polyhedron P can be presented as the union of its 2-dimensional and its 1-dimensional parts. The 1-dimensional part (the closure of the set of points with 0-dimensional links) is a graph, the 2-dimensional part consists of points whose links contain an arc. If P cannot be collapsed onto a smaller subpolyhedron, then its 2-dimensional part is a simple polyhedron (maybe disconnected).

The notions of a true vertex, singular graph, 2-component of an almost simple polyhedron are introduced in the same way as for simple polyhedra, see Sect. 1.1.3. A *true vertex* of an almost simple polyhedron P is a point with the link $L = \Gamma_4$, the *singular graph SP* consists of points whose links contain a circle with a diameter, and *2-components* are the connected components of the set of all the points whose links contain a circle but do not contain a circle with a diameter. Note that the 1-dimensional part does not affect these notions. For instance, a 2-component may contain a point of the 1-dimensional part, and this point is not a true vertex of P.

Almost simple spines are easier to work with than special spines, since we may puncture cells and stay within the realm of almost simple spines. So, for example, the process we used to construct a special spine for a given manifold may be simplified to give an almost simple spine; there is no longer need for the arch construction, see Fig. 1.8.

2.1.2 Definition and Estimation of the Complexity

The complexity function adverted to in the introduction to this chapter can now be defined.

Definition 2.1.3. *The* complexity $c(P)$ *of a simple polyhedron P is equal to the number of its true vertices.*

Definition 2.1.4. *The* complexity $c(M)$ *of a compact 3-manifold M is equal to k if M possesses an almost simple spine with k true vertices and has no almost simple spines with a smaller number of true vertices. In other words, $c(M) = \min_P c(P)$, where the minimum is taken over all almost simple spines of M.*

Let us give some examples. The complexity of S^3, of the projective space RP^3, of the lens space $L_{3,1}$, and the manifold $S^2 \times S^1$ is equal to zero, since they possess almost simple spines without true vertices: the point, the projective plane, the triple hat, and the wedge of S^2 with S^1, respectively. Recall that by the triple hat we mean the quotient space of D^2 by a free action of the group Z_3 on ∂D^2. Among compact manifolds with boundary, zero complexity is possessed by all handlebodies, I-bundles over surfaces, as well as some other manifolds such as the complement of the trefoil knot. Indeed, any handlebody collapses to a graph that (being considered as an almost simple polyhedron) has no true vertices. The I-bundles collapse to surfaces, and the complement of the trefoil collapses to the quotient space of the Möbius band by a free action of the group Z_3 on the boundary.

In general, the problem of calculating the complexity $c(M)$ is very difficult. Let us start with a simpler problem of estimating $c(M)$. To do that it suffices to construct an almost simple spine P of M. The number of true vertices of P will serve as an upper bound for the complexity. Since an almost simple spine can be easily constructed from practically any presentation of the manifold, the estimation problem does not give rise to any difficulties. Let us describe several estimates of the complexity based on different presentations of 3-manifolds. It is convenient to start with an observation that removing an open ball does not affect the complexity.

Proposition 2.1.5. *Suppose that B is a 3-ball in a 3-manifold M. Then $c(M) = c(M \setminus Int\ B)$.*

Proof. If M is closed, then $c(M) = c(M \setminus Int\ B)$ since M and $M \setminus Int\ B$ have the same spines by definition of the spine of a closed manifold. Let $\partial M \neq \emptyset$, and let P be an almost simple spine of $M \setminus Int\ B$ possessing $c(M \setminus Int\ B)$ true vertices. Denote by C the connected component of the space $M \setminus P$ containing B. Since M is not closed, there exists a 2-component α of P that separates C from another component of $M \setminus P$. Removing an open 2-disc from α and collapsing yields an almost simple spine $P_1 \subset P$ of M. The number of true vertices of P_1 is no greater than that of P, since puncturing α and collapsing results in no new true vertices. Therefore, $c(M) \leq c(M \setminus Int\ B)$.

To prove the converse inequality, consider an almost simple spine P_1 of M with $c(M)$ true vertices. Let us take a 2-sphere S in M such that $S \cap P_1 = \emptyset$. Join S to P_1 by an arc ℓ that has no common points with $P_1 \cup S$ except the endpoints. Clearly, $P = P_1 \cup S \cup \ell$ is an almost simple spine of $M \setminus Int\ B$. New true vertices do not arise. It follows that $c(M) \geq c(M \setminus Int\ B)$. □

In Sect. 1.1.5 we described a relation between singular triangulations of closed 3-manifolds and special polyhedra. The same method works for estimating the complexity.

Proposition 2.1.6. *Suppose a 3-manifold M is obtained by pasting together n tetrahedra by affine identifications of their faces. Then $c(M) \leq n$.*

Proof. Recall that any tetrahedron Δ contains a canonical copy $P_\Delta = \cup|\operatorname{lk}_i(v_i, \Delta')|$ of the standard butterfly E, where $v_i, 0 \leq i \leq 3$, are the vertices of Δ. When pasting together the tetrahedra, these copies are glued together into a simple polyhedron $P \subset M$ that may have a boundary if M is not closed. P has n true vertices and is a spine of M with several balls removed from it. These balls are the neighborhoods of the points which are obtained by gluing the vertices of the tetrahedra and lie in the interior of M. It follows from Proposition 2.1.5 that $c(M) \leq n$. □

Remark 2.1.7. It follows from Corollary 1.1.27 that a closed 3-manifold M possesses a special spine with n true vertices if and only if it can be obtained by pasting together n tetrahedra. Further, we shall see that any minimal (in the sense of the number of true vertices) almost simple spine of a closed orientable irreducible 3-manifold M which differs from the "exceptional" manifolds $S^3, RP^3, L_{3,1}$, is special. Therefore, the complexity of such a manifold may be defined as the minimal number of tetrahedra that is sufficient to obtain M.

Proposition 2.1.8. *Suppose $M = H_1 \cup H_2$ is a Heegaard splitting of a closed 3-manifold M such that the meridians of the handlebody H_1 intersect the ones of H_2 transversally at n points. Suppose also that the closure of one of the components into which the meridians of H_1, H_2 decompose the Heegaard surface $\partial H_1 = \partial H_2$ contains m such points. Then $c(M) \leq n - m$.*

Proof. Denote by P the union of the Heegaard surface $F = \partial H_1 = \partial H_2$ with the meridional discs of the two handlebodies. Then P is a simple polyhedron whose true vertices are the crossing points of the meridians. Since the complement of P in M consists of two open 3-balls, P is a spine of M punctured twice. Removing from P the 2-component $\alpha \subset F$ whose closure contains m true vertices, we fuse together the balls and get an almost simple spine of M which has $n - m$ true vertices, since the vertices in the closure of α will cease to be true vertices, see Fig. 2.3.   □

Proposition 2.1.9. *Suppose $\widetilde{M}$ is a k-fold covering space of a 3-manifold M. Then $c(\widetilde{M}) \leq kc(M)$.*

Proof. Let P be an almost simple spine of M having $c(M)$ true vertices. Consider the almost simple polyhedron $\widetilde{P} = p^{-1}(P)$, where $p \colon \widetilde{M} \to M$ is the covering map. Since the degree of the covering is k, the polyhedron $\widetilde{P}$ has $kc(M)$ true vertices. If $\partial M \neq \emptyset$, then $\widetilde{P}$ is an almost simple spine of $\widetilde{M}$, since the collapse of M onto P can be lifted to a collapse of $\widetilde{M}$ onto $\widetilde{P}$. Therefore, $c(\widetilde{M}) \leq kc(M)$.

If M is closed, $\widetilde{P}$ is a spine of the manifold $\widetilde{M} \setminus \pi^{-1}(V)$, where V is an open 3-ball in M. The inverse image $p^{-1}(V)$ consists of k open 3-balls, hence, by Proposition 2.1.5, we have $c(\widetilde{M}) = c(\widetilde{M} \setminus p^{-1}(V)) \leq kc(M)$.   □

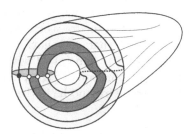

Fig. 2.3. Special spine of $L_{4,1}$ obtained from the standard Heegaard diagram of $L_{4,1}$

Remark 2.1.10. If M in the above proof is closed, then one can get an almost simple spine of $\widetilde{M}$ by puncturing those 2-components of $\widetilde{P}$ that separate different balls in $p^{-1}(V)$. To fuse k balls together, we must make $k-1$ punctures, and each of them decreases the total number of true vertices by the number of true vertices in the boundary of the 2-component we are piercing through. Thus, as a rule, $c(\widetilde{M})$ is significantly less than $kc(M)$.

Now we turn our attention to link complements and surgery presentations of 3-manifolds. Assume that a link L in the space $R^3 = S^3 \setminus \{*\}$ with coordinates x, y, z is in a general position with respect to the projection of R^3 onto the plane R^2 with the coordinates x, y. We will use the generally accepted way of presenting L by its projection $\bar{L}$, disconnecting it at the lower double points. The words *lower* and *higher* are understood in the sense of the value of the coordinate z. Connected components of the projection cut up in this way will be called *overpasses*. Each overpass is bounded by two lower points, and contains several upper crossing points. Their number will be called the *overpass degree*. We may look at the link from below and disconnect it at upper double points. Then we get *underpasses*. The number of lower points on an underpass is called the *underpass degree*. Let us call an overpass and an underpass *independent,* if the corresponding sets of double points (including the endpoints) are disjoint.

Often it is convenient to think of L as being contained in $S^3 = R^3 \cup \{*\}$ rather than in R^3. Then the projection $\bar{L}$ of L is in the sphere $S^2 = R^2 \cup \{*\}$. In this case the complement space $C(L) = S^3 \setminus \mathrm{Int}\, N(L)$, where $N(L)$ is an open tubular neighborhood of L in S^3, is a compact 3-manifold.

Proposition 2.1.11. *Suppose a link $L \subset S^3$ is given by a projection $\bar{L}$ with n crossing points so that there are an overpass of degree k and an independent underpass of degree m. Then the complexity of the complement space $C(L)$ of L is no greater than $4(n - m - k - 2)$.*

Proof. Let us attach the annulus $S^1 \times I$ along the projection $\bar{L}$ to S^2 and to the other parts of the annulus previously pasted on. We get a "tunnel," see Fig. 2.4, where the attaching procedure is shown in the neighborhood of a crossing point.

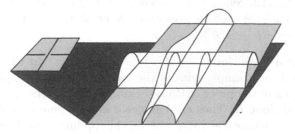

Fig. 2.4. Attaching a tunnel to S^2 produces a simple spine of the twice punctured link complement

The result will be a simple polyhedron P with $4n$ true vertices: each crossing point produces four of them. The complement to a regular neighborhood $N(P)$ of P in S^3 is the union of a tubular neighborhood $N(L)$ of L and two balls B_1, V_2 that lie inside and outside S^2, respectively. In order to get a spine of $C(L)$, one should fuse the balls with $N(L)$ by puncturing two 2-components of P that separate $N(L)$ from the balls. Choose for the puncture the 2-components $\alpha \subset S^1 \times I$ and $\beta \subset S^2$ that correspond to the overpass of degree k and the underpass of degree m, respectively. When we remove α, then $4k + 4$ true vertices disappear (two pairs correspond to the endpoints of the overpass, and $4k$ are related with k crossing points). Removing β, we destroy $4m + 4$ true vertices. It follows that after collapsing we get an almost simple spine of $C(L)$ with no more than $4(n - m - k - 2)$ vertices.   □

Remark 2.1.12. It can be shown that if the projection $\bar{L}$ has $n \geq 6$ crossings, then one can always find an overpass and independent underpass satisfying $k + m \geq 2$. The complexity of $C(L)$ can then be estimated by $4n - 16$. If there are no independent overpasses and underpasses, then one can use dependent ones or, alternatively, puncture a 2-component that lies on S^2 and separates the balls B_1, B_2. The number of disappearing true vertices in this case may be smaller, since the same true vertex may be taken into account twice.

Consider now the surgery presentation of 3-manifolds [72]. For simplicity, we restrict ourselves to the case when M is presented by a framed knot K. Recall that the *writhe* $w(\bar{K})$ of a projection $\bar{K}$ may be defined as the framing number of the "vertical" framing of K by the vector field orthogonal to R^2. To get an arbitrary framing s, one should twist the vertical framing $|s - w(\bar{K})|$ times in the appropriate direction.

Denote by ℓ the *preferred longitude* of K, i.e., the simple closed curve in $\partial N(K)$ that intersects a meridian m of $\partial N(K)$ at one point and is homologous to 0 in the complement to $N(K)$. Let K have the framing s. To convert S^3 to M, one should make two steps:

(1) Cut $N(K)$ out of S^3
(2) Glue in the solid torus $D^2 \times S^1$ so that the meridian $\partial D^2 \times \{*\}$ winds once around the longitude ℓ and s times around the meridian m

Proposition 2.1.13. *Suppose M is obtained by Dehn surgery along a knot K with framing s such that the projection $\bar{K}$ of K has $n \geq 1$ crossing points. Then $c(M) \leq 5n + |s - w(\bar{K})|$.*

Proof. First we assume that $s = w(\bar{K})$ or, equivalently, that the framing of K is vertical. Let P be a simple spine of the twice punctured complement $C(K)$ of K constructed in the proof of Proposition 2.1.11, i.e., the sphere S^2 with a tunnel attached along $\bar{K}$. Then one can get a simple spine P_1 of M punctured three times by attaching the disc D^2 along the top line of the tunnel. The disc plays the role of the meridional disc of the solid torus that is glued in instead of $N(K)$. Each time when the tunnel climbs onto itself, there appear two new

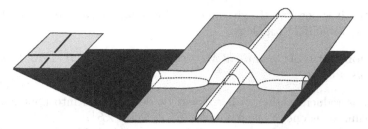

Fig. 2.5. An alternative construction of an almost simple spine of the link complement. The top line of the tunnel contains a smaller number of triple points, and each its winding around the meridian produces only one new true vertex

true vertices (where the base lines of the upper tunnel intersect the top line of the lower one). Thus P_1 possesses $6n$ true vertices (n is the number of crossing points of $\bar{K}$). To decrease the number of true vertices, we modify the construction of P as shown in Fig. 2.5. The new spine P of $C(K)$ punctured twice has the same number of true vertices, but the corresponding new spine P_1 of trice punctured M will have only $5n$ true vertices. The explanation is simple: if the tunnel climbs onto itself, then in the top line of the lower part of the tunnel there appears only one new true vertex.

If $s \neq w(\bar{K})$, one should force the top line of the tunnel to make $|s - w(\bar{K})|$ additional rotations. Each of them produces a new true vertex, so the total number of true vertices would be $5n + |s - w(\bar{K})|$. It remains to puncture two 2-components of P_1 that separate different balls and get an almost simple spine of M with a smaller number of true vertices. □

2.2 Properties of Complexity

2.2.1 Converting Almost Simple Spines into Special Ones

We have already stated the advantages of using almost simple spines, yet there are important downsides too. In general, almost simple spines determine 3-manifolds in a nonunique way, and cannot be represented by regular neighborhoods of their singular graphs alone. Since special spines, as has been mentioned before, are free from such liability, we would like to go from almost simple polyhedra to special ones whenever possible. So the question is: when is it possible? We shall study it in this section.

Let P be an almost simple spine of a 3-manifold M that is not a special one. Then P either possesses a 1-dimensional part or has 2-components not homeomorphic to a disc. Our aim is to transform P into a special spine of M without increasing the number of true vertices. In general this is impossible. For example, if M is reducible or has compressible boundary, any minimal almost simple spine of M must contain a 1-dimensional part. Nevertheless, in some cases it is possible. To give an exact formulation, we need to recall a few notions of 3-manifold topology.

Definition 2.2.1. *A 3-manifold M is called* irreducible *, if every 2-sphere in M bounds a 3-ball.*

If M is reducible, then either it can be decomposed into nontrivial connected sum, or is one of the manifolds $S^2 \times S^1$, $S^2 \tilde{\times} S^1$.

Recall that a compact surface F in a 3-manifold M is called *proper*, if $F \cap \partial M = \partial F$.

Definition 2.2.2. *A 3-manifold M is* boundary irreducible, *if for every proper disc $D \subset M$ the curve ∂D bounds a disc in ∂M.*

Definition 2.2.3. *Let M be an irreducible boundary irreducible 3-manifold. A proper annulus $A \subset M$ is called* inessential, *if either it is parallel rel ∂ to an annulus in ∂M, or the core circle of A is contractible in M (in the second case A can be viewed as a tube possessing a meridional disc). Otherwise A is called* essential.

Of course, these notions will be considered in more detail later.

Theorem 2.2.4. *Suppose M is a compact irreducible boundary irreducible 3-manifold such that $M \neq D^3, S^3, RP^3, L_{3,1}$ and all proper annuli in M are inessential. Then for any almost simple spine P of M there exists a special spine P_1 of M having the same or a fewer number of true vertices.*

Proof. Identify M (or M with a 3-ball removed, if M is closed) with a regular neighborhood of P. We will assume that P cannot be collapsed to a smaller subpolyhedron. We convert P into P_1 by a sequence of transformations (moves) of three types. To control the number of steps, we assign to any almost simple polyhedron P the following three numbers:

1. $c_2(P)$, the number of 2-components of P.
2. $-\chi_2(P) = -\sum_\alpha \chi(\alpha)$, where the sum is taken over all 2-components α of P and $\chi(\alpha)$ is the Euler characteristic.
3. $c_1(P) = \min e(X_P)$, where the 1-dimensional part X_P of P (i.e., the union of points having 0-dimensional links) is presented as a graph with $e(X_P)$ edges and the minimum is taken over all such presentations.

The triples $(c_2(P), -\chi_2(P), c_1(P))$ will be considered in the lexicographic order.

MOVE 1. Suppose that the 1-dimensional part X_P of P is nonempty. Consider an arc $\ell \subset X_P$ and a proper disc $D \subset M$ which intersects ℓ transversally at one point. Since M is irreducible and boundary irreducible, D cuts a 3-ball B out of M. Removing $B \cap P$ from P and collapsing the rest of P as long as possible, we get a new almost simple spine $P' \subset M$. If $B \cap P$ contains at least one 2-component of P, then $c_2(P') < c_2(P)$. If

$B \cap P$ is 1-dimensional, then the 2-dimensional parts of P, P' coincide and thus $c_2(P') = c_2(P), -\chi_2(P') = -\chi_2(P)$. Of course, $c_1(P') < c_1(P)$.

Assume that a 2-component α of P contains a nontrivial simple closed curve l so that the restriction to l of the normal bundle ν of α is trivial. If α is not D^2, S^2 or RP^2, then l always exists. It follows that one can find a proper annulus $A \subset M$ that intersects P transversally along l. Since all annuli are inessential, either A is parallel to the boundary or its core circle is contractible.

MOVE 2. Suppose that A is parallel to the boundary. Then it cuts off a solid torus V from M so that the remaining part of M is homeomorphic to M. Removing $V \cap P$ from P, we obtain (after collapsing) a new almost simple spine $P' \subset M$. This move annihilates α, so $c_2(P') < c_2(P)$.

MOVE 3. Suppose that the core circle of A is contractible. Then both circles of ∂A are also contractible. Choose one of them. By Dehn's Lemma [106], it bounds a disc in M and, since M is boundary irreducible, a disc D in ∂M. It follows that there is a disc $D \subset \mathrm{Int}\, M$ such that $D \cap P = \partial D = l$. Since $M \setminus P$ is homeomorphic to $\partial M \times (0, 1]$, D cuts a proper open 3-ball B out of $M \setminus P$, see Definition 1.2.12. If we puncture D, collapse B and then collapse the rest of D, we return to P. However, if we get inside the ball B through another 2-component of the free boundary of B (see Fig. 2.6), we get after collapsing a new almost simple spine $P' \subset M$.

Let us analyze what happens to α under this move. If l does not separate α, then the collapse eliminates α completely together with D. In this case we have $c_2(P') < c_2(P)$.

Suppose that l separates α into two parts, α' and α'' (the notation is chosen so that the hole is in α''). Then the collapse destroys α'', and we are left with $\alpha' \cup D$. In this case either $c_2(P') < c_2(P)$ (if the collapse destroys some other 2-components of P), or $c_2(P') = c_2(P)$ and $-\chi_2(P') < \chi_2(P)$ since $-\chi(\alpha' \cup D) < -\chi(\alpha)$.

Now let us perform Steps 1, 2, 3 as long as possible. The procedure is finite, since each step strictly decreases the triple $(c_2(P), -\chi_2(P), c_1(P))$ and hence any monotonically decreasing sequence of triples is finite. Let P_1 be the resulting almost simple spine of M. By construction, P_1 has no 1-dimensional

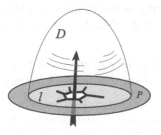

Fig. 2.6. Attaching D^2 along l and puncturing another 2-component produces a simpler spine

part and no 2-components different from D^2, S^2, and RP^2. The following cases are possible:

1. P_1 has no 2-components at all. Since it also has no 1-dimensional part, P_1 is a point and thus $M = S^3$ or $M = D^3$.
2. P_1 contains a 2-component which is not homeomorphic to the disc. In this case P_1 is either RP^2 or S^2. Suppose that $P_1 = RP^2$. Then $M = RP^2 \times I$ or RP^3. We cannot have $M = RP^2 \tilde{\times} I$, since this manifold is a punctured projective space and hence is reducible. For the same reason we cannot have $P_1 = S^2$: the manifold $S^2 \times I$ is reducible.
3. All the 2-components of P_1 are discs and P_1 has no true vertices but contains triple points. Denote by k the number of 2-components of P_1. We cannot have $k = 3$, since the union of three discs with common boundary is a spine of S^3 with three punctures, which is a reducible manifold. The simple polyhedron obtained by attaching two discs to a circle is unthickenable, see Example 1.1.18. We may conclude that P_1 has only one 2-component, which is homeomorphic to the disc. In this case M is homeomorphic to $L_{3,1}$.
4. There remains only one possibility: P_1 has true vertices and all its 2-components are discs. In this case P_1 is special.

$\square$

2.2.2 The Finiteness Property

Theorem 2.2.5. *For any integer k, there exists only a finite number of distinct compact irreducible boundary irreducible 3-manifolds that contain no essential annuli and have complexity k.*

Proof. Follows immediately from Theorems 2.2.4 and 2.1.1. $\square$

Restricting ourselves to the most interesting case of closed orientable irreducible 3-manifolds, we immediately get Corollary 2.2.6.

Corollary 2.2.6. *For any integer k, there exists only a finite number of distinct closed orientable irreducible 3-manifolds of complexity k.*

Recall that a compact 3-manifold M is *hyperbolic* if Int M admits a complete hyperbolic metrics of a finite volume. It is known (see [136]) that any hyperbolic 3-manifold is irreducible, has incompressible boundary, and contains no essential annuli.

Corollary 2.2.7. *For any integer k, there exists only a finite number of distinct orientable hyperbolic 3-manifolds of complexity k.*

Both corollaries follow immediately from Theorem 2.2.5. Let $n_c(k)$ and $n_h(k)$ be the numbers of all closed orientable irreducible 3-manifolds of complexity k and all orientable hyperbolic 3-manifolds of complexity k, respectively. Then for small k the exact values of these numbers are listed in the table below.

k	0 1 2 3
$n_c(k)$	3 2 4 7
$n_h(k)$	0 0 2 9

Remark 2.2.8. To show that the assumptions of Theorem 2.2.5 are essential, let us describe three infinite sets of distinct 3-manifolds of complexity 0. The sets consist of manifolds that are either reducible (1), or boundary reducible (2), or contain essential annuli (3).

(1) For any integer n the connected sum M_n of n copies of the projective space RP^3 is a closed manifold of complexity 0. To construct an almost simple spine of M_n without true vertices, one may take n exemplars of the projective plane RP^2 and join them by arcs. Alternatively, one can start with $L_{3,1}$ and the triple hat instead of RP^3 and RP^2.
(2) The genus n handlebody H_n is irreducible, but boundary reducible. Since it can be collapsed onto a 1-dimensional spine, $c(H_n) = 0$.
(3) Manifolds $\partial H_n \times I$ are irreducible and boundary irreducible, but contain essential annuli. They have complexity 0 since can be collapsed onto the corresponding surfaces.

2.2.3 The Additivity Property

Recall that the *connected sum* $M_1 \# M_2$ of two compact 3-manifolds M_1, M_2 is defined as the manifold $(M_1 \backslash \text{Int } B_1) \cup_h (M_2 \backslash \text{Int } B_2)$, where $B_1 \subset \text{Int } M_1$, $B_2 \subset \text{Int } M_2$ are 3-balls, and h is a homeomorphism between their boundaries. If the manifolds are orientable, their connected sum may depend on the choice of h. In this case $M_1 \# M_2$ will denote any of the two possible connected sums. Alternatively, one can use signs and write $M_1 \#(\pm M_2)$

To define the boundary connected sum, consider two discs $D_1 \subset \partial M_1$, $D_2 \subset \partial M_2$ in the boundaries of two 3-manifolds. Glue M_1 and M_2 together by identifying the discs along a homeomorphism $h: D_1 \to D_2$. Equivalently, one can attach an index 1 handle to $M_1 \cup M_2$ such that the base of the handle coincides with $D_1 \cup D_2$. The manifold M thus obtained is called the *boundary connected sum* of M_1, M_2 and is denoted by $M_1 \amalg M_2$. Of course, M depends on the choice of the discs (if at least one of the manifolds has disconnected boundary), and on the choice of h (homeomorphisms that differ by a reflection may produce different results). Thus the notation $M_1 \amalg M_2$ is slightly ambiguous, like the notation for the connected sum. When shall use it to mean that $M_1 \amalg M_2$ is one of the manifolds that can be obtained by the above gluing.

Theorem 2.2.9. *For any 3-manifolds M_1, M_2 we have:*

1. $c(M_1 \# M_2) = c(M_1) + c(M_2)$
2. $c(M_1 \amalg M_2) = c(M_1) + c(M_2)$

Proof. We begin by noticing that the first conclusion of the theorem follows from the second one. To see that, we choose 3-balls $V_1 \subset \operatorname{Int} M_1, V_2 \subset \operatorname{Int} M_2$, and $V_3 \subset \operatorname{Int} (M_1 \# M_2)$. It is easy to see that $(M_1 \setminus \operatorname{Int} V_1) \perp\!\!\!\perp (M_2 \setminus \operatorname{Int} V_2)$ and $(M_1 \# M_2) \setminus V_3$ are homeomorphic, where the index 1 handle realizing the boundary connected sum is chosen so that it joins ∂V_1 and ∂V_2. Assuming (2) and using Proposition 2.1.5, we have: $c(M_1 \# M_2) = c((M_1 \# M_2) \setminus V_3) = c(M_1 \setminus \operatorname{Int} V_1) + c(M_2 \setminus \operatorname{Int} V_2) = c(M_1) + c(M_2)$.

Let us prove the second conclusion. The inequality $c(M_1 \perp\!\!\!\perp M_2) \le c(M_1) + c(M_2)$ is obvious, since if we join minimal almost simple spines of M_1, M_2 by an arc, we get an almost simple spine of $M_1 \perp\!\!\!\perp M_2$ having $c(M_1) + c(M_2)$ true vertices.

The proof of the inverse inequality is based on Haken's theory of normal surfaces (see Chap. 3). So we restrict ourselves to a reference to Corollary 4.2.10, which states that attaching an index 1 handle preserves complexity. $\square$

2.3 Closed Manifolds of Small Complexity

2.3.1 Enumeration Procedure

It follows from the finiteness property that for any k there exist finitely many closed orientable irreducible 3-manifolds of complexity k. The question is: *how many?* The constructive proof of Theorem 2.1.1 allows us to organize a computer enumeration of special spines with k true vertices. Of course, the list of corresponding 3-manifolds can contain duplicates as well as nonorientable, nonclosed, or reducible manifolds. All such manifolds must be removed.

Let us briefly describe the enumeration results in historical order. First, Matveev and Savvateev tabulated closed irreducible orientable manifolds up to complexity 5, see [91]. The manifolds were listed with the help of a computer and recognized manually. This was the first paper on computer tabulation of 3-manifolds. It contained all basic elements of the corresponding theory, which much later have been rediscovered by various mathematicians. This table was extended to the level of complexity 6 in [80,83]. The same approach was used by Ovchinnikov [102,103] in composing the table of complexity 7. The manifolds were still recognized manually, although by an improved method (by distinguishing and using elementary blocks). Later Martelli wrote a computer program which is based on the same principle, but tabulates 3-manifolds in two steps. First, it enumerates some special building blocks (bricks), and only then assembles bricks into 3-manifolds. An interesting relative version of the complexity theory (see [74]) serves as a theoretical background for the program. We describe it in Sect. 7.7.

Let us present the results of these enumeration processes for $k \le 7$ (see Sect. 7.5 for the similar results for $k \le 12$).

Theorem 2.3.1. *The number $n_c(k)$ of closed orientable irreducible 3-manifolds of complexity k for $k \le 7$ is given by the following table:*

k	0	1	2	3	4	5	6	7
$n_c(k)$	3	2	4	7	14	31	74	175

Closed orientable irreducible 3-manifolds of complexity 0 are the following ones: the sphere S^3, the projective space RP^3, and the lens space $L_{3,1}$. Their almost simple spines without true vertices were described in Sect. 2.1.2. The complexity of $S^2 \times S^1$ is also equal to 0, but this manifold is reducible. Closed orientable irreducible 3-manifolds of complexity 1 are lens spaces $L_{4,1}$ and $L_{5,2}$. There are four 3-manifolds of complexity 2. They are the lens spaces $L_{5,1}$, $L_{7,2}$, $L_{8,3}$, and the manifold S^3/Q_8, where $Q_8 = \{\pm 1, \pm \mathbf{i}, \pm \mathbf{j}, \pm \mathbf{k}\}$ is the quaternion unit group (the action of Q_8 on S^3 is linear). See Sect. 2.3.3 and the Appendix for the description and the complete table of manifolds of complexity $k \le 6$.

Let us give a nonformal description of the computer program that was used for creating the table up to complexity 7. The computer enumerates all the regular graphs of degree 4 with a given number of vertices. The graphs may be considered as work-pieces for singular graphs of special spines. For each graph, the computer lists all possible gluings together of butterflies that are taken instead of true vertices (see the proof of Theorem 2.1.1). Note that if the graph has k vertices, then there are $2k$ edges, and thus potentially 6^{2k} different gluings of the triodes. Not all of them produce spines of orientable manifolds: it may happen that we get a special polyhedron which is not a spine or is a spine of a nonorientable manifold. To avoid this, we supply each copy of E with an orientation (in an appropriate sense), and use orientation reversing identifications of the triodes. This leaves us with no more than $2^{k-1}3^k$ spines of orientable manifolds. One may decrease this number by selecting spines of closed manifolds, but it still remains too large. The problem is that we get a list of spines, while it is a list of manifolds we are interested in (as we know, any 3-manifold has many different special spines). Also, some manifolds from the list thus created would be reducible. A natural idea to obtain a list of manifolds that does not contain duplicates and reducible manifolds consists in considering minimal spines, i.e., spines of minimal complexity. Unfortunately, there are no general criteria of minimality. The good news here is that there are a lot of partial criteria of nonminimality. In Sect. 2.3.2 we present two of them that appeared to be sufficient for casting out all reducible manifolds and almost all duplicates up to $k = 6$.

The completion of the table of closed orientable irreducible 3-manifolds up to complexity 6 was made by hand. It was a big job indeed: for each pair of spines that had passed the minimality tests one must decide whether or not they determine homeomorphic manifolds. In practice we calculated their invariants: homology groups and, in worst cases, fundamental groups [83, 91]. Later, after Turaev–Viro invariants had been discovered, we used them to verify the table. If the invariants did not help to distinguish the manifolds, we tried to transform one spine into the other by different moves that preserve the manifold. In all cases a definitive answer was obtained.

We point out that the Turaev–Viro invariants are extremely powerful for distinguishing 3-manifolds. In particular, invariants of order ≤ 7 distinguish all orientable closed irreducible 3-manifolds up to complexity 6 having the same homology groups. The only exception are lens spaces, for which there is no need to apply these invariants.

2.3.2 Simplification Moves

We describe here only two types of moves. The moves have the following advantage: It is extremely easy to determine whether or not one can apply them to a given special spine.

Definition 2.3.2. *Let P be a special polyhedron and c a 2-component of P. Then we say that the boundary curve of c has a* counterpass, *if it passes along one of the edges of P twice in opposite directions. We say that the boundary curve is* short *, if it passes through no more than 3 true vertices of P and through each of them no more than once.*

For instance, Bing's House contains two 2-components with boundary curves of length 1 while the boundary curve of the third 2-component has a few counterpasses (see Fig. 1.6).

Proposition 2.3.3. *Suppose that P is a special spine of a 3-manifold M such that either:*

1. *P has a 2-component with a short boundary curve.*
2. *M is closed, orientable, and the boundary curve of one of the 2-components of P has a counterpass.*

Then M possesses an almost simple spine with a smaller number of true vertices.

Proof. Assume that P has a 2-component c with a short boundary curve. A regular neighborhood of $\mathrm{Cl}(c)$ in P can be presented as a lateral surface of a cylinder with $k \leq 3$ wings and the 2-component c as a middle disc. Attach to P a disc parallel to c and drill a hole in a lateral face of the cylinder thus obtained, see Fig. 2.7. Collapsing the resulting polyhedron, we get a new

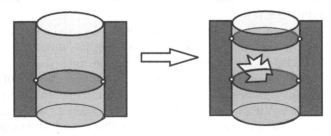

Fig. 2.7. Attaching a new 2-cell and making a hole decreases the number of true vertices

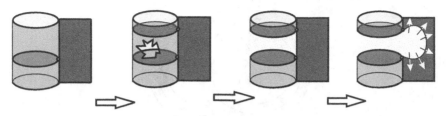

Fig. 2.8. Collapsing the unique wing

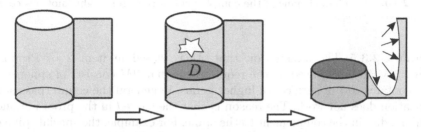

Fig. 2.9. Simplification by a counterpass

almost simple spine of M. It has a smaller number of vertices, since attaching the disc creates k new true vertices, and piercing the lateral face and collapsing destroys at least four of them if $k > 1$, and at least two if $k = 1$. It may be illuminating to note that the above transformation of P coincides with the move L^{-1} if $k = 2$, and with the move T^{-1} if $k = 3$. For $k = 1$ the result is drastic: We collapse not only the pierced 2-component, but also the unique wing of the cylinder. See Fig. 2.8.

Assume now that M is closed and orientable, and the boundary curve of a 2-component c of P has a counterpass on an edge e. Then there exists a simple closed curve $l \subset \mathrm{Cl}(c)$ that intersects e transversally at exactly one point. It decomposes c into two 2-cells c', c''. Since M is closed and orientable, one can easily find a disc $D \subset M$ such that $D \cap P = \partial D = l$. To construct D, one may push l by an isotopy to the boundary of a regular neighborhood of P and span it by a disc in the complementary ball. The polyhedron $P \cup D$ is a special spine of the twice punctured M, that is, of M with two balls B_1, B_2 cut out of it. To get a spine of M, we make a hole in c' or c'' depending on which of them is a common face of these balls. After collapsing we get an almost simple spine of M having a smaller number of true vertices, see Fig. 2.9.   □

Remark 2.3.4. Suppose P has a 2-component such that its boundary curve visits four true vertices, and each of them exactly once. If we apply the same trick (glue in a parallel 2-cell and puncture a lateral one), we get another spine of M having the same number of true vertices. Sometimes this transformation is useful for recognition of duplicates.

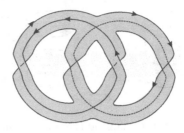

Fig. 2.10. The minimal spine of the complement of the figure eight knot has counterpasses

Remark 2.3.5. The assumption that M is closed in item 2 of Proposition 2.3.3 can be replaced by the requirement that ∂M consists of spheres. If ∂M contains tori or surfaces of higher genus, in general the counterpass simplification does not work. The reason is that the curve l in the proof may not bound a disc in the complement to the spine. For example, the special spine of the complement to the figure eight knot shown in Fig. 2.10 has counterpasses but cannot be simplified since it is minimal.

2.3.3 Manifolds of Complexity ≤ 6

The list of all closed orientable irreducible 3-manifolds up to complexity 6 contains 135-manifolds, see Sect. A.2 and its description in the Appendix. Each manifold is presented by a regular neighborhood of the singular graph of its minimal special spine. If the manifold has several minimal spines, all of them are included in the table. Let us comment on which kinds of 3-manifolds can be found in the table.

A. All closed orientable irreducible 3-manifolds up to complexity 6 are Seifert manifolds. All the manifolds of complexity ≤ 5 and many manifolds of complexity 6 have finite fundamental groups. They are elliptic, that is, can be presented as quotient spaces of S^3 by free linear actions of finite groups. Groups which can linearly act on S^3 without fixed points are well known (see [94]). They are:

1. The finite cyclic groups
2. The groups $Q_{4n}, n \geq 2$
3. The groups $D_{2^k(2n+1)}, k \geq 3, n \geq 1$
4. The groups P_{24}, P_{48}, P_{120}, and $P'_{8(3^k)}, k \geq 2$
5. The direct product of any of these groups with a cyclic group of coprime order

Lower indices show the orders of the groups. Presentations by generators and relations are given in the preliminary to Sect. A.2, see the Appendix.

B. The list contains representatives of all the five series of elliptic manifolds. In particular, the manifolds $S^3/P_{24}, S^3/P_{48}$, and the Poincaré homology sphere S^3/P_{120} have complexities 4,5, and 5, respectively. The first manifold with a nonabelian fundamental group is S^3/Q_8, where Q_8 is the quaternion unit group. It has complexity 2. More generally, for $2 \leq n \leq 6$ the manifolds S^3/Q_{4n} have complexity n. The simplest manifold of the type $S^3/D_{2^k(2n+1)}$, that is, S^3/D_{24}, has complexity 4 while the simplest manifold of the type $S^3/P'_{8(3^k)}$, the manifold S^3/P'_{72}, has complexity 5. There also occur quotient spaces of S^3 by actions of direct products of the above-mentioned groups with cyclic groups of relatively prime orders. The simplest of these (the manifold $S^3/Q_8 \times Z_3$) has complexity 4.

C. All six flat closed orientable 3-manifolds have complexity 6, among them the torus $S^1 \times S^1 \times S^1$ and the Whitehead manifold obtained from S^3 by Dehn surgery on the Whitehead link with trivially framed components. The last two are the only closed orientable irreducible manifolds of complexity ≤ 6 having the first homology group of rank ≥ 2. Recall that the Whitehead manifold coincides with the mapping torus of a homeomorphism $S^1 \times S^1 \to S^1 \times S^1$ which is the Dehn twist along nontrivial simple closed curve.

D. Among the manifolds of complexity ≤ 6 there is just one nontrivial homology sphere S^3/P_{120}. It has a unique minimal special spine with five true vertices. The singular graph of the spine is the complete graph on five vertices.

E. If the complexity of the lens space $L_{p,q}$ with $p > 2$ does not exceed 6, then it can be computed by the formula $c(L_{p,q}) = S(p,q) - 3$, where $S(p,q)$ is the sum of all partial quotients in the expansion of p/q as a regular continued fraction. Most probably, the formula $c(L_{p,q}) = S(p,q) - 3$ holds for all lens spaces, but we know only how to prove the inequality $c(L_{p,q}) \leq S(p,q) - 3$: it follows from Remark 2.3.8.

In practice, it is more convenient to calculate $c(L_{p,q})$ by the following empirical rule: if $p > 2q$, then $c(L_{p,q}) = c(L_{p-q,q}) + 1$. For example, $c(L_{33,10}) = c(L_{23,10}) + 1 = c(L_{13,10}) + 2 = c(L_{13,3}) + 2 = c(L_{10,3}) + 3 = c(L_{7,3}) + 4 = c(L_{4,3}) + 5 = c(L_{4,1}) + 5 = c(L_{3,1}) + 6 = 6$ since $c(L_{3,1}) = 0$ (we have used twice that lens spaces $L_{p,q}$ and $L_{p,p-q}$ are homeomorphic). This shows once again how natural the notion of complexity is.

It should be noted that the number of true vertices of a special spine as a measure of complexity of 3-manifolds was implicitly used by numerous authors. Ikeda proved that any simply-connected manifold having a simple spine with ≤ 4 vertices is homeomorphic to S^3 [49]. Together with Yoshinobu [50] he listed all closed 3-manifolds which in our terminology possess complexity ≤ 2. A complete list of all closed orientable irreducible 3-manifolds of complexity ≤ 5 was obtained by means of a computer as early as 1973 by Matveev and Savvateev [91]. Gillman and Laszlo, who were interested only in homology spheres [35], with the help of a computer proved that among manifolds of complexity ≤ 5 only S^3/P_{120} and S^3 have trivial homology. Actually, this fact can be extracted easily from the Matveev and Savvateev list. A list of closed orientable irreducible 3-manifolds of complexity 7 was obtained by

Ovchinnikov [102, 103]. It consists of 175-manifolds and is too large to be presented in full. Fortunately, a major part of the manifolds can be divided into four series admitting clear descriptions. In Appendix we present these descriptions and list the remaining exceptional manifolds.

It is interesting to note that not all regular graphs can be realized as singular graphs of minimal special spines of 3-manifolds. Let us try to single out several types of graphs that produce the majority of 3-manifolds up to complexity 6.

Definition 2.3.6. *A regular graph G of degree 4 is called a* nonclosed chain *if it contains two loops, and all the other edges are double. G is a* closed chain, *if it has only double edges. Finally, G is called a* triangle with a tail, *if it is homeomorphic to a wedge of a closed chain with three vertices and a nonclosed chain such that the base point of the tail (i.e., the common point of these two chains) lies on a loop of the nonclosed chain. See Fig. 2.11.*

We will say that a special spine of a closed orientable 3-manifold is *pseudo-minimal* if it has no counterpasses and short boundary curves. In particular, any minimal special spine is pseudominimal. For brevity we will say that a special spine P is *modeled* on a graph G if G is homeomorphic to the singular graph of P.

Proposition 2.3.7. *A closed orientable 3-manifold M has a pseudominimal special spine modeled on a nonclosed chain if and only if M is a lens space $L_{p,q}$ with $p > 3$.*

Proof. Let P be a pseudominimal special spine of M modeled on a closed chain G with n vertices. Denote by i an involution on G having $n+2$ fixed points: n vertices and one additional point on each loop. The involution permutes edges having common endpoints. Since the boundary curves have no counterpasses, they are symmetric with respect to i. Moreover, there is a boundary curve that passes a loop of G twice. Remove from P the corresponding 2-component, and denote by P_1 the resulting polyhedron. Note that P_1 is a spine of $M \setminus \mathrm{Int} H_1$, where H_1 is a solid torus in M.

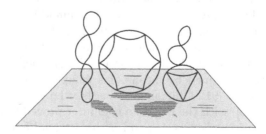

Fig. 2.11. Three useful types of singular graphs: a *nonclosed chain, closed chain,* and a *triangle with a tail*

Let us collapse P_1 for as long as possible by removing other 2-components together with their free edges, and free edges together with their free vertices. Using the above-mentioned symmetry of the boundary curves, one can easily show that we get a circle (actually, the second loop of G). To visualize this, one may take the spine of any lens space from Sect. A.2 and carry out the collapsing by hand. It follows that $M \setminus \mathrm{Int} H_1$ is a regular neighborhood of a circle, that is, a solid torus. Thus M is a lens space. $\square$

Remark 2.3.8. Let us describe a simple method for calculating parameters of the lens space presented by a picture that shows a regular neighborhood of the singular graph of its pseudominimal special spine. The correctness of this method can be easily proved by induction on the number of true vertices of the spine. Assign to each double edge and to each loop of the singular graph a letter ℓ or r as shown in Fig. 2.12. We get a string w of letters that we will consider as a composition of operators $r, \ell \colon Z \oplus Z \to Z \oplus Z$ given by $r(a, b) = (a, a + b)$ and $\ell(a, b) = (a + b, b)$. Then the lens space has parameters $p = m + n, q = m$, where $(m, n) = w(1, 1)$. For example, for the lens space shown in Fig. 2.12 we have $w = rrrr\ell\ell\ell$, $(m, n) = (4, 17)$, and $(p, q) = (21, 4)$, since by our interpretation of r, ℓ we have

$$(1, 1) \xrightarrow{\ell} (2, 1) \xrightarrow{\ell} (3, 1) \xrightarrow{\ell} (4, 1) \xrightarrow{r} (4, 5) \xrightarrow{r} (4, 9) \xrightarrow{r} (4, 13) \xrightarrow{r} (4, 17).$$

The same method can be used for constructing a pseudominimal special spine of a given lens space $L_{p,q}$: One should apply to the pair $(p - q, q)$ operators r^{-1}, ℓ^{-1} until we get $(1, 1)$, and then use the string of letters r, ℓ thus obtained for constructing the spine.

Proposition 2.3.9. *A closed orientable 3-manifold M has a pseudominimal special spine modeled on a triangle with a tail if and only if M is an orientable Seifert fibered manifold of the type $(S^2, (2, 1), (2, -1), (n, \beta))$, where $\beta, n > 0$, and $(n, \beta) \neq (1, 1)$.*

Proof. Let P be a pseudominimal special spine of M modeled on a triangle with a tail. Since the boundary curves have no counterpasses, they pass over

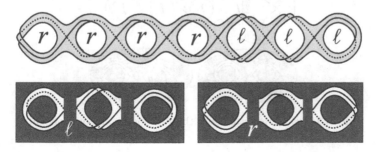

Fig. 2.12. How to write down the developing string for a nonclosed chain

the tail in a symmetric way with respect to the involution of the tail that permutes the double edges and reverses the orientation of the loop. For the same reason one of the boundary curves passes the loop twice. Remove from P the corresponding 2-component and denote by P_1 the resulting polyhedron. Of course, P_1 is a spine of $M_1 = M \setminus \mathrm{Int} H$, where H is a solid torus in M. Let us collapse P_1 as long as possible. Using the above-mentioned symmetry of the boundary curves, one can easily show that the tail disappears completely, together with all 2-cells that have common points with it, including all the 2-components whose boundary curves pass through the base point of the tail. It follows that we get a simple spine without singular points, that is, a closed surface K. Since the boundary of M_1 is a torus, K is the Klein bottle, and $M_1 = K \tilde{\times} I$. It is well known that $K \tilde{\times} I$ can be presented as the Seifert fibered manifold over the disc with two exceptional fibers of types $(2,1)$ and $(2,-1)$, see Example 6.4.14. Thus attaching the solid torus H converts M_1 to a Seifert fibered manifold $(S^2, (2,1), (2,-1), (n,\beta))$. The parameters (n,β) show how H is glued to $K \tilde{\times} I$. We can always get $n, \beta > 0$ by reversing the orientation of the manifold.

Just as in the proof of Proposition 2.3.7, let us describe a simple method for calculating the parameters (n,β) starting from a pseudominimal special spine modeled on a triangle with a tail. The correctness of the method can be easily proved by induction on the number of vertices of the tail. Assign to the loop and the double edges of the tail and to the pair of edges adjacent to it letters ℓ and r as shown in Fig. 2.13. We get a string w of letters which, as above, can be considered as a composition of operators $r, \ell \colon Z \oplus Z \to Z \oplus Z$ given by $r(a,b) = (a, a+b)$ and $\ell(a,b) = (a+b, b)$. Then $(n,\beta) = w(1,1)$. For example, for the spine shown in Fig. 2.13 we have $w = \ell r \ell \ell$ and $w(1,1) = (n,\beta) = (7,4)$.

The same method can be used for constructing a pseudominimal special spine of a given Seifert fibered manifold $(S^2, (2,1), (2,-1), (n,\beta))$: one should recover the string of r, ℓ by transforming (n,β) into $(1,1)$, and then use it for choosing the correct tail. Since $n \neq 0$ and $(n,\beta) \neq (1,1)$, the string and the tail exist.   □

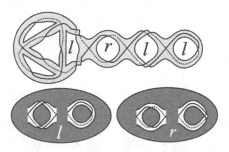

Fig. 2.13. How to write down the developing string for a tail

Note that for any pair of coprime positive integers (n, β) with $n \geq 1$ the fundamental group of the manifold $M_{n,\beta} = (S^3, (2,1), (2,-1), (n,\beta))$ is finite and has the presentation

$$\langle c_1, c_2, c_3, t | c_1^2 = t, c_2^2 = t^{-1}, c_3^n = t^\beta, c_1 c_2 c_3 = 1 \rangle$$

The order of the homology group $H_1(M_{n,\beta}; Z)$ is 4β. Using this, it is not hard to present $M_{n,\beta}$ as the quotient space of S^3 by a linear action of a group from the Milnor list [94] presented above. It turns out that the following is true:

(1) If $n > 1$ and β is odd, then $M_{n,\beta} = S^3/Q_{4n} \times Z_\beta$
(2) If $n > 1$ and β is even, then $M_{n,\beta} = S^3/D_{2^{k+2}n} \times Z_{2m+1}$, where k and m can be found from the equality $\beta = 2^k(2m+1)$
(3) If $n = 1$ and $\beta \neq 0, 1$, then $M_{n,\beta} = L_{4\beta, 2\beta+1}$

If $n = 1$ or $\beta = 1$, the pseudominimal special spine of $M_{n,\beta}$ modeled on the triangle with a tail is not minimal. An easy way to see that is to apply the transformation described in Remark 2.3.4. This is possible since the spine possesses a boundary curve that passes through four true vertices, and visits each of them exactly once. After the transformation we get a spine that has the same number of true vertices but possesses a boundary curve of length 3. Therefore, one can simplify the spine. In the case $n = 1$ we get a spine of a lens space modeled on a nonclosed chain with smaller number of true vertices. If $\beta = 1$, we get a simple spine of the manifold S^3/Q_{4n}.

Proposition 2.3.10. *A closed orientable 3-manifold M has a pseudominimal special spine modeled on a closed chain with $n \geq 2$ vertices if and only if M is S^3/Q_{4n}.*

Proof. One can easily show that any pseudominimal spine of a closed manifold modeled on a closed chain contains a boundary curve that goes twice around the chain and passes through all the edges. For any $n \geq 2$ there is only one such spine (see Fig. 2.14), and its fundamental group is Q_{4n}. Removing the

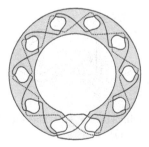

Fig. 2.14. The unique pseudominimal special spine modeled on a closed chain with n vertices is a spine of S^3/Q_{4n}

corresponding 2-component and collapsing, we get the Klein bottle. Therefore, M is a Seifert fibered manifold over S^2 with three exceptional fibers of
degree $2, 2$, and n. Among such manifolds only S^3/Q_{4n} has the fundamental
group Q_{4n}.   □

The following conjectures are motivated by Propositions 2.3.7–2.3.10 and
the results of the computer enumeration.

Conjecture 2.3.11. Any lens space $L_{p,q}$ with $p \geq 3$ has a unique minimal
special spine. This spine is modeled on a nonclosed chain.

Conjecture 2.3.12. For any $n \geq 2$ the manifold S^3/Q_{4n} has a unique minimal
special spine. This spine is modeled on a closed chain with n links.

Conjecture 2.3.13. Manifolds of the type $S^3/Q_{4n} \times Z_\beta, n > 1, \beta \neq \pm 1$ and
$S^3/D_{2^{k+2}n} \times Z_{2m+1}$ have minimal special spine modeled on triangles with a
tail.

Section A.2 shows that the conjectures are true for manifolds of
complexity ≤ 7.

Remark 2.3.14. One can prove that any pseudominimal special spine modeled on a triangle with three tails is a spine of a Seifert fibered manifold M
over S^2 with three exceptional fibers. Let $w_i, 1 \leq i \leq 3$ be the developing rl-
strings of the tails. Then $M = (S^2, (n_1, \beta_1), (n_2, \beta_2), (n_3, \beta_3), (1, -1))$, where
$(n_i, \beta_i) = w_i(1, 1)$ for $1 \leq i \leq 3$. We have inserted the regular fiber of the type
$(1,-1)$ to preserve the symmetry of the expression. Certainly, one may write
$M = (S^2, (n_1, \beta_1), (n_2, \beta_2), (n_3, \beta_3 - n_3))$. The formula works also for triangles
with < 3 tails, if we adopt the convention that the developing string for the
empty tail is ℓ and produces the exceptional fiber of type $(2,1)$. See Fig. 2.15.

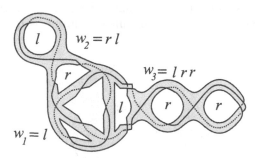

Fig. 2.15. The developing strings are ℓ (for the empty tail), $r\ell$, and ℓrr. Thus
$M = (S^2, (2, 1), (2, 3), (4, 3), (1, -1))$

2.4 Graph Manifolds of Waldhausen

Our discussion in Sect. 2.3.3 shows that all closed orientable irreducible 3-manifolds of complexity ≤ 6 belong to the class $\mathcal{G}$ of graph manifolds of Waldhausen. $\mathcal{G}$ contains all Seifert manifolds and all Stallings and quasi-Stallings 3-manifolds with fiber $S^1 \times S^1$. Its advantage is that it is closed with respect to connected sums. It follows that all closed orientable (not necessarily irreducible) 3-manifolds of complexity ≤ 6 are also graph manifolds. In Sect. 2.4.2 we will show that the same is true for 3-manifolds of complexity ≤ 8, but first we should study $\mathcal{G}$ in more detail.

2.4.1 Properties of Graph Manifolds

Graph manifolds have been introduced and classified by Waldhausen in two consecutive papers [129]. They turned out to be very important for understanding the structure of 3-manifolds. Indeed, it follows from JSJ-decomposition theorem (see [55, 57] and Sect. 6.4) that for any orientable closed irreducible 3-manifold M there exists a finite system $\mathcal{T}$ of disjoint incompressible tori $T_1, T_2, \ldots, T_n$ in M (unique up to isotopy) such that the following holds:

(1) $\mathcal{T}$ decomposes M into Seifert manifolds and manifolds which are not Seifert and contain no essential tori. We will call these submanifolds *JSJ-chambers*.
(2) $\mathcal{T}$ has the minimal number of tori among all systems possessing (1).

If $\mathcal{T} \neq \emptyset$, then all the JSJ-chambers are sufficiently large (see Sect. 4.1.6). Therefore, one can apply Thurston's results [62, 111, 122, 123] and prove that every non-Seifert JSJ-chamber is a hyperbolic manifold. The union of all Seifert JSJ-chambers is not necessarily a Seifert manifold, but it is composed of Seifert manifolds. In particular, it can have Stallings and quasi-Stallings components (see Definitions 6.4.16 and 6.5.12) with fiber $S^1 \times S^1$. Manifolds which can be obtained from Seifert manifolds by gluing their boundary tori are known as *graph manifolds of Waldhausen*.

Roughly speaking, the role of graph manifolds in 3-manifold topology may be expressed by the informal relation

$$\mathcal{M} = (\mathcal{G} + \mathcal{H}) \cup (?).$$

Here $\mathcal{M}, \mathcal{G}$, and $\mathcal{H}$ are the classes of all closed irreducible 3-manifolds, graph manifolds, and hyperbolic manifolds, respectively. The class $\mathcal{G} + \mathcal{H}$ consists of manifolds that can be decomposed by incompressible tori into graph and hyperbolic manifolds. The additional term (?) stands for the class of closed irreducible manifolds which contain no essential tori and are neither hyperbolic nor graph manifolds. If Thurston's Geometrization Conjecture [111] is true, then (?) is empty.

The class of graph manifolds was rediscovered by Fomenko [30]. It turned out that there is a close relationship between the integrability of Hamiltonian mechanical systems on symplectic 4-manifolds and the topological structure of level surfaces of the Hamiltonian: If the system is integrable, then each nonsingular level surface is a graph manifold. See [12] for further development of the theory.

To give a rigorous formal definition of graph manifolds, we prefer to compose them from Seifert manifolds of two very simple types. Denote by N^2 the disc D^2 with two holes. Then the manifold $N^2 \times S^1$ can be presented as the solid torus $D^2 \times S^1$ with two drilled out solid tori H_1, H_2 that are parallel to the core circle $\{*\} \times S^1$ of $D^2 \times S^1$. A more general way to view $N^2 \times S^1$ is to cut out a regular neighborhood of $c \cup \ell$ from $D^2 \times S^1$, where c is a core circle of $D^2 \times S^1$ and ℓ is any simple closed curve in Int $(D^2 \times S^1)$ that is parallel to a nontrivial curve in $\partial D^2 \times S^1$. The result does not depend on the choice of ℓ since all nontrivial simple closed curves in the boundary of $S^1 \times S^1 \times I$ are equivalent up to homeomorphisms of $S^1 \times S^1 \times I$. We will call the manifolds $D^2 \times S^1$ and $N^2 \times S^1$ *elementary blocks*.

Definition 2.4.1. *A compact 3-manifold M is called a* graph manifold *if it can be obtained by pasting together several elementary blocks $D^2 \times S^1$ and $N^2 \times S^1$ along some homeomorphisms of their boundary tori.*

It is often convenient to present the gluing schema by a graph having vertices of valence 1 and 3. The vertices of valence 3 correspond to blocks $N^2 \times S^1$. Vertices of valence 1 correspond either to blocks $D^2 \times S^1$ or to free boundary component of M. In Fig. 2.16 we represent them by black and white fat dots, respectively.

We next recall some well-known properties of graph manifolds, accompanying them with short explanations or informal proofs.

Proposition 2.4.2. *The class $\mathcal{G}$ contains all orientable Seifert manifolds.*

Proof. Suppose M is a Seifert manifold fibered over a surface F. Note that any surface can be decomposed by disjoint circles into the following elementary

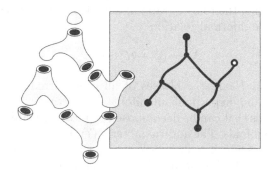

Fig. 2.16. A graph structure of a graph manifold

pieces: discs, copies of N^2, and Möbius bands. One may assume that all exceptional fibers correspond to the centers of the discs. Then the decomposition of F induces a decomposition of M into inverse images of elementary pieces. It remains to note that the inverse image of each piece P is either an elementary block (if $P = D^2, N^2$), or can be decomposed into three elementary blocks (if P is a Möbius band). The latter is true since the twisted product of a Möbius band and S^1 admits an alternative Seifert structure: it fibers over D^2 with two exceptional fibers of types (2,1), (2,-1), see Example 6.4.14.   □

Proposition 2.4.3. *The class $\mathcal{G}$ is closed with respect to connected sums, that is, $M_1 \# M_2 \in \mathcal{G} \iff M_1, M_2 \in \mathcal{G}$.*

Proof. To prove the implication $\Leftarrow$, it suffices to find a graph presentation of $D^2 \times S^1 \# D^2 \times S^1$. Let c be the core circle of $D^2 \times S^1$ and m a circle obtained from a meridian of $D^2 \times S^1$ by pushing it inward $D^2 \times S^1$. Denote by $N(c), N(m)$ regular neighborhoods of the circles. Then the manifold $D^2 \times S^1 \setminus \mathrm{Int} N(m)$ is homeomorphic to the connected sum of two solid tori. On the other hand, it can be obtained from the manifold $D^2 \times S^1 \setminus (\mathrm{Int}\ N(c) \cup \mathrm{Int} N(m))$ homeomorphic to $N^2 \times S^1$ by pasting back the torus $N(c)$.

To prove the inverse implication, assume that a graph manifold M contains a nontrivial 2-sphere S. Consider a decomposition of M into *extended elementary blocks*, where each extended block is the union of an elementary block $N^2 \times S^1$ and all the solid tori adjacent to it. Applying the innermost circle argument to the intersection of S with the boundaries of extended blocks, we locate an extended block B with compressible boundary. Recall that the boundary of any Seifert manifold is incompressible unless it is the solid torus. It follows that for B we have the following possibilities:

1. *B is a solid torus (presented as a union of smaller elementary blocks).* We consolidate the initial structure of a graph manifold by considering B as a new block.
2. *B is not a Seifert manifold.* This can happen only in the case when B is composed of $N^2 \times S^1$ and solid tori so that the meridian of one of the solid tori is isotopic to a fiber $\{*\} \times S^1$ of $N^2 \times S^1$. Then B can be presented as $B_1 \# B_2$, where each B_i is either a solid torus or a lens space. Thus we can decompose M into a connected sum of either simpler graph manifolds or a simpler graph manifold and $S^2 \times S^1$.

Continuing this process for as long as possible, we get a decomposition of M into a connected sum of prime graph manifolds. Since the topological types of the summands are determined by M, the prime decomposition summands for M_1, M_2 have the same types. It follows that M_1 and M_2 are graph manifolds.   □

Now we investigate the behavior of $\mathcal{G}$ with respect to boundary connected sums and, more generally, to cutting along discs. Of course, the statement $M_1 \amalg M_2 \in \mathcal{G} \iff M_1, M_2 \in \mathcal{G}$ is not true anymore. For example, let

V be a solid torus and B a 3-ball. Then $V \amalg V \notin \mathcal{G}$ although $V \in \mathcal{G}$ and $V \amalg B \in \mathcal{G}$ although $B \notin \mathcal{G}$. To formulate the correct corresponding statement, it is convenient to introduce the following notation: If M is a 3-manifold, then $\hat{M}$ denotes the 3-manifolds obtained from M by attaching 3-balls to all the spherical components of ∂M. In particular, if ∂M contains no spherical components, then $\hat{M} = M$. Recall also that if D is a proper disc in M, then M_D denotes the 3-manifold obtained from M by cutting along D.

Corollary 2.4.4. *Let D be a proper disc in a connected 3-manifold M such that ∂M consists of tori. Then M is a graph manifold if and only if so is $\hat{M}_D$.*

Proof. Denote by T the torus component of ∂M containing ∂D and by $N = N(D \cup T)$ a regular neighborhood of $D \cup T$ in M. Then M_D is homeomorphic to the manifold $M_1 = \mathrm{Cl}(M \setminus N)$. It is easy to see that $M_1 \cap N$ is either a sphere S (if ∂D does not decompose T) or the union of a sphere S and a torus T_1 (if ∂D decomposes T). There are three different cases, see Fig. 2.17. Let us list all of them together with the corresponding relation between $\hat{M}$ and $\hat{M}_D$ we wish to prove.

    Case 1. *D decomposes M into two components M', M''. Then $M = \hat{M}' \# \hat{M}''$.*

    Case 2. *∂D decomposes T, but D does not decompose M. Then $M = \hat{M}_D \# (S^2 \times S^1)$.*

    Case 3. *∂D does not decompose T. Then $M = \hat{M}_D \# (D^2 \times S^1)$.*

    Let us prove that. Suppose $M_1 \cap N = S \cup T_1$ is as in Cases 1, 2. Denote by M_S the 3-manifold obtained from $M = M_1 \cup N$ by cutting along S. If we attach N to M_1 along T_1, we get M_S. It follows that $\hat{M}_S$ can be obtained by attaching $\hat{N}$ to $\hat{M}_1$ along T_1. On the other hand, $\hat{N}$ is homeomorphic to $T_1 \times I$, so $\hat{M}_S$ and $\hat{M}_1$ (hence $\hat{M}_S$ and $\hat{M}_D$) are homeomorphic. We can conclude that M can be obtained from $\hat{M}_D$ by cutting out two 3-balls and identifying the two boundary spheres of the manifold thus obtained. This operation is equivalent to taking the connected sum with $S^2 \times S^1$ (if $\hat{M}_D$ is connected) or to taking the connected sum of its components (if not).

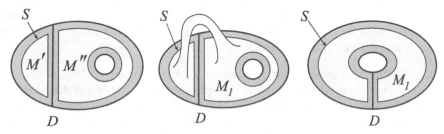

Fig. 2.17. M is either $\hat{M}' \# \hat{M}''$ (*on the left*), or $\hat{M}_D \# (S^2 \times S^1)$ (*in the middle*), or $\hat{M}_D \# (D^2 \times S^1)$ (*on the right*)

If $M_1 \cap N = S$ as in Case 3, then M_S is the disjoint union of M_1 (which is homeomorphic to M_D) and N (which is a punctured solid torus). It follows that $M = \hat{M}_D \# (D^2 \times S^1)$.

To conclude the proof of the corollary, it remains to recall that $S^2 \times S^1$ and $D^2 \times S^1$ are graph manifolds and apply Proposition 2.4.3.   □

Remark 2.4.5. One can easily generalize Corollary 2.4.4 as follows. Let D be a proper disc in a connected 3-manifold M so that ∂M consists of spheres and tori. Then $\hat{M}$ is a graph manifold if and only if so is $\hat{M}_D$. Indeed, if ∂D lies on a torus of ∂M, then the same proof works. Let ∂D lies on a sphere $S \subset \partial M$. Then either $\hat{M} = \hat{M}_D \# (S^1 \times S^2)$ (if D does not separate M), or $\hat{M} = \hat{M}' \# \hat{M}''$ (if it decomposes M into two components M', M'').

By Definition 2.4.1, any graph manifold can be decomposed onto elementary blocks by a finite system of disjoint tori. Our next goal is to decrease the number of tori by amalgamating the elementary blocks into Seifert manifolds called *Seifert blocks*. We will restrict ourselves to considering graph manifolds which are irreducible and boundary irreducible. This restriction is not very important. Indeed, the behavior of graph manifolds with respect to connected sums is already known (Proposition 2.4.3), and the only connected graph manifold which is irreducible but boundary reducible is the solid torus.

Definition 2.4.6. *A system $\mathcal{T} = \{T_1, T_2, \ldots, T_n\}$ of disjoint incompressible tori embedded into an irreducible boundary irreducible graph manifold M is called* canonical *if:*

(1) *$\mathcal{T}$ decomposes M into a collection of Seifert blocks (that is, Seifert manifolds).*
(2) *For any torus $T_i \subset \mathcal{T}$ and for any choice of Seifert fibrations on the adjacent blocks, the two S^1-fibrations on T induced from the both sides are not isotopic.*

Proposition 2.4.7. *Let a system $\mathcal{T} = \{T_1, T_2, \ldots, T_n\}$ of disjoint incompressible tori in an irreducible boundary irreducible graph manifold M decompose it into Seifert blocks. Then $\mathcal{T}$ contains a canonical subsystem.*

Proof. We introduce two moves that decrease the number of blocks for $\mathcal{T}$.

(1) If the two S^1-fibration on $T \subset \mathcal{T}$ induced by some Seifert fibrations of the adjacent blocks are isotopic, we remove the torus T from $\mathcal{T}$. The new block arising in this way is a Seifert manifold, the Seifert structure being composed from the Seifert structures of the old blocks.
(2) Suppose a torus $T \subset \mathcal{T}$ is compressible but is not the boundary of a block $D^2 \times S^1$. By irreducibility of M, it bounds a solid torus B in M. We amalgamate all the blocks lying in B into the new block B by erasing all the tori of $\mathcal{T}$ contained in B.

Let us now apply the moves to $\mathcal{T}$ as long as possible. Evidently, the resulting system (still denoted by $\mathcal{T}$) is canonical. Indeed, since the first move is impossible, any torus $T \in \mathcal{T}$ can inherit only distinct S^1-fibrations from the neighboring blocks. Also, all the tori in $\mathcal{T}$ are incompressible, because all moves of the second type are performed.   □

At first glance a graph manifold M can contain many canonical systems. Indeed, the initial decomposition into, say, elementary blocks is not unique, and the blocks can be amalgamated into larger blocks in many different ways. Nevertheless, if M is irreducible and boundary irreducible, then the canonical system is unique up to isotopy. This result follows from the JSJ-decomposition theorem (see Corollary 6.4.30 and Theorem 6.4.31) and, having been obtained 10 years earlier, can be considered as its infant stage. As a matter of fact, the Waldhausen classification of graph manifolds is nothing more than the JSJ-decomposition theorem for them. To supply a graph manifold M with a unique "name" which distinguishes it from all other graph manifolds, we simply describe its canonical Seifert blocks and the way how they are glued together. The gluing schema can be most naturally presented by a graph. This explains once more why graph manifolds are called so.

Proposition 2.4.8 shows that the class $\mathcal{G}$ is closed with respect to cutting along essential annuli. For simplicity, we formulate and prove it for irreducible manifolds. Recall that M_F denotes a manifold obtained from a manifold M by cutting along a surface $F \subset M$.

Proposition 2.4.8. *Let A be an essential annulus in an irreducible 3-manifold M. Then $M \in \mathcal{G} \iff M_A \in \mathcal{G}$.*

Proof. Let us prove that if M_A is a graph manifold, then so is M. Denote by Y^3 the connected component of a regular neighborhood of $\partial M \cup A$ that contains A. Since the boundary curves of A are nontrivial, $\partial M \cup A$ fibers onto circles. It follows that Y^3 also fibers onto circles over a surface F (one can prove that either $Y^3 = N^2 \times S^1$ or $Y^3 = K^2 \tilde{\times} S^1$). By construction, $M = M_1 \cup Y^3$, where the manifold $M_1 = M \setminus Y^3$ is homeomorphic with M_A, and $M_1 \cap Y^3$ consists of one or two boundary tori. It follows that $M_A \in \mathcal{G} \Rightarrow M \in \mathcal{G}$.

To prove the inverse implication, we construct a canonical system $\mathcal{T}$ of essential tori in M. As we have mentioned earlier, it coincides with the JSJ-system for M. One of the properties of $\mathcal{T}$ is that A is isotopic to another annulus (still denoted by A) which lies in the complement to $\mathcal{T}$, see Sect. 6.4.4. This means that A is contained in a JSJ-chamber Q of $\mathcal{T}$. In our case all the JSJ-chambers are Seifert manifolds. Since any essential annulus in a Seifert manifold Q is saturated (with respect to a Seifert structure on Q), the manifold Q_A is Seifert. It follows that M_A, being composed of Q_A and all the other JSJ-chambers of $\mathcal{T}$, is a graph manifold.   □

2.4.2 Manifolds of Complexity ≤8

As we know from Sect. 2.3.3, all closed orientable irreducible 3-manifolds of complexity ≤ 6 are graph manifolds. By Proposition 2.4.3, the class $\mathcal{G}$ is closed with respect to connected sums. It follows that all (not necessarily irreducible) closed orientable 3-manifolds of complexity ≤ 6 are graph manifolds. The following question arises: what is the complexity of the simplest closed orientable 3-manifold not contained in the class $\mathcal{G}$? In this section we show that the first nongraph closed orientable 3-manifold has complexity 9.

Theorem 2.4.9. *All closed orientable 3-manifolds of complexity no greater than 8 are graph manifolds.*

This was initially proved by computer. Later, a purely theoretical proof was found (see [32]). The computer program is based on the following observation.

Proposition 2.4.10. *Let M be an orientable 3-manifold with $\partial M = S^1 \times S^1$. Suppose that M has an almost simple spine P whose singular graph SP is either empty or consists of one or a few disjoint nonclosed chains with ≤ 2 vertices each. Then $M \in \mathcal{G}$.*

Proof. We apply to P the same simplification moves as in the proof of Theorem 2.2.4, with the following modifications:

(1) Since M may be reducible or boundary reducible, removing an arc ℓ from the 1-dimensional part of P may produce not only another spine of M, but also a spine of a new 3-manifold M_1. Let D be a proper disc in M intersecting ℓ transversally at one point. Then M_1 can be viewed as the manifold M_D, obtained by cutting M along D. It follows from Corollary 2.4.4 (see also Remark 2.4.5) that $M \in \mathcal{G} \iff \hat{M}_D \mathcal{G}$ is a graph manifold.

(2) It may happen that the proper annulus $A \subset M$ that intersects P along a nontrivial simple closed curve l in a 2-component α of P is essential. In this case we cannot apply moves 2 or 3 from the proof of Theorem 2.2.4, but simply cut P along l and obtain a spine of the manifold M_A. By Proposition 2.4.8, $M \in \mathcal{G} \iff M_A \in \mathcal{G}$.

At any step of the simplification procedure the above assumption concerning the singular graph is preserved: We obtain an almost simple spine whose singular graph consists of nonclosed chains with ≤ 2 vertices. After terminating the procedure, we get a collection of special spines modeled on closed chains with ≤ 2 vertices such that the boundaries of the corresponding manifolds are either empty or consist of tori. There are only a few such spines. It is easy to enumerate them and verify that in all cases they determine graph manifolds. Since our simplification moves preserve the property of a manifold to belong to $\mathcal{G}$, M is also a graph manifold. □

The computer works in the following way. It first looks through all the regular graphs of degree 4 with ≤ 8 vertices and, for each graph, lists all the possible spines modeled on it (see the proof of Theorem 2.1.1). Each spine P is tested for the following questions:

1. Is there a short boundary curve?
2. Is there a counterpass?
3. Is the corresponding manifold closed and orientable ?

If it obtains a positive answer to one of the first two questions, or a negative answer to the third question, the computer leaves aside P and goes on to the next spine. Otherwise it tests P for the following property:

4. Does there exist a 2-component α of P such that $P \setminus \alpha$ collapses to an almost simple polyhedron whose singular graph is either empty or consists of nonclosed chains with ≤ 2 vertices?

The main result of the computer experiment is that in all cases the answer to the last question turned out to be positive. By Proposition 2.4.10, this implies the conclusion of Theorem 2.4.9.

The complete text of the above-mentioned theoretical proof of Theorem 2.4.9 takes up nearly a 100 pages and therefore we will limit ourselves to a brief outline. The proof naturally splits up into three stages. First, we prove that any closed irreducible orientable 3-manifold of complexity ≤ 8 is obtained by attaching a solid torus to a 3-manifold of complexity ≤ 3 whose boundary is a torus. We then find out that all such 3-manifolds are graph manifolds except 14 remarkable manifolds $Q_i, 1 \leq i \leq 14$, which are hyperbolic and hence do not belong to the class $\mathcal{G}$ of graph manifolds. (In fact, Q_{12}, Q_{13}, Q_{14} are homeomorphic to Q_6, Q_1, Q_2, respectively. We distinguish them, since they have different special spines with three true vertices). This implies that all closed irreducible orientable 3-manifolds of complexity ≤ 8 are in the class $\mathcal{G}$ except possibly manifolds of the form $(Q_i)_{p,q}, 1 \leq i \leq 14$ (p, q are coprime integers) obtained by pasting solid tori to Q_i. Finally, a more specific analysis shows that any $(Q_i)_{p,q}$ is still in $\mathcal{G}$, provided that its complexity is ≤ 8.

Now let us comment on each step of the proof separately.

STEP 1. Let P be a minimal special spine of a closed irreducible orientable 3-manifold M of complexity ≤ 8. We wish to prove that P has a 2-component α such that after puncturing α and collapsing we get a spine with ≤ 3 true vertices. To simplify the notation, we restrict ourselves to the case when P has exactly eight true vertices. Recall that puncturing a 2-component of P corresponds to removing a solid torus from M.

Let us study in more detail what happens to P when we puncture and collapse its 2-component α. In the collapsing process α disappears completely. Suppose α is adjacent to an edge e of P twice. Then the 2-component β that is adjacent to e the third time also disappears completely. One can easily show that the boundary curves of α, β, and of all the other 2-components

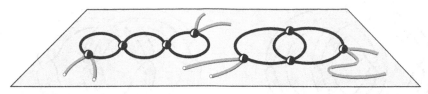

Fig. 2.18. Fragments containing boundary curves that pass through six edges and only four true vertices

that disappear under collapsing contain ≥ 5 true vertices of P together. This means that we get a spine with ≤ 3 true vertices.

Suppose now that no boundary curve passes through an edge twice. Let us call the *length* of a 2-component α of P (or of its boundary curve $c(\alpha)$) the total number of passages of $c(\alpha)$ through edges (with multiplicity taken into consideration). Since P has 16 edges, and since each of them is incident to exactly three 2-components, the total length of the 2-components is equal to 48. On the other hand, P has nine 2-components. It follows that there is a 2-component α adjacent to ≥ 6 different edges. If α contains ≥ 5 different true vertices, we may puncture it and get a spine with ≤ 3 true vertices. If not, then the singular graph SP of P contains one of the fragments shown in Fig. 2.18.

Analyzing the ways in which the boundary curves can pass through each of the fragments, one can always find another boundary curve that contains ≥ 5 different true vertices of P.

STEP 2. Let us introduce 14 remarkable special spines $P_i, 1 \leq i \leq 14$, with ≤ 3 true vertices that determine manifolds Q_i with tori as boundaries. It is convenient to do this by using Figs. 2.19 and 2.20. The manifolds Q_i are the complement spaces of knots in 3-manifolds of genus ≤ 1. For example, Q_2 and Q_{14} are homeomorphic to the complement space of **figure eight knot in S^3. One can show that Q_1 is homeomorphic to Q_{13} and Q_6 is homeomorphic to Q_{12}. All other manifolds Q_i are distinct.

Proposition 2.4.11. *Suppose that the boundary of a compact orientable 3-manifold Q is a torus and that Q has a special spine P with ≤ 3 true vertices. Then either Q is a graph manifold, or P is homeomorphic to one of the spines $P_i, 1 \leq i \leq 14$.*

The proof consists, roughly speaking, of going through all the possible special spines with three or less true vertices and analyzing the corresponding 3-manifolds. There are seven different regular graphs with ≤ 3 vertices. Only 3 of them (the closed chains with 2 and 3 vertices, and the chain with 2 vertices and with an additional loop) may produce manifolds that are not in $\mathcal{G}$. By using the symmetry of the three suspicious graphs and certain artificial tricks, the process can be kept within reasonable limits, which, however, are too large to be presented here. See [79] for details.

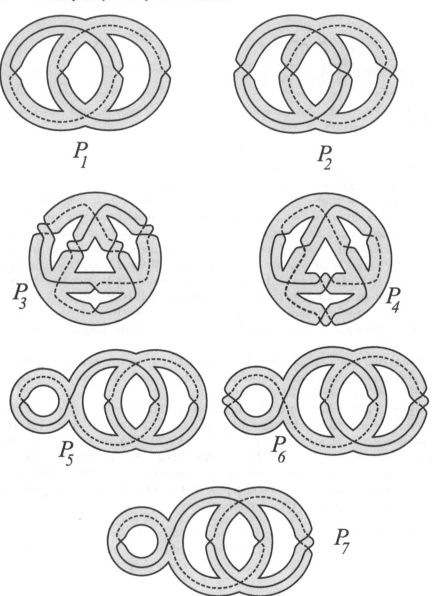

Fig. 2.19. Seven remarkable special spines with ≤ 3 true vertices

STEP 3. Now we prove that if a 3-manifold M of complexity $c(M) \leq 8$ is obtained by a Dehn filling of one of $Q_i, 1 \leq i \leq 14$, then $M \in \mathcal{G}$. Let P be a minimal special spine of M having ≤ 8 vertices. According to Step 1, one can puncture a 2-component α of P such that after collapsing we get a special polyhedron P' with ≤ 3 true vertices. By construction, P' is

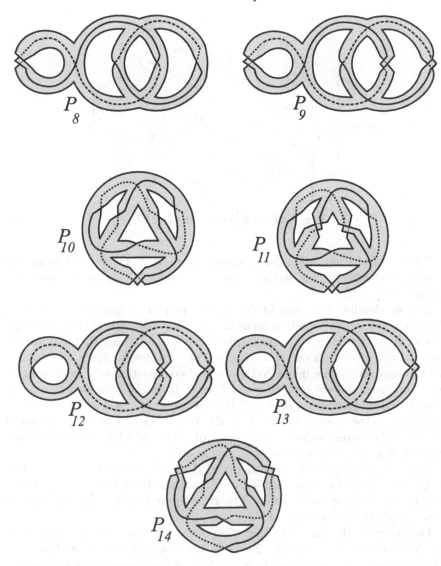

Fig. 2.20. Another seven remarkable special spines with ≤ 3 true vertices

a special spine of a 3-manifold Q such that M is a Dehn filling of Q. It follows from Proposition 2.4.11 that if P' is not homeomorphic to a polyhedron $P_i, 1 \leq i \leq 14$, then Q and M are graph manifolds. It remains to investigate the case when P' is one of P_i.

Proposition 2.4.12. *Suppose a special spine P of a closed orientable 3-manifold M has no more than eight true vertices and suppose that after puncturing one of its 2-components and collapsing we obtain the spine $P_i, 1 \leq i \leq 14$. Then M is a graph manifold.*

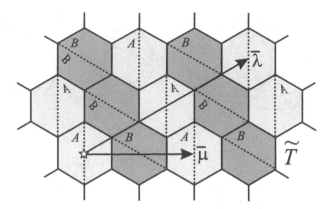

Fig. 2.21. Cell decomposition of $\tilde{T}$

The proof of Proposition 2.4.12 should be carried out in all 14 cases but it follows the same outline and uses the same tricks. Let us carry it out once for the case $i = 1$.

Let us identify the manifold Q_1 with a regular neighborhood of P_1 in M. Denote by T the boundary torus of Q_1. Then the natural collapse of Q_1 onto P_1 induces a locally homeomorphic map $T \to P_1$ such that the inverse image of each 2-component of P_1 consists of two 2-cells. Since P_1 contains two 2-components, this map determines a decomposition of T onto four 2-cells. Construct the universal covering $\tilde{T}$ of T. It can be presented as a plane decomposed into hexagons, see Fig. 2.21. The group of covering translations is isomorphic to the group $\pi_1(T) = H_1(T, Z)$. We choose a basis $\bar{\mu}, \bar{\lambda}$ as shown in Fig. 2.21. The corresponding elements μ, λ of $\pi_1(T)$ (which can be also viewed as oriented loops) form a coordinate system on T.

Since Q_1 is a regular neighborhood of P_1 in M, the difference $V = M \setminus \text{Int } Q_1$ is a solid torus. This means that M has the form $M = (Q_1)_{p,q}$, where coprime integers p, q are determined by the requirement that the curve $\mu^p \lambda^q$ is homotopic to the meridian of V.

Denote by X the part of P_1 that disappears after puncturing and collapsing. Assume that X is an open 2-cell. In other words, the spine P of M is obtained from P_1 by attaching the 2-cell $\bar{X}$ that disappears under puncturing and collapsing. Denote by ℓ the boundary curve of $\bar{X}$. All the intersection points of ℓ with the graph SP_1, as well as all the self-intersection points of ℓ, are true vertices of P. The number of such points must not be greater than 6, since the total number of true vertices is ≤ 8, and two of them are the true vertices of P_1.

Recall that if we factor this covering by the translations $\bar{\mu}, \bar{\lambda}$ corresponding to μ and λ, we recover T. If we additionally identify the hexagons marked by the letter A with respect to the composition of the symmetry in the dotted diagonal of the hexagon and the translation by $-\bar{\mu} + \bar{\lambda}/2$, and do the same for hexagons marked by the letter B, we obtain P_1. T is shown in Fig. 2.22 as

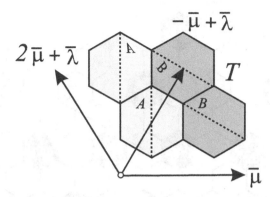

Fig. 2.22. Cell decomposition of T

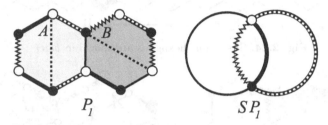

Fig. 2.23. Cell decomposition of P_1 and decorated SP_1

a polygonal disc D composed of four hexagons. Each side of D is identified with another one via the translation along one of the three vectors $\bar{\mu}, -2\bar{\mu}+\bar{\lambda}$, and $-\bar{\mu}+\bar{\lambda}$. P_1 can be presented as the union of two hexagons, see Fig. 2.23. The edges of the hexagons are oriented and decorated with four different patterns. To recover P_1, one should identify the edges having the same pattern. Figure 2.23 shows also the singular graph SP_1 of P_1 equipped with the same decoration.

To the curve ℓ on P_1 (the boundary curve of the attached 2-cell) there corresponds a curve $\bar{\ell}$ of type (p,q) on the torus T and an arc $\tilde{\ell}$ on $\tilde{T}$. One end of $\tilde{\ell}$ is obtained from the other by translation on the vector $p\bar{\mu}+q\bar{\lambda}$. Since ℓ crosses the edges of P_1 in ≤ 6 points, $\tilde{\ell}$ does the same with respect to the edges of $\tilde{T}$. Choosing one hexagon in $\tilde{T}$ as the initial one, and successively marking off those cells which may be reached at the expense of 1, 2, 3, 4, 5, or 6 intersections, one can select all the possible pairs of coprime parameters (p,q) that potentially may produce a spine with ≤ 8 true vertices. In our case they are the following: (1,0), (0,1), (1,1), (-1,1), (-2,1), (-3,1), (-4,1), (-1,2), (-3,2), (-5,2), (-4,3), (-5,3) (up to simultaneous change of signs). See Fig. 2.24.

Let us investigate these pairs. The pairs (-4,1), (-5,2), (-5,3), (-4,3), (-1,2), (1,1) are actually impossible, since in all these cases ℓ intersects at least six edges of P_1 and has at least one self-intersection. For example, if $\tilde{\ell}$ joins the hexagons (0,0) and (5,-2) as shown in Fig. 2.25a, it crosses the edges six times

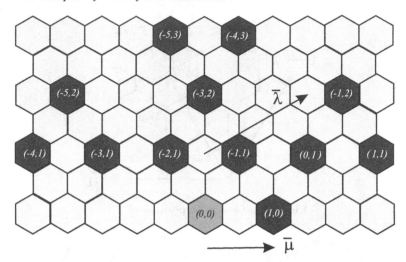

Fig. 2.24. Suspicious hexagons are shown in *black*

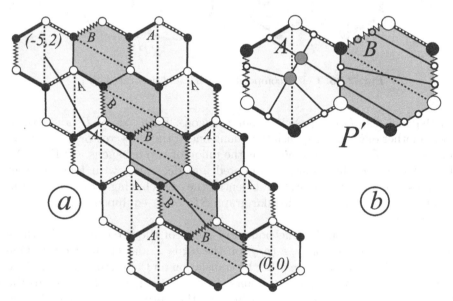

Fig. 2.25. A spine P' of $(Q_1)_{-5,2}$ having ten true vertices

and its projection $\ell \subset P_1$ has two self-intersections, see Fig. 2.25b, where the self-intersections are indicated with gray fat dots.

It can be checked directly that for all the remaining pairs (p,q) (i.e., for $(-3,1)$, $(-2,1)$, $(-3,2)$, $(-1,1)$, $(1,0)$, $(0,1)$) we get graph manifolds. Indeed, let us attach a 2-cell to P_1 so as to obtain a special spine P' of $(Q_1)_{p,q}$ with ≤ 8 true vertices. It turns out that in all these cases one can find a 2-component of P' so that after puncturing and collapsing we get a spine satisfying the assumption of Proposition 2.4.10. It follows that the corresponding manifold $(Q_1)_{p,q}$ belongs to the class $\mathcal{G}$.

However, the part X of P which disappears after puncturing and collapsing is a priori not necessarily a cell. One can represent it as a simple polyhedron $\bar{X}$ attached to P_1 along $\partial \bar{X}$. In this case $\partial \bar{X}$ is a regular graph of degree 3 and hence has a nonzero even number of vertices.

Suppose it has two vertices, which are joined by three edges, i.e., it is what is usually called a θ-curve. The case of *spectacles* (two circles joined by a segment) is excluded since P would have a counterpass and hence could be simplified. The case of four or more vertices is even simpler. So we restrict ourselves to considering only $\bar{X}$ such that $\partial \bar{X}$ is a θ-curve.

Denote by $\bar{a}, \bar{b}, \bar{c}$ the edges of $\bar{X}$. We can think of $\bar{X}$ as being contained in the solid torus $V = M \setminus \text{Int } Q_1, \partial V = T$, such that $\partial \bar{X} \subset T = \partial Q_1$ and the complement of $\partial V \cup X$ in V is an open 3-ball. When we attach $\bar{X}$ to P_1, the vertices of $\partial \bar{X}$ become true vertices of P. Since P has not more than eight true vertices, the images a, b, c of $\bar{a}, \bar{b}, \bar{c}$ under gluing may intersect the edges of the singular graph SP_1 in ≤ 4 points, and if in 4, then they cannot intersect each other or possess self-intersections. If we lift $\bar{a}, \bar{b}, \bar{c}$ to the arcs $\tilde{a}, \tilde{b}, \tilde{c}$ on $\tilde{T}$ having a common endpoint, we get a triode (a wedge of three arcs) intersecting the edges in ≤ 4 points. The free ends of the wedge lie on three different hexagons which are obtained from each other by translations on nontrivial integer linear combinations of $\bar{\mu}, \bar{\lambda}$. A simple analysis of the covering shows that there exist a few such triodes, but the projections onto P_1 of their edges have at least one additional intersection point. This finishes the proof of Proposition 2.4.12 and Theorem 2.4.9.

2.5 Hyperbolic Manifolds

2.5.1 Hyperbolic Manifolds of Complexity 9

As we have shown in the earlier section, all closed irreducible orientable 3-manifolds up to complexity 8 are graph manifolds. Is this result sharp? The answer is affirmative. We describe here a remarkable closed orientable 3-manifold M_1 of complexity 9 which is hyperbolic and thus does not belong to the class $\mathcal{G}$ of all graph manifolds. This manifold was discovered by Weeks [46, 133] and independently by Fomenko and Matveev [32]. We have called it "remarkable," since it is twice minimal. First, it has the minimal complexity among all closed orientable hyperbolic 3-manifolds. Second, its hyperbolic volume $V(M_1) \approx 0.94272$ is also minimal among all the *known* 3-manifolds of the same class. Conjecture 2.5.1 is motivated by these facts, together with an as experimental observation made in [32] that the growth of the volume correlates (in some sense) with the growth of the complexity.

Conjecture 2.5.1 ([32]). M_1 has the least volume of any closed orientable hyperbolic 3-manifold.

No counterexamples to this conjecture have appeared and the difference between the known lower estimates of the volumes and $V(M_1)$ remains substantial. Personally, I believe that the conjecture is true.

Let Q_1 be the 3-manifold represented by its special spine P_1 with two true vertices, see Fig. 2.19. Its boundary $T = \partial Q_1$ is a torus with coordinate system (μ, λ). One can think of μ and λ as oriented simple closed curves on T which are images of the oriented segments $\bar{\mu}, \bar{\lambda}$ in $\tilde{T}$ under the universal covering projection map $\tilde{T} \to T$, see Fig. 2.21. We define the above-mentioned remarkable manifold M_1 as the manifold $(Q_1)_{5,-2}$ obtained from Q_1 by attaching a solid torus such that the image of its meridian has the type $\mu^5 \lambda^{-2}$.

To show that M_1 is hyperbolic, we recall briefly the Thurston method, one of the most important and successful methods for understanding finite-volume hyperbolic 3-manifolds by considering their decomposition into ideal tetrahedra. This technique was introduced in [120] and was used for computation of volumes in [46,133] and [32]. It forms also the basis for the SNAPPEA computer program, written by Weeks, which allows one to determine the hyperbolic structure and volume of a large number of hyperbolic manifolds.

First, we use Corollary 1.1.28 to decompose Int Q_1 into two topological ideal tetrahedra. Next to each ideal tetrahedron we associate a complex variable that determines its geodesic shape. A system of complex polynomial equations is generated, the equations coming from the need for the tetrahedra to glue together correctly at the edges. This system of *consistency conditions* for Q_1 has many solutions parameterized by elements of U, where U is the set of all complex numbers z such that $\mathrm{Im}\, z > 0$ and $z \neq 1/2(1+t\mathbf{i})$, $\sqrt{15} < t < \infty$. If both ideal tetrahedra are regular, then the hyperbolic structure on Q_1 thus obtained is complete. This means that Q_1 is hyperbolic. By the way, all other manifolds Q_2–Q_{14} are also hyperbolic and, as shown in [32], are the only orientable hyperbolic 3-manifolds of complexity ≤ 3 having one cusp.

It turns out that the parameter z and hence the geometric shape of the ideal tetrahedra can be chosen so that z satisfies the consistency conditions and that the completion of the corresponding hyperbolic structure on Q_1 is a closed hyperbolic manifold homeomorphic to M_1, see [46,133] and [32]. This means that M_1 is hyperbolic.

Remark 2.5.2. The value of z that produces the hyperbolic structure of M_1 has irrational real and imaginary parts. So there may arise the question whether approximate values of z that can be found by computer (and that satisfy the consistency and the completion conditions only approximately) are sufficient for proving that M_1 is hyperbolic. This difficulty can be overcome, since z lies strictly inside U. On the other hand, M_1 is arithmetic [22,23], so one can prove the existence of a hyperbolic structure without using computers.

Theorem 2.5.3. *The complexity of M_1 is equal to 9.*

Proof. Let us construct a special spine of M_1 having nine true vertices. Since $c(M_1) > 8$ by Theorem 2.4.9, this would be sufficient for proving that $c(M_1) =$

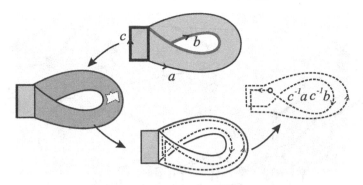

Fig. 2.26. Möbius triplet

9. The hexagon (5,-2) is one of "suspicious" hexagons, since it can be joined with the initial hexagon (0,0) by an arc $\tilde{l}$ in $\tilde{T}$ intersecting six edges, see Fig. 2.25. So it would be natural to look for $\tilde{l}$ such that its projection l onto P_1 has not more than one self-intersection point. If we find one, then a special spine of M_1 with 9=2+6+1 vertices can be obtained by attaching to P_1 a new disc 2-component along l. Unfortunately, the projections of all arcs on $\tilde{T}$ that join (0,0) with (5,-2) and intersect not more than six edges have at least two self-intersection points. Therefore, the maximum we can get by attaching a disc is a spine of $(Q_1)_{-5,2}$ having ten true vertices.

It turns out that one can save one true vertex by attaching to P_1 not a disc, but a so-called *Möbius triplet* Y, see Fig. 2.26. One can think of Y as being contained in the solid torus $V = M_1 \setminus \text{Int } Q_1$ such that $\partial Y \subset T = \partial Q_1$ and the complement of $\partial V \cup Y$ in V is an open 3-ball.

If we cut out a disc D from the Möbius 2-component of Y, then the rest collapses onto ∂Y. We need to track the behavior of ∂D under the collapse. Observe that ∂D is deformed into the curve $c^{-1}ac^{-1}b$, where a, b, c are the three coherently oriented edges of the θ-curve ∂Y. Let us attach Y to P_1 to obtain a special polyhedron $P' = P_1 \cup Y$ as shown in Fig. 2.27. The arcs $\tilde{a}, \tilde{b}, \tilde{c} \subset \tilde{T}$ that corresponds to a, b, c form a triode such that its branch point is in the hexagons $(-1/2, 0)$ while the free ends of $\tilde{a}, \tilde{b}, \tilde{c}$ are in the hexagons $(1/2, 0), (-1/2, 0), (-5/2, 1)$, respectively.

Therefore, the curve $c^{-1}ac^{-1}b$ has type (3,-1)+(2,-1)=(5,-2). It follows that P' is a special spine of $(Q_1)_{5,-2}$. The images in P_1 of the arcs a, b, c have only one intersection point (shown in Fig. 2.27 as a fat gray dot). Therefore, P' has nine vertices. A regular neighborhood of its singular graph is represented in Fig. 2.28. □

Remark 2.5.4. The manifold Q_2 (the complement of the figure eight knot) is a twin of Q_1: Just as Q_1, it admits a decomposition into two regular ideal tetrahedra. It was used by W. Thurston to illustrate his method [120]. Just as Q_1, it admits a decomposition into two regular ideal tetrahedra. The manifold $M_2 = (Q_2)_{5,1}$ has complexity 9 and its volume is the second one among all

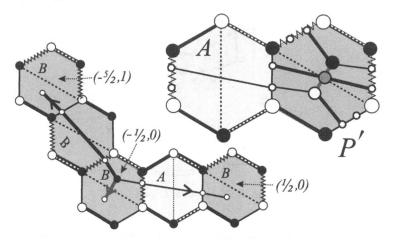

Fig. 2.27. A special spine P of $(Q_1)_{-5,2}$ obtained by attaching a Möbius triplet to P_1

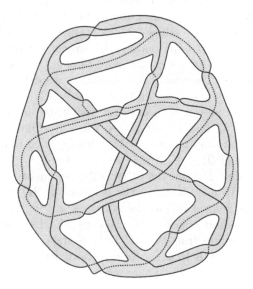

Fig. 2.28. A special spine of M_1

known volumes of closed hyperbolic 3-manifolds. A special spine of M_2 is shown on Fig. 2.29.

2.6 Lower Bounds of the Complexity

As we have shown in Sect. 2.1.2, it is relatively easy to obtain upper bounds for complexity. However, the problem of finding lower bounds is quite difficult. Of course, we know the exact value of the complexity for all the manifolds

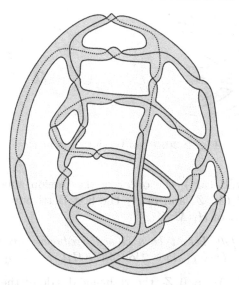

Fig. 2.29. A special spine of M_2

from the table (see Appendix), but there are only finitely many of them. In this section we present several lower bounds for the complexity of arbitrary 3-manifolds and describe two infinite series of hyperbolic 3-manifolds whose complexity is known exactly.

2.6.1 Logarithmic Estimates

The first bound is based on the evident observation that if the first homology group of a 3-manifold M is large, then $c(M)$ cannot be too small.

Lemma 2.6.1. *Suppose that a special spine P of a closed orientable 3-manifold M contains a 2-component α whose boundary curve passes along some edge e three times. Then M has an almost simple spine having a smaller number of true vertices.*

Proof. If the boundary curve $\partial\alpha$ of α has a counterpass on the edge e, then by Proposition 2.3.3 P can be simplified. Suppose $\partial\alpha$ passes along e all three times in the same direction. Then P contains a simple closed curve ℓ such that ℓ intersects SP at two points $A, B \in e$ and visits all three wings adjacent to e, see Fig. 2.30. ℓ can be easily constructed by thinking of α as an attached disc and considering two disjoint proper arcs in the disc that join points in distinct preimages of e. One can easily see that ℓ can be shifted away from P into the boundary of a regular neighborhood N of P. Since M is closed, ∂N is a 2-sphere. It follows that l bounds a disc $D \subset M$ such that $D \cap P = \ell$, and we can simplify the spine by adding D to P and piercing another 2-cell. $\square$

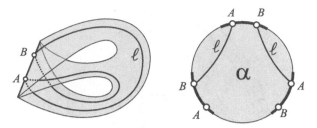

Fig. 2.30. ℓ can be shifted from P

Denote by $|\mathrm{Tor}(H_1(M))|$ the order of the torsion subgroup of the first homology group $H_1(M; Z)$ and by β_1 the first Betti number of M, i.e., the rank of the free part of $H_1(M; Z)$.

Theorem 2.6.2. *[90] Let M be a closed irreducible orientable 3-manifold different from $L_{3,1}$. Then $c(M) \geq 2\log_5 |\mathit{Tor}(H_1(M))| + \beta_1 - 1$.*

Proof. Since for $H_1(M) = 0, Z_2$ the right-hand side of the above inequality is negative, the conclusion of the theorem holds for $M = S^3$ and $M = RP^3$. So we can assume that M is not one of these manifolds. Choose an almost simple spine P of M having $c(M)$ true vertices. By Theorem 2.2.4, we may assume that P is special. Let $A(P)$ be the relation matrix of the presentation corresponding to P and n be the number of generators in that presentation. Then $n = c(M) + 1$, and, as M is closed, $A(P)$ is a square matrix of order n. Since P has the smallest number of true vertices, Lemma 2.6.1 implies that it has no edges along which some component passes three times. There are no counterpasses either, therefore each column of the matrix contains either two nonzero elements (one of them is equal to ±2, and the other is equal to ±1), or three elements (each is equal to ±1).

The matrix $A(P)$ has a minor A' of order $n - \beta_1$ whose determinant is nonzero and is divisible by $|\mathrm{Tor}(H_1(M))|$. On the other hand, the absolute value of the determinant is equal to the volume of the parallelepiped whose base vectors are the columns of A'. The volume does not exceed the product of the lengths of those vectors. It is clear that the length of each vector is not greater than $\sqrt{5}$. Hence $|\det A'| \leq (\sqrt{5})^{n-\beta_1}$, which implies that $n \geq 2\log_5 |\det A'| + \beta_1$. Since $|\det A'|$ is greater than any its divisor, and $n = c(M) + 1$, we have $c(M) \geq 2\log_5 |\mathrm{Tor}(H_1(M))| + \beta_1 - 1$. □

In some cases (for instance, for $L_{5,2}$) this bound is sharp. Let us show that for an infinite series of lens spaces this bound is *almost sharp* (in certain sense). Let $u_i, 1 \leq i < \infty$, be the Fibonacci numbers given by the initial values $u_1 = u_2 = 1$ and the recurrence relation $u_{i+1} = u_i + u_{i-1}$. Denote by L_n the lens space $L(p,q)$ with parameters $p = u_n, q = u_{n-2}$.

Corollary 2.6.3. *If $n > 4$, then $nC_n - 2 \leq c(L_n) \leq n - 4$, where $C_n = (2/n)\log_5(\sqrt{5}u_n)$.*

Proof. It follows from Theorem 2.6.2 that $c(L_n) \geq 2\log_5 u_n - 1 = nC_n - 2$, so we get the first inequality. To get the second one, we recall that all partial quotients in the expansion of u_n/u_{n-2} as a regular continued fraction are 1, so their sum $S(u_n, u_{n-2})$ is $n-1$. Therefore, by item E of Sect. 2.3.3, we have $c(L_n) \leq S(u_n, u_{n-2}) - 3 = n - 4$ for all $n > 4$. □

Remark 2.6.4. One can easily show that $C_n > 0.5$ for all $n > 4$ and that C_n tends to $2\log_5(\frac{1+\sqrt{5}}{2}) \approx 0.59798$ as $n \to \infty$.

Corollary 2.6.3 shows that for an infinite series of 3-manifolds complexity depends logarithmically on the order of the torsion subgroup of the first homology group. This remarkable fact was first observed by Pervova.

The bound in Theorem 2.6.2 has the following shortcoming: It is trivial for closed manifolds having zero first homology group, i.e., for homology spheres. It would be natural to attempt to find a bound depending on the fundamental group.

Definition 2.6.5. *Let a group G be given by a presentation*

$$G = \langle g_1, \ldots, g_n \mid r_1, \ldots, r_m \rangle.$$

Then the length *of that presentation is the number $|r_1| + \ldots + |r_m|$, where $|r_i|$ denotes the length of the word r_i with respect to $g_1, \ldots, g_n$. The presentation complexity $\hat{c}(G)$ of G is the minimum of the lengths of all its finite presentations.*

Let us consider some examples of estimates of complexity of groups. Evidently, the complexity of the cyclic group Z_n does not exceed n. It is interesting to note that it may be much less than n. For example, the presentation $\langle a, b, c \mid a^4 b, b^5 c, c^2 \rangle$ of Z_{40} has length 13. It may appear that this small value (compared with the order of the group) is due to the fact that 40 has nontrivial divisors. However, the group Z_{47}, which has prime order, can be given by the presentation $\langle a, b, c \mid a^4 b, b^4 c, c^3 a^{-1} \rangle$ of length 14.

Proposition 2.6.6. *Let M be a closed irreducible orientable 3-manifold, different from S^3, RP^3, and $L_{3,1}$. Then $c(M) \geq -1 + \hat{c}(\pi_1(M))/3$.*

Proof. Let P be a special spine of M having $k = c(M)$ true vertices. Then the length of the presentation of $\pi_1(M)$ that corresponds to P is $3(k+1)$. It follows that $\hat{c}(\pi_1(M)) \leq 3(k+1)$ and thus $k \geq -1 + \hat{c}(\pi_1(M))/3$. □

Sometimes this proposition allows to obtain better bounds than those given by Theorem 2.6.2. For example, it can be shown that the complexity of any nontrivial finitely presented group G that coincides with its commutator subgroup is at least 10. It follows from Proposition 2.6.6 that the complexity of any homology sphere cannot be less than 3. This agrees with the fact that the complexity of the first nontrivial homology sphere (the dodecahedron space) equals 5, see [35, 91].

2.6.2 Complexity of Hyperbolic 3-Manifolds

Now we turn our attention to hyperbolic 3-manifolds. As we have mentioned earlier, there is a correlation between their complexities and volumes. A very nice partial case of this observation was found by Anisov [6]. Recall that all regular ideal tetrahedra in the hyperbolic space H^3 are congruent and have the same volume $V_0 \approx 1.0149$. The volumes of all other ideal tetrahedra in H^3 are less than V_0.

Lemma 2.6.7. *Let M be a hyperbolic 3-manifold with nonempty boundary. Then $c(M) \geq V(M)/V_0$, where $V(M)$ is the hyperbolic volume of M.*

Proof. Since M is hyperbolic, it is irreducible and boundary irreducible, and contains no essential annuli. By Theorem 2.2.4, its minimal almost simple spine is special. It follows from Corollary 1.1.28 that M can be decomposed into $k = c(M)$ topological ideal tetrahedra $\Delta_i, 1 \leq i \leq k$. Further we follow Thurston's arguments [120]. These tetrahedra can be lifted to H^3, straightened inside H^3 to hyperbolic ideal tetrahedra and projected back into M. The new tetrahedra Δ_i' can overlap, but they still cover M. It follows that $V(M) \leq \sum_{i=1}^{k} V(\Delta_i) \leq kV_0$, where $V(\Delta_i)$ is the volume of Δ_i. We can conclude that $k \geq V(M)/V_0$. $\qquad\square$

In some cases Lemma 2.6.7 is sufficient for exact computation of complexity [6].

Corollary 2.6.8. *Suppose that a hyperbolic 3-manifold M can be decomposed into k straight regular ideal tetrahedra. Then $c(M) = k$.*

Proof. Since $V(M) = kV_0$, we have $c(M) \geq k$ by Lemma 2.6.7. The inequality $c(M) \leq k$ follows from Corollary 1.1.28. $\qquad\square$

There are not many 3-manifolds satisfying the assumption of Corollary 2.6.8. Q_1 and Q_2 (see Remark 2.5.4) as well as all their finite coverings are among them. Since $H_1(Q_1; Z) = Z \oplus Z_5$ and $H_1(Q_2; Z) = Z$, there are infinitely many such coverings. These manifolds form the first nontrivial infinite series of 3-manifolds with known complexities: if a 3-manifold M is a k-sheeted covering of Q_1 or Q_2, then $c(M) = 2k$.

Remark 2.6.9. Q_2 can be represented as a Stallings manifold fibered into punctured tori over S^1 with the monodromy matrix $\begin{pmatrix} 2 & 1 \\ 1 & 1 \end{pmatrix}$.

Let W_1 be a closed Stallings manifold with fiber $S^1 \times S^1$ and the same monodromy matrix. Consider the k-sheeted covering W_k of W_1 corresponding to the kernel of the superposition of the abelinization map $\pi_1(W_1) \to H_1(W_1; Z) = Z$ and the mod k reduction $Z \to Z_k$. One can easily construct a special spine P of W_k with $2k + 5$ true vertices. Indeed, it suffices to take a k-sheeted covering of the spine P_2 of Q_2 and attach an additional 2-cell D which fills up the puncture of the fiber. It follows that $c(W_k) \leq 2k+5$ (see [5]).

On the other hand, a short calculation shows that $|\text{Tor}(H_1(W_n))| = u_{2n+1} + u_{2n-1} - 2$. Taking into account that the first Betti number of W_n is 1 and applying Theorem 2.6.2, we get $c(W_n) \geq 2C'_n n$, where $C'_n = (1/n)\log_5(u_{2n+1} + u_{2n-1} - 2)$. It follows that $2C'_n n \leq c(W_n) \leq 2n + 5$, so we have another example of a logarithmic growth of the complexity (compare with Corollary 2.6.3). It is interesting to note that C'_n has exactly the same limit as C_n from Corollary 2.6.3.

2.6.3 Manifolds Having Special Spines with One 2-Cell

We describe for every $n \geq 2$ an interesting class $\mathcal{M}_n$ of orientable 3-manifolds having complexity n. The manifolds from $\mathcal{M}_n, 2 \leq n < \infty$ form the second infinite set of manifolds with known complexity (the first such set was described in the earlier section). This class was introduced in [33], see also [40].

Definition 2.6.10. *An orientable 3-manifold M belongs to the class $\mathcal{M}_n$, if it has a special spine with n true vertices and exactly one 2-cell.*

Examples of spines with one 2-cell are shown in Fig. 2.31. Each $\mathcal{M}_n$ contains a manifold presented either by the upper spine (if $n \neq 3k + 1$) or by the lower one (if $n = 3k + 1$). The only exception is the case $n = 1$ when $\mathcal{M}_n$ is empty. As proved in [33], the number of manifolds in $\mathcal{M}_n$ grows exponentially as $n \to \infty$.

Theorem 2.6.11. *[33] Let $M \in \mathcal{M}_n$. Then $c(M) = n$.*

Proof. By definition of the class $\mathcal{M}_n$, the manifold M has a special spine P with n vertices. Therefore, $c(M) \leq n$. To prove that $c(M) = n$, consider a handle decomposition ξ_P of M that corresponds to P. Since P has only one 2-cell, the set of all normal surfaces in M can be easily described. All closed normal surfaces are normally parallel to ∂M. Since $\chi(\partial M) = 2(1 - n)$ and $n > 0$, there are no normal spheres among them. Therefore, M is irreducible. All nonclosed normal surfaces are contained in the union of all balls and

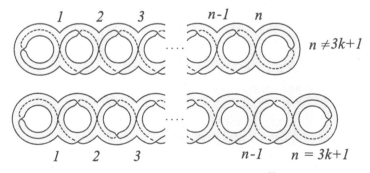

Fig. 2.31. Spines with one 2-cell

beams of ξ_P. The set of such surfaces contains no discs, thus M is boundary irreducible. This set can contain annuli, but all of them are compressible. It follows from Theorem 2.2.4 that M has a special spine P' with $m = c(M)$ true vertices.

Let us prove that $m \geq n$. Denote by k the number of 2-cells of P'. Counting the Euler characteristic of M, we get $2(k - m) = \chi(M) = 2(1 - n)$ and $n + k - 1 = m$. It follows that $m \geq n$. $\square$

Manifolds from $\mathcal{M}_n$ possess many other good properties. They are hyperbolic manifolds with totally geodesic boundary and have Heegaard genus $n + 1$. Moreover, each manifold $M \in \mathcal{M}_n$ has a unique special spine with n vertices, which is homeomorphic to the cut locus of M (the set of points of M having more than one shortest geodesic to ∂M). See [33] for the proof.

3

Haken Theory of Normal Surfaces

Normal surfaces were introduced by Kneser in 1929 [66]. The theory of normal surfaces was further developed by W. Haken in the early 1960s [38]. Its fundamental importance to the algorithmic topology cannot be overestimated. Most of the work on 3-manifolds since then is based on or related to it.

3.1 Basic Notions and Haken's Scheme

Haken is one of the first topologists who realized that the right strategy to investigate 3-manifolds is to look over the set of surfaces that are contained in them. It turned out that this infinite set has an algorithmically constructible finite basis, and thus admits an explicit description. The main idea of Haken's method consists in decomposing a given manifold M into simple pieces and considering how the surfaces intersect the pieces. There are several versions of the method, but the main steps are common to all of them. Here are the steps.

STEP 1. We choose a type of decomposition. By a decomposition of a 3-manifold M we mean presenting M as the union of simple (in some sense) 3-dimensional pieces with disjoint interiors. The pieces are called *elements* of the decomposition. Most common types of decompositions are triangulations and handle decompositions. As a rule, the triangulation approach is easier to describe, while handle decompositions are more flexible and efficient.

STEP 2. Let M be a manifold with a fixed decomposition ξ and F a proper surface in M. Then ξ generates a decomposition of F into *elementary pieces*, where each elementary piece is a connected component of the intersection of F with an element of ξ. We define a class of *normal surfaces* in M by specifying certain allowed types of elementary pieces. As a rule, normal surfaces are considered up to isotopies of M that preserve the decomposition, i.e., take every element of ξ to itself. Such an isotopy is called *normal*. Every allowed type is just an equivalence class of elementary pieces contained in the corresponding element of ξ (with respect to normal isotopy). From the viewpoint of the algorithmic approach, it is important to have only a finite number of allowed types in any element of ξ. Another important requirement: if a finite collection of elementary pieces in an element of ξ is realizable by disjoint pieces, then the realization is unique up to normal isotopy.

STEP 3. Let $\mathcal{N}$ denote the set of all normal surfaces in M (considered up to normal isotopy). At this step, we further refine the notion of elementary piece so as to ensure that $\mathcal{N}$ possesses the following two informal properties:

1. $\mathcal{N}$ is informative meaning that it contains representatives of all interesting classes of surfaces in M. In other words, we do not lose important information if we restrict ourselves to considering only normal surfaces. Certainly, the notions of interesting class and important information depend heavily on the problem we are trying to solve. Sometimes this item can be reduced to a motivating example, postponed, or skipped.
2. $\mathcal{N}$ admits a more or less explicit description. The next two steps give a general method for obtaining such a description.

STEP 4. Let $E_1, E_2, \ldots, E_n$ denote all the allowed types of elementary pieces in all the elements of ξ. A normal surface $F \subset M$ can intersect the elements along several pieces of each type E_i. Let $x_i = x_i(F)$ denote the number of these pieces. We get an n-tuple $\bar{x}(F) = (x_1, x_2, \ldots, x_n)$ of non-negative integers, that is, a vector with non-negative integer coordinates. We now need to verify that two normal surfaces with the same vector are normally isotopic (it is obvious from the above definitions that normally isotopic surfaces have the same vector). Having done that, we can identify the set of all normal surfaces in M with a subset of the set of integer points in R^n.

STEP 5. We have established on the previous step that to any normal surface there corresponds an n-tuple $\bar{x} = (x_1, x_2, \ldots, x_n)$ of non-negative integers. Conversely, we could begin with an n-tuple $\bar{x}$ of non-negative integers and try to build the corresponding normal surface. This is possible only if we subject the n-tuple to some constraints.

The first necessary condition is that, for every $x_i, x_j \neq 0$, the corresponding types E_i, E_j must be *compatible*, that is, they must contain disjoint representatives. Such n-tuples will be called *admissible*. As a rule, the compatibility condition guarantees us that for any admissible n-tuple not only two, but any collection of types E_i with $x_i \neq 0$ admits a collection of disjoint representatives.

In order to describe the second necessary condition, for each i we put x_i disjoint pieces of type E_i into the corresponding element of ξ. If the n-tuple is realizable by a normal surface, then the pieces must give normally isotopic patterns in the intersection of every pair of neighboring decomposition elements. This condition can be described by linear homogeneous equations. The equations together with the inequalities $x_i \geq 0$ form the so-called *matching system* $E(\xi)$. The main goal of this step consists in proving the following basic conclusion of Haken's theory:

(A) *The set of equivalence classes of normal surfaces in M can be parameterized by the set of admissible solutions to the matching system.*

STEP 6. This algebraic step is of general nature and does not depend on a specific version of the normal surface theory. Let E be a system of linear homogeneous equations with integer coefficients.

Definition 3.1.1. *A non-negative integer solution $\bar{x}$ to the system E is called fundamental, if it cannot be presented in the form $\bar{x} = \bar{y} + \bar{z}$, where $\bar{y}, \bar{z}$ are nontrivial non-negative integer solutions to E.*

The second basic conclusion of Haken's theory is:

(B) *The set of fundamental solutions to any system of linear homogeneous equations with integer coefficients is finite and can be constructed algorithmically. Any non-negative integer solution to the system can be presented as a linear combination of the fundamental solutions with non-negative integer coefficients.*

In particular, conclusion (B) is applicable to the matching system from Step 5. Let us return to geometry.

Definition 3.1.2. *A normal surface $F \subset M$ is called fundamental, if it corresponds to an admissible fundamental solution of the matching system.*

To get an explicit description of the set $\mathcal{N}$ of normal surfaces in M, it remains now to select admissible fundamental solutions, realize them by fundamental surfaces, and reveal a geometric interpretation of the algebraic summation of solutions. The desired description looks as follows: $\mathcal{N}$ is the set $\{\sum_{i=1}^{k} \alpha_i F_i\}$ of all linear combinations such that the coefficients α_i of each combination $\sum_{i=1}^{k} \alpha_i F_i$ are non-negative integers, and fundamental surfaces F_i with $\alpha_i > 0$ are compatible, i.e., consist of pairwise compatible pieces.

Thus, each normal surface in M can be constructed from a finite set of fundamental surfaces. This fact has essential applications in algorithmic problems. Indeed, in many cases the problem in hand can be reduced to the question of whether a given manifold M contains a surface with some specific property P. Quite often it turns out that if M contains a surface possessing P, then it contains a fundamental surface with that property. If P is algorithmically recognizable, then by checking each fundamental surface against possessing P, we find out in a finite number of steps whether M contains a required surface. That gives us an *algorithmic* solution of our initial problem.

In what follows we apply the above scheme to describing two basic versions of the Haken method. To facilitate understanding, we descend 1-dimension down and start with a review of *the theory of normal curves on surfaces*. It is remarkable that almost all key ideas of Haken's theory can be presented already at this level.

3.2 Theory of Normal Curves

3.2.1 Normal Curves and Normal Equations

Let F be a triangulated surface (later on we assume that the triangulation is fixed). By a *curve* we mean a 1-dimensional proper submanifold of F. It may consist of several disjoint components, each being either a simple closed curve or a simple arc with the endpoints on ∂F. We would like to describe the set of all curves on F. To be more precise, we want to find an algorithmically constructible list that contains nearly all isotopy classes of curves (possibly with duplicates). Saying "nearly" we mean that we might neglect trivial components, which cut off discs.

Let C be a curve on F. We may assume that C is in general position with respect to the triangulation, i.e., it does not pass through the vertices and intersects the edges transversally. Then the intersection of C with any triangle consists of connected components that can have one of the following three types:

(1) An arc with the endpoints on different edges of the triangle.
(2) An arc with the endpoints on the same edge e (we call this situation an *interior return* if e is inside F, and a *boundary return* if $e \subset \partial F$).
(3) A circle in the interior of the triangle.

Definition 3.2.1. *A general position curve C on F is called* normal, *if the intersection of C with any triangle contains no returns and no circles.*

So far we have carried out the first two steps of the general scheme. Theorem 3.2.2 shows that we do not loose anything by restricting ourselves to normal curves.

Theorem 3.2.2. *Any curve C on F can be transformed into a normal one by an isotopy and crossing out boundary returns and circle components contained in the interiors of the triangles.*

Proof. We will simplify the intersection of C with the triangles of the triangulation by a process called *normalization*. Assume that a triangle Δ contains an interior return of C. Any such return cuts off a half-disc D from Δ. Using D, we construct an isotopy that pushes the return across the edge into the adjacent triangle together with all the other interior returns and circle components that may be contained in D. This move decreases the total number of crossing points of C with edges by two or more (if D contains other returns). Repeating it for as long as possible, we get a curve without interior returns. It remains to cross out all boundary returns and circle components inside the triangles. $\square$

Remark 3.2.3. It may happen that normalization destroys the curve C completely, that is, reduces it to the empty set. On the other hand, if C contains no trivial component, which cuts a disc out of F, then we cross out nothing, and the normalization process is an isotopy.

Denote by $\mathcal{N}$ the set of all normal curves on F considered up to normal isotopy, which preserves the triangulation. To compose the matching system (see Step 5 above), in every angle of each triangle we put a variable x_i. Thus we get n variables $x_1, x_2, \ldots, x_n$ numbered in some order, where n is the total number of angles. For each interior edge e of the triangulation, there are four angles adjacent to it. Let x_i, x_j, x_k, x_l be the corresponding variables such that x_i, x_j are in Δ_1 and x_k, x_l are in Δ_2, where Δ_1, Δ_2 are the two triangles adjacent to e. Then we write the equation $x_i + x_j = x_k + x_l$.

Doing this for all the interior edges, we obtain a system of equations. The number of the equations coincides with the number of the interior edges of the triangulation. The matching system of linear homogeneous equations and inequalities is obtained by adding n inequalities $x_i \geq 0, 1 \leq i \leq n$:

$$\textit{Matching system}$$
$$x_i + x_j = x_k + x_l$$
$$x_i \geq 0, 1 \leq i \leq n$$

Let us relate integer solutions of the matching system (all of them admissible) and normal curves considered up to normal isotopy. Note that the set of such solutions is closed with respect to taking sums. Theorem 3.2.4 is just the first basic conclusion of Haken's theory for the case of triangulated surfaces.

Theorem 3.2.4. *There exists a natural bijection between the set of integer solutions of the matching system for F and the set of equivalence classes of normal curves on F.*

Proof. First, let us associate with a given normal curve C on F an n-tuple $\bar{x}(C) = (x_1, x_2, \ldots, x_n)$ of non-negative integers. For each $i, 1 \leq i \leq n$, let Δ be the triangle whose angle is labeled by x_i. We put x_i equal to the number of arcs in $C \cap \Delta$ that join the sides of the angle (the arcs are 1-dimensional counterparts of elementary pieces considered in Step 4 of Sect. 3.1). It follows that if the angles labeled by x_i, x_j are adjacent to an edge e of the triangulation from the same side, then $x_i + x_j$ equals the number of points in $e \cap C$. The same is true for the angles adjacent to e from the other side: $e \cap C$ consists of $x_k + x_l$ points, where x_k, x_l correspond to these angles. We may conclude that $x_i + x_j = x_k + x_l$. It follows that the n-tuple $\bar{x}(C) = (x_1, x_2, \ldots, x_n)$ is an integer solution of the matching system.

Conversely, given an integer solution $\bar{x} = (x_1, x_2, \ldots, x_n)$ of the system, draw the corresponding number of arcs in each angle. It is obvious that these arcs can be chosen so as to be disjoint. Thus all solutions are admissible. Since $\bar{x}$ is a solution, for each edge e the number of endpoints of arcs coming to e from one side equals the number coming from the other. Adjusting the arcs by an isotopy, we may assume that both sets of endpoints do coincide on each edge. This means that the union of all the arcs forms a normal curve (neither returns nor circles inside triangles can appear). $\qquad\square$

Remark 3.2.5. One can show that two normal curves are isotopic in the complement of the vertices if and only if they are normally isotopic. However, as the ensuing example shows, two curves can be isotopic without being normally isotopic.

Remark 3.2.6. Let T be a finite set of triangles. Divide some of their edges into pairs and identify the edges of each pair via a homeomorphism. We get a surface F. By construction, F is decomposed into *singular triangles* (the images of the triangles of T). Such a decomposition of F is called a *singular triangulation* of F. Although the intersection of two singular triangles may consist of more than one vertex or edge, the normal curve theory works for singular triangulations as well.

In the following educational example we apply Theorem 3.2.4 to describing the set of normal curves on the Klein bottle. As a corollary we show that there are only four nonisotopic nontrivial circles. See Fig. 3.1, where the bottle is presented as two Möbius strips glued together along the boundary.

Example 3.2.7. Let K be the Klein bottle. To decrease the number of variables, we consider a singular triangulation of K into only two triangles, see Fig. 3.2a. The matching system has the form

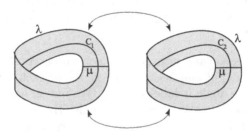

Fig. 3.1. Four simple *closed curves* on the Klein bottle: the meridian μ, the longitude λ, and core circles c_1, c_2 of the Möbius bands

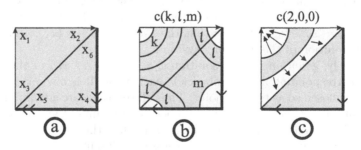

Fig. 3.2. Curves on the Klein bottle

$$x_1 + x_2 = x_1 + x_3$$
$$x_4 + x_5 = x_4 + x_6$$
$$x_2 + x_3 = x_5 + x_6$$
$$x_i \geq 0, 1 \leq i \leq 6,$$

and we immediately get the solution

$$x_1 = k, x_2 = x_3 = x_5 = x_6 = l, x_4 = m,$$

where k, l, m are arbitrary non-negative integers. By Theorem 3.2.4, the set of all normal curves on K is parameterized by 3-tuples k, l, m of non-negative integers and thus consists of curves of the type $c = c(k, l, m)$, see Fig. 3.2b.

Let us analyze the curves. If $k, l, m > 0$, then the curve $c(k, l, m)$ contains a trivial component that encircles the only vertex of K. Neglecting such components, we may assume that at least one of the parameters k, l, m equals 0.

CASE $l = 0$. The curve $\lambda = c(2, 0, 0)$ (the longitude of K) is connected and decomposes K into two Möbius strips with the core lines $c(1, 0, 0)$ and $c(0, 0, 1)$. The curve $c(0, 0, 2)$ is isotopic (but not normally!) to $c(2, 0, 0)$, since both of them are isotopic to the diagonal of the square, see Fig. 3.2c. Any other curve of the type $c = c(k, 0, m)$ is disconnected and consists of several parallel copies of the longitude to which the core lines of one or both Möbius strips may be added.

CASE $m = 0$. The curve $c(0, l, 0)$ consists of l parallel copies of the meridian $\mu = c(0, 1, 0)$ of the Klein bottle. If $k > 0$ and $l \geq 2$, then the isotopy shown in Fig. 3.3 transforms the curve $c(k, l, 0)$ into $c(k, l - 2, 0)$.

After a few steps we get either $c(k, 0, 0)$ (and we return to the Case $l = 0$) or $c(k, 1, 0)$. A similar isotopy transforms $c(k, 1, 0)$ into $c(k - 1, 0, 1)$, and we get the Case $l = 0$ again. Case $k = 0$ is similar.

We may conclude that any curve on K without trivial components consists either of several parallel copies of the longitude, to which core lines of the Möbius strips may be added, or of several parallel copies of the meridian. In particular, K contains exactly four nontrivial connected simple closed curves shown in Fig. 3.1. Another corollary: If a curve contains no trivial and no parallel components, then the number of its components is no greater than 3.

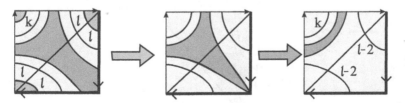

Fig. 3.3. Isotopy across the vertex

3.2.2 Fundamental Solutions and Fundamental Curves

We begin with a digression to the theory of linear equations. Consider a system E of linear homogeneous equations

$$a_{i1}x_1 + a_{i2}x_2 + \ldots + a_{in}x_n = 0, 1 \leq i \leq m,$$

where a_{ij} are integer.

Recall that a non-negative integer solution $\bar{x}$ to E is fundamental, if it cannot be presented in the form $\bar{x} = \bar{y} + \bar{z}$, where $\bar{y}, \bar{z}$ are nontrivial non-negative integer solutions to E (see Definition 3.1.1).

Let us prove the second conclusion (B) of Haken's theory (see Sect. 3.1).

Theorem 3.2.8. *The set of fundamental solutions to any system E of linear homogeneous equations is finite and can be constructed algorithmically. Any non-negative integer solution to E can be presented as a linear combination of the fundamental solutions with non-negative integer coefficients.*

Proof. Let R^n be a Euclidean space with coordinates $(x_1, \ldots, x_n)$. Denote by σ^{n-1} the simplex in R^n with vertices $(1, 0, \ldots, 0), \ldots, (0, 0, \ldots, 1)$. Let S be the set of all non-negative solutions to E over real numbers, L the support plane for σ^{n-1}, and $P = S \cap L$. Then we have:

(1) P is the intersection of L with the hyperplanes given by the above equations and with the half-spaces $x_i \geq 0$.
(2) P is contained in σ^{n-1} and hence is bounded.

It follows that P is a convex polyhedron of dimension $m \leq n - 1$. S can be considered as the union of straight rays that start at the origin and pass through points of P, see Fig. 3.4. The vertices of P have rational coordinates. Multiplying each vertex $\bar{v}$ by the smallest number $k > 0$ such that the coordinates of $k\bar{v}$ are integer, we get the set $\mathcal{V}$ of so-called *vertex* solutions. The vertex solutions are necessarily fundamental.

Since P, as any convex polyhedron of dimension m, can be decomposed into m-simplices without introducing new vertices, S can be presented as the

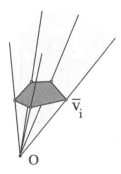

Fig. 3.4. The solution space is the cone over P

union of the cones over m-simplices with vertices in $\mathcal{V}$. It is sufficient to prove that each such cone contains only finitely many fundamental solutions.

Let δ be an m-simplex with vertices $\bar{V}_0, \bar{V}_1, \ldots, \bar{V}_m \in \mathcal{V}$ and $S_\delta \subset S$ the cone over δ. Since any point of δ is a non-negative linear combination of its vertices, any integer point $\bar{x} \in S_\delta$ can be presented in the form $\bar{x} = \sum_{i=0}^m \alpha_i \bar{V}_i$, where all α_i are non-negative. If one of the coefficients (say, α_i) is greater than 1, then $\bar{x}$ is not fundamental, since it can be presented as the nontrivial sum $\bar{x} = (\bar{x} - \bar{V}_i) + \bar{V}_i$ of non-negative integer solutions. We can conclude that all the fundamental solutions in S_δ are contained in the compact set $U = \{\sum_{i=0}^m \alpha_i \bar{V}_i : 0 \leq \alpha_i \leq 1\}$. Since they are integers, there are only finitely many of them.

To prove the second conclusion of Theorem 3.2.8, we successively decompose a given solution $\bar{x}$ into nontrivial sums of other solutions until only fundamental summands remain. The number of the summands is bounded by $\sum_{i=1}^n x_i$ (since each summand contributes to the sum at least 1), hence the decomposition process is finite. Note that the presentation of a solution as a sum of fundamental ones is, in general, not unique. $\square$

Remark 3.2.9. Theorem 3.2.8 remains true if we replace linear homogeneous equations by inequalities. The proof remains the same. For example, the system

$$-x + 4y \geq 0$$
$$3x - y \geq 0$$

has exactly six fundamental solutions see Fig. 3.5. The set U (shown shaded) contains 14 integer points (including the origin).

3.2.3 Geometric Summation

Let F be a triangulated surface. Theorem 3.2.4 tells us that there is a natural bijection between normal curves on F and non-negative integer solutions to the matching system for F. On the other hand, the set of all non-negative integer solutions possesses an evident additive structure. It follows that the

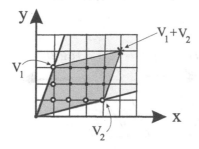

Fig. 3.5. Six fundamental solutions of the system $-x + 4y \geq 0, 3x - y \geq 0$

set of all normal curves is also additive: For any two normal curves $C_1, C_2 \subset F$ one can define their sum $C_1 + C_2$ as a normal curve realizing the algebraic sum of the corresponding solutions.

Let us describe the sum $C_1 + C_2$ geometrically. First, by means of normal isotopies, we straighten the intersection of C_1, C_2 with each triangle into a union of segments. We assume that the curves are in general position with respect to the triangulation and to each other. In particular, this means that the segments in the intersection of C_1, C_2 with each triangle have no common endpoints. At each crossing point exactly two cut-and-paste operations are possible: We cut both segments at the crossing point and glue the ends together in one of the two possible ways. We will call these operations *switches*.

Definition 3.2.10. *Let I_1, I_2 be two intersecting straight segments in a triangle Δ such that the endpoints of each segment lie on different sides of Δ. Let A be the crossing point of I_1, I_2. A switch at A is called* regular *if it produces two arcs such that the endpoints of each arc lie in different sides of Δ. Otherwise the switch is called* irregular, *see Fig. 3.6.*

Remark 3.2.11. It is easy to see that the regular switch at A produces two arcs that join the same sides of Δ as the original segments.

Lemma 3.2.12. *Let C_1, C_2 be two normal curves in a triangulated surface F such that they intersect each triangle along straight segments without common endpoints. Let C be obtained from C_1, C_2 by performing switches at all crossing points of C_1, C_2. Then the following holds:*

1. *If all the switches are regular, then C is a normal curve that realizes the sum of the corresponding solutions.*
2. *If at least one switch is irregular, then C has a return and hence is not normal.*

Proof. Let Δ be a triangle of the triangulation. The segments $C_1 \cap \Delta$ divide Δ into several polygonal regions. One of them (denote it by Ω_1) is adjacent to all three sides of Δ. Analogously, there is a region $\Omega_2 \subset \Delta$ bounded by segments from $C_2 \cap \Delta$ and three segments contained in different sides of Δ. Since both regions have common points with all three sides of Δ, their intersection is nonempty.

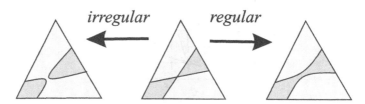

Fig. 3.6. Regular and irregular switches

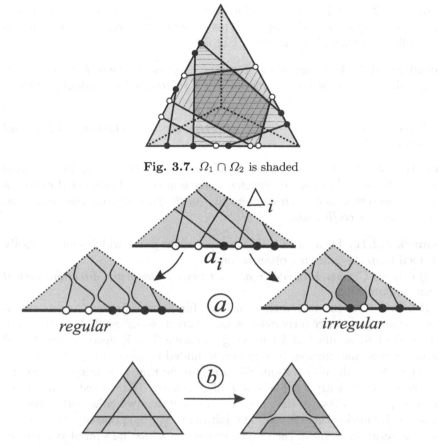

Fig. 3.7. $\Omega_1 \cap \Omega_2$ is shaded

Fig. 3.8. (a) Irregular switches of two curves produce returns; (b) Regular switches of three curves can also produce returns

Let us divide Δ into three triangles $\Delta_1, \Delta_2, \Delta_3$ by joining an interior point $A \in \Omega_1 \cap \Omega_2$ with the vertices of Δ. See Fig. 3.7, where the ends of the segments $C_1 \cap \Delta$ and $C_2 \cap \Delta$ are shown as black and white dots, respectively.

We see that the intersection of $C_1 \cup C_2$ with each Δ_i consists of segments joining the common side a_i of Δ_i and Δ with two other sides of Δ_i. Regular switches at all crossing points of these segments produce arcs of the same type: they join a_i with two other sides of Δ_i. This implies the first statement of the lemma.

Suppose that the switches at some points of $C_1 \cap C_2 \cap \Delta_i$ are irregular. Among all such points we choose a point x having the smallest distance to a_i. Then the irregular switch of $C_1 \cap C_2$ at x produces a return, and this return survives the switches at all other points of $C_1 \cap C_2$. It follows that the resulting curve is not normal. See Fig. 3.8a. $\square$

Remark 3.2.13. It is worth noticing that switches at all crossing points of three sets of segments in a triangle can produce returns, even if all the switches are regular. An example is shown in Fig. 3.8b.

Definition 3.2.14. *A normal curve on a triangulated surface F is called* fundamental, *if it corresponds to a fundamental solution of the matching system.*

Theorem 3.2.15 is a direct consequence of Theorem 3.2.4 and Theorem 3.2.8:

Theorem 3.2.15. *For any triangulated surface F, the set of fundamental curves is finite and can be constructed algorithmically. Any normal curve on F can be presented as a linear combination of the fundamental ones with non-negative integer coefficients.*

Remark 3.2.16. Theorem 3.2.15 has a simpler proof which can be easily obtained from the following observation:
If a curve C crosses an edge e at least twice in the same direction, then it is not fundamental.

It follows from the observation that any fundamental curve C crosses each edge no more than twice (otherwise at least two crossings would have the same direction). This is sufficient for proving Theorem 3.2.15, since the number of normal curves that cross each edge in a bounded number of points is finite.

Let us prove the observation. We can assume that C is oriented and the intersection of C with a collar $e \times I$ of e consists of oriented segments of the type $\{*\} \times I$. Assume that $C \cap (e \times I)$ contains two coherently oriented segments. Removing them from C and joining the initial point of each segment with the terminal point of the other, we present C as the sum of two curves $C_i, i = 1, 2$, maybe with self-intersections. Then we switch each C_i regular at all self-intersection points and get a collection of ≥ 2 normal curves whose sum is C. Therefore, C is not fundamental. See Fig. 3.9.

Example 3.2.17. Let us describe the set $\mathcal{N}$ of normal curves for the boundary of a tetrahedron Δ^3 considered as a triangulated 2-sphere. First, we show that there are exactly seven fundamental curves. Four curves (denote them by $X_i, 1 \leq i \leq 4$) are the boundaries of triangles that cut off the corners of Δ^3.

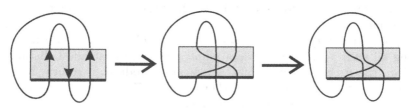

Fig. 3.9. Fundamental curve cannot cross an edge twice in the same direction

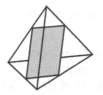

Fig. 3.10. Fundamental curves on the boundary of Δ^3

Each of the other three (denote them by $X_i, 5 \le i \le 7$) is isotopic to the boundary of a quadrilateral plane section parallel to a pair of opposite edges of Δ^3, see Fig. 3.10.

Indeed, let C be a fundamental curve and e an edge. If the endpoints of e lie in different components of $\partial \Delta^3 \setminus C$, then C crosses e in an odd number of points, hence exactly once by Remark 3.2.16. Of course, C decomposes $\partial \Delta^3$ into two discs containing at least one vertex each. Since $\partial \Delta^3$ has four vertices, there are two cases: 1+3 and 2+2. It is easy to see that in the first case we get four triangles $X_i, 1 \le i \le 4$, in the second one three quadrilaterals $X_i, 5 \le i \le 7$.

By Theorem 3.2.15, we have $\mathcal{N} = \{\sum_{i=1}^{k} \alpha_i X_i\}$, where α_i are non-negative integers. Moreover, one can show that any normal curve can be decomposed into the sum of fundamental ones in a unique way up to relation $X_1 + X_2 + X_3 + X_4 = X_5 + X_6 + X_7$.

The set of fundamental curves in F depends heavily on the triangulation. For example, any subdivision of F increases the number of fundamental curves, since there appear new vertices and hence small normal curves surrounding them. There can also appear other normal curves. In general, the growth is exponential. Therefore it would be reasonable to consider triangulations with the number of triangles being as small as possible. Another option is to consider decompositions of surfaces into polygons or other simple pieces. Section 3.2.4 is devoted to curves normal with respect to handle decompositions.

3.2.4 An Alternative Approach to the Theory of Normal Curves

Let ξ be a handle decomposition of a compact surface F. In case $\partial F \ne \emptyset$ we will assume that ξ contains no handles of index 2. Handles of indexes 0, 1, and 2 are called *islands*, *bridges*, and *lakes*, respectively. Any bridge can be presented as a strip (homeomorphic image of a rectangle) so that two opposite sides of the rectangle are adjacent to islands. Thus the intersection of any bridge with the union of the islands consists of two arcs which we call *the ends of the bridge*.

Let l be a proper arc in an island D. We say that l is a *return* if one of the following holds:

(1) (*Bridge return*) Both endpoints of l lie on the same end of a bridge.
(2) (*Half-return*) One endpoint of l lies on an end of a bridge, while the other is on the adjacent arc of $\partial F \cap \partial D$.
(3) (*Boundary return*) The endpoints of l lie on the same arc of $\partial F \cap \partial D$.

Definition 3.2.18. *A proper curve* $C \subset F$ *is called* normal (with respect to ξ), *if the following conditions hold:*

1. C *does not intersect lakes* (*handles of index 2*).
2. *The intersection of* C *with any bridge consists of simple arcs joining different ends of the bridge.*
3. (No returns) *The intersection of* C *with any island contains no returns of any kind.*
4. (No circles) *The intersection of* C *with any island contains no closed curves.*

See Fig. 3.11 for allowed and forbidden types of arcs in islands.

It is easy to see that just as in the case of triangulated surfaces (see Theorem 3.2.2), any curve on F can be transformed into a normal one by isotopies and crossing out boundary returns and closed curves inside the islands. Isotopies that destroy half- and bridge returns by moving them across corresponding bridges are shown in Fig. 3.12.

Fig. 3.11. Allowed and forbidden types of arcs are shown as *bold* and *dotted* lines, respectively

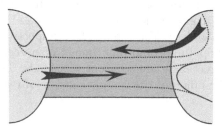

Fig. 3.12. Destroying returns

Note that if the valence v of an island D is 0 or 1, then all arcs in D are forbidden. If $v \geq 2$ and F is closed, then there are exactly $C_v^2 = v(v-1)/2$ types of allowed arcs that join different ends of the bridges. Let $\partial F \neq \emptyset$. Then there are exactly $v(2v-3)$ allowed types: C_v^2 of them join different bridges, just as many join different boundary arcs, and $v(v-2)$ types join the ends of the bridges with non-neighboring boundary arcs.

Let us begin to fulfill Steps 4, 5 of the Haken's scheme. We wish to compose a linear system and parameterize normal curves by its solutions. Since we are 1-dimension down, instead of elementary pieces we will consider elementary arcs, which are allowed by Definition 3.2.18. Lakes contain no arcs at all, while any bridge contains only one type of elementary arcs. So only elementary arcs in islands are of interest.

Let $E_1, E_2, \ldots, E_n$ be all the equivalence classes of elementary arcs in all the islands of ξ. Assign to them integer variables $x_1, x_2, \ldots, x_n$. To write the matching system, consider a bridge b with ends s_1, s_2. Denote by $E_{i_1}, E_{i_2}, \ldots, E_{i_l}$ the elementary arcs having an endpoint on s_1, and by $E_{j_1}, E_{j_2}, \ldots, E_{j_r}$ those arcs that have an endpoint on s_2. Then we write the equation

$$x_{i_1} + x_{i_2} + \ldots + x_{i_l} = x_{j_1} + x_{j_2} + \ldots + x_{j_r},$$

see Fig. 3.13 to the left. The matching system consists of such equations written for all the bridges of ξ, and of the inequalities $x_i \geq 0, 1 \leq i \leq n$. It turns out that any normal curve $C \subset F$ can be described by an integer solution $\bar{x}(C) = (x_1, x_2, \ldots, x_n)$ of the system, that shows how many arcs of each type are contained in the intersection of C with the islands.

In contrast to the case of triangulations, not every solution corresponds to a normal curve. For example, let D be an island of valence ≥ 4. Choose two arcs $e_i \in E_i, e_j \in E_j$ in D joining two pairs of bridges such that e_i crosses e_j transversally exactly once. Then the crossing point of e_i, e_j would survive any normal isotopy, and hence no solution with $x_i, x_j > 0$ has a chance to be realized by an embedded curve, see Fig. 3.13 to the right.

We say that a non-negative integer solution $(x_1, x_2, \ldots, x_n)$ is *admissible* if any pair $x_i, x_j > 0$ can be realized by disjoint arcs (compare with Step 5 of Sect. 3.1). Let us show that every admissible solution corresponds to a

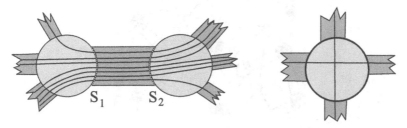

Fig. 3.13. Bridges produce equations (*left*). There exist nonadmissible solutions (*right*)

normal curve. Indeed, the admissibility allows us to realize the given solution $(x_1, x_2, \ldots, x_n)$ by disjoint arcs in islands, and the equations tell us why these arcs can be completed to a normal curve by adding parallel arcs inside the corresponding bridges.

The remaining part of the handle decomposition version of the theory of normal curves is similar to the one for triangulated surfaces. We restrict ourselves to a few comments:

1. There exists a natural bijection between admissible solutions of the matching system for ξ and equivalence classes of normal curves on F. The proof is actually the same as for Theorem 3.2.4.
2. In general, not every fundamental solution is admissible. Thus the fundamental curves correspond only to admissible fundamental solutions.
3. The sum of two admissible solutions may not be admissible. On the other hand, the sum of two solutions is admissible, then so are the summands.
4. To construct geometrically the sum of two normal curves with the admissible algebraic sum, we make them disjoint inside the islands and straight inside the bridges (with respect to a presentation of each bridge as a rectangle). Then we make regular switches at all crossing points of the curves. Similar to Definition 3.2.10, we call a switch between two curves in a bridge *regular*, if it produces two arcs that join different ends of the bridge.
5. Just as in the case of triangulated surfaces, the set of fundamental curves on F is finite and can be constructed algorithmically. Any normal curve can be presented as a linear combination of the fundamental ones with non-negative integer coefficients.
6. Any triangulation T of a closed surface F generates a handle decomposition ξ of F, which is called *dual*. It can be easily obtained by inserting an island into every triangle of T, and joining the islands by bridges across the edges, see Fig. 3.14. One can easily see that the matching systems for T and ξ do coincide. Thus the handle decomposition version of the normal curve theory is at least as powerful as the triangulation version. Moreover, the handle decomposition approach is more flexible and gives

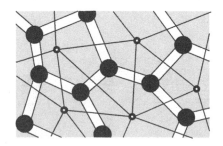

Fig. 3.14. Dual handle decomposition

us more freedom, especially in case $\partial F \neq \emptyset$. Hence in many cases it is preferable.

7. Let a surface F be decomposed into handles of index ≤ 1. As in the case of a triangulated surface (see Remark 3.2.5), two closed normal curves on F are isotopic if and only if they are normally isotopic. For arcs this is not true. Nevertheless, if there are no islands of valence ≤ 2, then any two isotopic normal arcs are either normally isotopic or parallel to the ends of the same bridge.

3.3 Normal Surfaces in 3-Manifolds

3.3.1 Incompressible Surfaces

As was pointed out in the beginning of this chapter, knowledge of surfaces contained in a given 3-manifold is useful for understanding its structure. Let us think a little about surfaces that are contained in R^3. An example of such surface (a cube with two knotted tubes) is shown in Fig. 3.15. This example turned out to have a general nature: any connected closed surface F in R^3 can be obtained from a 2-sphere by successive addition of tubes. The tubes may be knotted and linked, and run inside each other.

Now let us return to the case of an arbitrary 3-manifold M. The same procedure (adding tubes) works here as well, but all surfaces obtained in this way seem to be not very interesting. One possible explanation is that all surfaces that are contained in R^3 are contained in every 3-manifold M and thus carry no information on M. Therefore surfaces without tubes are the most interesting. To give a formal description of surfaces without tubes, note that if tubes are present, then the meridional disc of the last attached tube meets the surface only along its boundary. We come naturally to the following definitions.

Definition 3.3.1. *A compressing disc for a surface F in a 3-manifold M is an embedded disc $D \subset M$ which meets F along its boundary, i.e., $D \cap F = \partial D$.*

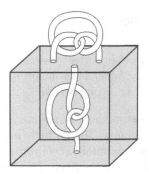

Fig. 3.15. A surface in R^3

The disc D is inessential *if the curve ∂D is trivial on F, i.e., bounds a disc in F. Otherwise D is* essential.

Definition 3.3.2. *A surface $F \subset M$ is called* incompressible *if it admits no essential compressing discs.*

Remark 3.3.3. According to our definition, any 2-sphere in any 3-manifold is incompressible. The same holds for any disc. Sometimes topologists call a sphere incompressible only if it does not bound a ball (see [43, 130]). We prefer to call such spheres *essential* or *nontrivial* without overloading the term "incompressible" with extra conditions.

Note that any essential compressing disc determines a nontrivial tube. To visualize the tube, we may squeeze the surface along the disc and get an isotopic surface with a thin tube. Therefore, incompressible surfaces contain no nontrivial tubes and thus have a chance to be informative.

Let us compare the notion of incompressible surface with a close notion of injective one.

Definition 3.3.4. *A connected surface $F \overset{i}{\subset} M$ is called* injective, *if the kernel of the induced homomorphism $i_*: \pi_1(F) \to \pi_1(M)$ is trivial.*

Clearly, every injective surface is incompressible. The converse is in general not true. For example, consider an orientable twisted I-bundle $K \tilde{\times} I$ over the Klein bottle K. We will think of K as being contained in $K \tilde{\times} I$ as a cross-section of the bundle. Let a 3-manifold M_h be obtained from $K \tilde{\times} I$ by pasting a solid torus $D^2 \times S^1$ via a homeomorphism $h: \partial D^2 \times S^1 \to \partial(K \tilde{\times} I)$. Since K contains only two nontrivial orientation-preserving simple closed curves (see Example 3.2.7), K is compressible in M_h only when $h(D^2) \times \{*\}$ is isotopic to one of them. On the other hand, K is not injective for any h.

Recall that a surface F in a 3-manifold M is two-sided in M, if the normal line bundle $F \tilde{\times} I$ of F in M is trivial. Otherwise F is one-sided. The surface $F \tilde{\times} \partial I$ will be denoted by $\tilde{F}$. In case F is two-sided, $\tilde{F}$ consists of two parallel copies of F.

Lemma 3.3.5. *Let F be a proper connected surface in a 3-manifold M. Then the following holds:*

(1) *If F is two-sided, then F is injective $\iff$ F is incompressible.*
(2) *If F is one-sided, then we have: $\tilde{F}$ is incompressible $\iff$ $\tilde{F}$ is injective $\iff$ F is injective $\Rightarrow$ F is incompressible.*

Proof. (1) Part $\Rightarrow$ of the statement is trivial. Part $\Leftarrow$ follows from the Loop Theorem and Dehn Lemma [106], which imply that any noninjective surface in the boundary of a 3-manifold is compressible. Indeed, suppose that F is incompressible and fails to be injective. Then at least one of the surfaces $F \times \{0, 1\}$ in the boundary of the 3-manifold $M_F = M \setminus \text{Int}\,(F \times I)$ is not injective. By the Loop Theorem and Dehn Lemma, this surface is compressible in M_F and hence in M, which contradicts our assumption that the isotopic surface $F \subset M$ is incompressible.

(2) Let F be one-sided. Since $\tilde{F}$ is always two-sided, the first equivalence follows from (1). The last implication is easy. Let us prove the middle equivalence. Since the restriction $p_{|\tilde{F}} \colon \tilde{F} \to F$ of the normal bundle projection is a covering, it induces an injection $p_* \colon \pi_1(\tilde{F}) \to \pi_1(F)$, which allows us to identify $\pi_1(\tilde{F})$ with a subgroup of $\pi_1(F)$. It follows that if F is injective, then so is $\tilde{F}$.

Let us prove the inverse implication. Suppose that $\tilde{F}$ is injective. First, we assume that $F \neq RP^2$. This assumption guarantees us that $\pi_1(F)$ contains no elements of order ≤ 2 except the unit. Consider an arbitrary element α of $\pi_1(F) = \pi_1(F \tilde{\times} I)$. Then α^2 lies in the subgroup $\pi_1(\tilde{F})$ of $\pi_1(F)$. The following implications are evident: $\alpha = 1$ in $\pi_1(M) \Longrightarrow \alpha^2 = 1$ in $\pi_1(M) \Longrightarrow \alpha^2 = 1$ in $\pi_1(\tilde{F})$ (since $\tilde{F}$ is injective) $\Longrightarrow \alpha^2 = 1$ in $\pi_1(F) \Longrightarrow \alpha = 1$ in $\pi_1(F)$ by the above assumption. Therefore F is injective.

The case $F = RP^2$ is simple, since $F \tilde{\times} I$ is a punctured projective space. Since $\partial F \tilde{\times} I$ is a 2-sphere, $F \tilde{\times} I$ and hence F are contained in M injectively. $\qquad \square$

Corollary 3.3.6. *Any incompressible proper connected surface in a 3-ball is either a 2-sphere or a proper disc.*

Proof. Let F be a connected proper incompressible surface in a 3-ball B^3. Then F is two-sided (as any other proper surface in B^3) and injective (by Lemma 3.3.5). It follows that $\pi_1(F) = 1$, but the only connected surfaces with the trivial fundamental group are S^2 and D^2. $\qquad \square$

To complete our discussion of the idea of interesting surfaces, let us introduce a relative counterpart of incompressibility.

Definition 3.3.7. *Let F be a surface in a 3-manifold M. An embedded disc $D \subset M$ is called a* boundary compressing disc *for F if D meets F along a proper arc $l \subset \partial D$ and meets ∂M along the remaining arc of ∂D. The disc D is* inessential, *if l cuts off a disc D' from F so that $\partial D'$ consists of a copy of l and an arc on ∂M. Otherwise D is* essential.

Definition 3.3.8. *A surface $F \subset M$ is called* boundary incompressible *if it admits no essential boundary compressing discs.*

We point out that Definitions 3.3.7 and 3.3.8 make sense also for nonproper surfaces. Just as compressing discs determine tubes, boundary compressing discs determine half-tubes, which we call *tunnels*.

Recall that a 3-manifold M is irreducible, if every 2-sphere in it is inessential, i.e., bounds a 3-ball. Also, M is *boundary irreducible*, if the boundary curve of any proper disc $D \subset M$ bounds a disc on ∂M. In other words, all proper discs in M must be inessential. It follows that M is boundary irreducible if and only if ∂M is incompressible. So the statements "M is boundary irreducible" and "∂M is incompressible" are completely equivalent. Also, it is worth mentioning that if M is irreducible, then a proper disc $D \subset M$ is inessential if and only if it cuts off a 3-ball from M.

3.3.2 Normal Surfaces in 3-Manifolds with Boundary Pattern

We consider the theory of normal surfaces in triangulated 3-manifolds in slightly more general setting for manifolds with boundary pattern, which we will need later. The notion was introduced by Johannson [57].

Definition 3.3.9. *A manifold (M, Γ) with boundary pattern Γ is a 3-manifold M with a fixed graph (1-dimensional polyhedron) $\Gamma \subset \partial M$ containing no isolated vertices.*

In particular, every manifold can be considered as a manifold with the empty boundary pattern. A homeomorphism between manifolds with boundary pattern is a homeomorphism $M_1 \to M_2$ which takes Γ_1 to Γ_2, i.e., a homeomorphism of pairs $(M_1, \Gamma_1) \to (M_2, \Gamma_2)$. Let (M, Γ) be a manifold with boundary pattern.

Definition 3.3.10. *We say that a surface $F \subset M$ is proper and write $F \subset (M, \Gamma)$, if $F \cap \partial M = \partial F$ and ∂F is in general position with respect to Γ. This means that ∂F does not pass through vertices of Γ and intersects its edges transversally.*

Definition 3.3.11. *A subset X of (M, Γ) is called* clean *if it does not intersect Γ. An isotopy $f_t \colon X \to M$ is* clean *if $f_t(X) \cap \Gamma = \emptyset$ for all t.*

Definition 3.3.12. *If $X \overset{i}{\subset} M$ is a closed subpolyhedron of M, then an isotopy $f_t : X \to M$ is called* admissible *if there is an ambient isotopy $g_t \colon (M, \Gamma) \to (M, \Gamma)$ of pairs such that $g_t i = f_t$ for all t. In particular, any clean isotopy is admissible.*
   Similarly, a homeomorphism $h \colon M \to M$ is called admissible, *if $h(\Gamma) = \Gamma$, i.e., if it determines a homeomorphism $(M, \Gamma) \to (M, \Gamma)$ of pairs.*

The presence of a boundary pattern does not influence the notions of incompressible surface and irreducible 3-manifold. The notions of boundary incompressible surface and boundary irreducible 3-manifold admit straightforward generalizations to the case of manifolds with boundary pattern. Let F be a surface in a manifold (M, Γ) with boundary pattern.

Definition 3.3.13. *A* boundary compressing disc *for F is a clean disc $D \subset M$ which meets F along an arc $l \subset \partial D$ and meets ∂M along the remaining arc of ∂D. The disc D is* inessential *if l cuts off a clean disc D' from F so that $\partial D'$ consists of a copy of l and a clean arc on ∂M. Otherwise, D is* essential.

Definition 3.3.14. $F \subset (M, \Gamma)$ *is called* boundary incompressible *if it admits no essential boundary compressing discs.*

Just as in the absolute case, Definitions 3.3.13 and 3.3.14 make sense also for nonproper surfaces. It is worth mentioning that the property of a surface in a 3-manifold (M, Γ) to be incompressible does not depend on the boundary pattern Γ. On the contrary, boundary incompressibility does. For example, let A be a clean annulus in (M, Γ) which is parallel rel ∂ to an annulus $A' \subset \partial M$ such that $A' \cap \Gamma$ contains a copy of the middle circle of A'. Then A is boundary compressible in $(M, \emptyset)$ and incompressible in (M, Γ).

Similarly, boundary irreducibility of (M, Γ) depends on Γ. For example, a solid torus with boundary pattern Γ is boundary reducible if and only if the complement to Γ contains a meridian of the torus.

Let (M, Γ) be a compact triangulated 3-manifold. We will always assume that the boundary pattern Γ is a 1-dimensional subcomplex of the triangulation T of M, i.e., it consists of vertices and edges. The case $\Gamma = \emptyset$ is also allowed.

Definition 3.3.15. *A proper surface $F \subset M$ is called* normal *if F is in general position with respect to the triangulation and the following holds:*

1. *The intersection of F with every tetrahedron consists of discs. Those discs are called* elementary *(they play the role of elementary pieces in the Haken's scheme, see Sect. 3.1).*
2. *The boundary of every elementary disc crosses at least one edge and crosses each edge at most once.*

We point out that the property of a surface to be normal does not depend on the boundary pattern Γ: F is normal in (M, Γ) if and only if F is normal in $M = (M, \emptyset)$ (provided that Γ consists of edges).

It is easy to show that in each tetrahedron there are seven types of allowed elementary discs: four triangles and three quadrilaterals. Their boundary curves are just fundamental curves on $\partial \Delta^3$, see Fig. 3.10. Recall that elementary discs as well as normal surfaces are usually considered up to normal isotopy, which preserves the triangulation.

3.3.3 Normalization Procedure

We are now ready to fulfill Step 3 of Haken's scheme by proving that all interesting (in our case, incompressible and boundary incompressible) surfaces are isotopic to normal ones. Consider a 3-manifold M equipped with a triangulation T and a boundary pattern Γ such that Γ consists of edges. Recall that a surface in M is in general position, if it does not pass through vertices and intersects the edges and triangles of the triangulation transversally. We introduce eight *normalization moves* N_1–N_8 on general position surfaces in (M, Γ). As we will see later, these moves are sufficient for converting every

surface into a normal one. To control the number of moves we introduce an
integer characteristic called the *edge degree* (or the *weight*) of a surface.

Definition 3.3.16. *Let F be a general position proper surface in a 3-manifold
M equipped with a fixed triangulation T and boundary pattern Γ. Then the
edge degree $e(F)$ of F is the number of points in the intersection of F with
all the edges of T.*

Let $F \subset M$ be a proper general position surface in (M, Γ) and Δ^3 a
tetrahedron of the triangulation. Then $F \cap \Delta^3$ is a proper surface in Δ^3, maybe
disconnected. Normalization moves N_1–N_8 can be described as follows:

N_1: Suppose that $F \cap \Delta^3$ admits an essential compressing disc $D \subset \Delta^3$, which
meets F along ∂D. Then we compress F along D, i.e., cut F along ∂D
and fill in two new boundary circles by two parallel copies of D.

N_2: Suppose that $F \cap \Delta^3$, considered as a proper surface in Δ^3, admits an
essential boundary compressing disc D such that $D \cap \partial\Delta^3$ is a closed
subinterval of an edge e not contained in ∂M. Then we use D to eliminate
two points in $F \cap e$ by an isotopy of F, see Fig. 3.16.

N_3: Suppose that a component of $F \cap \Delta^3$ is a 2-sphere. Then we remove it.

N_4: Suppose that F has a spherical component intersecting a triangle Δ^2
of the triangulation along a circle and consisting of two discs in the
tetrahedra adjacent to Δ^2. Then we remove this component.

As we shall see later, these moves are sufficient for normalizing closed
surfaces. For surfaces with boundary we need additional moves.

Definition 3.3.17. *A proper disc D in a triangulated 3-manifold M is called
edge-linked, if D intersects an edge $e \subset \partial M$ of the triangulation in exactly
two points $A, B \subset e$ and the intersection of D with any triangle Δ^2 either is
empty (if $e \not\subset \Delta^2$), or is an arc joining A and B (if $e \subset \Delta^2$). Of course, any
edge-linked disc is inessential in M, see Fig. 3.17.*

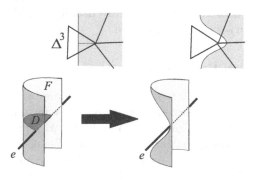

Fig. 3.16. Reducing $e(F)$ by an isotopy

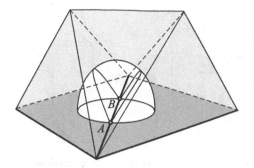

Fig. 3.17. An example of an edge-linked disc

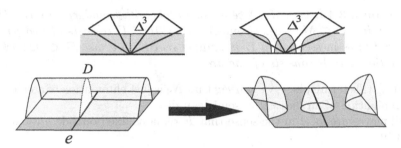

Fig. 3.18. Compression along two parallel discs

N_5: Suppose that $F \cap \Delta^3$ admits a boundary compressing disc D such that $D \cap \partial\Delta^3$ is a closed subinterval of an edge $e \subset \partial M$ and the connected component of F containing the complementary arc of ∂D is not an edge-linked disc. Then we compress F along two parallel copies D', D'' of D such that D', D'' intersect no edge of the triangulation. This move converts F into an edge-linked disc and a surface having a smaller edge degree, see Fig. 3.18.

N_6: Suppose a component of $F \cap \Delta^3$ is a proper disc $D \subset \Delta^3$ such that ∂D lies inside a triangular face $\Delta^2 \subset \partial M$ of Δ^3. Then we remove D.

N_7: Suppose that a component D of $F \cap \Delta^3$ is a disc linked to an edge $e \subset \partial M$ such that $e \not\subset \Gamma$. Then we remove D.

N_8: The same as N_7, but e is in Γ. Suppose that a component D of $F \cap \Delta^3$ is a disc linked to an edge $e \subset \Gamma$. Then we remove D.

Remark 3.3.18. Let us analyze the behavior of F under moves N_1–N_8 from a purely topological point of view, without taking into account the triangulation. N_1 compresses a tube, N_2 is an admissible isotopy, N_3 and N_4 consist in removing an inessential S^2. Further, N_5 compresses a tunnel while N_6, N_7 eliminate inessential discs. All these moves preserve $\partial F \cap \Gamma$. In contrast to this, move N_8 changes $\partial F \cap \Gamma$, but it is applicable only in a very specific situation.

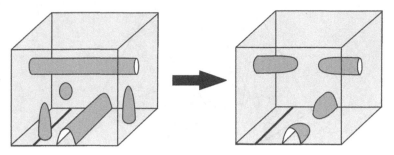

Fig. 3.19. Local moves keep the surface fixed outside a 3-ball

Definition 3.3.19. *Let (M, Γ) be a 3-manifold with boundary pattern. Then a proper disc $D \subset M$ is called* semiclean *if $\partial D \cap \Gamma$ consists of two points. D is said to be* inessential, *if D is parallel rel ∂D to a disc $D' \subset \partial M$ whose intersection with Γ consists of one arc.*

Adopting this terminology, we can say that N_8 is just elimination of an inessential elimination of an inessential semiclean disc.

All moves are local in the sense that F remains fixed outside a 3-ball, see Fig. 3.19.

Lemma 3.3.20. *Let $F_\Delta \subset \Delta^3$ be a proper general position surface in a tetrahedron Δ^3 such that the following holds:*

1. *Each component of F_Δ is a disc.*
2. *Each component of ∂F_Δ crosses at least one edge of Δ^3 and at least one component crosses an edge e more than once.*

Then F_Δ admits a boundary compressing disc D such that $D \cap \partial \Delta^3$ is a closed subinterval of e.

Proof. Consider the set $\mathcal{A}$ of all discs in $\partial \Delta^3$ such that for every disc $A \in \mathcal{A}$ the boundary ∂A of A is a component of ∂F_Δ and at least one of the connected components of $A \cap e$ is a closed subinterval in the interior of e.

The choice of e guarantees us that $\mathcal{A}$ is not empty. Let A be an innermost disc in $\mathcal{A}$, which contains no other discs from $\mathcal{A}$. Denote by l a connected component of $A \cap e$ that lies in the interior of e, and by B the disc component of F_Δ which spans ∂A. Since A is innermost, l has no common points with F except the endpoints.

Now let us slightly push A relative ∂A inward Δ^3. We get a proper disc A'. Since $\partial A' = \partial A = \partial B$, and since any two discs in Δ^3 with the same boundary are isotopic rel ∂, we may think of A' and B as being identical. On the other hand, pushing can be made so subtle that the trace of l would form a boundary compressing disc D for B and thus for $F \cap \Delta^3$ such that $D \cap \partial \Delta^3$ is a closed subinterval of e. See Fig. 3.20. □

Theorem 3.3.21 is a 3-dimensional version of Theorem 3.2.2.

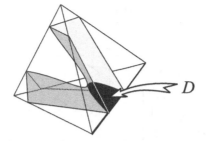

Fig. 3.20. Finding a compressing disc

Theorem 3.3.21. *Any general position proper surface F in a triangulated 3-manifold (M, Γ) can be transformed into a normal surface F' by a sequence of moves N_1–N_8.*

Proof. Let us apply to F moves N_1–N_8 as long as possible. We claim that the procedure is finite and that the resulting surface is normal.

To prove the first claim, we associate to $F \subset M$ the following numerical characteristics:

(1) The edge degree $e(F)$.
(2) The *reduced edge degree* $\hat{e}(F) = e(F) - 2k$, where k is the number of edge-linked disc components of F. Equivalently, $\hat{e}(F)$ is the edge degree of the surface obtained from F by removing all edge-linked disc components.
(3) $\gamma(F) = \sum_{i=1}^{m}(1 - \chi(F_i))$, where $F_1, \dots, F_m$ are those connected components of the intersection of F with all the tetrahedra that are not homeomorphic to S^2.
(4) $n(F)$, the total number of connected components of F.

We will measure the complexity of general position surfaces in M by the 4-tuples $\bar{t}(F) = (e(F), \hat{e}(F), \gamma(F), n(F))$ considered in the lexicographical order.

Let us investigate the behavior of $\bar{t}$ under the moves N_1–N_8. Obviously, moves N_2, N_7, N_8 decrease $e(F)$. Move N_5 preserves $e(F)$, but decreases $\hat{e}(F)$. Of course, moves N_3, N_4, N_6 preserve $e(F), \hat{e}(F), \gamma(F)$, and decrease $n(F)$ while move N_1 preserves $e(F)$ and does not increase $\hat{e}(F)$.

Let us prove that N_1 decreases $\gamma(F)$. Suppose that N_1 transforms a connected component F_i of the intersection of F with a tetrahedron Δ^3 into a surface $G \subset \Delta^3$. In other words, N_1 replaces F by the surface $F' = (F \backslash F_i) \cup G$. Then $\chi(F') = \chi(F) + 2$ and G consists of ≤ 2 components. Let us compare the contributions of F_i and G to $\gamma(F)$, respectively, $\gamma(F')$:

1. Suppose that G consists of two components G', G''. Since G is obtained from F_i by compression along an essential disc, neither G' nor G'' is a 2-sphere. It follows that

$$\gamma(G) = (1 - \chi(G')) + (1 - \chi(G'')) = -\chi(F_i) < 1 - \chi(F_i) = \gamma(F_i).$$

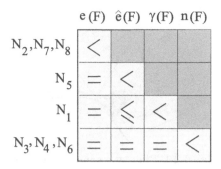

Fig. 3.21. Each line of the table describes the behavior of $e, \hat{e}, \gamma$, and n under corresponding moves

2. Suppose that G is connected and different from S^2. Then

$$\gamma(G) = 1 - \chi(G) = -1 - \chi(F_i) = \gamma(F_i) - 2 < \gamma(F_i).$$

3. Suppose that $G = S^2$. Then F_i is a torus, $\gamma(F_i) = 1 - \chi(F_i) = 1$, and $\gamma(G) = 0$ (since spherical components are neglected).

In all three cases we have $\gamma(G) < \gamma(F_i)$, which implies $\gamma(F') < \gamma(F)$.

The information on the behavior of $e(F), \hat{e}(F), \gamma(F)$, and $n(F)$ is summarized in the table (see Fig. 3.21). It follows from the table that all the moves strictly decrease $\bar{t}(F)$. On the other hand, $\bar{t}(F)$ is bounded from below by $(0,0,0,0)$. Therefore, the process of applying moves N_1–N_8 is finite.

Let us prove the second claim, which states that the surface F' obtained from F by applying all possible moves N_1–N_8 is normal. Indeed, the intersection of F' with any tetrahedron Δ^3 consists of incompressible surfaces (otherwise we could apply move N_1). Any proper incompressible surface in Δ^3 is the union of discs and spheres, but we cannot have spheres because of move N_3. Also, all components of $\partial(F' \cap \Delta^3)$ must cross edges, since otherwise we could apply move N_6.

It remains to show that any component of $\partial(F' \cap \Delta^3)$ crosses any edge no more than once. On the contrary, suppose that a component of $\partial(F' \cap \Delta^3)$ crosses an edge e at least twice. Then there exists a boundary compressing disc D for $F' \cap \Delta^3$ such that $D \cap \partial\Delta^3$ is a subinterval of e, see Lemma 3.3.20. It follows that one can apply either move N_2 (if $e \not\subset \partial M$) or one of moves N_5, N_7, N_8 (if $e \subset \partial M$), a contradiction.   □

Remark 3.3.22. We will refer to the process of applications of moves N_1–N_8 used in the proof of Theorem 3.3.21 as *normalization procedure*. It is important to note that, for any edge e, moves N_1, N_3, N_4–N_6 preserve its individual degree (the number of points in $F \cap e$), and each of the remaining moves N_2, N_7, N_8 either preserves it or decreases it exactly by two. It follows that the normalization procedure never increases the edge degree. Moreover, if the intersection of F with an edge e consists of no more than one point, then the normalization procedure preserves $F \cap e$.

Remark 3.3.23. Suppose that the intersection of F with the boundary of a tetrahedron of the triangulation contains a curve which crosses an edge more than once. Then at least one of moves N_2, N_7, N_8 must be performed. It follows that the normalization procedure strictly decreases $e(F)$. In particular, if the intersection of F with a triangle of the triangulation contains a return (an arc having the endpoints in the same edge), then $e(F)$ becomes smaller.

Other properties of the normalization are collected in Proposition 3.3.24.

Proposition 3.3.24. *Let T be a triangulation of a 3-manifold M whose boundary pattern Γ consists of edges of T, and let $F \subset (M, \Gamma)$ be a general position proper connected surface. Suppose that a normal surface F_1 is obtained from F by the normalization procedure. Then the following holds:*

1. *If F is an essential 2-sphere, then so is at least one connected component F' of F_1.*
2. *If F is a clean essential disc, then so is at least one connected component F' of F_1.*
3. *Suppose ∂F crosses each edge of T no more than once and no component of ∂F is a circle inside a triangle of T. Then $\partial F' = \partial F$. If, in addition, F is incompressible and different from S^2, then there exists a unique connected component F' of F_1 such that F' is homeomorphic to F. Furthermore, if M is irreducible, then F' is admissibly isotopic to F.*
4. *Suppose that (M, Γ) is irreducible and boundary irreducible, F is incompressible and boundary incompressible, and F is neither a 2-sphere, nor a clean disc, nor an inessential semiclean disc. Then there exists a unique connected component F' of F_1 such that F' is admissibly isotopic to F.*
5. *If F is orientable, then so is F_1.*
6. *If the intersection of F with a triangle of the triangulation contains a return (an arc with the endpoints contained in the same edge), then $e(F_1) < e(F)$.*

Proof. 1. Evident, since the property of a surface of containing an essential 2-sphere is preserved under moves N_1–N_4, which normalize a closed surface.

2. Also evident, since the property of a surface to contain a clean essential disc is preserved under all eight normalization moves.

3. We can apply moves N_5, N_7, N_8 only when ∂F crosses an edge twice. Also, move N_6 is possible only when ∂F contains a circle lying strictly inside a triangle of T. Since both situations are forbidden by the assumption and cannot be brought about by moves N_1–N_4, the normalization procedure does not use moves N_5–N_8 at all. Obviously, moves N_1–N_4 preserve ∂F. What can happen with F? Moves N_2–N_4 leave F essentially unchanged: N_2 is an isotopy, and N_3, N_4 consist in removing an inessential sphere. Let us investigate the behavior of F with respect to N_1. If F is incompressible,

then N_1 transforms F into a 2-sphere and a homeomorphic copy of F. Moreover, if M is irreducible, then this 2-sphere bounds a 3-ball, which can be used for constructing an admissible isotopy from F to its homeomorphic copy. This is sufficient for proving item 3.

4. Just as above, one can easily show that the property of a surface to contain an incompressible boundary incompressible connected component which is admissibly isotopic to F is preserved under all eight normalization moves.
5. Evident, since each of N_1–N_8 preserves the orientability.
6. Follows from Remark 3.3.23.   □

Corollary 3.3.25. *Let an incompressible boundary incompressible surface F in an irreducible boundary irreducible 3-manifold (M, Γ) is neither a 2-sphere, nor a clean disc, nor an inessential semiclean disc. Then F is admissibly isotopic to a normal surface of the same or a smaller edge degree.*

Proof. Follows from conclusion 4 of Proposition 3.3.24 and Remark 3.3.22.   □

Remark 3.3.26. Corollary 3.3.25 tells us that the class of normal surfaces is informative in the sense that quite often it contains representatives of interesting classes of surfaces (here we have in mind surfaces which are incompressible and boundary incompressible).

3.3.4 Fundamental Surfaces

To give an explicit description of the set of all normal surfaces in a given 3-manifold, we do the same as in the case of normal curves: We compose the matching system of linear equations and prove that normal surfaces correspond to admissible solutions. Note that the boundary pattern is irrelevant here. Let M be a triangulated 3-manifold, and let $E_1, E_2, \ldots, E_n$ denote all the equivalence classes of elementary discs in all the tetrahedra (we consider the discs up to normal isotopy). Since each tetrahedron contains seven equivalence classes of elementary discs (four triangles and three quadrilaterals), we have $n = 7t$, where t is the number of tetrahedra in T. We assign to these classes integer variables $x_1, x_2, \ldots, x_n$. To write the equations, consider an angle of a common triangular face of two tetrahedra Δ_1^3 and Δ_2^3. Each of the tetrahedra contains two types of elementary discs that intersect both sides of the angle: one triangle and one quadrilateral. Let $E_i, E_j \subset \Delta_1^3$ and $E_k, E_l \subset \Delta_2^3$ be the corresponding equivalence classes. Then we write the equation $x_i + x_j = x_k + x_l$. Doing so for all the angles, we get a system of $3m$ equations, where m is the number of triangles in the interior of M (if M is closed, then $m = 2t$). The matching system is obtained by adding n inequalities $x_i \geq 0, 1 \leq i \leq n$:

$$
\begin{gathered}
\textit{Matching system} \\
x_i + x_j = x_k + x_l \\
x_i \geq 0, 1 \leq i \leq n
\end{gathered}
$$

To each normal surface F we assign an n-tuple $\bar{x}(F) = (x_1, x_2, \ldots, x_n)$ of non-negative integer numbers in the following natural way: We take the number of elementary discs (triangles or quadrilaterals) of each type E_i in the intersection of F with the tetrahedra. It is obvious that the obtained tuple will be a solution to the matching system.

On the other hand, just as for curves in handle decompositions, not all solutions correspond to normal surfaces. Indeed, being embedded, our surface has no self-intersections, yet any two quadrilateral discs that lie in the same tetrahedron and have different types do intersect. Therefore, we should restrict ourselves to considering *admissible* solutions such that for all three types of quadrilateral discs in the same tetrahedron, no more than one of the corresponding variables is positive. Certainly, we have in mind only non-negative integer solutions. Let us prove the basic conclusion (A) of Haken theory for this case.

Theorem 3.3.27. *There exists a natural bijection between the set of admissible solutions to the matching system for a triangulated 3-manifold M and the set of equivalence classes of normal surfaces in M.*

Proof. Let $E_{i_1}, \ldots, E_{i_7}$ be the seven equivalence classes of elementary discs in a tetrahedron Δ^3 such that $E_{i_1}, \ldots, E_{i_4}$ are triangular while $E_{i_5}, E_{i_6}, E_{i_7}$ are quadrilateral. Let $(x_{i_1}, \ldots, x_{i_7})$ be a tuple of seven non-negative integers. The proof is based on the following evident observations:

(1) $(x_{i_1}, \ldots, x_{i_7})$ can be realized by a collection of disjoint elementary discs in Δ^3 if and only if at least two of the three numbers $E_{i_5}, E_{i_6}, E_{i_7}$ are zeros.
(2) The realization is unique up to normal isotopy.

Using these observations, one can realize any admissible solution $\bar{x} = (x_1, \ldots, x_n)$ by the corresponding set of elementary discs so that the discs match along all triangle faces and thus form a normal surface, which is unique up to normal isotopy. □

By Theorem 3.2.8, the matching system has only finitely many fundamental solutions. Some of them are admissible and thus correspond to *fundamental surfaces*. Just as earlier, all admissible solutions can be obtained as sums of fundamental ones. Note that not every sum of admissible solutions is admissible, though if the sum of solutions is admissible, then so are the summands.

3.3.5 Geometric Summation

Let us describe a geometric interpretation of algebraic summation of solutions. Consider two normal surfaces F_1, F_2 such that the sum $\bar{x} = \bar{x}(F_1) + \bar{x}(F_2)$ of the corresponding admissible solutions is also admissible. Shift F_1, F_2 by a normal isotopy so that their intersection inside each tetrahedron Δ^3 consists of double arcs with endpoints in the interiors of faces. Moreover, we require that the intersection of any two elementary discs of F_1, F_2 in Δ^3 consists of no more than two arcs. It is possible to achieve, because we may apply to each surface a normal isotopy that afterward the following holds:

1. Each surface intersects all the triangle faces of the triangulation along straight segments.
2. All triangular elementary discs in the intersection of the surfaces with tetrahedra are planar.
3. In general, we cannot achieve that the quadrilaterals are planar, since not every four points in edges of a tetrahedron are contained in a planar section. Nevertheless, we may achieve that any quadrilateral elementary disc in the intersection of the surfaces with tetrahedra is the union of two planar triangles having a common edge. Two such quadrilaterals can have no more than two common arcs.

Consider a double line c of $F_1 \cap F_2$ and decompose it into arcs which are connected components of the intersection of c with tetrahedra of the triangulation. Let $l \subset c \cap \Delta^3$ be one of these arcs. It belongs to the intersection of two elementary discs D_1, D_2 (pieces of F_1 and F_2) in Δ^3. Like in the case of normal curves on surfaces (see Sect. 3.2.1), there are two cut-and-paste operations along l: we cut the discs along l and glue them again in one of the two possible ways. The operations are called *exchange moves* or *switches*. The *regular switch* along l produces two elementary discs of the same types as D_1, D_2. In the case of *irregular switch* we get at least one disc that crosses an edge twice, i.e., does not satisfy condition 2 of Definition 3.3.15 of normal surface. See Fig. 3.22, where we illustrate the case of a triangle and a quadrilateral.

Note that any switch of elementary discs induces switches at all crossing points of their boundary curves (considered as normal curves on $\partial\Delta^3$). The switch is regular if and only if it induces regular switches of the curves. It follows that regular switches along all arcs in c agree in the triangle faces and thus give a global *regular switch* along c. In particular, if c is closed, then c is either orientation preserving on both F_1, F_2 or orientation reversing on both F_1, F_2.

Now let us perform regular switches along all double lines of $F_1 \cap F_2$. By construction, the resulting surface F is normal and $\bar{x}(F) = \bar{x}(F_1) + \bar{x}(F_2)$.

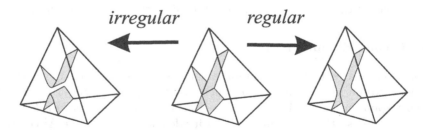

Fig. 3.22. Regular and irregular switches

We will write $F = F_1 + F_2$ and call F the *geometric sum* of F_1 and F_2. Note that $\bar{x}(F)$ depends only on $\bar{x}(F_1), \bar{x}(F_2)$. Thus the normal isotopy class of F is defined correctly although the relative position of F_1, F_2 (in particular, the number of curves in $F_1 \cap F_2$) can vary when we shift the surfaces by normal isotopies. If they are disjoint, then the sum is their union.

For convenience of future references we formulate the following direct consequence of Theorem 3.3.27 and Theorem 3.2.8.

Theorem 3.3.28. *For any triangulated 3-manifold M, the set of fundamental surfaces is finite and can be constructed algorithmically. Any normal surface in M can be presented as a linear combination of the fundamental ones with non-negative integer coefficients.*

Remark 3.3.29. Evidently, the edge degree of surfaces (see Definition 3.3.16) is additive with respect to geometric sums. Let us show that the Euler characteristic possesses the same property, i.e., $\chi(F_1 + F_2) = \chi(F_1) + \chi(F_2)$. Indeed, any normal surface F has a natural cell decomposition into connected components of the intersection of F with the edges, faces, and tetrahedra of the triangulation. Since the numbers of i-dimensional cells of F_1, F_2 sum up to those of $F, 0 \leq i \leq 2$, the same happens with the Euler characteristics.

There are many ways to present F as a geometric sum of two surfaces F_1, F_2. In general, $F_1 \cap F_2$ consists of several connected components (circles and arcs). The number of the components will be denoted by $\#(F_1 \cap F_2)$. Let us describe a simple trick that helps us to decrease $\#(F_1 \cap F_2)$.

Lemma 3.3.30. *Let a connected normal surface F be presented in the form $F = F_1 + F_2$ such that at least one of the following holds:*

1. *One of the surfaces F_1, F_2 is disconnected while the other is nonempty.*
2. *There is a curve $C \subset F_1 \cap F_2$ which decomposes both F_1, F_2.*

Then there is another presentation $F = F_1' + F_2'$ such that $\#(F_1' \cap F_2') < \#(F_1 \cap F_2)$.

Proof. Assume that one of the surfaces F_1, F_2 (say, F_1) consists of two or more components. Denote by $\tilde{F}_1$ one of them, and by F_1' the union of the remaining ones. By making regular switches along all curves in $\tilde{F}_1 \cap F_2$, we get a new surface F_2'. Then $F = F_1' + F_2'$ and $\#(F_1' \cap F_2') < \#(F_1 \cap F_2)$, see Fig. 3.23a.

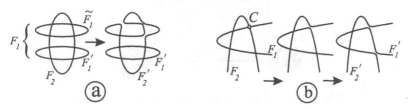

Fig. 3.23. How regular switches produce a more economic presentation

Assume now that F_1, F_2 are connected and a double curve $C \subset F_1 \cap F_2$ decomposes both F_1, F_2. Then $F_1 \setminus C$ and $F_2 \setminus C$ consist of two pieces each. The regular switch along C pastes together only two pairs of them, therefore we get two new surfaces, maybe with self-intersections. Switching all self-intersection curves regular, we get F_1', F_2'. By construction, $F = F_1' + F_2'$, see Fig. 3.23b.   □

3.4 Normal Surfaces in Handle Decompositions

In this section we describe a handle decomposition version of the theory of normal surfaces. All the basic facts remain the same, but sometimes, especially when investigating special spines, the handle decomposition version is more convenient. For simplicity we consider manifolds without boundary pattern.

Every 3-manifold M with nonempty boundary can be decomposed into handles of indexes 0, 1, and 2 called *balls*, *beams*, and *plates*, respectively. In case M is closed, its handle decomposition must contain at least one handle of index 3. Each beam and each plate has a direct product structure $D^2 \times I$. The difference is that the beam $D^2 \times I$ is attached to balls along two discs $D^2 \times \{0, 1\}$, while the plate $D^2 \times I$ is attached to the union of balls and beams along the annulus $\partial D^2 \times I$. The intersection of the balls with the beams is the union of discs called *islands*. Connected components of the intersection of the balls with the plates are called *bridges*. In the boundary of every ball, the complement to the union of islands and bridges consists of several regions, called *lakes*. See Fig. 3.24.

Let ξ be a handle decomposition of a 3-manifold M.

Definition 3.4.1. *A proper surface $F \subset M$ is called* normal *(with respect to ξ) if the following conditions hold:*

1. *F does not intersect handles of index 3.*
2. *F intersects each plate $D^2 \times I$ in a number of parallel copies of the form $D^2 \times \{*\}$ of the disc D^2.*

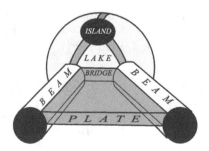

Fig. 3.24. A fragment of a handle decomposition

3. *The intersection of F with each beam $D^2 \times I$ has the form $L \times I$, where L is a finite system of disjoint simple proper arcs in D^2. Here the disc D^2 can be identified with the island $D^2 \times \{0\}$ as well as with the island $D^2 \times \{1\}$.*

4. *None of the systems L contains a closed curve or a* lake *return (an arc $l \subset L$ such that the endpoints of l lie in the same connected component of the intersection of an island with a lake).*

5. *The intersection of F with each lake contains no closed curves which are trivial in the lake and no island returns (an island return is an arc in a lake whose endpoints lie in the same connected component of the intersection of the lake with an island).*

6. *The intersection of F with each ball consists of discs (these discs are called elementary).*

7. *The boundary curve C of each elementary disc crosses each bridge and each lake at most once. Moreover, if a bridge and a lake are adjacent, then C can intersect only one of them.*

Remark 3.4.2. Condition 7 implies further restrictions on L: The endpoints of any arc l of L lie neither in the same end of the same bridge nor in a bridge and in an adjacent lake. It follows from condition 1 that the intersection of F with a lake A can be nonempty only if $A \subset \partial M$. By condition 5, $\partial F \cap A$ cannot contain trivial circles and island returns. Together with condition 7, it forbids once more lake returns from condition 4. See Fig. 3.25, where we the dotted lines show all types of curves that are forbidden by conditions 4, 5, and 7.

Remark 3.4.3. The intersection of a normal surface F with a lake can contain a closed curve l, but this can happen only if F is a disc in a ball B of ξ and each connected component of $\partial B \setminus l$ contains at least one island.

Just as in the case of triangulated manifolds, normal surfaces have the advantage that they may be presented algebraically. To describe this, let us consider the set of all elementary discs. Let ξ be a handle decomposition of a 3-manifold M.

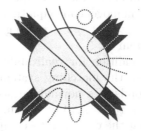

Fig. 3.25. Different types of curves in an island and lakes

Definition 3.4.4. *Two elementary discs* D_1, D_2 *in a ball* B *of* ξ *are called* equivalent *if they are normally isotopic, that is, there exists an isotopy of the ball that takes one disc onto the other and is invariant on the islands and bridges.*

Using condition 7, one can easily show that every ball of ξ contains only a finite number of nonequivalent elementary discs. Let $E_1, E_2, \ldots, E_n$ denote elementary discs representing without repetitions all the equivalence classes in all the balls of ξ. A normal surface $F \subset M$ can intersect the balls in several parallel copies of each disc E_i. Let $x_i = x_i(F)$ denote the number of these copies. We get an n-tuple $\bar{x}(F) = (x_1, x_2, \ldots, x_n)$ of non-negative integers, that is, a vector with non-negative integer coordinates.

Two normal surfaces F_1, F_2 are called *equivalent* if there exists a normal isotopy $h_t : M \to M$ which takes F_1 to F_2 (an isotopy is *normal* if it is invariant on all the handles of the decomposition). It is easy to see that F_1, F_2 are equivalent if and only if $\bar{x}(F_1) = \bar{x}(F_2)$. Thus to any equivalence class of normal surfaces there corresponds an n-tuple $\bar{x}$. Conversely, we could begin with an n-tuple $\bar{x}$ of non-negative integers and try to build the corresponding normal surface. This is possible only if we subject the n-tuple to two constraints.

The first constraint is that, for every $x_i, x_j > 0$, the equivalence classes of discs E_i, E_j must have disjoint representatives. Such n-tuples will be called *admissible*. One can easily prove that if an n-tuple $(x_1, x_2, \ldots, x_n)$ is admissible, then not only pairs, but also all E_i with $x_i > 0$ can be chosen to be disjoint. In order to describe the second constraint, let us multiply each of the discs E_i in x_i parallel copies. In the beam $D^2 \times I$ we choose a simple arc $l \times \{0\}$ in the island $D^2 \times \{0\}$.

Let us count the total number of copies of $l \times \{0\}$ in the intersection of $D^2 \times \{0\}$ with all the parallel copies of all discs E_i. We get a linear combination of x_i with coefficients 0 and 1, where the coefficient at x_i is 1 if and only if ∂E_i contains an arc normally isotopic to $l \times \{0\}$ (it follows from the last condition in Definition 3.4.1 that this coefficient cannot be greater than 1). In exactly the same fashion, we calculate the number of copies of $l \times \{1\}$ in the intersection of the discs with the island $D^2 \times \{1\}$. If $\bar{x}$ corresponds to a normal surface, these two numbers must be equal for all possible choices of l. We get a system of linear homogeneous equations with integer coefficients. Adding to it n inequalities $x_i \geq 0, 1 \leq i \leq n$, we get a handle decomposition version of the matching system for ξ.

The second necessary condition for realizability is that the coordinates of $\bar{x}$ must form a solution of the matching system. Let us show that every admissible solution corresponds to a normal surface. Admissibility allows us to realize a given solution $(x_1, x_2, \ldots, x_n)$ by disjoint elementary discs in balls, and the equations tell us that the discs can be completed to a normal surface by adding strips in beams and discs in plates.

Therefore the first conclusion of Haken's theory holds also for the handle decomposition version:

The set of equivalence classes of normal surfaces in M can be parameterized by the set of non-negative integer admissible solutions of the system.

Just as in the previous version of the theory of normal surfaces, Theorem 3.2.8 tells us that *the set of fundamental solutions to the matching system is finite and can be constructed algorithmically.* Fundamental surfaces correspond to admissible fundamental solutions. They form a basis in the following sense: each normal surface in M can be presented as a sum (with non-negative integer coefficients) of fundamental ones.

To show the geometric meaning of the summation, suppose that normal surfaces F_1, F_2 correspond to admissible solutions $\bar{x}(F_1), \bar{x}(F_2)$. We will assume that the solutions (and the surfaces) are compatible, that is, for any positive coordinates $x_i(F_1), x_j(F_2)$ the equivalence classes of the corresponding elementary discs E_i, E_j have disjoint representatives. Then the solution $\bar{x} = \bar{x}(F_1) + \bar{x}(F_2)$ will also be admissible. Let a surface F correspond to $\bar{x}$. How can one geometrically reconstruct F from F_1, F_2?

The compatibility condition assures us that F_1, F_2 can be realized so that their elementary discs are disjoint. The intersections of these surfaces with each beam $D^2 \times I$ consist of bands isotopic to bands of the form $l \times I$, where l is an arc in D^2. Straightening the bands, we can make them intersect only along double lines joining their lateral sides. Any two bands must intersect along no more than one line, see Fig. 3.26a. Similarly, one can bring to an appropriate form the sheets along which the surfaces intersect the plates of ξ. We can arrange things so that the intersection of any two sheets consists of arcs which join the endpoints of double lines in beams. So we may assume that either the surfaces F_1, F_2 do not intersect at all or they intersect in a collection of disjoint simple curves. Then "geometric summation" of F_1 and F_2 can be achieved by *switching* along all the curves in $F_1 \cap F_2$. Each switch of the surfaces consists in cutting them along a common curve and gluing back. Clearly, along any curve in $F_1 \cap F_2$ there are two ways to switch, but we perform a *regular switch* which yields a surface $F = F_1 + F_2$ essentially in normal position without further isotopy, see Fig. 3.26b). Note that the Euler characteristic of surfaces is additive with respect to the summation.

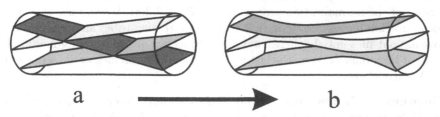

a ⟶ b

Fig. 3.26. Regular switches produce a normal surface

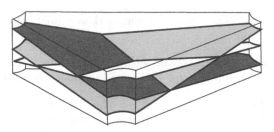

Fig. 3.27. The intersection of compatible normal surfaces can contain triple points, even after straightening bands and sheets

Having described geometric summation of two compatible normal surfaces, we can inductively describe the sum of a larger number of them. For example, one may consider $F_1 + F_2 + F_3$ as $(F_1 + F_2) + F_3$. The result does not depend on the order of summation, since in all cases we get the surface that realizes the algebraic sum of the corresponding n-tuples. One should point out that if there are more than two compatible normal surfaces, then their intersection can contain triple points, even after straightening. This means that in general we cannot make the switches along the double lines simultaneously; we have to perform them one after another. I am indebted to Benedetti, Lisca, and Petronio for showing me a simple example of an unavoidable triple point (see Fig. 3.27).

Let ξ be a handle decomposition of a 3-manifold M and F a proper surface in M disjoint from handles of index 3.

Definition 3.4.5. *The* beam degree $b(F)$ *of* F *is the total number of components in the intersection of* F *with the union of all beams.*

For each plate $D^2 \times I$ of ξ we choose an arc $l = \{*\} \times I$ which is transversal to F. The union $L_d(\xi)$ of all these arcs is called *a dual link* for ξ. Any dual link can be considered as a proper link in the union of all handles of indices ≤ 2. $L_d(\xi)$ is called *minimal* if the number $p(F) = \#(F \cap L_d(\xi))$ of points in $F \cap L_d(\xi)$ is as small as possible.

Definition 3.4.6. $p(F)$ *is called the plate degree of* F.

The plate degree is a handle decomposition analogue of the edge degree of a surface in a triangulated manifold. Let us describe a handle decomposition version of the normalization procedure (see Sect. 3.3.3). We will use the same normalization moves (see Remark 3.3.18): compressing tubes and tunnels, and crossing out inessential spheres and discs. Since we are considering manifolds without boundary pattern, the last move (crossing out an inessential semiclean disc) is irrelevant.

Theorem 3.4.7. *Any proper surface* F *in a handle decomposition* ξ *of a 3-manifold* M *can be transformed into a normal surface* F' *by a sequence of isotopic deformations and normalization moves.*

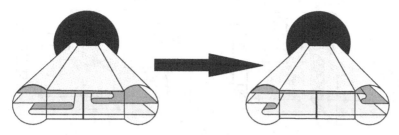

Fig. 3.28. Improving surfaces in plates

Proof. We divide the proof into several steps which we will take in turn.

STEP 1. Any 3-handle D^3 contains a small ball D_0^3 disjoint from F. Blow up D_0^3 by an isotopy to the size of D^3 and thus force out F to the outside of D^3.

Further on we will improve F by transformations described in Steps 2–6 below. After performing each of Steps 3–6 we start the procedure anew by returning back to Step 2. To control the total number of steps, we use the beam and plate degrees of F.

STEP 2. For each plate $D^2 \times I$ of ξ choose a dual arc $l = \{*\} \times I$ (transversal to F) which intersects F in the smallest possible number of points (among such arcs). Stretching a small regular neighborhood of l in $D^2 \times I$ by an isotopy to the whole $D^2 \times I$, we get condition 2 of Definition 3.4.1. See Fig. 3.28.

STEP 3. Similarly, for each beam $D^2 \times I$ choose a disc $D = D^2 \times \{*\}$ such that D is transversal to F and the number of components of $F \cap D$ is as small as possible. Condition 3 can be satisfied by stretching a regular neighborhood of D in the beam to the whole beam. $b(F)$ and $p(F)$ remain the same. After performing Steps 1–3, we get a surface satisfying conditions 1–3 of Definition 3.4.1.

STEP 4. Compress all tubes and tunnels in the beams of ξ, thus obtaining condition 4. Both transformations do not increase $p(F)$ and strictly decrease $b(F)$.

STEP 5. Suppose that the intersection of F with a lake contains a closed component l. Compress the tube outgoing from l and throw out the inessential proper disc obtained by the compression, thus diminishing the number of circles in ∂F. If there is an island return l, eliminate it by an isotopy which takes l into the adjacent beam, see Fig. 3.29. This transformation preserves $p(F)$ and strictly decreases $b(F)$.

STEP 6. Compressing nontrivial tubes in the intersection of F with all balls of ξ converts the intersection into the union of discs and maybe inessential 2-spheres. Remove the spheres. $b(F)$ and $p(F)$ do not increase.

STEP 7. To get condition 7, consider a disc D in the intersection of F with a ball. Suppose that D either crosses a bridge twice or crosses a bridge and an adjacent lake. Since all proper discs in a ball with the same boundary

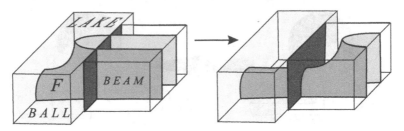

Fig. 3.29. Two cubes present a ball (*left*) and an adjacent beam (*right*). Plates and other beams are not shown

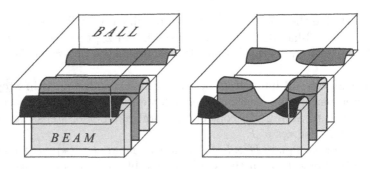

Fig. 3.30. Killing tunnels

curve are isotopic *rel* ∂, we may assume that near the bridge and the lake D looks like a tunnel. Destroy the tunnel by an isotopy. This transformation preserves $b(F)$ and strictly decreases $p(F)$. If D crosses a lake twice, compress the corresponding tunnel. $p(F)$ and $b(F)$ do not increase. All three cases are shown in Fig. 3.30.

Let us perform now Steps 2–7 for as long as possible. It follows from the description of the steps that we shall and up with a normal surface F'.   □

Corollary 3.4.8. *Let ξ be a handle decomposition of an irreducible boundary irreducible 3-manifold M. Then any incompressible boundary incompressible proper surface $F \subset M$ without inessential spherical or disc components is isotopic to a normal surface.*

Proof. We do the same as in the proof of Proposition 3.3.24: We substitute each tube or tunnel compression by an isotopic deformation of F. This is possible, since by assumption M is irreducible and boundary irreducible, and F is incompressible and boundary incompressible. Since the forbidden components do not appear under isotopy, the other two normalization moves are not needed at all.   □

Proposition 3.4.9. *Let ξ be a handle decomposition of an irreducible 3-manifold M. Suppose that M contains an essential proper disc D. Then M contains a normal essential proper disc D_1. Moreover, if D is nonseparating, then one can find a nonseparating normal disc D_1.*

Proof. Perform the same steps as in the proof of Theorem 3.4.7. The only difference is that each tunnel compression in Step 4 transforms an essential disc into two discs, at least one of which is essential. We discard the other. $\square$

4

Applications of the Theory of Normal Surfaces

4.1 Examples of Algorithms Based on Haken's Theory

The theory of normal surfaces is used extensively in algorithmic topology. Algorithms based on it most often follow the General Scheme described below. Suppose that we wish to solve a problem about a given 3-manifold M.

GENERAL SCHEME

1. Reduce the problem at hand to one of the existence in M of a surface with some specific characteristic property, which we denote by P. Let $\mathcal{P}$ be the class of all characteristic surfaces in M, i.e., the class of all surfaces that possess P.

2. Choose a triangulation of M and show that if M contains at least one characteristic surface F, then there exists a normal characteristic surface. Quite often it can be done by proving that P is stable with respect to the normalization procedure, i.e., with respect to isotopies and moves N_1–N_8 that bring surfaces in normal position, see Theorem 3.3.21. By stability we mean that if F_1 is obtained from $F \in \mathcal{P}$ by isotopies and moves N_1–N_8, then at least one connected component of F_1 is also in $\mathcal{P}$.

3. Show that if there is a normal characteristic surface, then there is a fundamental characteristic surface. One possible way to do that is to prove that if a characteristic surface F is not fundamental, then M contains a less complicated characteristic surface. Certainly, we should know how to measure the complexity of a surface in M. The edge degree $e(F)$, i.e., the total number of points in the intersection of F with the edges, may serve as a good candidate for the purpose.

4. Construct an algorithm to decide whether or not a given surface is characteristic.

Assume that all four steps of the General Scheme are carried out. Then the algorithm that solves the problem works as follows:

1. Choose a triangulation T of M.
2. Write down the corresponding matching system of linear equations.
3. Find the finite set of fundamental solutions.
4. Realize the fundamental solutions by normal surfaces.
5. Test each of the obtained fundamental surfaces for being characteristic. It follows that M contains a characteristic surface (i.e., that the problem in question has a positive answer) if and only if at least one of the fundamental surfaces is characteristic.

4.1.1 Recognition of Splittable Links

We will illustrate the above scheme by describing algorithms for recognizing splittable links in S^3. Recall that a link $L \subset S^3$ is a collection of disjoint simple closed curves in S^3. The curves are called the *components* of L. A link L is called *splittable*, if there is a 2-sphere $S \subset S^3$ such that $S \cap L = \emptyset$ and each connected component of the complement $S^3 \setminus S$ contains at least one component of L. We will call S *a splitting sphere*. For example, a splittable link of two components is nothing more than the union of two knots contained in disjoint balls. The boundary sphere of either ball can be taken as a splitting sphere.

Theorem 4.1.1. *There is an algorithm to decide if a given link $L \subset S^3$ is splittable.*

Proof. We will follow the General Scheme.

STEP 1. It is convenient to replace the noncompact 3-manifold $S^3 \setminus L$ by the compact manifold $M = S^3 \setminus \text{Int } N(L)$, where $N(L)$ is a regular neighborhood of L. The boundary of M is a collection of 2-dimensional tori. Let us formulate the following property P of closed surfaces in M: a surface $S \subset \text{Int } M$ possesses P if S is a splitting sphere for L. Clearly, P is characteristic for the problem at hand: L *is splittable* $\Longleftrightarrow$ M *contains a sphere* $S \in \mathcal{P}$.

Note that a sphere $S \subset M$ splits L if it is essential in M. Even more: S splits L if and only if it determines a nontrivial element $[S]$ of $H_2(M; Z_2)$. Further, a closed surface in M determines a nontrivial element of $H_2(M; Z_2)$ if and only if there exists a proper arc $a \subset M$ which crosses L transversally in an odd number of points.

STEP 2. Choose a triangulation T of M. The statement we need to prove here is that if M contains a splitting sphere, then it also contains a normal splitting sphere.

The proof is natural: Take the splitting sphere S which exists by assumption, and normalize it by a sequence of moves N_1–N_4. Note that moves N_5–N_8 (see Sect. 3.3.3) are irrelevant since S is closed. Each time we apply move N_1, we get two 2-spheres S', S'' such that $[S'] + [S''] = [S] \neq 0$ in $H_2(M; Z_2)$. Therefore at least one of them splits M. We cross out the other.

STEP 3. We wish to prove that if M contains a normal splitting 2-sphere S, then such a sphere can be found among fundamental surfaces. Our strategy is to show that if S is not fundamental, then there is another splitting sphere which is simpler than S. For measuring the complexity of S we use the edge degree $e(S)$.

Let S be presented as a geometric sum $S = F_1 + F_2$ of two surfaces. There are many ways to present S in such form. By Lemma 3.3.30, we may assume that F_1, F_2 are connected, and no component of $F_1 \cap F_2$ separates both surfaces. Taking into account that the Euler characteristic is additive with respect to geometric sums and $\chi(S) = 2$, $\chi(F_i) \leq 2$, we come naturally to two options: 2=1+1 and 2=2+0. The first option does not occur, since the only closed surface with $\chi = 1$ is the projective plane RP^2, which cannot be embedded into S^3.

In the second case we may conclude that one of the surfaces (say, F_1) is a sphere, while the other (F_2) is a torus. The Klein bottle, which also has zero Euler characteristic, does not embed into S^3 either. Since $[F_1]+[F_2] = [S] \neq 0$, at least one of elements $[F_1], [F_2]$ is not zero.

CASE 1. If $[F_1] \neq 0$, then F_1 splits M. Clearly, this new sphere is simpler than S.

CASE 2. Suppose $[F_1] = 0, [F_2] \neq 0$. We claim that there exists a proper arc $a \subset M$ which does not intersect F_1 and intersects F_2 transversally at an odd number of points. Indeed, consider an arc b that joins two points $x, y \in \partial M$ contained in different components of $M \setminus F_2$ and is transversal to F_1, F_2. Since $[F_1] = 0$, the intersection $F_1 \cap b$ consists of an even number of points, while the number of points in $F_2 \cap b$ is always odd. The desired arc a can be obtained by replacing the subarc of b between the first and the last points in $F_1 \cap b$ by an arc that runs near F_1 but does not intersect it.

Recall that each circle $c \subset F_1 \cap F_2$ does not separate at least one of the surfaces. Since all circles on the sphere F_1 do separate, c is a nonseparating curve on F_2 and thus is nontrivial. Note that any collection of disjoint non-trivial simple closed curves in the torus F_2 decomposes it into annuli. Since a intersects F_2 at an odd number of points, at least one of the annuli (denote it by A) contains an odd number of crossing points.

Denote by d_1, d_2 the boundary circles of A. They bound in F_1 disjoint discs D_1, D_2. Let us perform (not necessarily regular) switches along d_1, d_2 which adjoin the discs to A. The switches produce a sphere S' that corresponds to $D_1 \cup A \cup D_2$, and the remaining part R of $F_1 \cup F_2$ that corresponds to the union of $F_1 \setminus \mathrm{Int}\,(D_1 \cup D_2)$ and $F_2 \setminus \mathrm{Int}\,A$. Note that S' splits M, since a does not intersect D_1, D_2 and thus the number of intersection points of a and S' is odd, see Fig. 4.1.

Let us prove that either S' is simpler than S or S' is isotopic to a simpler sphere. Denote by R' the surface obtained from R by performing regular switches along all the curves in the self-intersection of R. Of course, at least one of the switches at d_1, d_2 is irregular, because otherwise S would consist of S' and R' and thus be disconnected. This irregular switch produces a return, i.e., an arc in a triangle of the triangulation having both endpoints on the same edge. If the return is contained in R', then $e(R') > 0$, and we obtain

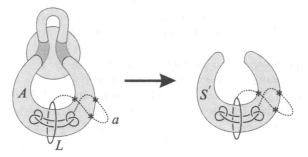

Fig. 4.1. Constructing a simpler splitting sphere

$e(S') = e(S) - e(R') < e(S)$. Let the return be in S'. Then we decrease the degree of S' by an isotopy which annihilates the return together with its endpoints. Applying the normalization procedure to the resulting surface, we get a new essential 2-sphere that has a strictly smaller edge degree.

Now we apply the simplification process as long as the sphere remains nonfundamental. Since each time we get a smaller edge degree, the process is finite and we end up with a fundamental splitting sphere. The last step of the General Scheme is easy. According to the scheme, we get a recognition algorithm for splittable links.

In conclusion we note that the proof works for any polyhedral subset of S^3. We have never used that L is the union of disjoint circles.   □

4.1.2 Getting Rid of Clean Disc Patches

Later we will describe other problems of 3-manifold topology whose algorithmic solutions are based on the theory of normal surfaces. Among they are:

1. Recognizing the Unknot.
2. Calculating the genus of a given simple closed curve on the boundary of a 3-manifold.
3. Recognizing irreducibility and boundary irreducibility of a 3-manifold.
4. Testing two-sided surfaces for incompressibility and boundary incompressibility.
5. Detecting sufficiently large 3-manifolds (see Sect. 4.1.6 for the definition of a sufficiently large 3-manifold).

Solutions of all these problems follow the same General Scheme. In all cases Step 3 (if M contains a surface having a specific characteristic property, then such a surface can be found among the fundamental ones) plays a crucial role. The right strategy for transforming a given normal surface into a fundamental one consists in considering the so-called *clean disc patches*; in many cases they are responsible for the existence of characteristic surfaces which are not fundamental.

It is surprising that four technical tricks, elaborated by Jaco and Oertel, [56] work in all five cases listed above. We present the tricks in the form of four lemmas, but before doing that we describe clean disc patches and related notions.

Let a normal surface F in a triangulated 3-manifold (M, Γ) be presented as the geometric sum of two normal surfaces, i.e., have the form $F = G_1 + G_2$. Then the union $G_1 \cup G_2$ is a 2-dimensional polyhedron of a very specific type. The set $G_1 \cap G_2$ of its singular points consists of double lines (intersection lines of the surfaces). If we cut $G_1 \cup G_2$ along all double curves, we get a collection of pieces called *patches*. Each patch P is a compact surface. If $P \cap \partial M \neq \emptyset$, P is said to be a *boundary patch*. Otherwise P is an *interior patch*.

Performing the regular switches along all double curves of $G_1 \cup G_2$, i.e., decomposing $G_1 \cup G_2$ into patches and gluing the patches together in the appropriate way, we restore F. It is convenient to think of F as being decomposed into the same patches by *trace curves* (images of the boundaries of the patches under the gluing). If a double curve l of $G_1 \cup G_2$ is two-sided on both G_1 and G_2, then it contributes two trace curves $l_1, l_2 \subset F$. They are called *twins*. If l is one-sided on G_1 and hence on G_2, it contributes only one trace curve, which is the twin of itself.

Let l be a double curve contained in two distinct patches P_1, P_2 of $G_1 \cup G_2$. We say that P_1, P_2 are *opposite* at l, if they lie in the same surface $G_i, i = 1, 2$. The patches are *adjacent*, if they are in different surfaces. The terminology is motivated by the actual position of patches near l, see Fig. 4.2. It may happen that the same patch P approaches l from opposite sides. Then we say that P is *self-opposite*.

By a *clean disc patch* we mean a patch P homeomorphic to a disc such that either ∂P is a closed double curve, or ∂P consists of a double arc and a clean arc on ∂M. In the first case P is an interior clean disc patch, in the second it is a boundary clean disc patch. We emphasize that if a clean patch P intersects ∂M along two or more disjoint arcs, then P is *not* a clean disc patch, even if it is homeomorphic to a disc. A boundary patch P is called *quadrilateral*, if ∂P consists of two trace arcs and two arcs on ∂M such that these pairs correspond to opposite sides of the quadrilateral.

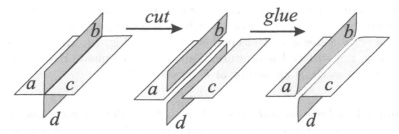

Fig. 4.2. Trace curves are shown by *dotted lines*. Pairs a, c and b, d consist of opposite patches, all other pairs consist of adjacent ones

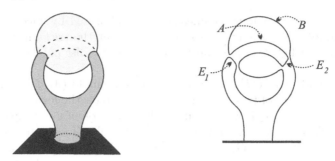

Fig. 4.3. Adjacent and opposite companions

Assume that $E, \partial E = s$, is a not self-opposite clean disc patch of F such that the twin curve s' of s cuts off a clean disc E' from F. We say that E' is *an adjacent* or *an opposite companion* of E, if the patch of E' containing s' is adjacent or opposite to E, respectively. For example, there are two clean disc patches in Fig. 4.3. Patch E_1 has the adjacent companion $A \cup E_2$, patch E_2 has the opposite companion $B \cup A \cup E_2$.

The next four lemmas show how the presence of clean disc patches helps us to simplify nonfundamental normal surfaces. Here and later on we will measure the complexity of a surface by its edge degree. So F is *simpler* than F' if $e(F) < e(F')$.

Lemma 4.1.2. *Let a normal surface F in a triangulated 3-manifold M be presented in the form $F = G_1 + G_2$. Suppose that $G_1 \cup G_2$ has a self-opposite disc patch E. Then the following holds:*

1. *M contains a normal projective plane P such that P is simpler than F.*
2. *M contains a surface F' such that $\partial F' = \partial F$, F' is homeomorphic to F, and F' is simpler than F.*

Proof. We may assume that E is contained in G_1. Then the connected component G_1' of G_1 containing E is a normal projective plane having a smaller edge degree. It intersects G_2 along a unique closed curve l, which corresponds to ∂E. This gives us the first conclusion.

To get the second conclusion, we replace the regular switch along l by the irregular one. We get a surface F' which is homeomorphic to F and has the same edge degree and the same boundary. See Fig. 4.4. Since the switch is irregular, this surface has a return. Thus we can eliminate the return and hence decrease $e(F)$ by an isotopy of F fixed on ∂F.   □

Lemma 4.1.3. *Let a normal surface F in a triangulated 3-manifold (M, Γ) be presented in the form $F = G_1 + G_2$. Suppose that there is a not self-opposed clean disc patch E of $G_1 \cup G_2$ having no clean companion discs. In other words, we suppose that the twin curve s' of $s = \partial E$ is nontrivial in F. Then F is*

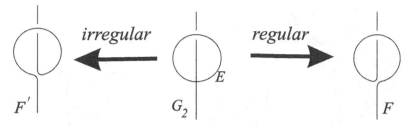

Fig. 4.4. Regular and irregular switches along l produce homeomorphic surfaces

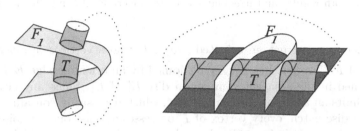

Fig. 4.5. T consists of annular or quadrilateral patches and therefore is a surface with $\chi(T) = 0$: a torus, a Klein bottle, an annulus, or a Möbius band

either compressible or boundary compressible, depending on whether E is an interior or a boundary patch.

Proof. Evident, since a parallel copy E' of E is an essential compressing or boundary compressing disc for F. □

Lemma 4.1.4. *Let a normal surface F in a 3-manifold (M, Γ) with boundary pattern be presented in the form $F = G_1 + G_2$. Assume that any clean disc patch E of $G_1 \cup G_2$ admits an adjacent companion disc E'. Then either there is a clean disc patch $E \subset F$ such that its adjacent companion disc E' is also a patch of F, or the following holds:*

1. *F can be presented in the form $F = F_1 + T$, where T is a torus, a Klein bottle, an annulus, or a Möbius band. Certainly, F_1 is simpler than F.*
2. *All patches of T are either annuli or quadrilaterals, and each double curve of $F_1 \cup T$ cuts off a clean disc patch from F_1, see Fig. 4.5. Here by a quadrilateral we mean a patch P whose boundary consists of two trace arcs and two arcs on ∂M such that these pairs correspond to opposite sides of P.*
3. *F_1 is homeomorphic to F as well as to the surface F' obtained from $F_1 \cup T$ by irregular switches along all the curves in $F_1 \cap T$.*
4. *If (M, Γ) is irreducible and boundary irreducible, then F is admissibly isotopic to F_1 and to F' (see Fig. 4.6).*

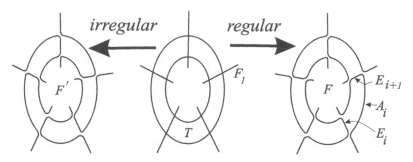

Fig. 4.6. Both regular and irregular switches convert $F_1 \cup T$ into homeomorphic surfaces

Proof. Let us construct an oriented graph Γ whose vertices are clean disc patches of F. Two patches E_1, E_2 are joined by an oriented edge $\overrightarrow{E_1 E_2}$ if E_2 is contained in the adjacent companion disc E_1' of E_1. Since any clean disc patch admits an adjacent companion disc, which necessarily contains at least one clean disc patch, every vertex of Γ possesses at least one outgoing edge. It follows that Γ contains a simple cycle Z (a subgraph of Γ which consists of coherently oriented edges and is homeomorphic to a circle).

Denote by $E_0, E_1, \ldots, E_{k-1}$ the successive vertices of Z. For each i, the clean disc patch E_{i+1} is contained in the adjacent companion disc E_i' of E_i (indices are taken modulo k). It may happen that $k = 2$ and $E_0' = E_1$. Then we get a pair of clean disc patches, each being a companion of the other. This corresponds to the first alternative of the conclusion of the lemma.

Assume that this situation never occurs. Then all $A_i = E_i' \setminus \mathrm{Int}\, E_{i+1}, 0 \leq i \leq k - 1$, are either annuli (if E_i are interior patches) or quadrilaterals (if they are boundary ones). Let us glue now each trace curve ∂E_i to its twin trace curve back (the same result can be obtained by making regular switches of $G_1 \cup G_2$ along all double curves except those that correspond to ∂E_i). We get a presentation $F = F_1 + T$, where T is obtained by gluing A_i while F_1 is the union of all patches of F that are not contained in T. Conditions 1, 2 are fulfilled by construction.

Let us prove that F_1, F, and F' are homeomorphic. Performing regular switches along all the curves in $F_1 \cap T$, which correspond to $\cup_{i=0}^{k-1} \partial E_i$, we actually replace each clean disc patch E_i of F_1 by the disc $A_i \cup E_{i+1}$. Similarly, irregular switches replace each disc E_i by the disc $A_{i-1} \cup E_{i-1}$. These replacements preserve the homeomorphism type of F_1. On the other hand, the result is F in the case of regular switches, and F' in the case of irregular ones. Thus F_1, F, and F' are homeomorphic.

To prove 4, consider a collection $\{S_i, 0 \leq i \leq k - 1\}$ of surfaces in M consisting either of 2-spheres or of clean proper discs. Each S_i is composed from A_i, E_{i+1}, and a copy of E_i, $0 \leq i \leq k - 1$. Since (M, Γ) is irreducible and boundary irreducible, each member of the collection cuts off a clean ball from M. These balls can be used for constructing an isotopy from F to F_1 and from F to F'. $\square$

Lemma 4.1.5. *Let F be an incompressible normal surface in a triangulated 3-manifold (M, Γ). Suppose that F can be presented in the form $F = G_1 + G_2$ such that a not self-opposite clean disc patch E of $G_1 \cup G_2$ admits an opposite companion disc E'. Then the following holds:*

1. *If E is an interior disc patch, than there exists a general position surface $F' \subset M$ such that F' is homeomorphic with F, has the same boundary, and $e(F') < e(F)$. If, in addition, M is irreducible, then F' is admissibly isotopic to F.*

2. *If E is a boundary clean disc patch and (M, Γ) is irreducible and boundary irreducible, then there exists a general position surface $F' \subset M$ such that $e(F') < e(F)$ and F' is admissibly isotopic to F.*

Proof. Denote by s the double curve of $G_1 \cup G_2$ which corresponds to the twin trace curves $s_1 \subset \partial E, s_2 \subset \partial E'$. Suppose that E is an interior disc patch.

CASE 1. *E' does not contain E.* Then the sphere $S = E \cup E'$ does not decompose M, thus M is reducible. Therefore this situation never occurs for an irreducible manifold. Replacing the regular switch along s by the irregular one, we get a homeomorphic surface F_1 with a return, see Fig. 4.7a. Removing the return by an isotopy, we get a surface F' with $e(F') < e(F_1) = e(F)$. This modification of F takes place far from ∂F, so $\partial F = \partial F'$.

CASE 2. *E' contains E.* Let us replace again the regular switch along s by the irregular one. We get the disjoint union $F_1 \cup T$ of two surfaces. The first surface F_1 is homeomorphic to F and has the same boundary. The second surface T is obtained from the annulus $E' \setminus \text{Int } E$ by identifying its boundary circles and thus is either a torus or a Klein bottle. See Fig. 4.7b. Since the switch is irregular, $F_1 \cup T$ has a return. If the return is in F_1, then we can remove it by an admissible isotopy and get a surface F' with $e(F') < e(F_1) = e(F)$. If it is in T, then $e(T) \neq 0$ and $e(F_1) = e(F) - e(T) < e(F)$. Therefore, we can take $F' = F_1$.

It remains to prove that F is admissibly isotopic to F_1 (and hence to F'), provided that M is irreducible. Indeed, assuming that, attach to F a parallel copy E_1 of E. Then the sphere $S = E' \cup E_1$ bounds a ball which can be used for constructing an isotopy from F to F_1.

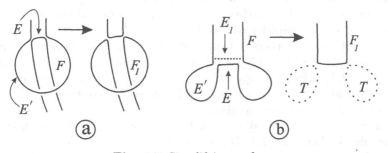

Fig. 4.7. Simplifying surfaces

If E is a boundary clean disc patch and (M, Γ) is irreducible and boundary irreducible, we apply the same tricks. The only difference is that in this case S is clean proper disc and T is either an annulus or a Möbius band.   □

As a first application of Lemmas 4.1.2–4.1.5 we describe an important situation when clean disc patches are impossible. Let (M, Γ) be a triangulated 3-manifold. As before, we will measure the complexity of a proper surface $F \subset M$ by its edge degree $e(F)$.

Definition 4.1.6. *A normal surface F in (M, Γ) is called* minimal, *if $e(F)$ is the minimum for the values $e(F')$, where F' ranges over all general position surfaces in M that are admissibly isotopic to F.*

Definition 4.1.7. *Let a normal surface F in (M, Γ) be presented in the form $F = G_1 + G_2$. Then the sum $G_1 + G_2$ is in* reduced form *if F cannot be written as $F = G'_1 + G'_2$, where G'_1, G'_2 are normal surfaces admissibly isotopic to G_1, G_2, respectively, and $G'_1 \cap G'_2$ consists of fewer components than $G_1 \cap G_2$.*

It follows from conclusion 4 of Proposition 3.3.24 that if (M, Γ) is irreducible and boundary irreducible, F is incompressible and boundary incompressible, and no component of F is an essential 2-sphere, an essential clean disc, or an essential semiclean disc, then F is admissibly isotopic to a minimal normal surface. Also, if a normal surface F is written in the form $F = G_1 + G_2$, then F can be written as a sum (of isotopic surfaces) in reduced form. Recall that it makes sense to speak about incompressible and boundary incompressible patches, though they are almost never proper.

Lemma 4.1.8. *Let (M, Γ) be an irreducible boundary irreducible 3-manifold and $F \subset (M, \Gamma)$ a minimal normal surface presented in a reduced form $F = G_1 + G_2$. If F is incompressible and boundary incompressible, then $G_1 \cup G_2$ has no clean disc patches. Moreover, all patches are incompressible and boundary incompressible.*

Proof. First, we note that since F is normal and M is irreducible, no component of F is a sphere. Otherwise the sphere would be inessential and we could decrease $e(F)$ by shifting the component into the interior of a tetrahedron. Similarly, no component of F is a clean proper disc or RP^2. Let us prove that $G_1 \cup G_2$ has no clean disc patches. To the contrary, suppose that such a patch E does exist. Let us consider all possible types of E.

If E is self-opposite, then by Lemma 4.1.2 M contains a normal projective plane P such that $e(P) < e(F)$. It follows that M, being irreducible, is homeomorphic to RP^3. Up to isotopy, RP^3 contains only one closed incompressible surface without spherical components. Therefore, P is isotopic to F and we get a contradiction with the minimality of F.

If E has no companion disc, then F is either compressible or boundary compressible by Lemma 4.1.3. This contradicts our assumption. Suppose that E has an opposite clean companion disc. Then Lemma 4.1.5 tells us that

F is admissibly isotopic to a general position surface F' such that $e(F') < e(F)$, in contradiction with our assumption. Finally, suppose that every clean disc patch of $G_1 \cup G_2$ has an adjacent companion disc. Then we can apply Lemma 4.1.4. Since F is minimal, conclusion 4 of Lemma 4.1.4 does not hold. Therefore, there is a clean disc patch E such that its adjacent clean companion disc E' is also a patch. Let us merely permute E and E' by performing the regular switch of $G_1 \cup G_2$ only along the double curve $\partial E \cap \partial E'$. This gives another presentation $F = G_1' + G_2'$ having a fewer number of double lines (since $\#(G_1' \cap G_2') = \#(G_1 \cap G_2) - 1$). On the other hand, $E \cup E'$ is a sphere or a clean proper disc and thus cuts off a clean ball from the irreducible boundary irreducible manifold (M, Γ). Deforming E to E' and E' to E through the ball, construct isotopies from G_1' to G_1 and from G_2' to G_2. But this contradicts the assumption that $G_1 + G_2$ is in reduced form.

Since all possibilities led to a contradiction, $G_1 \cup G_2$ has no clean disc patches. To prove that any patch P of F is incompressible, consider a compressing disc D for P. Our aim is to prove that D is inessential. Note that D can intersect F not only along ∂D. We will assume that the intersection is transversal.

Choose an innermost circle $c \subset F \cap D$, which bounds a disc $D' \subset D$ such that $F \cap \text{Int } D' = \emptyset$. Since F is incompressible, c bounds a disc $D'' \subset F$. It cannot happen that $D'' \supset P$; otherwise D'' would contain at least one clean disc patch of F. Therefore c can be eliminated by an isotopy of D' to the other side of F through the ball bounded by $D' \cup D''$. Doing so as long as possible, we get a new compressing disc for P (still denoted by D) such that D has the same boundary and $F \cap \text{Int } D = \emptyset$. Since F is incompressible, ∂D must bound a disc in F and hence in P (otherwise that disc would contain at least one clean disc patch). We may conclude that P is incompressible.

To see that P is boundary incompressible, consider a boundary compressing disc D for P and follow actually the same procedure for eliminating first all circles and then all arcs in $F \cap \text{Int } D$, and for proving that D must be inessential. The only difference is that we use boundary incompressibility of F, boundary irreducibility of (M, Γ), and an outermost arc argument for eliminating arcs.   □

4.1.3 Recognizing the Unknot and Calculating the Genus of a Circle in the Boundary of a 3-Manifold

Let K be a knot in S^3. We would like to know whether K is trivial, i.e., bounds a disc embedded in S^3. It is well known that K is always spanned by a connected orientable surface $F \subset S^3$. Any such surface is called *a Seifert surface* for K. The minimal possible genus of Seifert surfaces for K is called the *genus* of K. Since D^2 is the only connected nonclosed surface with $\chi = 1$, K is trivial if and only if its genus is 0. Hence the recognition problem for the unknot is a partial case of the genus calculation problem.

Denote by $N(K)$ a tubular neighborhood of K in S^3, and by M_K the complement space $S^3 \setminus \text{Int } N(K)$. Recall that a nontrivial simple closed curve $m \subset \partial N(K)$ is a *meridian* of K if it bounds a disc in $N(K)$.

Any simple closed curve $l \subset \partial N(K)$ which crosses m transversally exactly once is a *longitude* of K. A longitude l_0 is *principal* if there is an orientable surface $F \subset M_K$ such that $\partial F = l_0$. The principal longitude always exists and is unique up to isotopy. To construct it, one may take any longitude l and improve it by k negative twists along m, where $k = \text{lk}(l, K)$ is the linking number. Since the homology group $H_1(M_K, Z)$ is cyclic and generated by the meridional cycle $[m]$, we get a curve l_0 such that $[l_0] = [l] - k[m] = 0 \in H_1(M_K, Z)$. It follows that l_0 bounds an orientable surface F in M_K and hence is a principal longitude.

It is important to note that there is actually no difference between Seifert surfaces for K and surfaces that bound l_0. Indeed, l_0 is isotopic to the core circle of the torus $N(K)$, i.e., to K. Therefore any surface $F \subset M_K$ that spans l_0 is isotopic to a spanning surface for K. The converse is also true, since any Seifert surface is isotopic rel K to a surface $F \subset S^3$ such that $F \cap N(K)$ is an annulus. Then the surface $F \cap M_K$ spans l_0.

Theorem 4.1.9. *There exists an algorithm that calculates the genus of a knot.*

It seems reasonable to expect that one can prove a more general statement which deals with arbitrary curves in the boundary of arbitrary 3-manifolds. Let l_0 be a simple closed curve on the boundary of a 3-manifold M. Consider the set of connected orientable surfaces in M that are bounded by l_0. The minimal genus of such surfaces is called the *genus* of l_0. If l_0 bounds no orientable surface, then the genus is ∞.

Theorem 4.1.10. *There exists an algorithm that calculates the genus of any simple closed curve l_0 in the boundary of any 3-manifold M.*

Proof. Let us triangulate M so that l_0 crosses each edge of the triangulation no more than once and is not contained inside a triangle of the triangulation. We will use the asterisk to indicate that a general position surface is connected and bounded by l_0. So F^* is *a star surface*, if F^* is connected, intersects the edges transversally, and $\partial F^* = l_0$. Let us prove that if the genus g_0 of l_0 is finite, then there is an orientable star surface of genus g_0 which is fundamental. This is quite sufficient for algorithmic calculation of the genus. All what we have to do is to construct all fundamental surfaces, select among them orientable star surfaces, and take the minimum of their genera.

By the definition of the genus, there exists an orientable star surface of genus g_0. Note that it is incompressible (otherwise we could compress it and get an orientable star surface of a smaller genus). Among all such surfaces chose a *minimal* surface, which has the smallest edge degree, and apply to it the normalization procedure described in Sect. 3.3.3. It follows from item 3 of Proposition 3.3.24 that l_0 bounds a normal homeomorphic copy F^* of the minimal surface. Since the normalization procedure does not increase the edge degree, F^* is also minimal.

Let us prove that F^* is fundamental. Assuming the contrary, choose a presentation of F^* in the form $F^* = G_1^* + G_2$. By Lemma 3.3.30 we may suppose that G_1^*, G_2 are connected, and no circle from $G_1^* \cap G_2$ separates both surfaces. One of the surfaces (G_1^*) is bounded by l_0, the other one is closed (otherwise l_0 would intersect some edge more than once). It follows that there are no boundary disc patches. Let us show that the presence of an interior disc patch leads to a contradiction.

1. Assume that there is a self-opposite interior disc patch. Then we apply conclusion 2 of Lemma 4.1.2 to construct an orientable star surface which has a return and thus can be normalized into a simpler star surface by items 3, 6 of Proposition 3.3.24. This contradicts the minimality of F^*.
2. Since F^* is incompressible, every interior disc patch of $G_1^* \cup G_2$ has a companion disc by Lemma 4.1.3.
3. Assume that an interior disc patch of $G_1^* \cup G_2$ admits an opposite companion disc. Using conclusion 1 of Lemma 4.1.5, we can replace F^* by an orientable star surface which is simpler than F^*. This contradicts the minimality of F^*.
4. Assume that every interior disc patch E admits an adjacent companion disc E'. Since the presentation $F^* = G_1^* + G_2$ had been chosen so that no double curve of $G_1^* \cup G_2$ separates both surfaces, there is no pair of companion interior disc patches. Then conclusions 1–4 of Lemma 4.1.4 hold. In particular, F^* has the form $F^* = F_1 + T$, where F_1^* is an orientable star surface homeomorphic to F^*. Obviously, F_1^* has a smaller edge degree, in contradiction with the minimality of F^*.

In Cases 1–4 we have considered all the logical possibilities for interior disc patches and found that all of them lead to a contradiction. Therefore we can suppose that $G_1^* \cup G_2$ contains no clean disc patches at all. Observe the following important fact: The no-disc-patch condition implies that G_2 is neither a sphere nor a projective plane. Therefore, $\chi(G_2) \leq 0$ and $-\chi(G_1^*) \leq -\chi(F^*)$.

Let us prove that G_1^* is nonorientable. Suppose, on the contrary, that it is orientable. Then its genus $g(G_1^*) = (1 - \chi(G_1^*))/2$ does not exceed $g_0 = (1 - \chi(F))/2$ and hence equals g_0 because of the minimality of g_0. On the other hand, we have $e(G_1^*) = e(F) - e(G_2) < e(F)$, in contradiction with our choice of F.

To proceed further, choose an orientation of F^* and supply all the patches of $G_1^* \cup G_2$ with the inherited orientations. Decompose the set of all double curves of $G_1^* \cup G_2$ into two subsets A, B. A double curve l is in A, if by the irregular switch along l the orientations of the uniting patches match together (by the regular switch they always do). Otherwise l is in B, see Fig. 4.8. Note that l is in A if and only if the orientations of patches that are opposite at l do not match. Since G_1^* is nonorientable, $A \neq \emptyset$.

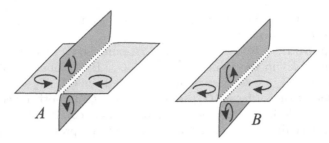

Fig. 4.8. In Case A opposite patches have opposite orientations, and the orientations of all patches match after any switch. In Case B opposite patches have coherent orientations

Now perform irregular switches at all curves in A and regular switches at all curves in B. We get an orientable star surface F'^*. Normalizing it, we decrease the edge degree (since at least one irregular switch has been made). Just as above, the no-disc-patch condition assures us that the star component of F'^* has the same or a bigger Euler characteristic. Therefore that component is simpler than F^*, in contradiction with the choice of F.

We may conclude that if the orientable star surface F^* is not fundamental, then there always exists a simpler orientable star surface. So the simplest orientable star surface must be fundamental. This completes Step 3 of the General Scheme.

In order to calculate the genus of l_0, it remains to construct all fundamental surfaces and choose the simplest orientable star surface among them. Its genus gives us the answer (if there are no orientable star surfaces at all, the genus of l_0 is ∞). $\qquad\Box$

Recall that the genus of a nonorientable surface F with boundary S^1 can be defined as $g(F) = (1 - \chi(F))/2$, i.e., by the same formula as for orientable surfaces. For example, if F is a punctured connected sum of m projective planes, then $g(F) = m/2$. The *nonorientable genus* of a simple closed curve $l_0 \subset \partial M$ can be defined as the minimum of genera of all connected surfaces which bound l_0, including nonorientable ones.

Theorem 4.1.11. *There exists an algorithm that calculates the nonorientable genus of any simple closed curve l_0 in the boundary of any 3-manifold M.*

The proof is similar to the one of Theorem 4.1.10. The only difference is that we do not need to care about orientability of surfaces and thus do not need the last step of the proof. Instead, we simply replace F^* by G_1^*.

4.1.4 Is M^3 Irreducible and Boundary Irreducible?

As we have indicated in the introduction to the book, Haken manifolds play an important role in 3-manifold topology. By definition, Haken manifolds are irreducible, boundary irreducible, and sufficiently large (see Sect. 4.1.6). But how do we know that a given 3-manifold M possesses these properties? Here we show that irreducible and boundary irreducible 3-manifolds can be recognized algorithmically (modulo algorithmic recognition of S^3, which we describe in Sect. 5).

Irreducibility. Recall that a 3-manifold M is irreducible, if every 2-sphere $S \subset M$ is inessential, i.e., bounds a 3-ball in M. Note that if M contains a one-sided projective plane P, then either M is reducible or $M = RP^3$. Indeed, the boundary S of a regular neighborhood of P is a 2-sphere (if P is normal, one can take $S = 2P$, having in mind the geometric summation). The sphere can be either essential or not. In the first case M is reducible, in the second it is RP^3.

Theorem 4.1.12. *Let M be an orientable triangulated 3-manifold. Then the following holds:*

1. *If M contains a projective plane, then at least one of the projective planes in M is fundamental.*
2. *If M is reducible, then there exists an essential sphere $S \subset M$ such that S either is fundamental or has of the form $S = 2P$, where P is a fundamental projective plane.*

Proof. Any projective plane $P \subset M$ is incompressible, since the only nontrivial simple closed curve on P is orientation-reversing, and thus cannot bound a compressing disc. By conclusion 3 of Proposition 3.3.24, P can be replaced by a normal projective plane (still denoted by P). Suppose that P is not fundamental. We claim that then M contains a normal projective plane having a smaller edge degree.

To prove this, present P as a nontrivial sum $G_1 + G_2$, where normal surfaces G_1, G_2 are connected and no double curve of $G_1 \cup G_2$ decomposes both G_1, G_2. Since the Euler characteristics of the patches of $G_1 \cup G_2$ sum up to $\chi(P) = 1$, there is at least one clean disc patch E. If E is self-opposite, then by Lemma 4.1.2 one of the surfaces G_1, G_2 is a simpler projective plane. "No companion disc" situation is impossible, since P is incompressible (see Lemma 4.1.3). If there is a clean disc patch with an opposite companion disc, or if every clean disc patch admits an adjacent companion disc, then a simpler normal projective plane in M can be found by conclusion 1 of Lemma 4.1.5 or conclusion 3 of Lemma 4.1.4, and normalization.

We have considered all logical possibilities and found that if P is not fundamental, then there is a simpler normal projective plane. It follows that any normal projective plane in M having the smallest edge degree must be fundamental.

Suppose now that M is reducible and contains no projective planes (otherwise one can take $S = 2P$, where P is a fundamental projective plane). By conclusion 3 of Proposition 3.3.24, M contains an essential normal sphere S. Assuming that S is not fundamental, present it as a nontrivial sum $G_1 + G_2$ such that normal surfaces G_1, G_2 are connected and no double curve of $G_1 \cup G_2$

decomposes both G_1, G_2. Our goal is to prove the existence of an essential normal sphere which is simpler than S, i.e., has a smaller edge degree. Indeed, since by the Jordan Theorem any circle in S bounds discs on both sides, every clean disc patch of $G_1 \cup G_2$ admits an adjacent companion disc. By Lemma 4.1.4, we can construct a presentation $S = S_1 + T$, where S_1 is a sphere and T is a torus or a Klein bottle. T consists of $k \geq 1$ annular patches $A_i, 0 \leq i \leq k-1$, while S_1 has k clean disc patches $E_i, 0 \leq i \leq k-1$, and one patch homeomorphic to a sphere with k holes.

If $k = 1$, then S_1 does not decompose M and hence is essential. Evidently, S_1 is simpler than S. Assume now that $k > 1$. We get a collection of spheres composed from $E_i \cup A_i$ and a copy of E_{i+1}, where $0 \leq i \leq k$ and indices are taken modulo k. Each of these spheres is simpler than S. If at least one of them is essential, then after normalization we get a simpler essential normal sphere (see conclusion 3 of Proposition 3.3.24). Assume that all of them bound balls. Then these balls can be used for constructing an isotopy from S to the simpler sphere S_1.

We may conclude that if S is not fundamental, then there is a simpler essential normal sphere. It follows that the simplest essential normal sphere in M must be fundamental. □

Theorem 4.1.12 is insufficient for recognition of irreducibility. It is true that if M is reducible, then an essential sphere can be found among a finite algorithmically constructible set of 2-spheres. But how can we be sure that a given 2-sphere $S \subset M$ is essential? If S does not decompose M or if it decomposes M into two parts, each containing a component of ∂M (as in the recognition of splittable links), then we are happy. In general, to make the last step of the General Scheme, we have to have a recognition algorithm for the standard 3-ball or, equivalently, a recognition algorithm for S^3. We describe both algorithms in Chap. 6.

Boundary irreducibility. It is easy to show that the connected sum of 3-manifolds is boundary irreducible if and only so are the summands. Therefore, it suffices to construct an algorithm that recognizes boundary irreducibility of irreducible manifolds.

Theorem 4.1.13. *There exists an algorithm to decide whether or not a given irreducible 3-manifold (M, Γ) is boundary irreducible. In case it is boundary reducible, the algorithm constructs an essential compressing disc.*

Proof. Let us triangulate (M, Γ). Suppose that ∂M admits an essential clean compressing disc D. By conclusion 2 of Proposition 3.3.24, after normalization we obtain a normal essential compressing disc (still denoted by D). Our goal is to show that then such a disc can be found among fundamental surfaces.

We claim that if D is not fundamental, then there is a simpler essential compressing disc. This is sufficient for proving the theorem, since then the simplest essential compressing disc must be fundamental. For proving the claim we apply the same procedure as in the proof of Theorem 4.1.12. Let D

be presented in the form $D = G_1 + G_2$, where G_1, G_2 are connected surfaces and each double curve does not separate at least one of them.

Assume that $G_1 \cup G_2$ contains an interior disc patch. Then one can construct a simpler disc D_1 with the same boundary by the same tricks as in the proof of Theorem 4.1.12. Indeed, if there exists a self-opposite disc patch E or if an interior disc patch E admits an opposite companion disc E', then it suffices to switch all curves in $G_1 \cap G_2$ regular except ∂E which we switch irregular. By conclusion 1 of Lemma 4.1.2 or conclusion 1 of Lemma 4.1.5 and normalization, we get a simpler disc with the same boundary. Since every circle in D bounds a disc, "no companion disc" situation is impossible. Thus we can use Lemma 4.1.4 to present D in the form $D = F_1 + T$ and replace D with the simpler disc $D_1 = F_1$ having the same boundary.x

Now assume that $G_1 \cup G_2$ has no interior disc patches and has a clean boundary disc patch. Observe that since every proper arc in D divides D into two discs, every boundary disc patch admits an adjacent clean companion disc. By Lemma 4.1.4, one may present D in the form $D = F_1 + T$, where F_1 is a proper disc and T is either an annulus or a Möbius band. Recall that T consists of $k \geq 1$ quadrilateral patches A_i and F_1 is the union of k boundary clean disc patches E_i plus one exceptional patch. If $k > 1$ and the collection of proper discs $E_i \cup A_i \cup E_{i+1}, 0 \leq i \leq k - 1$, contains at least one essential disc, then we replace D by that disc and normalize it, thus obtaining a simpler essential compressing disc. If all the disc in the collection are inessential or if $k = 1$, then we replace D by F_1. In both cases we get a simpler essential compressing disc, since F_1 is admissibly isotopic to D for $k > 1$ and does not separate ∂M for $k = 1$.   □

Corollary 4.1.14. *There is an algorithm to decide whether a given irreducible 3-manifold M is a solid torus.*

Proof. Evident, since the solid torus is the only irreducible boundary reducible 3-manifold whose boundary is homeomorphic to $S^1 \times S^1$.   □

4.1.5 Is a Proper Surface Incompressible and Boundary Incompressible?

Theorem 4.1.15. *There is an algorithm to decide if a given two-sided surface in a 3-manifold M is incompressible.*

Proof. Denote by M_F the manifold obtained from M by cutting along F. Then ∂M_F contains two copies F_1, F_2 of F. Triangulate M_F such that $\partial F_1 \cup \partial F_2$ is the union of edges and supply M_F with the boundary pattern Γ composed of all edges in ∂M_F that have no common points with Int $(F_1 \cup F_2)$. It follows from the construction that (M_F, Γ) is boundary reducible if and only if F is compressible. It remains to apply Theorem 4.1.13.   □

Remark 4.1.16. There is no algorithm known to decide if a one-sided surface F is incompressible. However, injectivity of F can be recognized: It suffices to test the double of F for incompressibility.

To describe a recognition algorithm for boundary incompressibility of surfaces we need some preparation.

Lemma 4.1.17. *Let an incompressible boundary incompressible connected normal surface F in an irreducible boundary irreducible triangulated 3-manifold $(M\Gamma)$ be presented in the form $F = G_1 + G_2$. Assume that there are two clean patches $P_1 \subset G_1$, $P_2 \subset G_2$ such that the following holds:*

1. *P_1, P_2 are quadrilateral discs, each having exactly two opposite sides in ∂M.*
2. *These four sides bound two clean biangles in ∂M.*
3. *$P_1 \cap P_2$ consists of the other two opposite sides of the quadrilaterals. Those sides are not in ∂M.*

Then either F is admissibly isotopic to a surface of a smaller edge degree or the sum $G_1 + G_2$ is not in reduced form.

Proof. We say that a common side l of P_1, P_2 is of type **a**, if by the regular switch at l the patches are pasted together, and of type b, if not. Then we have four possibilities: aa, ab, ba, bb, see Fig. 4.9. Since F is connected, case aa is impossible. In Cases ab and ba replace the regular switch along the b-type arc by the irregular one. We get a new surface F' consisting of a compressible annulus that cuts off a clean ball from M, and of a surface F'' that is admissibly isotopic to F. The isotopy consists in deforming a portion of F through the ball. Normalizing F'', we get a normal surface admissibly isotopic to F and having a smaller edge degree.

Consider the Case bb. Replacing G_1 by $G_1' = (G_1 \setminus P_1) \cup P_2$ and G_2 by $G_2' = (G_2 \setminus P_2) \cup P_1$, i.e., merely switching $G_1 \cup G_2$ along the two segments $P_1 \cap P_2$ irregular, we get another presentation $F = G_1' + G_2'$ such that G_i' is admissible isotopic to $G_i, i = 1, 2$, and $G_1' \cap G_2'$ consists of fewer components than $G_1 \cap G_2$. This contradicts the assumption that $G_1 \cup G_2$ is in REDUCED form.   □

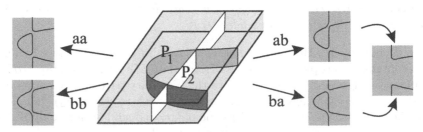

Fig. 4.9. (**aa**) F is not connected; (**ab, ba**) F is not minimal; (**bb**) F is not in reduced form

Recall that a proper disc $D \subset M$ is called *semiclean* if $\partial D \cap \Gamma$ consists of two points, see Definition 3.3.19. D is said to be *inessential*, if D is parallel rel ∂D to a disc $D' \subset \partial M$ whose intersection with Γ consists of one arc.

Lemma 4.1.18. *Let (M, Γ) be an irreducible boundary irreducible 3-manifold and L a collection of disjoint circles in Γ such that each edge of L separates two different components of $M \setminus \Gamma$. Then one can algorithmically decide, whether or not (M, Γ) contains an essential semiclean disc D such that both points of $\partial D \cap \Gamma$ lie in L.*

Proof. Denote by $\mathcal{D}$ the set of all essential semiclean normal discs in (M, Γ) whose boundaries intersect L in two points. Our goal is to construct an algorithm to decide whether or not $\mathcal{D}$ is nonempty. Let us choose a triangulation of M such that Γ is a subcomplex. We claim that $\mathcal{D} \neq \emptyset$ if and only if $\mathcal{D}$ contains a fundamental disc. Obviously, this is sufficient for proving the lemma.

To prove the claim, choose a disc $D \in \mathcal{D}$ having the minimal edge degree. Clearly, D is incompressible. Since L separates two different regions of $\partial M \setminus \Gamma$, D is boundary incompressible. Therefore we can normalize it by an admissible isotopy (see conclusion 4 of Proposition 3.3.24). Let us prove that the resulting essential semiclean normal disc D is fundamental.

Suppose, on the contrary, that there is a nontrivial presentation $D = G_1 + G_2$. By Lemma 3.3.30, we can assume that G_1, G_2 are connected and no double curve of $G_1 \cup G_2$ decomposes both surfaces. We can also assume that the presentation is in reduced form. Then by Lemma 4.1.8, $G_1 \cup G_2$ contains no clean disc patches.

Recalling that $\chi(D) = 1, \chi(G_i) \leq 2$, and the Euler characteristic is additive, we arrive at two options: 1=2+(-1) and 1=1+0. The first option is impossible, since then one of the surfaces G_i is a sphere, which must contain a disc patch. In the second case one of the surfaces (say, G_1) is a disc, the other one is either an annulus or a Möbius band. Moreover, both points in $D \cap L$ lie in G_1, since otherwise G_1 would contain a clean disc patch. It follows that G_1 is a semiclean disc, automatically inessential, since it is simpler than the minimal essential semiclean disc D.

The double curves decompose G_1 into two disc patches, each containing a point of $\partial G_1 \cap D$, and several clean quadrilaterals. Recall that by our assumption no double curve of $G_1 \cup G_2$ decomposes both surfaces. Since each proper curve on G_1 does decompose it, no double curve decompose G_2. It follows that all the patches of G_2 (which is either an annulus or a Möbius band) are clean quadrilaterals, each having two opposite sides in G_1, the other two in ∂M.

Let P_2 be a quadrilateral patch of G_2 contained in the ball B bounded by G_1 and a disc in ∂M. Suppose that P_2 is outermost with respect to B. Then the arcs $P_2 \cap G_1$ cut out a quadrilateral patch $P_1 \subset G_1$ from G_1, see Fig. 4.10. By construction, P_1, P_2 satisfy the assumption of Lemma 4.1.17. Applying it, we come to a contradiction with our assumption that $D = G_1 + G_2$ is minimal and in reduced form. This contradiction shows that D is fundamental. $\square$

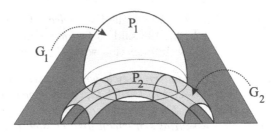

Fig. 4.10. P_1 and P_2 are two quadrilateral patches having a common pair of opposite sides

Theorem 4.1.19. *There is an algorithm to decide if a given incompressible two-sided surface in an irreducible boundary irreducible 3-manifold (M, Γ) is boundary incompressible.*

Proof. Denote by M_F the 3-manifold obtained from M by cutting along F. Let $F_1, F_2 \subset \partial M_F$ be two copies of F thus obtained. We supply M_F with the boundary pattern $\Gamma' = \Gamma_F \cup L'$, where $\Gamma_F \subset \partial M_F$ is obtained from Γ by cutting along ∂F, and $L' = \partial F_1 \cup \partial F_2$. Obviously, each edge of L' separates two different components of $\partial M_F \setminus \Gamma'$.

Let us show that F is boundary compressible if and only if (M_F, Γ') contains an essential semiclean disc D such that both points of $\partial D \cap \Gamma'$ lie in L'. Indeed, there is a natural map $\varphi \colon M_F \to M$ obtained by the reverse identification of F_1 with F_2. It is easy to see that if $D \subset M_F$ is as above, then $\varphi(D) \subset M$ is an essential boundary compressing disc for F. The converse implication is also easy, since cutting along F transforms any essential boundary compressing disc for F into an essential semiclean disc in M_F.

The desired algorithm for recognition of boundary incompressible surfaces can be now constructed as follows: we simply apply Lemma 4.1.18 to (M_G, Γ') and all the circles of $L' \subset \Gamma'$. It follows from the claim that F is boundary reducible if and only if we get at least one essential semiclean disc.   □

4.1.6 Is M^3 Sufficiently Large?

Definition 4.1.20. *A compact 3-manifold M is sufficiently large, if it contains a closed connected surface which is different from S^2, RP^2 and is incompressible and two-sided.*

Note that the exceptional surfaces listed above are the only closed surfaces having a finite fundamental group. Since every two-sided incompressible surface is injective (see Lemma 3.3.5), the fundamental group of every sufficiently large manifold is infinite.

The class of sufficiently large 3-manifolds is "sufficiently large." For example, every irreducible 3-manifold M with nonempty boundary is either a handlebody (orientable or not), or sufficiently large. To prove that, we define a *core* of a compact 3-manifold M. Recall that if F is a proper surface in M, then M_F denotes the 3-manifold obtained from M by cutting it along F. It is convenient to think of cutting as removing an open regular neighborhood of F and thus consider M_F as a submanifold of M.

Definition 4.1.21. *Let M be an irreducible 3-manifold with nonempty boundary. Then we call a compact 3-manifold $Q \subset \text{Int } M$ a core of M, if the following holds:*

1. *Q is boundary irreducible and no connected component of Q is a 3-ball.*
2. *Q is obtained from M by successive cutting along proper discs, removing all 3-ball components that might appear under this cutting, and removing an open collar of the resulting 3-manifold to get a submanifold of Int M.*

Remark 4.1.22. It follows from condition 1 that any continuation of the cutting process brings us nothing new. Since Q is boundary irreducible, each next cut results in the appearance of a new 3-ball and the removal of it. The isotopy class of Q remains the same.

Remark 4.1.23. Sometimes the following reformulation of condition 2 is more convenient: M can be obtained from Q by collaring ∂Q, adding disjoint 3-balls, and attaching handles of index 1, see Fig. 4.11. In particular, if Q is empty, then M is a handlebody (the converse is also true).

Remark 4.1.24. Let $D_1, D_2, \ldots, D_n$ be the sequence of discs that determines Q. Then each next disc is proper in the manifold obtained by cutting along the previous discs, but it may not be proper in M. Nevertheless, the discs can be modified by an isotopy so that afterward they are disjoint and proper in M. The core remains the same (modulo isotopy). Indeed, arguing by induction, we can assume that the first k discs $D_1, \ldots, D_k$ are already disjoint and proper in M. Then we simply shift the boundary ∂D_{k+1} of the next disc from the copies of $D_1, \ldots, D_k$ in the boundary of M_{k+1}, where M_{k+1} is the manifold obtained from M by cutting along $D_1, \ldots, D_k$.

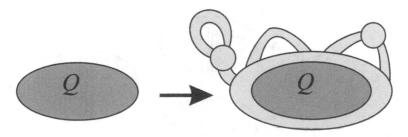

Fig. 4.11. Any 3-manifold can be obtained from its core Q by adding $\partial Q \times I$, disjoint 3-balls, and disjoint handles of index 1

Proposition 4.1.25. *Any irreducible 3-manifold has a core, which is unique up to isotopy.*

Proof. Let us construct a sequence $D_1, D_2, \ldots$ of disjoint proper discs in M. Each next disc D_k must be essential in the 3-manifold M_k obtained from $M_1 = M$ by cutting along $D_1 \cup \ldots \cup D_{k-1}$. If there appear 3-balls, we remove them at once. Let us observe that for each k we have the following:

1. $\chi(M_{k+1}) \geq \chi(M_k)$ and $\beta_1(M_{k+1}) \leq \beta_1(M_k)$, where χ is the Euler characteristic and β_1 is the first Betti number.
2. If $\chi(M_{k+1}) = \chi(M_k)$, then $\beta_1(M_{k+1}) < \beta_1(M_k)$.

Indeed, cutting M_k along D_k increases χ by one. We can get $\chi(M_{k+1}) = \chi(M_k)$ only if there appears a 3-ball, whose removal decreases χ by one. In this situation D_k, being essential, is a meridional disc of a component of M_k homeomorphic to a solid torus or to a solid Klein bottle and we have $\beta_1(M_{k+1}) = \beta_1(M_k) - 1$.

It follows that the process of constructing new essential discs and new 3-submanifolds without 3-ball components is finite and must stop with a boundary irreducible 3-manifold $M_n \subset M$, which admits no essential proper discs and contains no 3-ball components. Removing an open collar of ∂M_n in M_n, we get a core Q of M.

To prove that Q is unique up to isotopy, consider another core Q' with the defining system $D_1', D_2', \ldots, D_m'$ of discs which are disjoint and proper in M (see Remark 4.1.24). Let $\mathcal{D} = \cup_{i=1}^n D_i$ and $\mathcal{D}' = \cup_{j=1}^m D_j'$. Since M is irreducible, we can eliminate all circles in $\mathcal{D} \cap \mathcal{D}'$ and assume that the intersection consists of proper arcs. The arcs decompose the discs into smaller discs.

Let $M_{\mathcal{D}}, M_{\mathcal{D}'}$, and $M_{\mathcal{D} \cup \mathcal{D}'}$ be 3-manifolds obtained from M by cutting along, respectively, $\mathcal{D}, \mathcal{D}'$, and $\mathcal{D} \cup \mathcal{D}'$ (the latter can be considered as the union of those smaller discs). See Fig. 4.12. Since $M_{\mathcal{D}}$ is boundary irreducible, any disc in $\mathcal{D}' \cap M_{\mathcal{D}}$ is inessential. Therefore, cutting $M_{\mathcal{D}}$ along it preserves the isotopy class of Q. We can conclude that Q is isotopic to a submanifold $Q_0 \subset M$ obtained from $M_{\mathcal{D} \cup \mathcal{D}'}$ by throwing away 3-ball components and

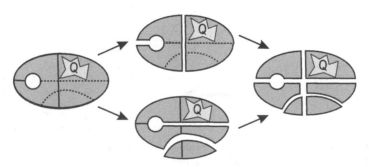

Fig. 4.12. Cutting M along $\mathcal{D}$ (*solid lines*) and along $\mathcal{D}'$ (*dotted lines*)

removing an open collar of the boundary of the resulting manifold. The same is true for Q'. Therefore, Q and Q' are isotopic. $\qquad\square$

Lemma 4.1.26. *Let $Q \subset Int\ M$ be the core of an irreducible 3-manifold M. Then ∂Q is incompressible in M.*

Proof. Consider an arbitrary compressing disc $\Delta \subset M$ for ∂Q. We wish to prove that Δ is inessential, that is, its boundary curve bounds a disc in ∂Q. If Δ lies in Q, then it is inessential by condition 1 of Definition 4.1.21.

Suppose that Δ lies outside Q. Let $\mathcal{D} = D_1 \cup \ldots \cup D_n$ be a collection of disjoint proper discs in M which decomposes M into 3-balls and a copy M_n of Q. Since M is irreducible, we can deform Δ isotopically away from $\mathcal{D}$. Then Δ is contained in the collar $M_n \subset Int\ Q = \partial Q \times I$ and hence must be inessential in this case as well. $\qquad\square$

Corollary 4.1.27. *Every irreducible 3-manifold M with nonempty boundary is either a handlebody or sufficiently large.*

Proof. Follows from Proposition 4.1.25 and Lemma 4.1.26: if M is not a handlebody, then the boundary of the core of M is an incompressible surface in M different from S^2 and RP^2. $\qquad\square$

Definition 4.1.28. *A compact 3-manifold M is called* Haken, *if it is irreducible, boundary irreducible, and sufficiently large.*

It follows from Corollary 4.1.27 that every irreducible boundary irreducible 3-manifold M with nonempty boundary is either a 3-ball or Haken. So all interesting non-Haken 3-manifolds are closed. Examples of such manifolds can be found among Seifert manifolds fibered over S^2 with three exceptional fibers. Some closed hyperbolic 3-manifolds also possess this property, among them the manifolds M_1, M_2 described in Sect. 2.5.1. In general, if you take a closed irreducible 3-manifold by chance, then most probably it would be Haken.

Lemma 4.1.29. *Let M be a closed irreducible 3-manifold such that the group $H_1(M; Z)$ is infinite and M contains no projective planes. Then M is Haken.*

Proof. Since $H_1(M; Z)$ is infinite, there exists a map $f: M \to S^1$ such that the induced homomorphism $f_*: \pi_1(M) \to H_1(M; Z)$ is surjective. Then the inverse image of any regular point $a \in S^1$ is a nonseparating surface. Compressing it as long as possible, we get an incompressible surface F. Since the property of a surface to be nonseparating is preserved under compressions, F is nonempty. It follows from our assumption that F is different from S^2, RP^2. $\qquad\square$

In this section we present the following important result of Jaco and Oertel [56].

Theorem 4.1.30. *There exists an algorithm to decide if an irreducible manifold is sufficiently large.*

Since irreducibility and boundary irreducibility of a 3-manifold can be recognized algorithmically (see Theorem 4.1.13), Corollary 4.1.31 is evident.

Corollary 4.1.31. *There exists an algorithm to decide if an irreducible manifold is Haken*

The ideas used in the proof of Theorem 4.1.30 influenced significantly the development of the normal surface theory and turned out to be very important for analyzing the complexity of algorithms based on it. The proof follows the same General Scheme, but before presenting it we prove several lemmas that are useful not only for this particular proof.

Let (M, Γ) be a triangulated 3-manifold. As before, we will measure the complexity of a general position proper surface $F \subset M$ by its edge degree $e(F)$. We slightly generalize the notion of compressing disc for a surface by extending it to the case of arbitrary subpolyhedra. Let X be a compact 2-dimensional subpolyhedron of a 3-manifold (M, Γ). By a *singular graph* $S(X)$ of X we mean the union of all points in X having no disc neighborhood.

Definition 4.1.32. *A disc* $D \subset M$ *is called* a compressing disc *for* X *if* $D \cap X = \partial D$ *and the curve* ∂D *is transversal to* $S(X)$. *Similarly, a clean disc* $D \subset M$ *is* a boundary compressing disc *for* X *if* $l = D \cap X$ *is an arc in* ∂D, l *is transversal to* $S(X)$, *and* $D \cap \partial M$ *is the complementary arc in* ∂D.

Let $F = G_1 + G_2$ be a normal surface in a 3-manifold M presented as a geometric sum of two normal surfaces, and let l be a component of $G_1 \cap G_2$. Locally, in a close neighborhood of every point $x \in l$, the surfaces G_1, G_2 look like two planes (or half-planes, if $x \in \partial M$) forming four dihedral angles. A dihedral angle is called *good*, if the patches forming its sides are pasted together under the regular switch at l. Otherwise the angle is *bad*.

Any compressing disc D for $G_1 \cup G_2$ can be considered as a curvilinear polygon whose angles are labeled by letters g or b depending on whether the corresponding dihedral angles are good or bad, respectively. Analogously, any boundary compressing disc D for $G_1 \cup G_2$ is a polygon whose all but two angles are labeled by letters g or b (each of the two angles without labels is formed by an arc in $D \cap \partial M$ and an arc in $D \cap G_1$ or $D \cap G_2$). Let us describe the behavior of D with respect to the regular switch of $G_1 \cup G_2$ and to the resulting surface $F = G_1 + G_2$. Near each double line l of $G_1 \cup G_2$, consider a strip A that runs along l and spans the trace curves of l. If l is closed, then A is either an annulus (if l is two-sided in both G_1, G_2) or a Möbius band (if l is one-sided). In case l is a proper arc, A is a disc band (homeomorphic image of a rectangle) such that two opposite sides of A are in ∂M while the other two coincide with the trace curves of l.

Consider the union of F with all strips obtained in this way. Then D determines a compressing or boundary compressing disc $\tilde{D}$ for the union. Near each good angle of D, the boundary curve of $\tilde{D}$ is contained in F while

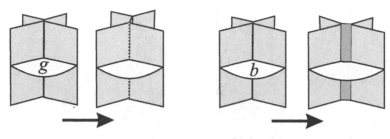

Fig. 4.13. Behavior of compressing discs near good and bad angles

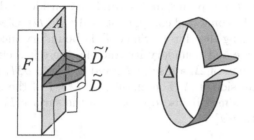

Fig. 4.14. Parallel compressing discs for $F \cup A$ and a compressing disc Δ for F

near each bad angle it crosses the corresponding strip. See Fig. 4.13. If all angles of D are good, then $\tilde{D}$ is a compressing or boundary compressing disc for F.

Lemma 4.1.33. *Let an incompressible normal surface F in an irreducible boundary irreducible manifold (M, Γ) be presented in the form $F = G_1 + G_2$ such that there are no clean disc patches. Assume that $G_1 \cup G_2$ admits a compressing disc D such that precisely one angle of D is bad. Then the edge degree of F can be decreased by an admissible isotopy.*

Proof. Denote by l the double line of $G_1 \cup G_2$ that passes through the vertex of the bad angle of D. Let A and $\tilde{D}$ be the corresponding band and compressing disc for $F \cup A$. Assume that l is closed. Denote by B' the ball bounded by $\tilde{D}$, a close parallel copy $\tilde{D}'$ of $\tilde{D}$ (see Fig. 4.14), and the portions A' of A and F' of F between them. Consider the disc $\Delta \subset M$ composed of $\tilde{D}$, $\tilde{D}'$, and the band $\mathrm{Cl}(A \setminus A')$, see Fig. 4.14 to the right. It is easy to see that Δ is a compressing disc for F. Since F is incompressible and M is irreducible, there is a ball $B \subset M$ bounded by Δ and a disc δ in F. It cannot happen that $B' \subset B$, otherwise each trace curve of l would be contained in δ and bound a disc in $F \cap \delta \subset F$, which contradicts the "no clean disc patch" assumption. Therefore $B' \cap B = \tilde{D} \cup \tilde{D}'$, and the union $B \cup B'$ is a solid torus T. This solid torus helps us to construct an isotopy of F to the surface $F_1 = (F \setminus \partial T) \cup A$ which is simpler than F (see Fig. 4.15).

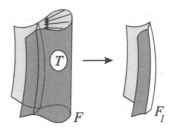

Fig. 4.15. Simplifying isotopy

The case when l is a proper arc is similar. Consider a disc $\Delta \subset (M, \Gamma)$ composed of $\tilde{D}$ and a connected component of $A \setminus (A \cap \tilde{D})$. Clearly, Δ is a clean boundary compressing disc for F. Since F is boundary incompressible and (M, Γ) is irreducible and boundary irreducible, there is a clean ball $B \subset M$ bounded by the union of Δ, a disc in F, and a disc in ∂M. A similar ball is placed on the other side of Δ. The union $T \sim \tilde{D} \times I$ of these two balls helps us to construct an admissible isotopy of F to the surface $F_1 = (F \setminus \partial T) \cup A$ which is simpler than F. □

Lemma 4.1.34. *Let an incompressible boundary incompressible normal surface F in a manifold (M, Γ) be presented in the form $F = G_1 + G_2$ such that there are no clean disc patches. Assume that $G_1 \cup G_2$ admits a compressing or boundary compressing disc D such that D has no bad angles and at least one good angle. Then for $i = 1, j = 2$ or for $i = 2, j = 1$ there exists a disc $D_0 \subset G_i$ such that ∂D_0 consists of an arc in G_j, an arc in ∂D, and maybe an arc in ∂M while $\text{Int}\, D_0$ has no common points with $G_j \cup D \cup \partial M$.*

Proof. Since there are no bad angles, D determines a compressing or boundary compressing disc $\tilde{D}$ for F. The curve $\alpha = \partial \tilde{D} \cap F$ must cut off a disc $\tilde{D}'$ from F. Recall that trace curves decompose F into patches. Consider the induced decomposition of $\tilde{D}'$. The boundary of every region of the induced decomposition consists of arcs contained in trace curves and of arcs in $\partial \tilde{D}'$. The "no clean disc patch" assumption assures us that $\tilde{D}'$ contains no closed trace curves and proper trace arcs having both endpoints on ∂M. Therefore there exists a region bounded by a trace arc, an arc in α, and maybe an arc in ∂M. Any such region (in Fig. 4.16 they are marked by stars) determines a disc D_0 satisfying the conclusion of the lemma. □

Let a proper normal surface F in a triangulated 3-manifold M be presented in the form $F = G_1 + G_2$ and let Δ be a compressing or boundary compressing disc for G_1. We will assume that Δ is in general position with respect to G_2. Then $\Delta \cap G_2$ is a proper 1-dimensional submanifold of Δ. Define *the weight* of Δ to be the number $c(\Delta) + c_\partial(\Delta)$, where $c(\Delta)$ is the number of components of $\Delta \cap G_2$ and $c_\partial(\Delta)$ is the number of points in $\partial \Delta \cap G_2$ which are not on ∂M. Our goal is to decrease the weight.

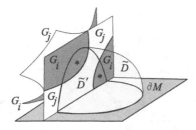

Fig. 4.16. Trace curves decompose F as well as $\tilde{D}'$

Lemma 4.1.35. *Let $F = G_1 + G_2$ be a proper normal surface in a triangulated irreducible boundary irreducible 3-manifold (M, Γ) such that all patches of F are incompressible and boundary incompressible. Let Δ be an essential compressing or boundary compressing disc for G_1. Assume that there is a disc $D_0 \subset M$ such that at least one of the following holds:*

(1) *$D_0 \subset \Delta$ and either $D_0 \cap G_2 = \partial D_0$ or $D_0 \cap G_2$ is an arc in ∂D_0 while the complementary arc of ∂D_0 is in ∂M.*
(2) *$D_0 \subset G_1$, $\partial D_0 = D_0 \cap (G_2 \cup \Delta \cup \partial M)$, and ∂D_0 consists of an arc in G_2, an arc in $\partial \Delta$, and maybe an arc in ∂M.*
(3) *$D_0 \subset G_2$ and ∂D_0 consists of an arc in G_1, an arc in $\partial \Delta$, and maybe an arc in ∂M, while Int D_0 is disjoint from $G_1 \cup \Delta \cup \partial M$.*

Then there is an essential compressing or boundary compressing disc Δ' for G_1 having a strictly smaller weight.

Proof. CASE 1. It follows from the assumptions that $\partial D_0 \cap G_2$ is contained in a patch $P \subset G_2$ of $G_1 \cup G_2$. Since P is incompressible and boundary incompressible, the curve $\partial D_0 \cap G_2$ cuts off a clean disc D from P. Recall that M is irreducible and boundary irreducible. It follows that $D \cup D_0$ cuts off a clean ball from M. Using the ball, construct an isotopy of D_0 to the other side of P and get a new disc Δ' having a smaller weight.

CASE 2. We use D_0 for constructing an isotopy of Δ that shifts the arc $l = D_0 \cap \Delta$ to the other side of G_2. Clearly, the isotopy decreases the weight. See Fig. 4.17a,b for the cases $\partial l \cap \partial M = \emptyset$ and $\partial l \cap \partial M \neq \emptyset$.

CASE 3. Compressing Δ along D_0, we get two simpler compressing or boundary compressing discs for G_1 with strictly smaller weights. It is clear that if Δ is essential, then so is at least one of the new discs. See Fig. 4.18. □

Theorem 4.1.36. *Let a minimal connected normal surface F in an irreducible boundary irreducible 3-manifold (M, Γ) be presented in the form $F = G_1 + G_2$. If F is incompressible and boundary incompressible, then so are G_1, G_2. Moreover, neither G_1 nor G_2 is a sphere, a projective plane, a clean disc or a disc whose boundary crosses Γ exactly once.*

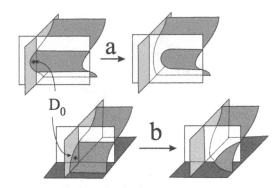

Fig. 4.17. Simplifying Δ by shifting $\partial\Delta$ along G_1

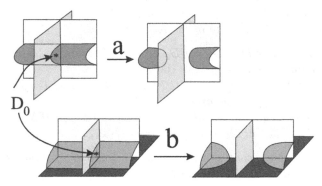

Fig. 4.18. Compressing Δ

Proof. It is sufficient to prove the theorem with the additional assumption that $F = G_1 + G_2$ is in reduced form. Then Lemma 4.1.8 tells us that $G_1 \cup G_2$ has no clean disc patches and all the patches of $G_1 \cup G_2$ are incompressible and boundary incompressible. Since any decomposition of S^2, RP^2, or D^2 with no more than one point in $\partial D^2 \cap \Gamma$ into patches contains a clean disc patch, G_1, G_2 are different from these surfaces. By the symmetry, we only need to prove that one surface, say, G_1, is incompressible and boundary incompressible.

On the contrary, suppose that G_1 admits an essential compressing or boundary compressing disc Δ. The idea is to decrease the weight $c(\Delta)+c_\partial(\Delta)$.

Assume that $\Delta \cap G_2$ contains a circle or an arc with the endpoints on ∂M. Using an innermost circle or an outermost arc argument, we find a disc $D_0 \subset \Delta$ satisfying condition (1) of Lemma 4.1.35. Applying the lemma, we find a simpler compressing or boundary compressing disc. So for the remainder of the argument we may assume that $\Delta \cap G_2$ contains no circles and no arcs with endpoints in ∂M. It follows that the arcs $\Delta \cap G_2$ cut Δ into regions which are homeomorphic to the standard disc and thus can be considered as curvilinear polygons. Every region is a compressing or boundary compressing disc for $G_1 \cup G_2$.

Let us count the number of good and bad angles in these polygons. Any point in $\partial \Delta \cap G_2$ is a common vertex of two angles belonging to different polygons. One of the angles is good, the other is bad. We relate the total number of polygons in Δ and the total number of their bad angles. Denote by m the number of arcs in $\Delta \cap G_2$ that are disjoint from ∂M, and by n the number of arcs having one endpoint in ∂M. Then there are $m+n+1$ polygons and $2m+n$ bad marks. Since $2(m+n+1) > 2m+n$, there is a polygon having ≤ 1 bad angles.

Note that there are no polygons having precisely one bad angle; otherwise F would be not minimal by Lemma 4.1.33. Therefore there is a polygon D that has no bad angles. By Lemma 4.1.35, there exists a disc $D_0 \subset M$ satisfying either assumption (2) or assumption (3) of Lemma 4.1.35. Applying the lemma, we get a simpler essential compressing or boundary compressing disc again.

Continuing this simplification procedure for as long as possible, we get an essential compressing or boundary compressing disc Δ for G_1 that has zero weight and thus does not intersect G_2. This contradicts the assumption that all patches of $G_1 \cap G_2$ are incompressible and boundary incompressible.   □

Corollary 4.1.37. *Let a minimal connected normal surface F in an irreducible boundary irreducible 3-manifold (M, Γ) be presented as a sum $F = \sum_{i=1}^{n} G_i$ of $n > 0$ nonempty normal surfaces. If F is incompressible and boundary incompressible, then so are all G_i. Moreover, no G_1 is a sphere, a projective plane, a clean disc or a disc whose boundary crosses Γ exactly once.*

Proof. Rewrite F in the form $F = G_1 + G'$, where $G' = \sum_{i=2}^{n} G_i$. By Theorem 4.1.36, G_1 is incompressible, boundary incompressible, and different from S^2, D^2, and RP^2. The same trick works for all other surfaces G_i.   □

Remark 4.1.38. If the surface F in the statements of Theorem 4.1.36 and Corollary 4.1.37 is closed, then the assumption that M is boundary irreducible is superfluous. Indeed, we have used it in the proofs of the theorem, corollary, and preceding lemmas only when $\partial F \neq \emptyset$ (if $\partial F = \emptyset$, then all events are going on strictly inside M).

Proof of Theorem 4.1.30 (On Recognizing Sufficiently Large Manifolds). Let us triangulate a given manifold M and construct the finite set $\{G_i, 1 \leq i \leq n\}$ of all closed fundamental surfaces. Suppose that M contains a two-sided closed incompressible surface $F \neq S^2, RP^2$. We may assume that F is minimal. Present F as a sum $F = \sum_{i=1}^{n} k_i G_i, k_i > 0$, of some fundamental surfaces. It follows from Theorem 4.1.36 that all surfaces $G_i, 1 \leq i \leq n$, are incompressible and different from S^2 and RP^2. However, this is not good enough to develop the desired algorithm. It may be not true that at least one of G_i is two-sided.

To overcome this obstacle, consider the presentation $2F = \sum_{i=1}^{n} k_i(2G_i)$. Since $2F$ is also minimal and incompressible, all surfaces $2G_i$ are incompressible by Corollary 4.1.37. Certainly, they are two-sided.

The desired algorithm can be described as follows. We apply the algorithm of Corollary 4.1.15 to the check whether the double $2G$ of each fundamental surface $G \neq S^2, RP^2$ is incompressible. If we get a positive answer at least once (i.e., if we find a fundamental surface $G \neq S^2, RP^2$ with an incompressible double $2G$), then M is clearly sufficiently large. If all the doubles turn to be compressible, then M is not sufficiently large.   □

Remark 4.1.39. The algorithm constructed above can be sharpened by reducing the set of all fundamental surfaces to a much smaller subset of so-called *vertex surfaces*. Every vertex surface G_j corresponds to an admissible vertex solutions $\bar{V}_j$ to the matching system E for the triangulation of M (see the proof of Theorem 3.2.8 for a description of vertex surfaces). Any integer solution $\bar{x}$ to E can be written in the form $\bar{x} = \sum_{j=1}^{N} \alpha_j \bar{V}_j$, where α_j are non-negative rational numbers. Obviously, if $\bar{x}$ is admissible, then so are all $\bar{V}_j$ with $\alpha_j > 0$. It follows that for every normal surface F an appropriate multiple $2kF$ can be presented in the form $2kF = \sum_{j=1}^{m} k_j(2G_j)$, where all k_j are integer and G_j are vertex surfaces. If F is minimal and incompressible, then Corollary 4.1.37 tells us that all these vertex surfaces are incompressible. Therefore M is sufficiently large if and only if the double of at least one vertex surface is incompressible.

4.2 Cutting 3-Manifolds along Surfaces

In this section we investigate what happens to the complexity of a 3-manifold when we cut the manifold along an incompressible surface.

4.2.1 Normal Surfaces and Spines

Let M be a 3-manifold with nonempty boundary. There is a close relationship between handle decompositions and spines of M. Indeed, let ξ be a handle decomposition of M into balls, beams, and plates, without handles of index 3. Collapsing the balls, beams, and plates of ξ onto their core points, arcs, and discs, we get a spine P of M. By construction, P is equipped with a natural cell decomposition into the core cells of the handles. Conversely, let P be a *cellular spine* of M, i.e., a spine equipped with a cell decomposition. Replace each vertex of P by a ball, each edge by a beam, and each 2-cell by a plate (we used a similar construction in Sect. 1.1.4). We get a handle decomposition ξ of a regular neighborhood $N(P)$ of P in M. Since $\partial M \neq \emptyset$, $N(P)$ can be identified with the whole manifold M, so ξ can be considered as a handle decomposition of M.

Consider a normal surface $F \subset M$ and denote by M_F the 3-manifold obtained by cutting M along F. Let us investigate the behavior of ξ and P under the cut.

Since F is normal, it decomposes the handles of ξ into handles of the same index. The new handles form a handle decomposition ξ_F of M_F. Denote by P_F the corresponding cellular spine of M_F. We can think of each handle of ξ_F as being contained in the corresponding handle of ξ. This inclusion relation induces a cellular map $\varphi\colon P_F \to P$ (a map is *cellular* if it takes cells to cells). For any vertex v of P_F, the map φ, being cellular, induces a map $\varphi_v\colon \mathrm{lk}(v, P_F) \to \mathrm{lk}(w, P)$ between links, where $w = \varphi(v)$.

Another way for describing φ_v consists in the following. Collapsing each island in ∂B_v to its core point and each bridge to its core edge, we get a graph isomorphic to $\mathrm{lk}(v, P_F)$. The same is true for the ball B of ξ containing w: the island-beam configuration in ∂B has the shape of $\mathrm{lk}(w, P)$. Then φ_v is induced by the inclusion relation between islands and bridges of ξ_F and ξ.

Lemma 4.2.1. *Suppose that a handle decomposition ξ of a 3-manifold M with nonempty boundary corresponds to an almost simple cellular spine P of M. Let F be a normal surface in M, and let ξ_F, P_F, and φ be, respectively, the induced handle decomposition of M_F, cellular spine of M_F, and cellular map $P_F \to P$. Then P_F is almost simple, and for any vertex v of P_F and the corresponding vertex $w = \varphi(v)$ of P the induced map $\varphi_v\colon \mathrm{lk}(v, P_F) \to \mathrm{lk}(w, P)$ is an embedding.*

Proof. Let v, $w = \varphi(v)$ be vertices of P_F, P, and $B_v \subset B_w$ the balls of ξ_F, ξ containing them. We observe the following:

(a) *Any bridge of B_v is contained in a bridge of B_w, and any bridge of B_w contains no more than one bridge of B_v.* The first statement is evident. Let us prove the second (it is true for all handle decompositions, not only for those arising from almost simple spines).

Suppose, on the contrary, that a bridge b of B_w contains two bridges b', b'' of B_v. We can assume that b', b'' are neighbors, that is, the strip $S \subset b$ between them contains no other bridges of B_v. Note that the region $U = \partial B_v \cap \partial B_w$, just as every connected region in $\partial B_w \approx S^2$ bounded by disjoint circles, has exactly one common circle with each connected component of $\mathrm{Cl}(\partial B_w \setminus U)$. It follows that the lateral sides of S lie in the same circle $C \subset \partial U$. Since C is the boundary of an elementary disc in $F \cap B_w$ and since the boundary of any elementary disc crosses each bridge no more than once, we get a contradiction.

(b) *Any island of B_v is contained in an island of B_w, which contains no other islands of B_v.* Since ξ corresponds to an almost simple spine and thus all islands of ξ have valence ≤ 3, the boundary curve of any elementary disc passes through any island no more than once (otherwise condition 7 of Definition 3.4.1 of a normal surface would be violated). Thus the same proof as in (a) does work.

Since φ_v is induced by the inclusion relation between islands and bridges of ξ_F and ξ, it follows from (a), (b) that φ_v is an embedding. Since it is true for all vertices of P_F and P is almost simple, P_F is also almost simple. $\square$

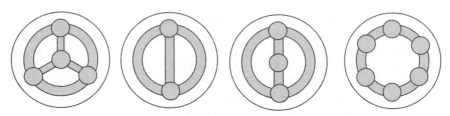

Fig. 4.19. Island-bridge-lake configurations in the boundaries of balls

Suppose P is a simple spine of a 3-manifold M with nonempty boundary. It is convenient to choose a cell decomposition σ of P such that any true vertex of P is incident to only four 1-cells, at any vertex of σ inside a triple line only two or three 1-cells meet together, and inside 2-components there are no vertices of σ incident to only one 1-cell. Denote by ξ the handle decomposition of M induced by σ.

et v be a vertex of σ. Denote by B_v the corresponding ball. The boundary of B_v is decomposed into islands, bridges, and lakes. If v is a true vertex of P, then ∂B_v contains 4 islands, 6 bridges, and 4 lakes such that any two islands are joined by one bridge. If v is a triple point, then ∂B_v contains two or three islands, and if v is a nonsingular point, then the islands and bridges compose an annulus, see Fig. 4.19.

Let us investigate the types of elementary discs in B_v.

Definition 4.2.2. *An elementary disc $D \subset B_v$ has type (k, m) if its boundary curve ℓ intersects k bridges and m lakes (recall that by definition of an elementary disc, ℓ passes through each bridge and each lake no more than once).*

Lemma 4.2.3. *Let a ball B_v of ξ corresponds to a vertex v of a simple spine P. Then any elementary disc in B_v has one of the following types:*

1. *Types $(4,0), (3,0), (2,1), (1,2), (0,4), (0,3), (0,2)$, if v is a true vertex.*
2. *Types $(2,0), (1,1), (0,2)$, if v lies on a triple line and ∂B_v contains two islands.*
3. *Types $(3,0), (2,1), (2,0), (1,1), (0,2), (0,3)$, if v lies on a triple line and ∂B_v contains three islands.*
4. *Types $(0,2)$ and $(k,0)$, if v is a nonsingular vertex and ∂B_v contains k islands.*

With a few exceptions, each type determines the corresponding elementary disc in a unique way up to homeomorphisms of B_v taking islands to islands, bridges to bridges, and lakes to lakes. The exceptions are:

1. *If v is a true vertex, then B_v contains two elementary discs of the type $(0,3)$.*
2. *If v is a triple vertex and B_v contains three islands, then there are two elementary discs of the type $(0,2)$.*

3. *If v is a nonsingular vertex of valence k, then there is one elementary disc of the type $(k, 0)$ and $[k/2]$ discs of the type $(0, 2)$, where $[k/2]$ is the integer part of $k/2$.*

Proof. Let ℓ be the boundary curve of a type (k, m) elementary disc. Then $n = k + m$ is the number of arcs in the intersection of ℓ with the union of all islands. Since each island has valence ≤ 3, ℓ visits each islands no more than once (otherwise condition 7 of Definition 3.4.1 would be violated). Moreover, at least two islands must be visited.

CASE 1. If v is a true vertex, then B_v contains four islands. Therefore, $2 \leq k + m \leq 4$. It remains to enumerate all possible pairs (k, m) with $2 \leq k + m \leq 4$ and verify that only the pairs listed in the statement of Lemma 4.2.3 are realizable by elementary discs, and that each of them admits a unique realization except the pair $(0,3)$ that admits two, see Fig. 4.20.

CASE 2. If v is a triple vertex such that B_v contains two islands, then $k + m = 2$. It is easy to verify that only the pairs listed in item 2 are realizable and that the realizations are unique.

CASE 3 is similar. The only difference is that there are two elementary discs of the type $(0, 2)$, see Fig. 4.21.

In the last CASE 4 of a nonsingular vertex v the proof is evident. For the annular island-bridge configuration with $k = 6$ islands presented on Fig. 4.21 to the right we show all three discs of type $(0, 2)$.   □

Our next goal is to show that in many cases cutting a 3-manifold M along an incompressible surface makes M simpler.

Lemma 4.2.4. *Suppose that a handle decomposition ξ of a 3-manifold M with nonempty boundary corresponds to a simple cellular spine P of M. Let F be a connected normal surface in M, and let ξ_F, P_F, and φ be, respectively, the induced handle decomposition of M_F, cellular spine of M_F, and cellular map $P_F \to P$. Then the following holds:*

1. *φ embeds the set of true vertices of P_F into the one of P. This embedding (denote it by φ_0) is bijective if and only if all the elementary discs of F in the balls around true vertices have type $(3, 0)$.*
2. *If φ_0 is bijective, then φ embeds the union of triple circles of P_F into the one of P. This embedding (denote it by φ_1) is bijective if and only if all elementary discs of F in the balls around triple vertices have type $(3, 0)$ or $(2, 0)$.*
3. *If φ_0 and φ_1 are bijective and $\partial F \neq \emptyset$, then F is either an annulus or a Möbius band intersecting only balls of ξ_P that correspond to nonsingular vertices. Moreover, F can intersect these balls only along elementary discs of type $(0, 2)$ and $F \cup P$ is a middle circle of F.*

Proof. Let us prove the first conclusion of the lemma. Let v_1 and $v = \varphi(v_1)$ be true vertices and $B_1 \subset B$ the corresponding balls. Then $\mathrm{lk}(v_1, P_F)$ and

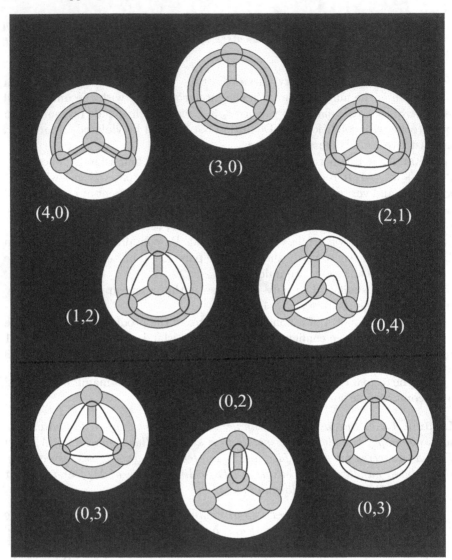

Fig. 4.20. Eight types of elementary discs in a ball neighborhood of a true vertex

$\mathrm{lk}(v, P)$ are homeomorphic. We claim that every elementary disc $D \subset F \cap B$ has type $(3, 0)$. Obviously, the first conclusion follows from the claim.

To prove the claim, suppose that D has type (k, m) with $m > 0$. Then cutting B along D destroys at least one three-valent vertex of $\mathrm{lk}(v, P)$, namely, the one corresponding to the island where ∂D crosses the coast of a lake. Since $\mathrm{lk}(v_1, P_F)$ has four three-valent vertices and $\mathrm{lk}(v_1, P_F)$ and $\mathrm{lk}(v, P)$ are homeomorphic, it is impossible. Cutting along type $(4, 0)$ disc preserves

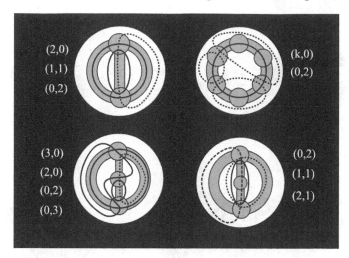

Fig. 4.21. The boundaries of elementary discs in ball neighborhoods of other vertices

three-valent vertices, but decomposes lk(v, P) into two graphs, each containing two three-valent vertices. Therefore this case is also impossible.

This observation can be easily verified by applying Lemma 4.2.3 and considering Fig. 4.22, which shows why type $(3, 0)$ discs preserve the true vertex and how discs of all other types destroy it.

The proof of the second conclusion of the lemma is similar. Denote by E the union of all true vertices of P with all the triple edges that join them. Let E_F be the similar union for P_F. Suppose that φ_0 is bijective. Then φ induces a homeomorphism between E_F and E. Let us investigate the behavior of φ on the set of triple circles. It follows from Lemma 4.2.3 that the image $\varphi(C)$ of a triple circle $C \subset P_F$ is a triple circle in P if and only if all the elementary discs in the balls around vertices of $\varphi(C)$ have types $(2, 0)$ or $(3, 0)$. Indeed, cutting along such discs preserves the triple circle while cutting along discs of type (k, m) with $m > 0$ destroys it.

Let us prove the last conclusion of the lemma. Suppose that φ_0 and φ_1 are bijective. It means that φ takes the union of singular points of P_F onto the one of P homeomorphically. Then ∂F cannot intersect the balls around true and triple vertices, since otherwise at least one of them would not survive the cut. It follows that ∂F is contained in the union of balls and beams of ξ that correspond to the vertices and edges of P inside a 2-component of P. Each such ball B can be presented as $D^2 \times I$ such that $D^2 \times \partial I$ are the lakes of B. Suppose that B has at least one common point with ∂F. Then any elementary disc $D \subset F \cap B$ has type $(2, 2)$. Therefore D is a quadrilateral having two opposite sides in the islands of B, the other two in the lakes. Similarly, if ∂F passes along a beam $D^2 \times I$, then each strip in $F \cap (D^2 \times I)$ has two opposite sides in the islands, the other two in ∂M.

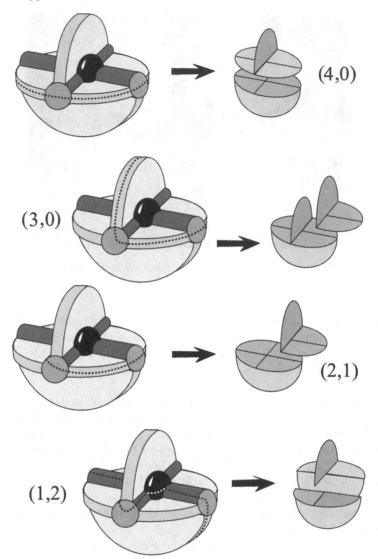

Fig. 4.22. Cutting along discs of all the types except (3,0) destroys the true vertex (see continuation)

Consider the union F_0 of all such quadrilaterals and strips. Since each quadrilateral intersects exactly two strips along its two island sides, F_0 is the disjoint union of annuli and Möbius bands. On the other hand, $\partial F_0 = \partial F$. Taking into account that F is connected, we conclude that $F = F_0$ and F_0 is either an annulus or a Möbius band intersecting P along its middle circle. □

Recall that if P is an almost simple polyhedron, then $c(P)$ denotes the complexity of P, i.e., the number of its true vertices.

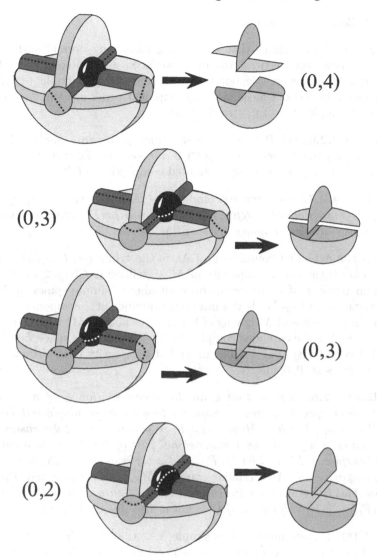

Fig. 4.22. Cutting along discs of the all types except (3,0) destroys the true vertex (continued from the previous page)

Corollary 4.2.5. *Suppose that a handle decomposition ξ of a 3-manifold M with nonempty boundary corresponds to an almost simple cellular spine P of M. Let F be a connected normal surface in M, and let ξ_F, P_F, and φ be, respectively, the induced handle decomposition of M_F, cellular spine of M_F, and cellular map $P_F \to P$. Then $c(P_F) \leq c(P)$ and $c(P_F) = c(P)$ if and only if all elementary discs of F in the balls around true vertices have type $(3,0)$.*

Proof. Follows from item 1 of Lemma 4.2.4.   □

Corollary 4.2.5 tells us that by cutting along any normal surface there appear no new true vertices. However, we need a more complete information. To any almost simple polyhedron we associate a triple (c, c_1, c_2) of non-negative integers, which, in lexicographical ordering, will measure the "extended complexity" of the polyhedron.

Definition 4.2.6. *Let P be an almost simple polyhedron having $c(P)$ true vertices, $c_1(P)$ triple circles, and $c_2(P)$ 2-components. Then the triple $\bar{c}(P) = (c(P), c_1(P), c_2(P))$ is called the* extended complexity *of P.*

Definition 4.2.7. *The* extended complexity *$\bar{c}(M) = (c(M), c_1(M), c_2(M))$ of a compact 3-manifold M is defined as $\bar{c}(M) = min_P \bar{c}(P)$, where the minimum is taken over all almost simple spines of M.*

Thus the extended complexity of M is the triple $(c(M), c_1(M), c_2(M))$, where $c(M)$ is the usual complexity of M as defined in Chap. 2, $c_1(M)$ is the minimum number of triple circles over all almost simple spines of M with $c(M)$ vertices, and $c_2(M)$ is the minimum number of 2-components over all almost simple spines of M having $c(M)$ vertices and $c_1(M)$ triple circles. For example, S^3 has the extended complexity $(0, 0, 0)$ and lens space $L_{3,1}$ has the extended complexity $(0, 1, 1)$. The extended complexity of all I-bundles over closed surfaces is $(0, 0, 1)$.

Corollary 4.2.8. *Suppose that a handle decomposition ξ of a 3-manifold M with nonempty boundary corresponds to an almost simple cellular spine $P = H \cup G$ of M, where H is a simple polyhedron (the 2-dimensional part of P) and G is a graph (the 1-dimensional part of P). Let F be a connected normal surface in M, and let ξ_F, P_F, and φ be, respectively, the induced handle decomposition of M_F, cellular spine of M_F, and cellular map $P_F \to P$. Suppose that $\partial F \neq \emptyset$. Then P_F can be collapsed onto a spine P'_F of M such that $\bar{c}(P'_F) \leq \bar{c}(P)$. Moreover, if F is not a disc, then $\bar{c}(P'_F) < \bar{c}(P)$.*

Proof. STEP 1. Assume that P is simple, i.e., that $P = H$. If $c(P_F) < c(P)$ or if $c(P_F) = c(P)$ and $c_1(P_F) < c_1(P)$, we are done. Otherwise we are in the situation of item 3 of Lemma 4.2.4 and hence can conclude that F is an annulus or a Möbius strip such that $F \cap P$ is its middle circle. This means that P_F is obtained from P by cutting along the circle contained in a 2-component α of P. Collapsing P_F, we eliminate α and get a spine $P'_F \subset P_F$ of M_F such that either $c(P'_F) < c(P)$ (if the boundary circles of α pass through at least one true vertex of P), or $c_1(P'_F) < c_1(P)$ (if they are triple circles), or $c_2(P'_F) = c_2(P) - 1$ (if P is a closed surface). In all three cases we get $\bar{c}(P'_F) < \bar{c}(P)$.

STEP 2. Assume that P is not simple, i.e., $G \neq \emptyset$. Since F is normal, it does not intersect the handles of ξ that correspond to edges of G. It follows that either:

(a) F lies in a ball of ξ around a vertex of G which is not in H.
(b) F is contained in the union $N(Q)$ of handles that correspond to the cells of H.

Evidently, in Case (a) F is a disc of the required type. Consider Case (b). Recall that H is a simple polyhedron. If F is normal in $N(H)$, then we apply Step 1 and get the desired inequality $\bar{c}(P'_F) < \bar{c}(P)$. Therefore, we may assume that F, being normal in M, is not normal in $N(H)$. The only reason for that phenomenon is violation of condition 5 of Definition 3.4.1. It follows that ∂F is contained in a lake and is nontrivial there. We may conclude that in this case F is a disc in a ball of ξ. □

Remark 4.2.9. Condition $\partial F \neq \emptyset$ in Corollary 4.2.8 is essential. Indeed, consider a normal surface $F \subset M$ which is normally parallel to ∂M. Then P_F consists of a copy of itself and a copy of ∂M. Therefore, $\bar{c}(P'_F) > \bar{c}(P)$.

Lemma 4.2.10. *If D is a proper disc in a 3-manifold M, then $c(M_D) = c(M)$ and $\bar{c}(M_D) = \bar{c}(M)$. In other words, the complexity and extended complexity of a 3-manifold are preserved under removing as well as under attaching a handle of index 1.*

Proof. It is sufficient to prove that $\bar{c}(M_D) = \bar{c}(M)$. First, we note that if P_D is a minimal almost simple spine of M_D, then an almost simple spine of M having the same extended complexity can be obtained from P_D by adding an appropriate arc. It follows that $\bar{c}(M_D) \geq \bar{c}(M)$. Iterating this argument, we can conclude that $\bar{c}(Q) \geq \bar{c}(M_D) \geq \bar{c}(M)$, where Q is the core of M (see Definition 4.1.21).

To prove the inverse inequality $\bar{c}(Q) \leq \bar{c}(M)$ we choose a minimal almost simple spine P of M. Denote by ξ_P the corresponding handle decomposition. If ∂M admits an essential boundary compressing disc, then, normalizing it, we get an essential normal compressing disc D'. Then $\bar{c}(P'_{D'}) \leq \bar{c}(P)$ by Corollary 4.2.8. It follows that $\bar{c}(M_{D'}) \leq \bar{c}(M)$.

Let us perform now boundary compressions along essential normal discs as long as possible. As it is explained in the proof of Proposition 4.1.25, after a finite number of compressions we end up with a core Q' of M. Since Q and Q' are isotopic by Proposition 4.1.25 and since each compression does not increase $\bar{c}$, we can conclude that $\bar{c}(Q) = \bar{c}(Q') \leq \bar{c}(M)$. Combining the inequalities $\bar{c}(Q) \geq \bar{c}(M)$ and $\bar{c}(Q) \leq \bar{c}(M_D) \leq \bar{c}(M)$, we get $\bar{c}(M_D) = \bar{c}(M)$. Therefore any compression along any proper disc preserves both $c(M)$ and $\bar{c}(M)$. □

Corollary 4.2.11. *Let S be a 2-sphere in a 3-manifold M. Then $c(M_S) = c(M)$.*

Proof. Cutting along S can be realized by removing a ball and cutting along a disc whose boundary lies in the boundary of the ball. Both preserve $c(M)$. Or vice-versa: Gluing two boundary spheres together means attaching an index 1 handle followed by attaching a 3-ball. Both preserve $c(M)$. □

Remark 4.2.12. Cutting along a 2-sphere can increase the extended complexity of a 3-manifold. For example, if M is a 3-ball, then $\bar{c}(M) = (0,0,0) < (0,0,1) = \bar{c}(M_S)$

Let ξ be a handle decomposition of a 3-manifold M and $F \subset M$ a proper incompressible surface. Then F can be normalized by the normalization procedure described in Theorem 3.4.7, which consists of tube and tunnel compressions, and eliminating trivial spheres and discs. Let us modify the procedure as follows. Since F is incompressible, each tube compression results in appearance of a 2-sphere. If this sphere is inessential, then we accomplish the compression by throwing it away. Other normalization moves remain the same. Of course, the modified normalization procedure transforms F into actually the same normal surface F' as the unmodified one. The only difference is that we get a fewer number of trivial spherical components. Note also that if F is a closed surface different from a sphere, then a connected component of F' is homeomorphic to F.

Lemma 4.2.13. *Suppose a handle decomposition ξ of a 3-manifold M corresponds to an almost simple cellular spine P of M. Let $F \subset M$ be a connected incompressible surface and let a surface $F' \subset M$ be obtained from F by the modified normalization procedure described above. Then $c(M_F) = c(M_{F'})$ and, if F is not a 2-sphere, $\bar{c}(M_F) < \bar{c}(M_{F'})$.*

Proof. Let us analyze the behavior of $\bar{c}(M_F)$ under the modified normalization moves.

1. Let us show that the modified tube compressions preserve $\bar{c}(M_F)$. Indeed, denote by D the compressing disc of a tube and by D' the disc bounded by ∂D in F. Then the surface F' resulting from the tube compression can be presented as $F' = (F \setminus \mathrm{Int}\ D') \cup D$. Let $W = (M_F)_D = (M_{F'})_{D'}$ be the manifold obtained from M by cutting along $F \cup D = F' \cup D'$. Then W is obtained from M_F by cutting along D, and simultaneously it is obtained from M'_F by cutting along D', see Fig. 4.23. It follows from Lemma 4.2.10 that $\bar{c}(M_F) = \bar{c}(W) = \bar{c}(M_{F'})$.

2. Let us show that tunnel compressions also preserve $\bar{c}(M_F)$. Denote by D a boundary compressing disc for F, which can be considered as a partition wall inside a tunnel. Let $W = (M_F)_D$ be the 3-manifold obtained from M_F by cutting it along D. Consider the surface F' obtained from F by compressing along D. Then $\bar{c}(M_F) = \bar{c}(W)$ by Lemma 4.2.10, and $\bar{c}(W) = \bar{c}(M_{F'})$, since W and $M_{F'}$ are homeomorphic. It follows that $\bar{c}(M_F) = \bar{c}(M_{F'})$, see Fig. 4.24.

3. Crossing out an inessential disc component of F preserves $\bar{c}(M_F)$ by Lemma 4.2.10.

4. We may conclude that the first three modified normalization moves preserve the extended complexity and hence the complexity of the manifold. Consider the last normalization move (removing an inessential 2-sphere).

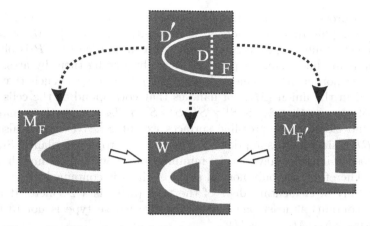

Fig. 4.23. W is obtained from M_F and $M_{F'}$ by cutting along discs

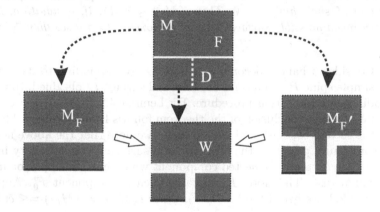

Fig. 4.24. W is homeomorphic to $M_{F'}$ and can be obtained from M_F and by cutting D

This move preserves $c(M_F)$ by Corollary 4.2.11, but can decrease $\bar{c}(M_F)$. Nevertheless, if F is not a sphere, then, thanks to the modification of the first normalization move, spherical components do not appear at all. So we do not need the last move at all.   □

Theorem 4.2.14. *Let F be a connected incompressible surface in a 3-manifold M. Then $c(M_F) \leq c(M)$. Moreover, if M is closed and contains no projective planes, and F is not a 2-sphere, then $c(M_F) < c(M)$.*

Proof. Consider a handle decomposition of M corresponding to its minimal almost simple spine P and a connected normal surface F' obtained from F by the modified normalization procedure. By Lemma 4.2.13, $c(M_F) = c(M_{F'})$. Therefore, the first conclusion of the theorem follows from Corollary 4.2.5.

Let us prove the second. Let F_0' be a connected component of F' which is homeomorphic to F. Consider a decomposition $M = \#_{i=1}^n M_i$ of M into the connected sum of prime summands. We can assume that P is obtained from minimal almost simple spines P_i of M_i by joining them by arcs. Then, since F_0' is normal and connected, there is $j, 1 \leq j \leq n$, such that F_0' is contained in the union $(M_j)_0$ of handles that correspond to the cells of P_j. It cannot happen that M_j is $S^1 \times S^2$, $S^1 \tilde{\times} S^2$ or $L_{3,1}$, since these manifolds contain no closed incompressible surfaces except S^2. We cannot also have $M_j = RP^3$, since by assumption M contains no projective planes. Since all other closed manifolds have special minimal spines, we can assume that P_j is special. Therefore, the only normal surface which is contained in $(M_j)_0$ and consists only of elementary discs of the type $(3, 0)$, is a 2-sphere. It follows that F_0' contains at least one elementary disc whose type is not $(3, 0)$. By Corollary 4.2.5, $c(M_{F_0'}) < c(M)$. □

Theorem 4.2.15. *Let F be a connected proper incompressible surface in a 3-manifold M such that $\partial F \neq \emptyset$. Then $\bar{c}(M_F) \leq \bar{c}(M)$. If, in addition, F is a boundary incompressible surface not homeomorphic to a disc, then $\bar{c}(M_F) < \bar{c}(M)$.*

Proof. Consider a handle decomposition of M corresponding to its minimal almost simple spine P and a connected normal surface F' obtained from F by the modified normalization procedure. By Lemma 4.2.13, $\bar{c}(M_F) = \bar{c}(M_{F'})$. Therefore, the first conclusion of the theorem follows from Corollary 4.2.5. To obtain the second conclusion of the theorem, we note that the above normalization procedure preserves the property of a surface to be boundary incompressible and contain a connected component which has nonempty boundary and is not a disc. Therefore, F' contains such a component F_0'. Applying Corollary 4.2.8, we get $\bar{c}(M_{F_0'}) < \bar{c}(M)$. Since $\bar{c}(M_F) = \bar{c}(M_{F'}) = \leq \bar{c}(M_{F_0'})$, we are done. □

4.2.2 Triangulations vs. Handle Decompositions

As the reader might have observed, the triangulation and handle decomposition versions of the normal surface theory are in a sense parallel. To make the observation precise, we note that the triangulation version works without any changes for closed manifolds equipped with singular triangulations. Recall that by Corollary 1.1.27 one-vertex singular triangulations of a closed manifold M correspond bijectively to special spines of M.

Let T be a one-vertex triangulation of a closed 3-manifold M and P the corresponding dual special spine of M. Recall that P has a natural cell decomposition into true vertices, edges, and 2-components. This decomposition induces a handle decomposition ξ_P of M such that ξ_P has only one handle of index 3, and balls, beams, and plates of ξ_P correspond naturally to the true vertices, edges, and 2-components of P. Since P is dual

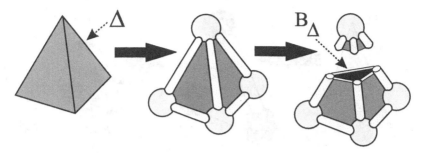

Fig. 4.25. Elementary discs in Δ correspond bijectively to the ones in the ball B_Δ shown *black*

to T, ξ_P is dual to the handle decomposition ξ_T of M obtained by thickening vertices, edges, and triangles of T to balls, beams, and plates, respectively, see Chap. 1. The duality of the handle decompositions means that every index i handle of ξ_T is considered as an index $(3 - i)$ handle of ξ_P, $0 \le i \le 3$.

Let T be a one-vertex singular triangulation of a closed 3-manifold M and ξ_P the corresponding handle decomposition of M.

Theorem 4.2.16. *The matching system for T (see Sect. 3.3.4) coincides with the one for ξ_P (see Sect. 3.4).*

Proof. We will think of ξ_P as being obtained from ξ_T by appropriate renumbering of indices of handles. Let Δ be a tetrahedron of T. Denote by B_Δ the ball of ξ_P which is placed inside Δ. See Fig. 4.25, where B_Δ is shown as a black core of Δ. It is evident that any elementary disc for Δ is normally isotopic to a disc that crosses B_Δ along an elementary disc for B_Δ. This gives us a one-to-one correspondence between the variables of the matching systems for T and ξ_P. It means that the matching systems have actually the same variables.

Similarly, each equation of the matching system for T appears by considering an arc l in a triangle having the endpoints in different sides of the triangle. Each such arc l determines a strip in the corresponding beam of ξ_P joining distinct islands, see Fig. 4.26. The strip is responsible for an equation of the matching system for ξ_P. It is easy to see that these equations of the matching systems for ξ_T and ξ_P are the same. This means that the systems are identical.   □

Corollary 4.2.17. *Let T be a one-vertex singular triangulation of a closed 3-manifold M and ξ_P the corresponding handle decomposition. Then any surface in M normal with respect to T is normally isotopic to a surface normal with respect to ξ_P. This correspondence determines a bijection between the sets of the normal isotopy classes of normal surfaces in T and ξ_P, and respects the summation.*

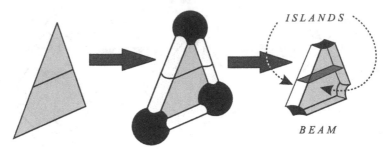

Fig. 4.26. Each arc in a triangle of T determines a strip in the corresponding beam of ξ_P

Proof. Follows directly from Theorem 4.2.16.   □

Later on we will use Corollary 4.2.17 to switch from triangulations to handle decompositions and back whenever we find it advantageous. Note that for manifolds with boundary $\neq \emptyset$, S^2 the matching systems are quite different.

5

Algorithmic Recognition of S^3

As we have mentioned earlier, recognizing irreducibility of 3-manifolds requires the existence of a recognition algorithm for the sphere S^3. Another motivation for constructing such an algorithm is the following. To the late 1970s topologists elaborated methods for proving an algorithmic classification theorem for Haken manifolds (though a complete proof appeared only in 1997, see Chap. 6). These methods play a crucial role in solving many other problems about Haken manifolds, but they do not work for manifolds which are not sufficiently large. What can one do with them?

It would have been natural to begin expansion into the world of non-Haken 3-manifolds by considering the simplest non-Haken manifold S^3. To that time it was generally thought that the recognition algorithm for S^3 does exist. This view was supported by existence of two different algorithms for recognizing S^3 in the class of manifolds of Heegaard genus 2 [48, 127, 128]. Also, Haken made a short announcement on the existence of a general S^3-recognition algorithm (publication of the proof was prevented by its huge size).

In 1992 Rubinstein also announced the existence of an algorithm. His proof was based on some ideas developed by him and Jaco, and never appeared in full. Finally, in 1994 Thompson succeeded in realizing Rubinstein's ideas in a beautiful and purely topological proof [119] (see also [81]). The algorithm is based on Rubinstein's theorem. Its handle decomposition version states that each special spine P of S^3 possesses the following property: The handle decomposition ξ_P corresponding to P contains a 2-normal 2-sphere with a quadrilateral or an octagon. Most of this chapter is devoted to the development of the necessary technique.

The above property enables us to construct an algorithm which successively decomposes each homology sphere into a connected sum of simpler homology spheres until we get either a collection of spheres with one-point spines or run into a homology sphere which does not possess that property and hence is not S^3. To complete the algorithm, we show that verifying whether ξ_P contains a 2-normal 2-sphere with a quadrilateral or an octagon can be done algorithmically.

5.1 Links in a 3-Ball

5.1.1 Compressing Discs and One-legged Crowns

Let S be a 2-sphere in the interior of a 3-ball B. It divides B into two pieces, the *upper* part, which contains ∂B, and the *lower* part. Anything in the upper part lies *above* S, anything in the lower one lies *below*.

Definition 5.1.1. *An arc α in B is called an* upper arc *for S, if the endpoints of α lie on S and small initial and terminal segments of α lie above S; α is a* lower arc, *if the segments lie below S.*

We emphasize that an upper or a lower arc does not necessarily lie entirely in the corresponding part of the ball. By a link in a 3-ball B we mean a finite collection of disjoint proper arcs in B.

Definition 5.1.2. *Let L be a link in B and $S \subset Int\, B$ a 2-sphere which intersects L transversally. A disc $D \subset B$ is called an* uppercompressing disc *for S (or, more precisely, for $S \cup L$) if ∂D can be represented as the union of an upper arc α and an arc $\beta \subset S$ such that $D \cap L = \alpha$ and $\partial \alpha = \partial \beta$. If the same conditions hold, but α is a lower arc, then D is a* lower compressing disc. *In both cases β is called the* base *of D.*

We stress that no conditions are imposed on the possible intersections of Int D with S. If $S \cap Int\, D = \emptyset$, the D is called a *strict* (upper or lower) compressing disc for S. An example of a nonstrict compressing disc for $S \cup L$ is shown in Fig. 5.1a.

Definition 5.1.3. *An upper and a lower compressing discs for S are called* independent *if the intersection of their base arcs either is empty or consists of their common endpoint.*

See Fig. 5.1b for examples of independent and dependent two-sided compressing discs.

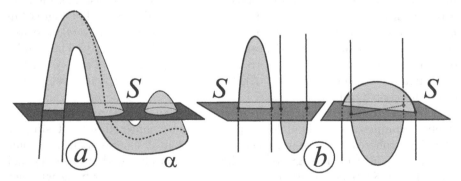

Fig. 5.1. (a) Nonstrict compressing disc; (b) independent and dependent compressing discs

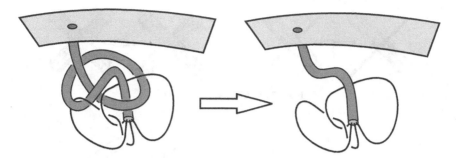

Fig. 5.2. One can improve the leg by an isotopy

Now we make a digression and describe peculiar geometric objects called *one-legged crowns*. A *crown* is a finite collection of disjoint arcs in the interior of a 3-ball B. By a *leg* we mean the image $G = \lambda(D^2 \times I)$ of an embedding $\lambda: D^2 \times I \to B$ such that $G \cap \partial B = \lambda(D^2 \times \{1\})$. The discs $\partial_0 G = \lambda(D^2 \times \{0\})$ and $\partial_1 G = \lambda(D^2 \times \{1\})$ are called the lower and upper *bases* of the leg. Curves of the form $\lambda(\{*\} \times I)$ are its *generators*.

Definition 5.1.4. *Let C and G be, respectively, a crown and a leg in B such that all the endpoints of C lie on $\partial_0 G$, and they are the only points in the intersection of C with G. Then the union $C \cup G$ is called a* one-legged crown.

Consider two one-legged crowns on Fig. 5.2. They have the same crowns but different legs. Each leg can be knotted and linked with the arcs of the crown in arbitrary way. It is surprising, but any two such one-legged crowns are isotopic (by an isotopy of $(B, \partial B)$)!

Lemma 5.1.5. *Let $C \cup G$, $C' \cup G'$ be two one-legged crowns such that $C \cup \partial_0 G$ and $C' \cup \partial_0 G'$ are isotopic. Then there is an isotopy $(B, \partial B) \to (B, \partial B)$ taking $C \cup G$ to $C' \cup G'$.*

Proof. Replacing $C \cup G$ by an isotopic one-legged crown (still denoted by $C \cup G$), we may assume that $\partial_1 G = \partial_1 G'$ and $C \cup \partial_0 G = C' \cup \partial_0 G'$. It follows from fairly general reasons borrowed from knot theory that $C \cup G$ can be transformed into $C' \cup G'$ by isotopic deformations and moves (*"crossing changes"*) of the following two types. Performing the first move, we pull a subarc l of the crown through a portion $\lambda(D^2 \times I_1)$ of the leg $G = \lambda(D^2 \times I)$, where I_1 is a subinterval of I. The second move consists in pulling one portion of the leg through another. Both moves are performed inside a small ball having no other common points with the crown and the leg. See Fig. 5.3.

Claim. Each of the crossing change moves can be realized by an isotopy of the one-legged crown.

Evidently, this is sufficient for proving the lemma. To prove the claim, we start with a rigorous description of the first move.

Fig. 5.3. Crossing changes

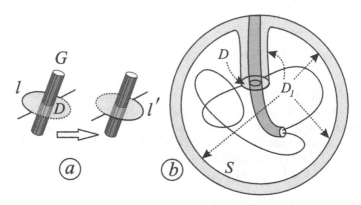

Fig. 5.4. Realizing the crossing change moves by isotopy

Let D be a disc in B such that D intersects the leg $G = \lambda(D^2 \times I)$ in a disc $\lambda(D^2 \times \{a\})$, $D \cap L$ is an arc in ∂D, and D has no other common points with $C \cup G$. Then the first move consists in replacing l by the complementary arc $l' = \partial D \setminus \text{Int } l$, see Fig. 5.4a.

Consider a sphere $S \subset B$ such that S is parallel and very close to ∂B, and intersects the leg G along a disc $\lambda(D^2 \times \{b\})$. If we look attentively at Fig. 5.4b, we find easily a disc $D_1 \subset B$ which has the same boundary as D and intersects $C \cup G$ only in l. It consists of the tube $\lambda(\partial D^2 \times [a, b])$ running along the leg from D to S, and of the disc $S \setminus \lambda(D^2 \times \{b\})$. The existence of an isotopy of $C \cup G$ that realizes the first move is now evident: We simply deform l to l' along D_1 and extend this deformation by the identity to an isotopic deformation of $C \cup G$.

The case of the second move is similar. We may assume that the moving portion of the leg is very thin, and operate with it in the same way. The only difference is that the portion $\lambda(D^2 \times [a, b])$ of the leg running from D to S should not contain the moving portion of the leg. This can be easily achieved by choosing correctly which of the two leg portions participating in the move should move.   □

Let $C \cup G \subset B$ be a one-legged crown. We associate with it the link $L \subset B$ consisting of the crown C and the generators of the leg G that join the endpoints of the crown with ∂B. We will refer to $C \cup G$ as to a *one-legged presentation* of L.

Definition 5.1.6. *Let L be a link in B and $S \subset Int\ B$ a 2-sphere which intersects L transversally. We say that L admits an inverse leg (with respect to S), if there is a one-legged presentation $C \cup G$ of L such that the lower base $\partial_0 G$ of the leg is contained in S and the leg approaches S from below.*

Recall that a link $L = l_1 \cup l_2 \cup \ldots \cup l_k$ in a 3-ball B is *trivial* if there are disjoint discs $D_1, D_2, \ldots, D_k$ in B such that $D_i \cap L = l_i$ is an arc in ∂D_i and $D_i \cap \partial B$ is the complementary arc of ∂D_i. Theorem 5.1.7 is the first step in the construction of the recognition algorithm for S^3.

Theorem 5.1.7. *For any nontrivial link L in B there exists a transversal 2-sphere $S \subset Int\ B$ such that the following holds:*

1. *$S \cup L$ admits an upper and a lower compressing discs.*
2. *$S \cup L$ admits no pair of independent compressing discs.*
3. *$S \cup L$ admits no inverse leg.*

The proof is based on the concept of thin position of links [34]. It will take Sect. 5.1.2 to develop the technique needed for the proof, which will be given at the end of the section.

5.1.2 Thin Position of Links

Denote by B^3 the standard unit ball in Euclidean space R^3 centered at the origin O. Let a link $L \subset B^3$ consist of smooth proper arcs in B^3 not passing through the center. Consider a foliation $\mathcal{F}_0$ of $B^3 \setminus \{O\}$ with round spheres $S_r, 0 < r \leq 1$, centered at O (r is the radius of S_r). It is convenient to divide these spheres into two types: *transverse spheres*, which intersect L transversally, and *singular spheres* containing tangency points. We shall refer to the spheres as *level* spheres, having in mind the *height function* $h : B^3 \to R$ given by the rule $h(S_r) = r$.

Definition 5.1.8. *A link $L \subset B^3$ is in* general position *(with respect to $\mathcal{F}_0$) if the following holds:*

1. *There is only a finite number of singular spheres in $\mathcal{F}_0$.*
2. *Each singular sphere in $\mathcal{F}_0$ contains exactly one tangency point.*
3. *All the tangency points are minimum or maximum points with respect to the restriction of h onto L. Here we mean local minima and maxima.*

Remove from B^3 all the singular spheres. We get a collection of 3-manifolds. One manifold is an open ball, another one is homeomorphic to $S^2 \times [0,1)$, and all the other are homeomorphic to $S^2 \times (0,1)$. The removal splits L into several open and half-open arcs.

Definition 5.1.9. *The total number $w(L)$ of the arcs into which the singular spheres decompose L is called the* width *of L.*

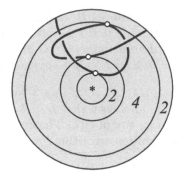

Fig. 5.5. Link in thin position of width 2+4+2=8

The width of a link is an unstable characteristic, since it may vary under an isotopy of the link. However, if the isotopy is $\mathcal{F}_0$-*regular*, that is, at each moment of time the link is in general position, then the width is preserved. Moreover, the width is preserved under an $\mathcal{F}_0$-regular homotopy, when self-intersections of L are allowed. The only requirement is that the minimum and maximum points neither appear nor disappear, and never come to the same level sphere.

Definition 5.1.10. *A link* $L \subset B^3$ *is in* thin position *if it has the minimum width over all links isotopic to L.*

An example of a one-component link in thin position is shown in Fig. 5.5. Its width is 2+4+2=8.

Obviously, any link is isotopic to a link in thin position. Our next goal is to investigate properties of links in thin position.

Lemma 5.1.11. *Let* $L \subset B^3$ *be a link,* $S_r \subset B^3$ *a transverse level sphere, and* $l \subset L$ *an upper arc for S_r. Replace l by an upper arc* $l' \subset B^3$ *with the same endpoints such that l' has only one maximum point, this maximum point lies below the global maximum of l, and the link* $L' = (L \setminus l) \cup l'$ *(not necessarily isotopic to L) is in general position. Then $w(L') \le w(L)$.*

Proof. Choose an upper arc $l'' \subset B^3$ such that $\partial l'' = \partial l$, l'' contains only one maximum point, and the height of the maximum equals the height of the global maximum of l. Then each singular sphere for the link $L'' = (L \setminus l) \cup l''$ is a singular sphere for L, and the number of arcs of L'' between any two neighboring singular spheres is no greater than the one of L between the same spheres. It follows that $w(L'') \le w(L)$.

We now deform l'' to l' keeping the endpoints fixed so that at each moment of time it has only one maximum point and this maximum moves strictly downward, see Fig. 5.6. During the deformation l'' may intersect the remaining part of L''. When the maximum point of l'' passes through another singular level sphere $S_{r'}$ of L'', the width of L'' is preserved if $S_{r'}$ contains a maximum

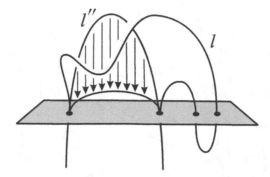

Fig. 5.6. Pulling the unique maximum down does not increase the width

point of L'', and decreases by four if it contains a minimum point. Hence $w(L') \leq w(L'')$ and thus $w(L') \leq w(L)$.   □

An analogous statement for lower arcs is also true. Indeed, let $l \subset L$ and $l' \subset B^3$ be two lower arcs with the same endpoints such that the only minimum point of l' is above the global minimum point of l. Then the replacement of l by l' does not increase the width.

We shall describe two methods for decreasing width. The first method works when there are independent compressing discs.

Lemma 5.1.12. *Let $L \subset B^3$ be a general position link and S_r a transverse level sphere. Suppose that S_r admits a pair of independent compressing discs with respect to L. Then L is isotopic to a link of a smaller width.*

Proof. Let D_u be an upper compressing disc for S_r with the upper arc α_u and the base β_u, and let D_l be an independent lower compressing disc with the lower arc α_l and the base β_l. Then L is isotopic to the link $L_1 = L \setminus (\alpha_u \cup \alpha_l)) \cup \beta_u \cup \beta_l$ (such an isotopy can be constructed by moving α_u to β_u and α_l to β_l along D_u and D_l, respectively). Slightly shift β_u upward to an arc α'_u having a single maximum point. Similarly, we shift β_l a little downward to an arc α'_l with a single minimum point. The shifts should be fixed at the endpoints of the arcs.

Consider the link $L' = L \setminus (\alpha_u \cup \alpha_l)) \cup \alpha'_u \cup \alpha'_l$. It follows from Lemma 5.1.11 that $w(L') \leq w(L)$. Since the shifts of β_u upward and of β_l downward are small, we may assume that there are no singular level spheres between the level of the maximum point of α'_u and the level of the minimum point of α'_l. Then we can decrease $w(L')$ by lowering α'_u below S_r and raising α'_l above S_r, see Fig. 5.7 for the case when α'_u and α'_l have no common points, and Fig. 5.8 for the case they have a common endpoint. The maximum of α'_u and the minimum of α'_l either change places or annihilate each other. In both cases we get a link of a strictly smaller width.   □

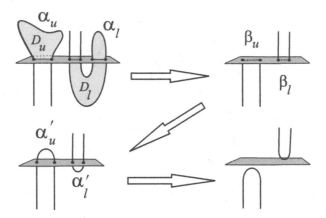

Fig. 5.7. The maximum and minimum change places

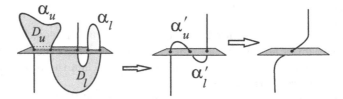

Fig. 5.8. The maximum and minimum annihilate each other

The second method of decreasing width works when there is an inverse leg.

Lemma 5.1.13. *Assume that a general position link $L \subset B^3$ admits an inverse leg with respect to a level sphere S. Then L is isotopic to a link of a smaller width.*

Proof. Let $C \cup G$ be a one-legged presentation of L such that the lower base $\partial_0 G$ of the leg G is contained in S and approaches S from below. We choose an arc of the crown C and move along it upward from one of its endpoints to the first maximum point, which we denote by A. The part traversed will be denoted by l. Let us now construct a new leg G' with the same lower base as follows. First, it must go down a little without intersecting singular level spheres, simultaneously becoming very thin. Next, it turns upward to follow very closely along l to the level of A. Finally, it goes monotonically upward to ∂B^3. Of course, the intersection $C \cap G'$ must consist of the endpoints of C. Joining these points with ∂B^3 by generators of G', we obtain a new link $L' \subset B^3$ having the same crown and the new leg G'. Then it follows from Lemma 5.1.11 that $w(L') \le w(L)$.

Let us prove that a transversal level sphere S_r, placed slightly below the level of A, admits a pair of independent compressing discs. Indeed, the upper compressing disc of this pair contains A, the lower one can be easily constructed by using the fact that G' turns upward immediately and goes along l. See Fig. 5.9.

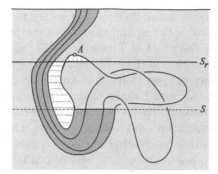

Fig. 5.9. Two independent compressing discs for S_r

It follows from Lemma 5.1.5 that L and L' are isotopic, while Lemma 5.1.12 ensures the existence of an isotopy of L' that strictly decreases its width. □

Proof of Theorem 5.1.7. It is sufficient to prove the theorem for the standard ball B^3. Put L in thin position. Any link in B^3 contains a minimum point, and any nontrivial link contains at least one maximum point (with respect to the height function h). It follows that there exists a transverse level sphere $S = S_r \subset B^3$ such that the first singular sphere S_{max} above S contains a maximum point of L, while the first singular sphere S_{min} below S contains a minimum point. We claim that S satisfies conclusions 1–3 of the theorem.

For proving conclusion 1 we note that a transverse sphere S' immediately below S_{max} admits an evident strict upper compressing disc containing the unique maximum point of L in S_{max}. Since the portion of L between S_{max} and S is the union of monotone arcs, there is an isotopy $B^3 \to B^3$ that takes S_{max} to S and is invariant on L. Thus S also admits a strict upper compressing disc. The same argument shows the existence of a strict lower compressing disc.

Since L is in thin position, conditions 2 and 3 are true for every transverse level sphere, in particular, for S (see Lemmas 5.1.12 and 5.1.13). □

5.2 The Rubinstein Theorem

5.2.1 2-Normal Surfaces

2-Normal surfaces differ from normal once by allowing the boundary of any elementary disc D in any tetrahedron Δ^3 to intersect some edges twice. Every normal surface is 2-normal, but not vice versa. Hence the class of 2-normal surfaces is larger. Nevertheless, it possesses the same properties and thus potentially is more useful. To give a rough idea which served as a motivation for the introduction of 2-normal surfaces, consider an arbitrary (say, closed)

surface F in a triangulated 3-manifold M. Let us try to normalize it by eliminating returns, compressing tubes, and throwing away trivial 2-spheres as was
done in Sect. 3.3.3, see the proof of Theorem 3.3.21. Since the edge degree is
decreasing, we end up with a surface F' that realizes a "local minimum" of
the edge degree. It may happen that F' is an interesting (i.e., informative)
normal surface, but it may also happen that F' is either empty or a union of
vertex spheres (small normal spheres surrounding vertices).

Assume that M contains no interesting surfaces which are normal. Then
the idea of trapping other interesting surfaces consists in keeping balance
between the latter two possibilities, i.e., between falling down to the empty
set and to a collection of vertex spheres. Surfaces that realize such "unstable
local maxima" of the edge degree may also be interesting and informative. As
a rule, they can be realized by 2-normal surfaces (but not by normal ones).

Definition 5.2.1. *Let T be a triangulation of a 3-manifold M, may be singular. A closed surface F is called* 2-normal *if F is in general position with
respect to T and the following holds:*

1. *The intersection of F with every tetrahedron consists of discs. Those discs
   are called* elementary.
2. *The boundary $\partial D \subset \partial\Delta^3$ of every elementary disc $D \subset \Delta^3$ is normal and
   crosses each edge at most twice.*

The weakening of the restrictions on elementary discs results in the
appearance of three new octagonal elementary discs in each tetrahedron Δ^3,
see Fig. 5.10. To prove this we recall that the set of fundamental curves in $\partial\Delta^3$
consists of four triangles and three quadrilaterals, and any connected normal
curve l in $\partial\Delta^3$ is either a triangle or a curve of the form $mX_i + nX_j$, where
X_i, X_j are quadrilaterals of different types, see Lemma 3.2.17. If l intersects
an edge twice, then $m = n = 1$ and by examining $X_i + X_j$ we can conclude
that l is an octagon.

Now we are ready to formulate the Rubinstein Theorem.

Theorem 5.2.2. *For every one-vertex singular triangulation of S^3 there is a
2-normal 2-sphere $S \subset S^3$ such that the intersection of S with at least one
tetrahedron contains either a quadrilateral or an octagon.*

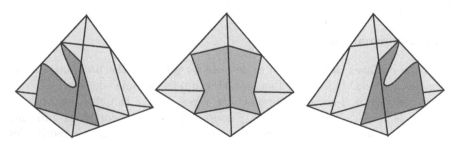

Fig. 5.10. Three octagons

Remark 5.2.3. The original Rubinstein's statement is stronger: There always exists an *almost normal* sphere, i.e., a 2-normal sphere containing exactly one octagon. For our version of the recognition algorithm this restriction is superfluous (although can be easily satisfied), so we have dropped it.

It is also worth mentioning that the condition that the intersection of S with all the tetrahedra contains at least one quadrilateral or octagonal elementary disc means that not all elementary disc in S are triangles or, equivalently, that S is not a small normal sphere surrounding the unique vertex of the triangulation.

We have mentioned earlier that the handle decomposition approach to the theory of normal surfaces is more flexible than the triangulation one. The same is true for 2-normal surfaces, so we shall use it for proving the Rubinstein Theorem. We shall consider handle decompositions generated by special spines. The theory of closed normal surfaces in such handle decompositions of closed manifolds coincides with the one in one-vertex triangulations (see Theorem 4.2.16 and Corollary 4.2.17). The same is true for the theory of 2-normal surfaces. For reader's convenience we give here a direct definition of a closed 2-normal surface in a handle decomposition. Let ξ be a handle decomposition of a 3-manifold M.

Definition 5.2.4. *A closed surface $F \subset M$ is called* 2-normal *(with respect to ξ) if the following conditions hold:*

1. *F does not intersect handles of index 3.*
2. *F intersects each plate $D^2 \times I$ in a collection of parallel sheets of the form $D^2 \times \{*\}$.*
3. *The intersection of F with each beam $D^2 \times I$ has the form $L \times I$, where L is a finite system of disjoint simple proper curves in D^2. Here the disc D^2 can be identified with the island $D^2 \times \{0\}$ as well as with the island $D^2 \times \{1\}$.*
4. *None of the systems L contains a closed curve.*
5. *None of the systems L contains an arc having both endpoints in the same end of the same bridge (such arcs are called* bridge returns*).*
6. *The intersection of F with each ball consists of discs (these discs are called* elementary*).*
7. *The boundary curve C of every elementary disc crosses each bridge at most twice.*

This definition is just Definition 3.4.1 restricted to closed surfaces with only one difference: We allow the boundary curve of any elementary disc to pass through any bridge twice. For the case of a handle decomposition ξ_P generated by a special spine P we have one more type of elementary discs (in addition to triangle and quadrilateral types (3,0) and (4,0), which appear in considering closed normal surfaces, see Definition 4.2.2 and Lemma 4.2.3). The boundary of such a disc (called an *octagon*) is shown in Fig. 5.11a; the other

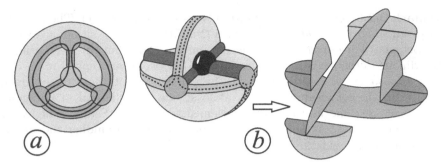

Fig. 5.11. (a) Octagon; (b) Octagon destroys the vertex

two octagons can be obtained by rotation by $\pm 120°$. If two octagons have different types and lie in the same ball, then their intersection is nonempty. The same is true for any quadrilateral and any octagon contained in the same ball. Note that cutting a ball of ξ_P along an octagon followed by collapsing destroys the corresponding true vertex of P, see Fig. 5.11b.

As we have mentioned earlier, all basic facts of the theory of normal surfaces for a handle decomposition ξ_P corresponding to a special spine P of a 3-manifold M remain true for 2-normal surfaces. In particular, we have the following:

1. Equivalence classes of 2-normal surfaces can be parameterized by admissible solutions of the corresponding matching system. Here the matching system for 2-normal surfaces is obtained in exactly the same way as for normal ones, see Sect. 3.4. The only difference consists in the appearance of three new types of elementary discs (octagons) in each ball of ξ_P. A non-negative integer solution to the matching system is *admissible*, if in each ball of ξ_P among quadrilateral and octagonal types of elementary discs no more than one type has a positive coefficient.
2. The set of fundamental solutions is finite and can be constructed algorithmically. It follows that the same is true for the set of fundamental surfaces, which correspond to admissible fundamental solutions.
3. Cutting M along a closed 2-normal surface produces a new manifold M_F with a natural almost simple spine P_F such that $c(P_F) \leq c(P)$ (i.e., the number of the true vertices of P_F is not greater than the one of P). Moreover, if F contains an elementary disc of the quadrilateral or octagonal type, then $c(P_F) < c(P)$.

Let P be a special spine of a closed 3-manifold M and ξ_P the corresponding handle decomposition of M.

We will prove the Rubinstein Theorem in the following equivalent formulation.

Theorem 5.2.5. *For every special spine of the standard ball B^3 there is a 2-normal 2-sphere $S \subset S^3$ such that the intersection of S with at least one ball of the handle decomposition ξ_P contains either a quadrilateral or an octagon.*

5.2.2 Proof of the Rubinstein Theorem

Let P be a special spine of the standard ball B^3 and ξ_P the corresponding handle decomposition. Consider a plate $D^2 \times I$ of ξ_P. An arc $l = \{*\} \times I$, where $\{*\}$ is a point on D^2, is called a *transversal arc* of the plate.

Definition 5.2.6. *A link $L(\xi_P) \subset M$ consisting of transversal arcs for the plates (one arc for each plate) is called a* dual link *of the decomposition ξ_P.*

Since each plate of ξ_P corresponds to an edge of the related one-vertex triangulation T of S^3, L is simply the union of truncated edges of T.

Example 5.2.7. Dual links for the Abalone and Bing House with two Rooms (see Sect. 1.1.4) are shown in Fig. 5.12.

The example above shows that the dual links for the Abalone and Bing house with two rooms are nontrivial. The same turned out to be true for all special spines of B^3.

Lemma 5.2.8. *For any special spine P of B^3 the dual link $L(\xi_P)$ is nontrivial.*

Proof. In fact, we prove more: $L = L(\xi_P)$ contains no trivial component. Assuming the contrary, choose a trivial component l. It is convenient to shift it into the intersection of the corresponding plate with a ball of the decomposition ξ_P, i.e., into a bridge b. Since l is trivial, there is a disc $D \subset B^3$ bounded by l and an arc $l_1 \subset \partial B^3$ such that D has no other common points with $L \cup \partial B^3$. Taking into account that $D \cap L = l$ and $l \subset b$, one can easily see that D can be isotoped *rel L* so that afterward it is contained in the union of all balls and beams.

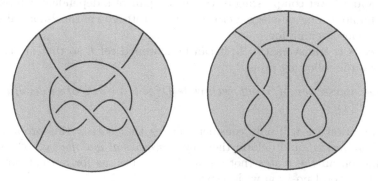

Fig. 5.12. Dual links for the Abalone (*left*) and Bing house with two rooms (*right*)

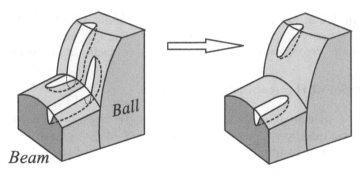

Fig. 5.13. Simplifying the intersection of D with the islands

Put D in general position with respect to the union U of all the islands. By an innermost circle argument we can destroy all closed curves in $D \cap U$ by an isotopy of D fixed on ∂D. Similarly, considering arcs in $D \cap U$ which are outermost in D and cut out from D discs not containing l, we can eliminate all arcs in $D \cap U$ by an isotopy of D which keeps l fixed and moves l_1 along ∂B^3. Here we use the following evident observation: If an arc in the intersection of a beam or a ball of ξ_P with ∂B^3 is disjoint from all plates and has both endpoints in the boundary of the same island, then it can be shifted by an isotopy to the adjacent ball or beam, respectively, see Fig. 5.13. Here the assumption that P is special is essential.

After the complete annihilation of $D \cap U$, the disc D and the arc l_1 will entirely lie in the ball containing b and in a lake, respectively. But this is impossible, since l_1 joins different sides of b, which lie in the coasts of different lakes. $\square$

Proof of Theorem 5.2.5. Let P be a special spine of B^3 and ξ_P the corresponding handle decomposition. By Lemma 5.2.8, the dual link $L = L(\xi_P)$ is nontrivial.

It follows from Theorem 5.1.7 that there is a sphere $S \subset B^3$ admitting an upper and a lower compressing discs, but no pair of independent compressing discs and no inverse one-legged crown. S is a prototype of the 2-normal sphere which we are looking for.

Since S is transversal to L, it can be isotoped rel L so that afterward S possesses the following property:

The intersection of S with each plate $D^2 \times I$ consists of sheets of the type $D^2 \times \{*\}$.

In this section we will only consider spheres that satisfy this property, call them *seminormal*, and deform them by *seminormal isotopies* of B^3 that are invariant on all the plates (not on all the handles, as for normal isotopies). Put S in general position with respect to the islands.

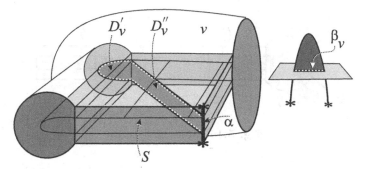

Fig. 5.14. Bridge return determines a compressing disc

The main problem is to eliminate bridge returns, i.e., arcs that lie in the intersection of S with the islands and have both endpoints at the end of the same bridge, see Item 5 of Definition 5.2.4. Let v be a bridge return which lies in an island J and has the endpoints in a bridge b. We call v an *upper return* if the arc $l_v \subset J \cap b$ between the endpoints of v is an upper arc for S, and a *lower return* if l_v is lower.

The following two observations are crucial for the proof:

1. Each upper or lower bridge return v determines an upper or, respectively, a lower compressing disc D_v for S, not necessarily strict.
2. S cannot have at the same time both an upper and a lower bridge returns. The same is true for every sphere seminormally isotopic to S.

Let us comment on the observations. The disc D_v corresponding to a bridge return $v \subset J$ consists of the disc $D_v' \subset J$ bounded by $v \cup l_v$, and of the straight strip $D_v'' \subset D^2 \times I$ joining l_v with the corresponding subarc α_v of $\{*\} \times I \subset L$. Here $D^2 \times I$ is the plate containing l_v, see Fig. 5.14. The arc $\beta_v = \partial D_v \setminus \text{Int } \alpha_v$ is the base of D_v. Since S is embedded, any two bridge returns v, w are disjoint. Therefore the base arcs β_v, β_w of the corresponding compressing discs D_v, D_w are either also disjoint or have at most one common endpoint. It follows that either both v, w are lower returns, or both upper. Otherwise $S \cup L$ would admit a pair of independent compressing discs, which contradicts the choice of S.

It is convenient to set out the continuation of the proof in several steps. We begin the elimination of bridge returns by setting out in the opposite direction: we create bridge returns.

STEP 1. *S is admissibly isotopic to a sphere S_l having a lower bridge return and to a sphere S_u having an upper bridge return.*

Indeed, consider a strict lower compressing disc D for S. Since D intersects only one component of L, D can be forced out from all the plates except the plate $P_1 = D^2 \times I$ containing the arc $c = D \cap L$. Moreover, we can arrange for $D \cap P_1$ to consist of a straight strip D'' joining c with an arc α in the boundary of an island J. The remaining part D' of D is a disc in the union

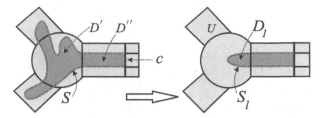

Fig. 5.15. Creating a return

of the balls and beams. Choose a seminormal isotopy of B^3 that shrinks D' along itself to a disc $D_1 \subset J$. This isotopy takes S to a seminormal sphere S_l that has a lower bridge return, see Fig. 5.15.

We may conclude that there is a seminormal isotopy taking S_l to S. Similarly, starting with a strict upper compressing disc for S, we can construct a seminormal isotopy of S to a sphere S_u having an upper bridge return.

STEP 2. *There exists a seminormal sphere $S' \subset B$ which is seminormally isotopic to S and has no bridge returns.*

The idea consists in considering a family $\{h_t(S_l), 0 \le t \le 1\}$ of seminormal spheres which joins S_l with S_u. The existence of such a family follows from Step 1. The sphere $S_l = h_0(S_l)$ has a lower bridge return, the sphere $S_u = h_1(S_l)$ has an upper one, and we never see simultaneously returns of both types. It means that for some t the sphere $h_t(S_l)$ has no bridge returns at all.

The rigorous proof of this claim is based on the well-known fact that any ambient isotopy of a manifold can be replaced by a composition of local shifts so that each shift is fixed outside a small ball. We replace the isotopy h_t taking S_l to S_u by a composition of seminormal local shifts such that the following holds:

(a) Before and after each shift the displaced sphere is in general position with respect to the islands.
(b) The support ball of each shift intersects no more than one island.

Consider the sequence $S_l = S_1, S_2, \ldots, S_n = S_u$ of seminormal spheres obtained from S_l by successive application of the shifts. There are two possibilities: At least one of the spheres has no bridge returns (then we are done), and there is $i, 1 \le i \le n-1$, such that S_i has a lower bridge return while S_{i+1} has an upper one.

Consider the second possibility. By construction, S_i and S_{i+1} coincide in a neighborhood of any island, except possibly an island J. All bridge returns of S_i are lower, all returns of S_{i+1} are upper. Therefore S_i can have bridge returns only in J. Otherwise a lower bridge return of S_i in any other island J' would be an upper bridge return of S_{i+1}, which is impossible. Indeed, the local shift taking S_i to S_{i+1} keeps a neighborhood of J' fixed and thus cannot convert a lower arc into an upper one.

To purify J from bridge returns, we present the beam containing J as $D^2 \times I$ such that $J = D^2 \times \{1\}$. Let $D^2 \times [0, \varepsilon]$ be a small neighborhood in the beam of the other island $D^2 \times \{0\}$, which contains no bridge return. Then we stretch linearly $D^2 \times \{0\}$ to the size of the whole beam $D^2 \times I$ and extend the stretching to a seminormal isotopy of B^3. This isotopy converts S_i into a seminormal sphere S' without bridge returns.

What should be done further to convert S' into a 2-normal sphere? Let us recall conditions 1–7 in Definition 5.2.4 of a 2-normal surface. The first condition is fulfilled automatically, since there are no handles of index 3. The second one holds since S' is seminormal. Condition 5 is the outcome of the previous step. It is very easy to get condition 3 by the same trick as above: for each beam $D^2 \times I$ we stretch a small neighborhood $D^2 \times [0, \varepsilon]$ of the island $D^2 \times \{0\}$ to the size of the whole beam.

Our next goal is to prove a version of condition 7. The remaining conditions 4, 6 will be considered later.

STEP 3. *Any curve in the intersection of S' with the boundary of any ball B_1 of ξ_P crosses each bridge at most twice.*

This property slightly differs from condition 7, since we do not know yet that S intersects tetrahedra along discs. Assuming the contrary , suppose that a closed curve C in the intersection of S' with ∂B meets a bridge b in more than two segments. From these we choose three adjacent (with respect to the bridge) segments I_1, I_2, I_3 such that I_2 lies between I_1 and I_3.

Let us choose a starting point and an orientation for C such that, traversing C, we meet consequently the segments I_1, I_2, I_3. For $i = 1, 2, 3$, denote by x_i^- and x_i^+ the initial and the terminal points of I_i with respect to the orientation. Then the arc of C between x_1^+, x_2^- and the arc l_1 in ∂b between the same points bound a disc D_1 in ∂B. Similarly, the arc in C between x_2^+, x_3^- and the arc l_2 in ∂b between the same points bound a disc D_2 in ∂B.

The interiors of D_1, D_2 may contain lakes and thus intersect ∂B^3. To avoid this, we slightly push them rel ∂ into the interior of B^3. Now we append to each $D_i, i = 1, 2$, a straight strip A_i contained in the plate $D^2 \times I$ such that $\partial A_i \subset l_i \cup L \cup S$, see Fig. 5.16.

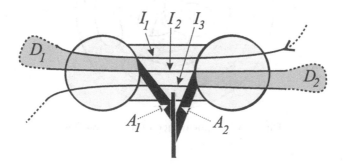

Fig. 5.16. Triple passing through a bridge produces independent compressing discs

We get two independent compressing discs whose base arcs have a common endpoint. This contradicts the property of S that it admits no independent compressing discs.

Now we turn our attention to conditions 4 and 6 of Definition 5.2.4. Violating any of them means that S' contains tubes. It is convenient to present S' as an interior connected sum of 2-normal spheres, i.e., as a collection $S_1, S_2, \ldots, S_k$ of 2-normal spheres connected by thin tubes. The tubes may be knotted and linked, and may run one inside another. More exactly, the spheres $S_1, S_2, \ldots, S_k$ can be obtained from S' in the following way. Replace the intersection of S' with every ball B of ξ_P by a collection of discs in B having the same boundary. Remove all spheres that are contained in the union of balls and beams. Then the remaining spheres form the collection.

STEP 4. *At least one of the spheres $S_1, S_2, \ldots, S_k$ contains either a quadrilateral or an octagon.*

Assuming the contrary, suppose that all spheres $S_1, S_2, \ldots, S_k$ intersect the balls of ξ_P only along triangle pieces. Then each S_i is normally parallel to ∂B^3. If $k = 1$, then $S_1 = S'$, and we get a contradiction with an obvious fact that a boundary-parallel sphere admits no upper compressing disc. Suppose that $k \geq 2$. Without loss of generality we can assume that S_1 is the outermost sphere from the collection. Let S_2 be the next one. Consider the upper part U of B^3 with respect to S_2. Present it as $S_2 \times I$ such that $U \cap L = E \times I$, where E is a finite set in S_2. Let D be a disc in S_2 that contains E and has no common points with tubes, i.e., $D \subset S'$. Note that a small neighborhood of D in the upper part U lies below S_1. It follows that the cylinder $D \times I$ is an inverse leg for S', see Fig. 5.17. This contradiction concludes the proof. □

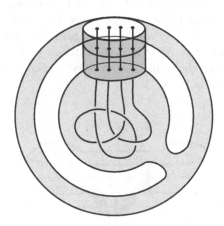

Fig. 5.17. Appearance of an inverse leg

5.2.3 The Algorithm

By a *homology sphere* we mean a closed manifold M with $H_1(M;Z) = 0$. A *homology ball* is a manifold M' with $\partial M' = S^2$ and $H_1(M';Z) = 0$. It is easy to show that all homology groups of a homology sphere or a homology ball coincide with the corresponding groups of S^3 and B^3, respectively.

Lemma 5.2.9. *Let ξ_P be a handle decomposition corresponding to a special spine P of a homology sphere M. Assume that there is a 2-normal sphere $S \subset M$ containing at least one quadrilateral or octagon. Then there exist homology spheres M_1, M_2 and their almost simple spines P_1, P_2 such that $M = M_1 \# M_2$ and $c(P_1) + c(P_2) < c(P)$. The spheres M_i and the spines P_i can be constructed algorithmically.*

Proof. To construct M_1, M_2, we fill with 3-balls the boundary spheres of the disconnected manifold $M_S = M_1' \cup M_2'$ obtained from M by cutting along S, see Fig. 5.18a. Almost simple spines P_1, P_2 of manifolds M_1, M_2, respectively, can be constructed essentially in the same way as in the proof of Lemma 4.2.1. Indeed, only one homology ball (say, M_2') contains the unique index 3 handle D^3 of ξ_P. Denote by M_2'' the manifold $M_2' \setminus D^3$, which has the same homology groups as $S^2 \times I$. Since S is 2-normal, it decomposes the handles of ξ_P into the handles of the same index. The new handles form a handle decomposition of $M_1' \cup M_2''$. Collapsing balls, beams, and plates of the new decomposition onto their core points, arcs, and 2-cells, respectively, we get an almost simple spine P_1 of M_1' (and hence of M_1) and a spine P_2'' of M_2''. Since S contains a quadrilateral or an octagon, the number of true vertices of $P_1 \cup P_2''$ is strictly less than the one of P, see Corollary 4.2.5 for the case S is normal. If S contains an octagon, then $c(P_1) < c(P_2)$ since cutting along an octagon destroys the corresponding true vertex, see Fig. 5.11.

To construct P_2, it suffices to puncture a 2-component of P_2'' which separates the two boundary components of M_2''. On the level of manifolds, this removal is equivalent to connecting the boundary components by a thin solid tube which intersects the 2-component in a meridional disc of the tube. This operation converts M_2'' (i.e., the twice punctured M_2) into M_2 punctured only once (i.e., into a homeomorphic copy of M_2'). See Fig. 5.18b.

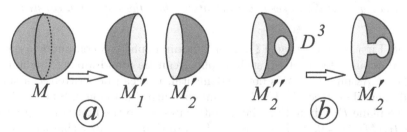

Fig. 5.18. (a) S decomposes M into two homology balls; (b) Removing the ball and the tube preserves the homeomorphism type

Clearly, puncturing a 2-component of P_2'' cannot increase the number of true vertices. Therefore, $c(P_1) + c(P_2) < c(P)$.   □

Lemma 5.2.10. *For any almost simple spine P of a homology sphere M there exists an algorithmically constructible finite collection of homology spheres M_i and their spines $P_i \subset M_i$, $1 \leq i \leq n$, such that M is the connected sum of $M_1, \ldots, M_n$, $\sum_{i=1}^{n} c(P_i) \leq c(P)$, and for each i the spine P_i is either a point or special.*

Proof. We shall follow the same procedure as in the proof of Theorem 2.2.4. For convenience we replace M by a regular neighborhood M' of P in M. Surely, M' is a homology ball.

STEP 1. Suppose that P possesses a 1-dimensional part. Consider an arc ℓ in it and a proper disc $D \subset M'$ which intersects ℓ transversely at one point. D decomposes M' into two homology balls, so that their boundary connected sum is M'. Their almost simple spines can be obtained by removing ℓ from P.

STEP 2. Assume that a 2-component of P contains a nontrivial orientation preserving simple closed curve l. Shifting l into $\partial M'$ and considering a disc bounded by it in ∂M, we construct a disc $D \subset M'$ such that $D \cap P = \partial D = l$ and D cuts a 3-ball V out of $M' \setminus P$. If we penetrate into V through a hole in another 2-component of the free boundary of V (see Fig. 2.6), we get after collapsing a new almost simple spine of M' with the same or a smaller number of true vertices.

STEP 3. Performing Steps 1, 2 for as long as possible, we get a finite collection of almost simple spines P_i of homology balls M_i' so that each P_i has no 1-dimensional part (i.e., is a simple polyhedron), and no 2-component containing a nontrivial orientation preserving simple closed curve, $1 \leq i \leq n$. Every homologically trivial polyhedron possessing these properties is either a point or special, see the discussion at the end of the proof of Theorem 2.2.4.

It remains to fill the boundaries of M_i' with 3-balls and get the desired collection M_i of homology spheres. By construction, M is the connected sum of M_i.   □

Lemma 5.2.11. *There is an algorithm which for a given homology 3-sphere M, a special spine P of M, and the corresponding handle decomposition ξ_P, determines whether there is a 2-normal sphere in M containing a quadrilateral or an octagon. If the answer is affirmative, then the algorithm constructs one of those spheres.*

Proof. First we show that if ξ_P has a 2-normal sphere S containing a quadrilateral or an octagon, then such a sphere can be found among fundamental surfaces. Assume that S is not fundamental, i.e., can be written as a sum of fundamental surfaces $F_i, 1 \leq i \leq n$ (maybe with repetitions). Since each projective plane RP^2 in a 3-manifold contributes Z_2 to the first homology group and $H_1(M; Z) = 0$, no F_i can be RP^2. Note that the Euler characteristics of F_i sum up to $\chi(S) = 2$. Therefore, at least one of the surfaces F_i (say, F_1) must be a 2-sphere.

We claim that F_1 contains at least one quadrilateral or octagon. Suppose, on the contrary, that all elementary discs in F_1 have type $(3,0)$. Let us shift them by a normal isotopy very close to the lakes contained in the boundaries of the corresponding balls of ξ_P such that afterward they compose a new sphere F_1 disjoint from the other surfaces $F_i, i \neq 1$. Then the surface $\sum_{i=1}^n F_i$ would be disconnected, a contradiction.

The desired algorithm can now be easily constructed by enumerating the fundamental surfaces and testing each of them for being a 2-sphere with a quadrilateral or an octagon. $\qquad\qquad\qquad\qquad\qquad\qquad\qquad\qquad\quad$ □

The Algorithm

Let M be a closed 3-manifold we wish to test. We calculate the first homology group. If it is nontrivial, then M is not a 3-sphere. If M is a homology sphere, we construct a special spine P of M and perform the following two steps.

STEP 1. Using Lemma 5.2.11, we look for a 2-normal sphere in M containing a quadrilateral or an octagon. If there are no such spheres, we stop.

STEP 2. If such a sphere does exist, then we apply Lemmas 5.2.9 and 5.2.10 and present M as a connected sum of homology spheres M_i. At the same time we construct spines $P_i \subset M_i$ such that each P_i is either a point or a special polyhedron.

We shall refer to M_i and P_i as the homology spheres and the spines of the first level. Then for each homology sphere M_i such that P_i is special we return to Step 1 and then to Step 2 taking M_i instead of M and P_i instead of P. After processing all M_i and P_i we obtain a collection of homology spheres and their spines of the second level, and so on. If all the spines of some level are points, we stop.

Since each performing of Step 1 strictly decreases the complexity of the spines, the process must end up after finitely many steps. If we stop at Step 1, i.e., if some M_i contains no 2-normal sphere with a quadrilateral or an octagon, then M_i (and hence M) is not S^3 by the Rubinstein theorem. If all the spines of some level are points, then all the corresponding homology spheres (and hence M) are homeomorphic to S^3.

6

Classification of Haken 3-Manifolds

6.1 Main Theorem

The following is known as *the recognition problem for 3-manifolds [39]:*

Does there exist an algorithm to decide whether or not two given 3-manifolds are homeomorphic?

Why is this problem important? The reason is that the positive answer would imply the existence of algorithmic classification of 3-manifolds. Indeed, one can easily construct an algorithm which enumerates step by step all compact 3-manifolds. Using it, we could create a list $M_1, M_2, \ldots$ of all 3-manifolds without duplicates by inquiring if each manifold has been listed before. It is this list that is considered the *classifying list* of 3-manifolds. Certainly, this is a classification in a very weak sense. The knowledge that the classifying list exists would not help to answer many possible questions, for example, whether the Poincaré conjecture is true. It is the proof of the existence that is important, since the search for it would inevitably lead one to deeper understanding of the intrinsic structure of 3-manifolds. Our goal is to present a positive solution of the above problem for Haken manifolds (see Definition 4.1.28). This case is especially important, since it implies a positive solution of the algorithmic classification problem for knots, one of the most intriguing problems of low-dimensional topology.

Recall that a *knot* is a circle embedded in S^3. Two knots are *equivalent,* if there is an isotopy $S^3 \to S^3$ taking one knot to the other. Knots are usually presented by *knot diagrams*, i.e., by generically immersed plane curves so that at each crossing point it is shown which strand goes over.

If we know at advance that two given knots are equivalent, then one can rigorously prove the equivalence by performing *Reidemeister moves* that transform one diagram into the other. If the knots are distinct, then sometimes one can prove that by calculating certain polynomial or numerical invariants. But what can we do, if both potentially infinite procedures (comparing the knots via Reidemeister moves or via knot invariants) do not stop? The right strategy consists in considering knot complement spaces, which happen to be Haken manifolds. The only exception is the complement of the unknot, which is a solid torus and hence boundary reducible.

This chapter is devoted to the proof of the following theorem and corollaries.

Theorem 6.1.1 (The Recognition Theorem). *There is an algorithm to decide whether or not two given Haken 3-manifolds are homeomorphic.*

To keep the proof within reasonable limits we restrict ourselves to considering orientable 3-manifolds. The proof for the nonorientable case is essentially the same. The only difference is that sometimes one has to consider additional special cases.

Corollary 6.1.2. *There exists an algorithmic classification of Haken 3-manifolds.*

Remark 6.1.3. The conditions that 3-manifolds under consideration must be irreducible and boundary irreducible are of technical nature. Both statements remain true for compact 3-manifolds which can be decomposed into connected sum of irreducible sufficiently large factors. The factors can be boundary reducible, but this is not an obstacle, since any irreducible sufficiently large 3-manifold can be decomposed into a boundary connected sum of Haken manifolds and handlebodies.

Corollary 6.1.4. *There exists an algorithmic classification of knots and links in S^3.*

Formally speaking, this corollary does not follow from Theorem 6.1.1. Of course, for the case of knots one can apply the Gordon and Luecke result that knots are determined by their complements [37]. But links are not determined by their complements, see an example in Fig. 6.1. Nevertheless, any link is determined by its complement equipped with the boundary pattern consisting of meridians of the link components. Therefore, Corollary 6.1.4 follows from Theorem 6.1.6, a stronger version of Theorem 6.1.1.

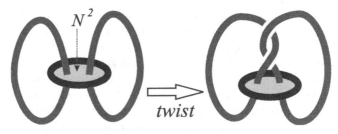

Fig. 6.1. Complements of two distinct 3-component links are related by a 2π-twist along the twice punctured disc N^2

Definition 6.1.5. *An irreducible boundary irreducible 3-manifold* (M, Γ) *with boundary pattern is called* Haken, *if either* M *is sufficiently large or* $\Gamma \neq \emptyset$ *and* M *is a handlebody, but not a 3-ball.*

In case $\Gamma = \emptyset$ this definition coincides with Definition 4.1.28. If $\Gamma \neq \emptyset$ and M is not a handlebody, then M is sufficiently large by Corollary 4.1.27. So the only essential new aspect here is that we include into the set of Haken manifolds boundary irreducible handlebodies with boundary patterns. It follows that any irreducible boundary irreducible 3-manifold (M, Γ) with nonempty boundary pattern is Haken, provided that M is not a 3-ball.

Theorem 6.1.6. *There is an algorithm to decide whether or not for two given Haken 3-manifolds* (M, Γ), (M', Γ') *with boundary patterns there is a homeomorphism* $(M, \Gamma) \to (M', \Gamma')$ *taking* Γ *to* Γ'.

Since boundary patterns appear already at the first step of the proof of Theorem 6.1.1, this generalization introduces no additional difficulties.

The history of the positive solution of the recognition problem for Haken 3-manifolds is very interesting. In 1962, Haken suggested an approach for solving the problem [39]. However, this approach contained a conceptual gap. Thanks to efforts of several mathematicians, by the early 1970s a crucial obstacle was singled out, and, when in 1978 Hemion overcame it [41], it was broadly announced that the problem has been solved [58,131]. Later on many topologists used extensively this result.

Nonetheless, trying to understand in detail how the theorem was proved, I discovered that there was no written complete proof at all. All papers and even books [42,58,59,131] devoted to this subject were written according to the same scheme: they contained informal descriptions of Haken's approach, of the obstacle, of Hemion's result, and the claim that these three ingredients provided a proof. But there has never appeared a paper containing a proof! I undertook an investigation of the question and came to the following conclusions:

1. The statement that the recognition problem for sufficiently large 3-manifolds is algorithmically solvable is true.
2. There is another obstacle of a similar nature that cannot be overcome by the same tools as the first one.
3. It can be overcome by using an algorithmic version of Thurston's theory of surface homeomorphisms that appeared only in 1995 in [9].

Thus for more than 20 years mathematicians relied on an unproven theorem. In this chapter we prove the theorem. The proof is based on the idea of Haken. We also make use of different ideas of Jaco, Johannson, Neumann, Shalen, Thurston, Waldhausen, and others. Nevertheless, the proof presented here is self-contained with the following two exceptions: We do not reprove the results of Hemion [41] and Bestvina and Handel [9] mentioned earlier.

6.2 The Waldhausen Theorem

In this section we prove the classical theorem of Waldhausen [130]. The proof is based on the notion of *hierarchy* (a specific decomposition of 3-manifolds), which is an important tool for solving many other problems. It is this approach that the proof of the Recognition Theorem is based on. We will demonstrate how the approach works by considering first its 2-dimensional version and then the 3-dimensional one. Nevertheless, neither the statement nor the proof of the Waldhausen Theorem will be used in the sequel.

It is well known that two closed surfaces are homeomorphic if and only if they are homotopy equivalent. Moreover, it suffices to require that they have isomorphic homology groups. This follows from the classification theorem for closed surfaces. For surfaces with boundary the above statements are not true. For example, the punctured torus and the disc with two holes, which are distinct surfaces, can be collapsed onto a wedge of two circles and hence are homotopy equivalent. Nevertheless, if a homotopy equivalence $f: F \to F'$ between compact surfaces takes ∂F to $\partial F'$ homeomorphically, then it can be deformed to a homeomorphism. A similar result for irreducible sufficiently large 3-manifolds is known as the Waldhausen Theorem.

Theorem 6.2.1 (Waldhausen). *Let $f: M \to M'$ be a homotopy equivalence between orientable irreducible 3-manifolds such that either $\partial M, \partial M'$ are nonempty and f takes ∂M onto $\partial M'$ homeomorphically, or M, M' are closed and M' is sufficiently large. Then f can be deformed to a homeomorphism $M \to M'$ by a homotopy fixed on ∂M.*

There is a more general form of this theorem. For the sake of completeness we present here its statement, but since the proof is based on the same ideas (see [43, 130]), we omit it.

Theorem 6.2.2. *Let M, M' be irreducible 3-manifolds which are sufficiently large. Suppose that $f: (M, \partial M) \to (M', \partial M')$ is a map such that $f_*: \pi_1(M) \to \pi_1(M')$ is injective and such that for each connected component F of ∂M the induced map $(f_{|F})_*: \pi_1(F) \to \pi_1(F')$ is injective, where F' is the connected component of $\partial M'$ containing $f(F)$. Then there is a homotopy $f_t: (M, \partial M) \to (M', \partial M')$ such that $f_0 = f$ and at least one of the following holds:*

1. $f_1: M \to M'$ is a covering map;
2. M is an I-bundle over a closed surface, and $f_1(M) \subset \partial M'$;
3. M' (hence also M) is a solid torus or a solid Klein bottle and $f_1: M \to M'$ is a branched covering whose branch set is a circle.

If $f_{|F}: F \to F'$ is already a covering map, we may assume that f_t is fixed on F.

The significance of the Waldhausen Theorem can hardly be overestimated. It belongs to the class of theorems that relate different categories, in our case the homotopy category and the topological one. For example, the famous

Poincaré Conjecture stating that any homotopy sphere is a topological one has the similar nature.

In a certain sense, the proof of the Waldhausen Theorem shows one reason why the Poincaré Conjecture and other problems for closed 3-manifolds containing no nontrivial incompressible surfaces are so difficult: there is no appropriate surface to start the decomposition process. We cannot get around this obstruction by puncturing the manifold, since this the punctured manifold is reducible.

6.2.1 Deforming Homotopy Equivalences of Surfaces

We start by recalling two well-known facts that will be used in the sequel.

Lemma 6.2.3. *Let $f: M \to M'$ be a map between connected manifolds of the same dimension n. Suppose that at least one of the following holds:*

1. *M, M' are closed and orientable, and the degree $\deg(f)$ of f is $\pm 1 \in Z$*
2. *M, M' are closed and nonorientable, and $\deg(f)$ is $1 \in Z_2$*
3. *$\partial M \neq \emptyset$ and f takes ∂M onto $\partial M'$ homeomorphically.*

Then the induced homomorphism $f_*: \pi_1(M) \to \pi_1(M')$ is surjective.

Proof. After an appropriate deformation of f we may assume that f takes an n-ball $B \subset M$ onto an n-ball $B' \subset M'$ homeomorphically and $f(M \setminus \mathrm{Int}\, B) = M' \setminus \mathrm{Int}\, B'$. Indeed, in the first two cases this is a well-known property of degree one maps. In the last case it suffices to deform f so that afterwards it takes a collar of ∂M onto a collar of $\partial M'$ homeomorphically, and choose B, B' inside the collars.

Choose basepoints $x_0 \in \mathrm{Int}\, B$ and $x_0' = f(x_0) \in \mathrm{Int}\, B'$. Let $C' \subset M'$ be a simple loop with endpoints at x_0'. We may assume that f is transversal to C'. Then the 1-dimensional submanifold $f^{-1}(C')$ contains a unique simple closed curve C passing through x_0. It follows that the element $[C']$ of $\pi_1(M', x_0')$ corresponding to C' lies in $f_*(\pi_1(M, x_0))$. Since $\pi_1(M', x_0')$ is generated by simple loops with endpoints at x_0', f_* is surjective.   □

Now we turn our attention to a 2-dimensional analogue of the Waldhausen Theorem.

Proposition 6.2.4. *Let $f: F \to F'$ be a homotopy equivalence between compact connected surfaces that takes ∂F onto $\partial F'$ homeomorphically. Then there exists a homotopy of f which keeps $f_{|\partial F}$ fixed and deforms f to a homeomorphism.*

Proof. CASE 1. F, F' *have nonempty boundaries.* Let C' be the union of disjoint proper arcs $C_1', C_2', \ldots, C_n'$ in F' which cut up F' into a disc D'. The existence of such arcs follows from the classification theorem for compact surfaces: arcs possessing the required property can be easily found on every model

Fig. 6.2. Three arcs which cut up the Klein bottle with two holes into a disc

surface. For example, if F' is a Klein bottle with two holes, then one can take the three arcs shown in Fig. 6.2. In general, exactly $1 - \chi(F')$ arcs are needed, since each cut along an arc increases the Euler characteristic by one.

We may assume that f is transversal to C'. Then the inverse image $f^{-1}(C')$ is a proper 1-dimensional submanifold of F. It consists of n disjoint proper arcs $C_1, C_2, \ldots, C_n$ that are mapped, respectively, onto $C'_1, C'_2, \ldots, C'_n$, and maybe of several closed components. Deforming f by a homotopy, we may assume that f takes each C_i to C'_i homeomorphically.

The arcs C_i cut F into a connected surface D, whose boundary is homeomorphic to $\partial D'$ and thus is a circle. Since the Euler characteristics of F and F' are equal, so are the ones of D and D'. It follows that D is a disc. Therefore, we can apply the cone construction to take D onto D' homeomorphically. The homeomorphism $F \to F'$ thus obtained is homotopic to f. Indeed, since $\partial F' \neq 0$ and thus $\pi_2(F') = 0$, there is no obstruction for constructing a homotopy between two maps of D into F'.

CASE 2. F, F' are closed. Since f is a homotopy equivalence, $\deg(f) = \pm 1$ if F, F' are orientable, and $\deg(f) = 1 \in Z_2$ if not. So we may assume that f takes a disc $D \subset F$ onto a disc $D' \subset F'$ homeomorphically and $f(F \setminus \text{Int } D) = F' \setminus \text{Int } D'$. To deform the map $f_{|(F \setminus \text{Int } D)} : (F \setminus \text{Int } D) \to (F' \setminus \text{Int } D')$ to a homeomorphism, we use the same method as in Case 1. The method does work, since the surfaces $F \setminus \text{Int } D$, $F' \setminus \text{Int } D'$ have the same Euler characteristic, and we have used only this property in the proof of Case 1.   □

6.2.2 Deforming Homotopy Equivalences of 3-Manifolds to Homeomorphisms

The proof of the 3-dimensional Waldhausen Theorem is based on the same idea as the above proof of the 2-dimensional one: we decompose the target 3-manifold M' into balls and then try to transfer the decomposition to the source 3-manifold M. There appear three essential differences.

First, instead of arcs, we take surfaces F'_i in M'. Their inverse images $F_i \subset M$ can be much more complicated, for example, have larger genera. To overcome this obstacle, we take surfaces F'_i that are incompressible. Then we compress their inverse images F_i by deforming the map $f: M \to M'$, thus converting them to incompressible surfaces. Since f is a homotopy equivalence, this allows us to deform f so that afterwards it takes F_i to F_i' homeomorphically.

Second, in general it is not possible to choose F_i' so that they are disjoint and cut up M' into a ball. An iterative procedure is needed here. The boundary curves of surfaces appearing at each next step must be allowed to be contained in the union of the previous surfaces.

Third, we should prove that this inductive procedure of decomposing M' into one or several 3-balls is finite.

To realize this program, we need several preparatory lemmas. The first two of them allow us to compress the inverse image of an incompressible surface at the expense of a homotopy deformation of the map.

Lemma 6.2.5. *Let M be a connected irreducible 3-manifold with infinite $\pi_1(M)$. Then $\pi_3(M) = 0$.*

Proof. Since M is irreducible, $\pi_2(M) = 0$ by the Sphere Theorem [106]. Therefore, the universal cover $\tilde{M}$ of M is a simply connected manifold with $\pi_2(\tilde{M}) = \pi_2(M) = 0$. Since $\pi_1(M)$ is infinite and hence $\tilde{M}$ is noncompact, we have $H_3(M; Z) = 0$. By the Hurewicz Theorem, $\pi_3(\tilde{M}) = H_3(\tilde{M}; Z) = 0$. It follows that $\pi_3(M) = 0$. $\qquad\square$

Lemma 6.2.6. *Let $f: B_0 \to M'$ be a map of a 3-ball into a connected irreducible 3-manifold M' with infinite $\pi_1(M')$. Suppose that f is transversal to a two-sided proper incompressible surface $F' \subset M'$ such that $F' \neq S^2$ and $F = f^{-1}(F')$ is a proper tube (i.e., a proper annulus) in B_0. Then there is a map $f_1: B_0 \to M'$ homotopic to f via a homotopy fixed on ∂B_0 such that f_1 is also transversal to F' and the surface $F_1 = f_1^{-1}(F')$ is the union of two discs. In other words, F_1 is obtained from F by compressing along a meridional disc of the tube.*

Proof. Choose two disjoint proper discs $D_1, D_2 \subset B_0$ which are bounded by ∂F. They decompose B_0 into three balls B_1, B_2, B_3. Since F' is injective, the curves $f(\partial D_1), f(\partial D_2)$, which bound singular discs $f(D_1), f(D_2)$ in M', bound singular discs D_1', D_2' in F'. We define $f_1: B_0 \to M'$ by setting $f_1 = f$ on ∂B_0, sending D_1, D_2 to D_1', D_2', and extending that map to the interiors of B_1, B_2, B_3 by taking them into $M' \setminus F'$. This is possible, since $F' \neq S^2$ and thus the manifold obtained by cutting M' along F' is irreducible and hence has trivial π_2. Evidently, the extension may be chosen so as to be transversal to F'. By construction, the surface $F_1 = f_1^{-1}(F')$ is the union of discs D_1, D_2 and the maps f and f_1 coincide on ∂B_0. Since $\pi_3(M') = 0$ by Lemma 6.2.5, f, f_1 are homotopic rel ∂. See Fig. 6.3. $\qquad\square$

Lemma 6.2.7. *Let $f: M \to M'$ be a homotopy equivalence between compact connected irreducible 3-manifolds with infinite fundamental groups such that*

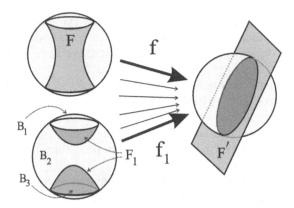

Fig. 6.3. Compressing the inverse image of F

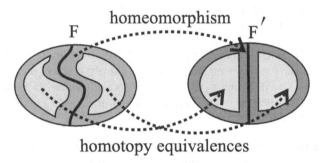

Fig. 6.4. Improving f to f_1

either M is closed or f takes ∂M onto $\partial M'$ homeomorphically. Then for any proper connected two-sided incompressible surface $F' \subset M', F' \neq S^2$, there exist a surface $F \subset M$ and a homotopy $f_t \colon M \to M'$ such that:

1. *$f_0 = f$ and f_t is fixed on ∂M*
2. *f_1 takes a regular neighborhood $N(F \cup \partial M)$ onto a regular neighborhood $N(F' \cup \partial M')$ homeomorphically*
3. *The restriction of f_1 onto $Q = M \setminus Int\,(N(F \cup \partial M))$ is a homotopy equivalence of Q onto $Q' = M' \setminus Int\,(N(F' \cup \partial M'))$, see Fig. 6.4*

Proof. STEP 1: *Improving f so that $f^{-1}(F')$ is an incompressible surface.* If $\partial M \neq \emptyset$, then after an appropriate deformation of f by a homotopy, we may assume that f is a homeomorphism on collars of $\partial M, \partial M'$ and is transversal to F'. If M' is closed, then $\deg f = 1$. Therefore, we can assume that there is a ball $B' \subset M'$ such that:

1. $B' \cap F'$ is a disc
2. $B = f^{-1}(B')$ is a ball in M and f takes B onto B' homeomorphically
3. $f(M \setminus Int\, B) = M' \setminus Int\, B'$

If the surface $G = f^{-1}(F')$ is compressible, we choose a compressing disc $D \subset M$ for G. Consider a ball neighborhood B_0 of D such that $G \cap B_0$ is a tube. By Lemma 6.2.6, one can deform f by a homotopy fixed outside B_0 so that afterwards the inverse image of F' is obtained from G by compressing G along D. Doing so as long as possible, we convert G to an incompressible surface.

STEP 2: *Improving f so that it takes a component of $f^{-1}(F)$ onto F' homeomorphically.* Let F be a connected component of G that contains either a component of ∂G (if $\partial G \neq \emptyset$) or the disc $B_0 \cap G$ (if G is closed). Since F is injective by Lemma 3.3.5 and $f_*: \pi_1(M) \to \pi_1(M')$ is an isomorphism, $(f_{|F})_*: \pi_1(F) \to \pi_1(F')$ is also injective. On the other hand, it is surjective (see Lemma 6.2.3). It follows that the map $f_{|F}: F \to F'$ is a homotopy equivalence, which by Proposition 6.2.4 can be deformed to a homeomorphism. This deformation can be easily extended to a regular neighborhood of F and then to the whole M. The resulting map $M \to M'$ is still denoted by f.

STEP 3: *Eliminating superfluous components of $G = f^{-1}(F')$.* Any other connected component F_1 of G is a closed incompressible surface. The map $(f_{|F_1})_*: \pi_1(F_1) \to \pi_1(F')$ is also injective. Therefore, if $\partial F' \neq \emptyset$, then the group $\pi_1(F_1)$, being a subgroup of the free group $\pi_1(F')$, is free. The same is true in the case $\partial F' = \emptyset$, since F_1 is mapped into $F' \setminus \mathrm{Int}\,(B' \cap F')$. We can conclude that F_1 is a 2-sphere, since S^2 is the only closed surface having a free fundamental group.

By irreducibility of M, the sphere F_1 bounds a ball V in M. If F' is closed, then V cannot contain B, since otherwise it would contain F, which is impossible. Let V_1 be a slightly larger ball whose boundary does not intersect G. Then the singular sphere $f(\partial V_1)$ lies in $M' \setminus F'$. Since M' is irreducible and $F' \neq S^2$ is incompressible, $M' \setminus F'$ is also irreducible and hence $\pi_2(M' \setminus F') = 0$. Therefore we may redefine f on V_1 by taking it into $M' \setminus F'$. This transformation eliminates F_1.

Doing so for as long as possible, we get a new map $f_1: M \to M'$ such that $f_1^{-1}(F') = F$, $f_1(M \setminus F) = M' \setminus F'$, and the restriction of f_1 onto $F \cup \partial M$ is a homeomorphism $F \cup \partial M \to F' \cup \partial M'$. Obviously, f_1 can be improved so that afterwards f_1 takes a regular neighborhood $N(F \cup \partial M)$ onto a regular neighborhood $N(F' \cup \partial M')$ homeomorphically and f_1 takes $Q = M \setminus \mathrm{Int}\, N(F \cup \partial M)$ onto $Q' = M' \setminus \mathrm{Int}\, N(F' \cup \partial M')$.

STEP 4: *Why is $Q \to Q'$ a homotopy equivalence?* To prove the last conclusion of the lemma, consider a connected component Q_j of Q, which is mapped onto the corresponding component Q'_j of Q'. By Lemma 6.2.3, the map $Q_j \to Q'_j$ induces a surjection $\pi_1(Q_j) \to \pi_1(Q'_j)$. On the other hand, since F is incompressible and two-sided, the embedding $Q_j \subset M$ induces an injection $\pi_1(Q_j) \to \pi_1(M)$. It follows that $\pi_1(Q_j) \to \pi_1(Q'_j)$ is also an injection and hence an isomorphism. Taking into account that $\pi_i(Q_j) = \pi_i(Q'_j) = 0$ for $i = 2, 3$, we conclude that the map $Q_j \to Q'_j$ is a homotopy equivalence.   □

Lemma 6.2.8. *Let M be an orientable connected 3-manifold. Suppose that a connected component F_0 of ∂M is not S^2. Then M contains an orientable proper incompressible boundary incompressible surface F such that $\partial F \neq \emptyset$ is homologically nontrivial in ∂M, i.e., $[\partial F] \neq 0$ in $H_1(\partial M; Z)$.*

Proof. Since F_0 is an orientable surface of genus ≥ 1, it contains a pair c_1, c_2 of oriented simple closed curves such that their intersection number $\lambda(c_1, c_2)$ is not 0. Let $\alpha_1, \alpha_2 \in H_1(M; Z)$ be the corresponding elements of the first homology group of M. We claim that the order of at least one of them is infinite.

Suppose that, on the contrary, there exist integers $k_1, k_2 \neq 0$ such that $k_1\alpha_1 = k_2\alpha_2 = 0$. Then there exists orientable surfaces $F_1, F_2 \subset M$ such that for $i = 1, 2$ the boundary ∂F_i consists of k_i parallel copies of c_i. Of course, F_1, F_2 can be singular, i.e., have self-intersections. The intersection $F_1 \cap F_2$ consists of arcs with endpoints in $\partial F_1 \cap \partial F_2$ such that the endpoints of each arc have opposite signs. Therefore, $\lambda(k_1\alpha_1, k_2\alpha_2) = k_1k_2\lambda(\alpha_1, \alpha_2) = 0$, which contradicts our assumption that $\lambda(c_1, c_2) \neq 0$.

Suppose that the order of α_1 is infinite. Then we do the same as in the proof of Lemma 4.1.29: we construct a surjective homomorphism $\varphi \colon H_1(M; Z) \to Z$ such that $\varphi(\alpha_1) \neq 0$. All homotopy groups $\pi_i(S^1), i > 1$, are trivial, so there is no obstruction for realizing φ by a map $f \colon M \to S^1$.

Now, let us take a point $a \in S^1$ such that f is transversal to a. Then $G = f^{-1}(a)$ is a proper orientable surface in M such that $\lambda(c_1, \partial G) \neq 0$. If G is compressible or boundary compressible, we compress it as long as possible. Since the compressions preserve the property $\lambda(c_1, \partial G) \neq 0$, the resulting surface F is incompressible, boundary incompressible, and $[\partial F] \neq 0$ in $H_1(\partial M; Z)$. □

Definition 6.2.9. *Let M be a compact 3-manifold. By an* hierarchy *in M we mean a finite sequence $M = P_0 \subset P_1 \subset \ldots \subset P_n$ of 2-dimensional subpolyhedra of M such that the following holds:*

1. *$P_0 = \partial M$ and each P_i splits M into connected 3-manifolds called chambers;*
2. *P_n is a skeleton of M, that is, all the chambers of P_n are 3-balls.*

Example 6.2.10. Let $M_0 = S^1 \times S^1 \times [0, 1]$. Then a hierarchy of M_0 can be constructed as follows:

1. $P_0 = S^1 \times S^1 \times \{0, 1\}$
2. $P_1 = P_0 \cup A_0$, where $A_0 = S^1 \times \{*\} \times [0, 1]$ is an annulus. P_1 has only one chamber, which is a solid torus
3. The terminal skeleton P_2 is obtained from P_1 by inserting a meridional disc D into the chamber. □

Proof of Theorem 6.2.1. The proof is based on the same idea as the proof of Proposition 6.2.4. We begin with constructing a hierarchy $P_0' \subset P_1' \subset \ldots \subset P_n'$ for M' as follows:

1. $P_0' = \partial M'$ if $\partial M' \neq \emptyset$ and $P_0' \neq S^2$ is an incompressible surface in M' if M' is closed (here is the only place where we use the assumption that M' is sufficiently large);
2. If $0 \leq i \leq n - 1$, then P_{i+1}' is obtained from P_i' by adding a surface F_i', where F_i' is a proper incompressible boundary incompressible two-sided nonseparating subsurface of a chamber Q_j' of P_i' such that $[\partial F_i'] \neq \emptyset$.

Let us prove that the process of erecting new "partition walls" F_i' described in item 2 above is finite and ends up with a skeleton of M'. For that we use the total extended complexity $(\sum_j c(Q_j), \sum_j c_1(Q_j), \sum_j c_2(Q_j))$ of the chambers, where the sums are taken over all chambers of P_i'. Here $c(Q_j), c_1(Q_j), c_2(Q_j)$ denote, respectively, the minimal possible numbers of true vertices, triple circles, and 2-components of almost simple spines of Q_j, see Definition 4.2.7. By Theorem 4.2.15, the erection of each next partition wall strictly decreases the extended complexity of the corresponding chamber or, if the wall is a disc, preserves it. Therefore, since each chamber admits only a finite number of nontrivial disc compressions, we complete the process in a finite number of steps. All chambers of the last polyhedron P_n' will automatically be 3-balls, since otherwise the process could be continued by Lemma 6.2.8.

Our next step consists in transferring the hierarchy from M' to M. Arguing by induction, we assume that there is a polyhedron $P_i \subset M$ such that f takes a regular neighborhood $N(P_i)$ onto a regular neighborhood $N(P_i')$ homeomorphically and the restriction of f onto each chamber of P_i is a homotopy equivalence onto the corresponding chamber of P_i'. The base $i = 0$ of the induction is given by the assumption if $\partial M \neq \emptyset$, and by Lemma 6.2.7 if M is closed. Applying the same lemma to the surface F_i' and the chambers $Q_j' \supset F_i'$, $Q_j = f^{-1}(Q_j')$, we perform the inductive step.

We have proved that f can be deformed so that afterwards f takes a regular neighborhood $N(P_n)$ onto a regular neighborhood $N(P_n')$ homeomorphically and the restriction of f onto each chamber Q_j of P_n is a homotopy equivalence onto the corresponding chamber Q_j' of P_n'. Since all Q_j' are 3-balls and all Q_j are irreducible, they are also 3-balls. Therefore, P_n is a skeleton of M. Our last step consists in replacing each map $f_{|Q_j} : Q_j \to Q_j'$ by a homeomorphism of the balls. $\square$

Remark 6.2.11. The original concept of a hierarchy (see [39,57,130]) differs from the one introduced above. The former referred to the process of splitting M along incompressible surfaces until one obtains a collection of 3-balls. If two 3-manifolds possess identical hierarchies, then they are homeomorphic. We prefer to insert surfaces into M instead of splitting M along them. The advantage is that (under appropriate restrictions, see the next section) there is no need to compare whole hierarchies. It suffices to compare the terminal skeletons: if the skeletons are homeomorphic, then so are the manifolds.

6.3 Finiteness Properties for Surfaces

We are now turning to the proof of the Recognition Theorem (Theorem 6.1.1). To start with, we state it in a more specific form, which suggests the direction the proof will take.

6.3.1 Two Reformulations of the Recognition Theorem

Recall that a 2-dimensional polyhedron is simple, if it has the simplest possible singularities: triple lines and true vertices (crossing points of triple lines). A simple polyhedron is special, if all its 2-components are 2-cells, and there is at least one true vertex (see Definitions 1.1.8 and 1.1.10). Let P be a simple subpolyhedron of a compact 3-manifold M. Then it decomposes M into connected 3-manifolds called *chambers*.

Definition 6.3.1. *A subpolyhedron P of a 3-manifold M is called a* simple skeleton *of M if P is special, contains ∂M, and all the chambers of P are 3-balls. In case M is equipped with a boundary pattern, we will always assume that the pattern is transversal to $SP \cap \partial M$.*

A simple skeleton can be viewed as a special spine of many times punctured M, one puncture for each chamber.

Lemma 6.3.2. *Let P, P' be simple skeletons of 3-manifolds M, M', respectively. Assume that the boundaries of M, M' contain no 2-spheres. Then any homeomorphism $h : P \to P'$ can be extended to a homeomorphism $H : M \to M'$.*

Proof. By Lemma 1.1.15, h can be extended to a homeomorphism H' between regular neighborhoods of the skeletons. Since the rest of M consists of balls, one can easily extend H' to a homeomorphism $H: M \to M'$.   □

Lemma 6.3.2 allows us to give the following reformulation of the Recognition Theorem.

Theorem 6.3.3. *There is an algorithm that, to any Haken manifold M, assigns a finite set $\mathcal{P}(M)$ of simple skeletons of M such that $\mathcal{P}(M)$ is characteristic in the following sense: Haken manifolds M, M' are homeomorphic if and only if $\mathcal{P}(M)$ and $\mathcal{P}(M')$ consist of the same number of pairwise homeomorphic polyhedra.*

Remark 6.3.4. Taking liberties with the language, one can state the conclusion of Theorem 6.3.3 as follows: $M = M' \iff \mathcal{P}(M) = \mathcal{P}(M')$. The part $\Leftarrow$ of the theorem is easy: if a polyhedron $P \in \mathcal{P}(M)$ is homeomorphic to a polyhedron $P' \in \mathcal{P}(M')$, then, according to Lemma 6.3.2, M is homeomorphic to M'.

Let us prove that the Recognition Theorem (Theorem 6.1.1) and Theorem 6.3.3 are equivalent. Indeed, Theorem 6.3.3 reduces the recognition problem for Haken 3-manifolds to the one for 2-dimensional polyhedra. Since the latter is easy, Theorem 6.3.3 implies the Recognition Theorem. On the other hand, if the Recognition Theorem is true, then the characteristic set of simple skeletons of a given Haken manifold M can be constructed as follows.

For each $k = 1, 2, \ldots$ we enumerate one after another all the special polyhedra with k true vertices (the number of such polyhedra being finite, see Chap. 2). Each subsequent polyhedron P we subject to the following tests:

1. Is P a spine of a 3-manifold N?
2. Is the manifold $\hat{N}$, obtained from N by filling all the spheres on ∂N with 3-balls, a Haken manifold?
3. Is $\hat{N}$ homeomorphic to M?

The verification of the first condition is easy, see Theorem 1.1.20. Also, if the answer is affirmative, then N and $\hat{N}$ can be effectively constructed. If $\hat{N}$ satisfies the second condition (that can be checked algorithmically, see Sect. 4.1.4), we apply our assumption that the Recognition Theorem is true to verify the last condition.

Since M is Haken, we inevitably meet the first polyhedron that withstands the above tests. Selecting all other polyhedra that have the same number of true vertices and also withstand the tests, we get the collection $\mathcal{P}(M)$ of simple skeletons of M with the minimal possible number of true vertices. Since the construction depends only on the topological type of M, the set $\mathcal{P}(M)$ is characteristic.

The theorem below is apparently the most precise reformulation of the Recognition Theorem.

Theorem 6.3.5. *There is an algorithm that associates with every Haken 3-manifold M its triangulation $T(M)$ in such a way that $T(M)$ is canonical in the following sense: two Haken manifolds are homeomorphic if and only if their associated triangulations are combinatorially isomorphic.*

Let us prove that Theorem 6.3.5 is also equivalent to the Recognition Theorem. Indeed, It is not difficult to verify whether or not two given simplicial complexes are combinatorially isomorphic. Thus Theorem 6.3.5 implies the Recognition Theorem. On the other hand, if the Recognition Theorem is true, then the canonical triangulation can be constructed as follows.

First, we introduce an order on the set of all 3-dimensional simplicial complexes. To do that, one can number the vertices, edges, triangles, and tetrahedra of a given complex K and represent K by the incidence matrices. Next, we combine the rows of the matrices into one line, obtaining a large integer that encodes the complex. Finally, among all such numbers we take the minimal number $n(K)$ (the minimum is taken over all possible numberings of simplices of K). The natural ordering of the numbers $n(K)$ induces the fixed ordering of simplicial complexes.

Then we take one by one all the 3-dimensional simplicial complexes in the fixed ordering. The first complex whose underlying space is homeomorphic with a given Haken manifold M can be taken as the canonical triangulation $T(M)$. Here we have used our assumption that the Recognition Theorem is true. Since the construction of $T(M)$ depends only on the homeomorphism type of M, homeomorphic 3-manifolds have combinatorially isomorphic canonical triangulations.

To prove the Recognition Theorem, reformulation 6.3.3 is most convenient. As we have mentioned earlier, we will only prove it for orientable 3-manifolds and allow the manifolds to be equipped with boundary patterns.

We describe the first steps towards constructing the characteristic set $\mathcal{P}(M)$.

Definition 6.3.6. *A simple subpolyhedron P of a 3-manifold (M, Γ) with boundary pattern is called* admissible *if the following holds:*

1. *P contains ∂M, and the singular graph SP of P is transversal to Γ;*
2. *Every 2-component $\alpha \subset$ Int M of P is incompressible and separates two different chambers.*
3. *Every 2-component $\alpha \subset \partial M$ is incompressible as a surface with the pattern $\Gamma \cap \alpha$. It means that α admits no clean essential compressing discs.*

Let P be an admissible subpolyhedron of a Haken 3-manifold (M, Γ) and Q_j a chamber of P. We shall consider Q_j as a 3-manifold with boundary pattern $\Delta_j = (SP \cup \Gamma) \cap \partial Q_j$, where SP is the singular graph of P. Since (M, Γ) is boundary irreducible and all the 2-components of P are incompressible, (Q_j, Δ_j) is boundary irreducible. It may happen that Q_j is a reducible manifold, but only when at least one connected component of P lies inside a 3-ball in M. We will always take care that this situation does not occur.

A fairly general method for constructing admissible subpolyhedra can be described as follows. Let (Q_j, Δ_j) be a chamber of an admissible subpolyhedron $P \subset (M, \Gamma)$. Consider a connected two-sided proper incompressible surface $F \subset (Q_j, \Delta_j)$. Certainly, we assume that ∂F is in general position with respect to Δ_j. It means that ∂F contains no vertices of Δ_j and intersects the edges transversally. If F does not separate Q_j, we replace F by two parallel copies of F. Then $P \cup F$ is an admissible subpolyhedron of M. In some sense, $P \cup F$ is obtained from P by inserting a new *partition wall* into Q_j. The wall consists of F or of two parallel copies of F, if F does not separate Q_j.

Definition 6.3.7. *A finite sequence $\partial M = P_0 \subset P_1 \subset \ldots \subset P_n$ of admissible subpolyhedra of a Haken 3-manifold (M, Γ) is called an* admissible hierarchy *of (M, Γ), if P_n is a simple skeleton of M.*

One can construct an admissible hierarchy just as in the proof of Theorem 6.2.1. We start with ∂M or with a closed incompressible surface if M is closed (such a surface exists, since M is sufficiently large). Then, subsequently applying the above construction, we erect new and new partition walls until getting a simple skeleton of M.

This methods always works. For example, one can construct an admissible hierarchy for $S^1 \times S^1 \times [0,1]$ as follows (compare with Example 6.2.10). We take $P_0 = S^1 \times S^1 \times \{0,1\}$, $P_1 = P_0 \cup 2A$, and $P_2 = P_1 \cup 2D_1 \cup 2D_2$, where $2A$ are two parallel copies of the annulus $A = S^1 \times \{*\} \times [0,1]$, and $2D_1, 2D_2$ are two pairs of parallel meridional discs of the two solid torus chambers of $P_1 \cup 2A$. Certainly, the boundaries of the discs must cross the boundaries of the annuli transversally (to be sure that P_2 is simple).

Sometimes, we will also use a more general method for constructing admissible hierarchies: we will allow to insert into chambers not only surfaces, but also more complicated 2-dimensional polyhedra.

Denote by $\mathcal{P}_\infty(M)$ the set of the terminal skeletons of all possible hierarchies of M considered, say, modulo self-homeomorphisms of M. Evidently, $\mathcal{P}_\infty(M)$ possesses the characteristic property $\mathcal{P}_\infty(M) = \mathcal{P}_\infty(M') \Leftrightarrow M = M'$ (compare with Theorem 6.3.3). A difficult problem is that the set $\mathcal{P}_\infty(M)$ thus defined can be infinite, which completely destroys our hope to use it for the algorithmic recognition of 3-manifolds.

6.3.2 Abstract Extension Moves

Fortunately, we do not have to take terminal skeletons of all hierarchies. It turns out that one can subject the process of erecting partition walls to such strong restrictions that for any Haken manifold we get a finite set of terminal skeletons which still is characteristic.

Let us formulate this idea more explicitly. We will describe below a few transformations of admissible subpolyhedra, which are called *extension moves*. Let P be an admissible subpolyhedron of a Haken 3-manifold (M, Γ). Each extension move transforms P into a larger admissible subpolyhedron, which is called *an extension* of P. Two extensions of P are *equivalent*, if there is an admissible (i.e., taking Γ to Γ) homeomorphism $M \to M$ taking one extension to the other. The following properties must hold:

C_1. For any admissible subpolyhedron P, the number of equivalence classes of all possible extensions of P is finite

C_2. There is an algorithm to construct a finite set $\mathcal{E}(P)$ of extensions of P which contains at least one representative of each equivalence class

C_3. Any sequence $P_0, P_1, P_2, \ldots$, in which each subsequent polyhedron P_{i+1} is an extension of P_i, must be finite

C_4. If P does not admit extension moves, then it must be a simple skeleton of M.

Proof of Theorem 6.3.3 (Under the Assumption that the Extension moves have Already been Described). Let (M, Γ) be a given Haken manifold. Denote by P_0 the boundary of M (if M is closed, P_0 is empty). Let us apply to P_0 step by step extension moves. Doing so, at each step we multiply the pair (M, P_i) in several number of exemplars to be able to realize separately all possible extensions from the set $\mathcal{E}(P_i)$. Properties C_3, C_2 guarantee that this branched process stops and that it is algorithmic. It follows from properties C_1, C_4 that we get a finite set of simple skeletons of M. This set can contain duplicates, i.e., homeomorphic skeletons (for example, at each step the set $\mathcal{E}(P_i)$ may contain more than one representative of each equivalence class). Removing the duplicates, we get a finite set $\mathcal{P}(M)$ of simple skeletons of M.

To show that $\mathcal{P}(M)$ is characteristic, we consider two Haken manifolds M, M'. Since at every step we apply all the extension moves (up to equivalence), the result of our branched process depends only on the homeomorphism type of the manifold. Therefore, if M, M' are homeomorphic, then $\mathcal{P}(M)$ and $\mathcal{P}(M')$ must consist of the same polyhedra. On the other hand, if $\mathcal{P}(M)$ and $\mathcal{P}(M')$ contain at least one pair of homeomorphic polyhedra, then M and M' are homeomorphic by Lemma 6.3.2.   □

6.3.3 First Finiteness Property and a Toy Form of the Second

It follows from the previous section that our only concern with proving Theorem 6.3.3 is to define extension moves satisfying C_1–C_4. How can we do that? Unfortunately, a whole bunch of preparatory work will be necessary. We will not actually define extension moves until Sect. 6.5, but will in the meantime establish three crucial finiteness properties for surfaces in 3-manifolds. The first property tells that any Haken manifold contains only a bounded number of disjoint incompressible boundary incompressible connected surfaces which are pairwise nonparallel (Theorem 6.3.10). The second property (Theorem 6.3.17) states that, up to a strong equivalence relation, any simple Haken manifold contains only finitely many incompressible boundary incompressible surfaces of bounded complexity. The third property is valid for all (not necessarily simple) Haken manifolds. It tells us that the number of equivalence classes of incompressible tori and annuli is finite for any Haken manifold, see Theorem 6.4.44.

We begin by describing an idea of getting properties C_1–C_3. As above, we consider each chamber as a 3-manifold with boundary pattern. As was indicated before, we wish to extend our polyhedra by adding surfaces. Let us look closer at this strategy keeping in mind our needed properties C_1–C_3.

It is natural to choose at each step of the construction of an hierarchy only surfaces that are minimal in certain sense. There are many different notions of the complexity of a proper surface F in a 3-manifold (M, Γ) with boundary pattern. All of them involve the Euler characteristic of F and $\#(\partial F \cap \Gamma)$, the number of points in the intersection of ∂F with the pattern. One can take any linear combination of $-\chi(F)$ and $\#(\partial F \cap \Gamma)$ with positive coefficients, or order pairs of these numbers lexicographically. We prefer the simplest combination.

Definition 6.3.8. *Let F be a proper surface in a 3-manifold (M, Γ). Then $\gamma(F) = -\chi(F) + \#(\partial F \cap \Gamma)$ is called the* pattern complexity *(or, in abbreviated form, p-complexity) of F.*

We emphasize that the pattern complexity does not depend on triangulation of M and is preserved under admissible isotopy of the surface. In this respect it drastically differs from the edge degree introduced in Definition 3.3.16.

The following proper connected surfaces have nonpositive p-complexity:

1. Spheres, projective planes, and clean discs (they have negative p-complexity)
2. Clean annuli, clean Möbius bands, tori, Klein bottles, and proper discs whose boundary circles cross the pattern at exactly one point (they have p-complexity 0).

One can easily verify that all other proper connected surfaces in 3-manifolds with boundary pattern have positive p-complexity.

We would like to use inserting p-minimal surfaces as extension moves. By a *p-minimal surface* we mean a surface having the smallest possible p complexity (among all surfaces of a given class). Let's see what our chances for obtaining properties C_1–C_3 are. Consider property C_3. The following lemma is of use here. Recall that two proper surfaces F_0, F_1 in a 3-manifold (M, Γ) are called *admissibly parallel*, if there is an embedding $\varphi \colon F_0 \times [0,1] \to M$ such that the following holds:

1. $\varphi(F_0 \times \{i\}) = F_i$, $i = 0, 1$
2. $\varphi(\partial F_0 \times [0,1]) \subset \partial M$ and $\varphi(\partial F_0 \times [0,1]) \cap \Gamma$ consists of segments of the type $\varphi(\{*\} \times [0,1])$.

Lemma 6.3.9. *Let T be a triangulation of a 3-manifold (M, Γ) with boundary pattern such that Γ is the union of some edges. Suppose that $\mathcal{F}$ is a finite family of disjoint two-sided proper connected normal surfaces in (M, Γ) such that no two of them are admissibly parallel. Then the number of surfaces in $\mathcal{F}$ is not greater than $10t$, where t is the number of tetrahedra in T.*

Proof. The surfaces intersect the tetrahedra of T along discs called patches. There are two types of patches: triangles and quadrilaterals. We call a patch P *good*, if it lies between two parallel patches of the same type. Otherwise P is *bad*. Since every tetrahedron contains at most 10 bad patches (see Fig. 6.5), the total number of bad patches does not exceed $10t$.

Fig. 6.5. Bad patches are shown *black*

Assume that $\mathcal{F}$ contains more than $10t$ surfaces. Then at least one of them (denote it by F) consists only of good patches. The surfaces from $\mathcal{F}$ decompose the tetrahedra of T into pieces. Consider the union U of all pieces which are adjacent to F. Since F contains no bad patches and is two-sided, U can be identified with $F \times I$ such that $F \times \{1/2\} = F$, the surfaces $F \times \{0,1\}$ also belong to $\mathcal{F}$, and $(F \times I) \cap \Gamma$ consists of edge segments of the type $\{*\} \times I$. This means that $\mathcal{F}$ contains even three admissibly parallel surfaces $F \times \{0, 1/2, 1\}$, a contradiction. □

Now we are ready to prove the following First Finiteness Property for surfaces in 3-manifolds, which is useful for proving property C_3.

Theorem 6.3.10 (First Finiteness Property). *Let (M, Γ) be an irreducible boundary irreducible 3-manifold with boundary pattern. Suppose that $\mathcal{F}$ is a finite set of disjoint two-sided proper connected surfaces in (M, Γ) such that no surface is a sphere, a clean disc, or a trivial semi-clean disc, all the surfaces are incompressible and boundary incompressible, and no two of them are admissibly parallel. Then the number of surfaces in $\mathcal{F}$ does not exceed $n(M, \Gamma)$, where $n(M, \Gamma)$ is a constant depending only on the topological type of (M, Γ).*

Proof. Triangulate M such that Γ is the union of edges and take $n(M, \Gamma) = 10t$, where t is the number of tetrahedra of the triangulation. By Corollary 3.3.25, the union of surfaces from $\mathcal{F}$ is admissibly isotopic to a normal surface, and we can apply Lemma 6.3.9. It follows that $\mathcal{F}$ contains not more than $n(M, \Gamma)$ surfaces. □

Now we consider properties C_1, C_2. Let (M, Γ) be a 3-manifold with boundary pattern. For property C_1, we will need two equivalence relations on the set of all 2-dimensional polyhedra in (M, Γ).

Definition 6.3.11. *Two subpolyhedra P_1, P_2 of (M, Γ) are equivalent, if there exists a homeomorphism $h : (M, \Gamma) \to (M, \Gamma)$ such that $h(P_1) = P_2$. In other words, equivalent polyhedra are related by an admissible homeomorphism of (M, Γ).*

Definition 6.3.12. *Two subpolyhedra P_1, P_2 of (M, Γ) are strongly equivalent, if there exists a homeomorphism $h : (M, \Gamma) \to (M, \Gamma)$ such that $h(P_1) = P_2$ and the restriction of h onto ∂M is admissibly isotopic to the identity.*

Evidently, strong equivalence occupies an intermediate position between equivalence and admissible isotopy: admissibly isotopic polyhedra are always strongly equivalent, and strongly equivalent polyhedra are always equivalent. We point out that it makes sense to speak about equivalent and strongly equivalent surfaces in (M, Γ) (since they are polyhedra).

It would be very nice if we could prove the following version of what we will call the Second Finiteness Property. Let (M, Γ) be a Haken 3-manifold and k a number. Then:

1. The set of all incompressible boundary incompressible surfaces of complexity $\leq k$ in (M, Γ) consists of only finitely many strong equivalence classes
2. Representatives of all the equivalence classes can be constructed algorithmically

Clearly, this version would imply properties C_1, C_2 for extension moves, if we define them as insertions of p-minimal incompressible boundary incompressible surfaces. Unfortunately, this version is not true: the number of inequivalent p-minimal surfaces can be infinite. For example, if we twist any surface in (M, Γ) along a clean annulus, we get another surface of the same complexity. Therefore, consecutive twists can produce infinitely many distinct surfaces of the same p-complexity.

We describe a toy situation when the Second Finiteness Property holds in an evident manner, even in a stronger form.

Proposition 6.3.13. *Let (M, Γ) be an irreducible boundary irreducible triangulated 3-manifold. Suppose that (M, Γ) contains no normal surfaces of p-complexity ≤ 0. Then for any k the number of admissible isotopy classes of incompressible boundary incompressible proper connected surfaces in (M, Γ) of p-complexity $\leq k$ is finite. Representatives of the classes can be constructed algorithmically.*

Proof. Since (M, Γ) is irreducible and boundary irreducible, all 2-spheres in M are isotopic and there are only finitely many clean discs (one disc for each connected component of $\partial M \setminus \Gamma$). The number of trivial semi-clean discs is bounded by the number of edges of Γ and thus is also finite. By Corollary 3.3.25, any other incompressible boundary incompressible proper surface in (M, Γ) is admissibly isotopic to a normal one. Therefore it can be presented as a linear combination of fundamental surfaces. Since all the fundamental surfaces have positive p-complexity and since p-complexity is additive, only a finite number of linear combinations of fundamental surfaces have p-complexity $\leq k$. $\square$

The assumption of Proposition 6.3.13 never occurs. For example, any triangulation contains either a normal sphere that surrounds an interior vertex or a normal disc that cuts off a boundary vertex. Also, if C is a simple loop consisting of interior edges such that no two consecutive edges belong to the same triangle, then a thin torus running along C is also normal. Moreover, (M, Γ) can contain other surfaces of complexity 0 such as incompressible tori or clean incompressible annuli. Nevertheless, the idea turns to be sufficient for proving the Second Finiteness Property for 3-manifolds which are simple.

6.3.4 Second Finiteness Property for Simple 3-Manifolds

The next step in our preparatory work will be to prove the Second Finiteness Property for surfaces in simple 3-manifolds. This property is stated in Theorem 6.3.17.

Definition 6.3.14. *Let* (M, Γ) *be a Haken 3-manifold. An incompressible annulus* $A \subset \partial M$ *is called* almost clean *if* $\partial A \cap \Gamma = \emptyset$ *and* $A \cap \Gamma$ *either is empty or consists of several parallel copies of the core circle of* A.

Definition 6.3.15. *An incompressible torus* T *in a Haken 3-manifold* (M, Γ) *is called* essential *if it is not parallel to a torus in* ∂M. *An incompressible boundary incompressible clean annulus* A *in* (M, Γ) *is called* essential *if it is not parallel to an almost clean annulus in* ∂M.

We emphasize that the property of a torus in a 3-manifold to be essential does not depend on the boundary pattern.

Definition 6.3.16. *An irreducible boundary irreducible 3-manifold* (M, Γ) *with boundary pattern is called* simple *if it contains no essential tori and annuli.*

Theorem 6.3.17 (Second Finiteness Property). *Let* (M, Γ) *be an orientable simple 3-manifold. Then for any number* k *there is an algorithmically constructible finite set* $\mathcal{F}_k$ *of surfaces in* (M, Γ) *such that any two-sided proper incompressible boundary incompressible connected surface* F *of p-complexity* $\leq k$ *is strongly equivalent to a surface in* $\mathcal{F}_k$.

The proof is based on the same idea as the proof of Proposition 6.3.13. We triangulate (M, Γ) and present a given surface F as the sum of fundamental surfaces. Then $F = G^+ + G^-$, where G^+, G^- are the sums of the summands that have positive, respectively, nonpositive p-complexity. To prove the theorem, it suffices to show that, up to replacement of F by a strongly equivalent surface, the number of summands of G^- is bounded by a constant depending only on k. Recall that surfaces of nonpositive p-complexity are:

1. Spheres, projective planes, clean discs (they have negative p-complexity)
2. Clean annuli, clean Möbius bands, tori, Klein bottles, and proper discs whose boundary curves cross the pattern at exactly one point (they have p-complexity 0)

In general, G^- can contain arbitrary many exemplars of each type. Let us analyze what happens in our situation.

We can quietly forget about projective planes, since the only irreducible orientable 3-manifold containing RP^2 is the projective space RP^3, which contains only two incompressible surfaces: RP^2 and S^2. We can always assume that F is minimal, i.e., it has the smallest edge degree among all general position surfaces admissibly isotopic to F. In addition, 2-spheres, clean discs, and discs intersecting Γ at exactly one point are irrelevant, since they cannot be summands of F by Corollary 4.1.37. Moreover, the same corollary tells us that those clean annuli and tori that are summands of F must be incompressible and boundary incompressible.

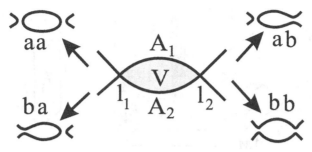

Fig. 6.6. (**aa**) F is not connected; (**ab, ba**) F is not minimal; (**bb**) F is not in reduced form

To estimate their number (see Propositions 6.3.20 and 6.3.21), we need two preparatory lemmas. Klein bottles and Möbius bands bring no additional difficulties and will be considered directly in the proof of Theorem 6.3.17. The proof of the following lemma is similar to that of Lemma 4.1.17.

Lemma 6.3.18. *Let an incompressible boundary incompressible connected normal surface F in a triangulated Haken manifold (M, Γ) be presented in the form $F = G_1 + G_2$. Assume that two annular patches $A_1 \subset G_1, A_2 \subset G_2$ bound a solid torus $V \subset M$ such that $V \cap (G_1 \cup G_2) = \partial V$ and two circles l_1, l_2 that form $\partial A_1 = \partial A_2$ are longitudes of V. Then F is either not minimal or not in reduced form.*

Proof. We shall say that $l_i, i = 1, 2$, has type **a**, if the annuli A_1, A_2 are pasted together by the regular switch of $G_1 \cup G_2$ along l_i. Otherwise l_i has type **b**. There are four possibilities: aa, ab, ba, bb, see Fig. 6.6. Let us consider them.

(aa) By the regular switch of $G_1 \cup G_2$ along l_1, l_2 the patches A_1, A_2 are glued together and give us a connected component of F which does not coincide with F. That contradicts the assumption that F is connected.

(ab) and (ba) Let us replace the regular switch along the **b**-type circle l_i by the irregular one. We get a new surface F' consisting of the boundary of a solid torus and a surface F'' isotopic to F (the isotopy consists in deforming a portion of F'' through the solid torus). Since F'' has a return, $e(F'')$ can be decreased by an admissible isotopy, see Remark 3.3.23. Therefore, F is not minimal.

(bb) Replacing G by $G_1' = (G_1 \setminus A_1) \cup A_2$ and G_2 by $G_2' = (G_2 \setminus A_2) \cup A_1$, i.e., merely switching A_1 and A_2, we get another presentation $F = G_1' + G_2'$ such that G_i' is admissibly isotopic to $G_i, i = 1, 2$, and $G_1' \cap G_2'$ consists of fewer components than $G_1 \cap G_2$. This means that $G_1 \cup G_2$ is not in reduced form. □

Lemma 6.3.19. *Let an incompressible boundary incompressible connected normal surface F in a triangulated Haken manifold (M, Γ) be presented in*

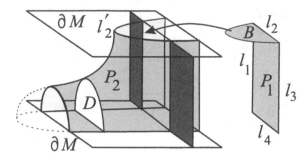

Fig. 6.7. $B \cup P_1$ determines a clean boundary compressing disc

the form $F = G_1 + G_2$. Assume that there are two patches $P_1 \subset G_1, P_2 \subset G_2$ such that

1. P_1 is a clean quadrilateral disc with the sides l_1, l_2, l_3, l_4 such that l_2, l_4 are opposite and lie in ∂M while the interiors of the other two opposite sides l_1, l_3 are in $Int\ M$.
2. Among the sides of P_2 there are three consecutive sides l_1', l_2', l_3' such that $l_1' = l_1, l_3' = l_3$, and l_2', l_2 bound a biangle B in ∂M.

Then F is either not minimal or not in reduced form.

Proof. Suppose, on the contrary, that F is minimal and in reduced form. Then P_2, just as any other patch of $G_1 \cup G_2$, is boundary incompressible by Lemma 4.1.8. It follows that ∂B is clean, since otherwise we could slightly shift B inward M (keeping l_2 fixed and letting l_2' slide along P_2) and get a nontrivial boundary compressing disc for P_2. Also, the disc $B \cup P_1$ determines a clean boundary compressing disc D for P_2: we simply shift $B \cup P_1$ inward M such that l_4 slides along ∂M while the remaining part of ∂D slides along P_2, see Fig. 6.7. Since D must be trivial, P_2 is a clean quadrilateral. Therefore, we can apply Lemma 4.1.17 and conclude that F is either not minimal or not in reduced form, in contradiction with our assumption.   □

Now we are ready to investigate the role of clean annuli. Fortunately, they contribute nothing.

Proposition 6.3.20. *Let F be a two-sided incompressible boundary incompressible connected surface in a triangulated Haken manifold (M, Γ). Suppose that F can be presented in the form $F = G + A$, where $G \neq \emptyset$ and A is a normal clean annulus admissibly parallel to an almost clean annulus A' in ∂M. Then F is not minimal.*

Proof. Suppose, on the contrary, that F is minimal. We may assume that $G + A$ is in reduced form. By Lemma 4.1.8, $G \cup A$ has no clean disc patches and all the patches of $G \cup A$ are incompressible and boundary incompressible. Recall that two-sided incompressible patches are injective. Denote by V the

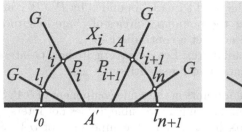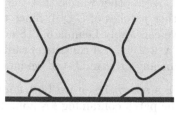

Fig. 6.8. If every patch $P_i \subset V$ is an annulus running from A to A', then any choice of switches leads to a disconnected surface

solid torus in M bounded by A and A'. Since $\pi_1(V) = Z$, each patch of G in V has a cyclic fundamental group and thus is either an incompressible annulus or a disc.

Let us investigate how the curves $G \cap A$ lie in A. Since there are no clean disc patches, either

1. $G \cap A$ consists of core circles of A
2. $G \cap A$ consists of radial arcs, which join different boundary components of A

CASE 1. $G \cap A$ consists of core circles of A. Then all the patches of G contained in V are incompressible annuli.

CASE 1.1. Suppose that $G \cap V$ is the union $\cup_{i=1}^{n} P_i$ of annuli such that each annulus P_i has exactly one boundary circle l_i in A and the other in A'. The circles l_i decompose A into annuli $X_i, 0 \leq i \leq n$. Let l_0, l_{n+1} be the boundary circles of A. We may assume that the notation is chosen so that each $X_i, 0 \leq i \leq n$, is bounded by l_i and l_{i+1}, see Fig. 6.8(left).

We will say that for $0 < i \leq n$ the circle l_i is *positive* or *negative*, if the regular switch at l_i pastes together respectively P_i and X_i or P_i and X_{i-1}. It cannot happen that some l_i is positive while l_{i+1} is negative. Otherwise F would contain a connected component composed from P_i, X_i, and P_{i+1}.

Assume that l_1 is negative. Then X_0 and P_1 also produce a connected component of F. If l_1 is positive, then so are $l_2, \ldots, l_n$, and we get a connected component $X_n \cup P_n$ of F again. All three possibilities of getting a connected component of F are shown in Fig. 6.8(right). Since F is assumed to be connected, Case 1.1 cannot occur.

Another proof of the same fact can be obtained by the following counting argument. Cutting $G \cup A$ along all n circles in $G \cap A$, we get a collection of $\geq 2n + 2$ connected surfaces (patches of $G \cup A$). Each switch decreases the number of components by 2 or less. Therefore, having performed all the switches, we get a disconnected surface.

CASE 1.2. Now we can assume that at least one annular patch $P \subset G$ has both boundary circles in A. Then P is parallel rel ∂P to an annulus $P' \subset A$. By an outermost annulus argument, we may assume that P' is also a patch of $G \cup A$, in other words, that the solid torus V' bounded by $P \cup P'$ contains no other patches of $G \cup A$. Since the boundary circles of P are longitudes of V', we can apply Lemma 6.3.18 and get a contradiction.

CASE 2. $G \cap A$ consists of radial arcs. Let us investigate how G intersects A'. For any arc l in $G \cap A'$ we have two possibilities:

1. l is a *return*, i.e., it has both endpoints in the same component of $\partial A'$. Since the patch containing l is boundary incompressible, l does not intersect Γ.
2. l is *radial*, i.e., its endpoints lie in different components of $\partial A'$. Since $\Gamma \cap A'$ consists of core circles of A', l crosses each circle, and since the patch containing l is boundary incompressible, l crosses each circle exactly once.

CASE 2.1. There is a return. It is convenient to denote it by l_2'. Denote also by a_1, a_3 its endpoints, by l_2 the arc in ∂A between them, by l_1, l_3 two radial arcs in $G \cap A$ outgoing from a_1, a_3, and by l_4 the arc in ∂A between the other two ends of these arcs. Let $P_1 \subset A, P_2 \subset G$ be the patches adjacent to l_2, l_2', respectively, see Fig. 6.9. Then P_1 is bounded by $l_1 \cup l_2 \cup l_3 \cup l_4$, and we are in the situation of Lemma 6.3.19. According to this lemma, F is either not minimal, or not in reduced form, a contradiction.

CASE 2.2. Now we may assume that all the arcs in $\partial G \cap A'$ are radial. Then the union of those arcs and the radial arcs in $G \cap A$ consists either of meridional circles of V or of circles that go several times along the meridian and at least once along the longitude of V. In the first case all the patches of G in V are meridional discs, in the second they are annuli.

CASE 2.2.1. All the patches of G in V are meridional discs. Denote them by $P_0', P_1', \ldots, P_{n-1}'$. Let $l_i \subset G, 0 \le i \le n-1$, be the corresponding radial arcs in A that cut P_i' out of G. Then the arcs decompose A into quadrilateral patches $P_0, P_1, \ldots, P_{n-1}$. We can choose notation so that each P_i is adjacent to l_i and l_{i+1} (indices are taken modulo n). As in Case 1.1, we call l_i positive, if the regular switch at l_i pastes P_i' to P_i, and negative otherwise. It cannot

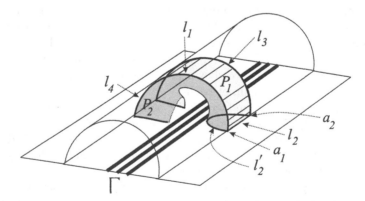

Fig. 6.9. Returns in A' lead to parallel quadrilaterals

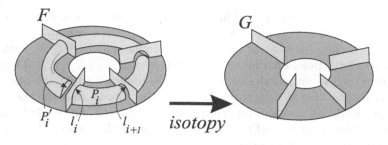

Fig. 6.10. $F = G + A$ is admissibly isotopic to G

happen that l_i is positive while l_{i+1} is negative, since in that case F would contain a component $P_i' \cup P_i \cup P_{i+1}'$ and hence be disconnected. Therefore all l_i have the same sign and F is admissibly isotopic to the simpler surface G, see Fig. 6.10, where all l_i are negative.

CASE 2.2.2. All the patches of G in V are annuli. Let P be one of them. Since any incompressible annulus in a solid torus is parallel to the boundary, any part of P adjacent to A' can be considered as a tunnel with radial base arcs. The tunnel has an evident meridional disc, which does not intersect Γ and hence is a nontrivial boundary compressing disc for P. Therefore case 2.2.2 does not occur. We have now considered all logical possibilities.   □

In contrast to annuli, addition of inessential tori can produce new surfaces, but only finitely many of them up to strong equivalence. Recall that strong equivalence of surfaces in (M, Γ) is generated by homeomorphisms $(M, \Gamma) \to (M, \Gamma)$ whose restrictions onto ∂M are admissibly isotopic to the identity, see Definition 6.3.12.

Proposition 6.3.21. *Let F be a connected two-sided incompressible boundary incompressible normal surface in a triangulated Haken manifold (M, Γ). Suppose that a torus $T_0 \subset \partial M$ contains exactly $m \geq 0$ boundary circles of F and F can be presented in the form $F = G + G'$, where $G \neq \emptyset$ and G' is the union of n disjoint tori parallel T_0. Let either $m = 0$ and $n = 1$ or $m \geq 1$ and $n \geq m$. Then there is a surface $F' \subset (M, \Gamma)$ such that F' is strongly equivalent to F and has a strictly smaller edge degree.*

Proof. We can assume that F is minimal (i.e., it has the smallest edge degree among all admissibly isotopic surfaces) and that $F = G + G'$ is in reduced form. Let $T_1, \ldots, T_n$ be the tori forming G'. It is convenient to number them from ∂M inwards so that T_i lies between T_j and T_0 for all $0 < i < j$.

Denote by V the submanifold of M bounded by T_n and ∂M. Then V is homeomorphic with $S^1 \times S^1 \times I$. By Lemma 4.1.8, $G \cup G'$ has no disc patches, and all the patches of $G \cup G'$ are incompressible. It follows that if G crosses a torus T_i, then it crosses it along circles which are nontrivial and hence mutually parallel in T_i. The same lemma tells us that every patch of $G \cup G'$ contained in V is incompressible and hence is an annulus.

CASE 1. Suppose that G contains an annular patch $A \subset V$ such that both boundary circles of A are contained in a torus $T_i, 1 \leq i \leq n$. Then A is parallel rel ∂A to an annulus $A' \subset T_i$. By an outermost annulus argument, we may assume that A' is a patch of T_i, in other words, that the solid torus bounded by $A \cup A'$ contains no other patches of $G \cup G'$. Since the boundary circles of A are longitudes of the torus, we can apply Lemma 6.3.18 and get a contradiction with our assumptions.

Note that if $m = 0$, i.e., if $\partial F \cap T_0 = \emptyset$, then we always are in Case 1. Hence for $m = 0$ the proof is finished. Further we will assume that $m \geq 1$ and $m = n$ (this is sufficient, since we can ensure this by replacing G by $G + T_{m+1} + \ldots + T_n$.

CASE 2. Suppose that we are not in Case 1, i.e., no annulus from $G \cap V$ has both boundary circles in the same torus T_i. Then $G \cap V$ is the union of annuli $A_j, 0 \leq j \leq m - 1$ such that each annulus A_j has one boundary circle in ∂M, the other one in T_m. We assume that the annuli are numbered in a cyclic order such that each A_j lies between A_{j-1} and A_{j+1} (indices are taken modulo m).

The circles $l_{ij} = T_i \cap A_j$ decompose G' and A_j, respectively, into annuli X_{ij} and Y_{ij} such that $\partial X_{ij} = l_{ij} \cup l_{ij+1}$ and $\partial Y_{ij} = l_{ij} \cup l_{i+1j}$ for $0 \leq i, j < m - 1$, see Fig. 6.11. We will say that l_{ij} is *positive*, if the regular switch at l_{ij} pastes together the annuli Y_{ij} and X_{ij-1}, and *negative*, if it pastes Y_{ij} and X_{ij}. It cannot happen that for some j the curve l_{1j} is positive while l_{1j+1} is negative, otherwise F would contain a connected component composed from Y_{0j}, X_{1j}, and Y_{0j+1}, which is shown in Fig. 6.11 as a dotted line. Therefore all curves l_{1j} have the same sign. Similar considerations show that all curves l_{2j} on T_2 have the same sign as well as the curves on $T_3, \ldots, T_m$. So we may divide the tori T_j into positive and negative. Further, any two neighboring tori $T_j, T_{j+1}, 1 \leq j \leq m - 1$ have the same signs. Otherwise F would not be minimal, since it would be isotopic to the surface $G + (G' \setminus (T_j \cup T_{j+1}))$, which has the smaller edge degree. See Fig. 6.12(left). So we may assume that all T_j

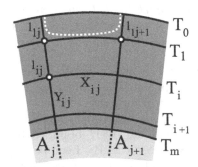

Fig. 6.11. The lattice of T_i and A_j

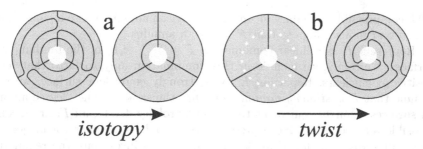

$$\xrightarrow{\quad}$$

isotopy   *twist*

Fig. 6.12. (a) Canceling opposite tori by isotopy; (b) canceling same sign tori by twists

have the same sign. Then F is isotopic rel ∂ to a surface obtained from G by one twist along any of T_i, see Fig. 6.12(right) for the case $m = 3$. Since G has a smaller edge degree, it satisfies the conclusion of the lemma. So we can take $F' = G$.   □

Proof of Theorem 6.3.17. Let k be given. Choose a triangulation of M such that Γ is the union of edges. Construct the finite set of all fundamental surfaces. Let $G_1, \ldots, G_m$ be those of them that have positive p-complexity, $T_1, \ldots, T_n$ all incompressible fundamental tori (necessarily boundary-parallel), $K_1, \ldots, K_r$ all fundamental Klein bottles, and $M_1, \ldots, M_s$ all fundamental Möbius bands. Consider the set $\mathcal{F}'_k$ of normal surfaces in M that can be presented in the form

$$\sum_{i=1}^{m} \gamma_i G_i + \sum_{i=1}^{n} \tau_i T_i + \sum_{i=1}^{r} \varepsilon_i K_i + \sum_{i=1}^{s} \mu_i M_i,$$

where $\gamma_i \geq 0, 0 \leq \tau_i \leq k+2, 0 \leq \varepsilon_i < 2(k+3)$, and $\mu_i = 0, 1$ for all i. Certainly, we take only one surface from every normal isotopy class. Since all G_i are of positive complexity and the coefficients at all other surfaces are bounded, $\mathcal{F}'_k$ is finite (see the proof of Proposition 6.3.13). Let $\mathcal{D}$ be a set of trivial discs and trivial semi-clean discs that represent all the admissible isotopy classes of such discs. This set is also finite, since the number of the discs is bounded by the number of edges of Γ plus the number of connected components of $\partial M \setminus \Gamma$. We add to $\mathcal{D}$ an embedded 2-sphere (necessarily trivial) and all clean incompressible boundary incompressible annuli. The annuli are considered up to admissible isotopy. All of them are parallel to almost clean annuli in ∂M, thus the number of such annuli is also finite.

Let us set $\mathcal{F}_k = \mathcal{F}'_k \cup \mathcal{D}$. We will show that any two-sided incompressible boundary incompressible connected surface F of p-complexity $\leq k$ is strongly equivalent to a surface in $\mathcal{F}_k$. If F is a sphere, or a clean disc, or a trivial semi-clean disc, then, according to the definition of $\mathcal{D}$, F is admissibly isotopic to a surface from $\mathcal{F}_k$. So we may assume that F is not one of the above surfaces and thus is admissibly isotopic to a normal surface by Corollary 3.3.25.

All incompressible boundary incompressible clean annuli are also included into $\mathcal{F}_k$, so we can also suppose that F is not an annulus.

Represent F as a linear combination $F = \sum_{i=1}^{N} k_i F_i$ of the fundamental surfaces. Certainly, we include in this combination only surfaces that have positive coefficients. Replacing F by a strongly equivalent surface, we can assume that F is *strongly minimal*, i.e., has the smallest edge degree among all surfaces strongly equivalent to F. By Corollary 4.1.37, all F_i are incompressible and boundary incompressible, and no F_i is a sphere, a projective plane, or a disc crossing Γ at ≤ 1 points. Let us investigate the remaining surfaces of p-complexity 0.

If F_i is an incompressible torus (necessarily boundary parallel), then $k_i \leq k+2$. Indeed, suppose that $k_i \geq k+3$. Since p-complexity of F does not exceed k, ∂F consists of not more than $k+2$ circles. It follows from Proposition 6.3.21 that in this case F is not strongly minimal, a contradiction.

Let F_i be a Klein bottle. Consider the torus $T = 2F_i$. Suppose that T is compressible. Then M is a closed 3-manifold obtained from a thick Klein bottle $K^2 \tilde{\times} I$ by attaching a solid torus. Any incompressible surface in such a manifold is composed of an incompressible boundary incompressible annulus in $K^2 \tilde{\times} I$ and two meridional discs of the solid torus. Therefore, it is a 2-sphere and thus this case cannot occur (see Remark 6.4.14 for the description of $K^2 \tilde{\times} I$).

If T is incompressible, then it is parallel to ∂M. This can occur only if M is homeomorphic to the thick Klein bottle $K^2 \tilde{\times} I$. We claim that $\varepsilon_i < 2(k+3)$. Suppose, on the contrary, that $\varepsilon_i \geq 2(k+3)$. Then F can be presented as the sum $F' + 2(k+3)F_i = F' + (k+3)T$, where $F' = F - 2(k+3)F_i$ and $T = 2F_i$ is a torus. Just as above, we may conclude that F is not strongly minimal, a contradiction.

It follows from Proposition 6.3.20 that if F_i is a clean incompressible boundary incompressible annulus, then $F = F_i$. Since all such annuli are in $\mathcal{F}_k$, so is F. Suppose that F_i is a clean incompressible Möbius band. Then $2F_i$ is a clean annulus, necessarily parallel to a clean or semi-clean annulus in ∂M. This can occur only if M is a solid torus and Γ consists of disjoint simple closed curves. If $\mu_i \geq 2$, then F can be presented as the sum $(F - 2F_i) + 2F_i$, which is impossible by Proposition 6.3.20 and the minimality of F. Therefore, $\mu_i = 0, 1$.

We have considered all F_i of p-complexity ≥ 0 and found that their coefficients satisfy the required inequalities. Therefore, F is strongly equivalent to a surface from $\mathcal{F}_k$. $\square$

6.4 Jaco–Shalen–Johannson Decomposition

Throughout this prolonged section, our main goal will be to establish one more "finiteness" result, Theorem 6.4.44. The advantage of this theorem compared to Theorem 6.3.17 is that it works for all (not necessarily simple) irreducible boundary irreducible 3-manifolds. However, its statement is weaker, since only tori and annuli are considered. Together with Theorem 6.3.17, it will play a central role in carrying out our task. To prove the theorem, we need to establish a version of the usual JSJ-decomposition theorem [54, 57] for 3-manifolds with boundary pattern. In the case of the empty boundary pattern our approach is very close to that of Neumann and Swarup. In fact, reading the paper [97] helped me significantly to improve the exposition presented in [85].

The JSJ-Decomposition Theorem states that any irreducible boundary irreducible 3-manifold (M, Γ) with boundary pattern contains a canonical system of disjoint essential tori and clean essential annuli that cut (M, Γ) into simple manifolds, Seifert manifolds and I-bundles. In order to obtain an algorithmic version (see Theorem 6.4.42), we need to develop some technique for working with tori and annuli constructively.

6.4.1 Improving Isotopy that Separates Surfaces

Let F, G be two-sided surfaces in an irreducible boundary irreducible 3-manifold (M, Γ) with boundary pattern such that each is either an incompressible torus or a clean incompressible boundary incompressible annulus. Assume that G is admissibly isotopic to a surface that is disjoint from F. Our first aim is to present a very accurate isotopy of G that shifts it from F and consists of simple steps such that each step strictly decreases the number of circles in $F \cap G$ (Theorem 6.4.1). To this end we introduce five *elementary separating moves*. Each move strictly decreases the number of curves in $F \cap G$.

MOVE 1 (*Disc shift*). Assume that $F \cap G$ contains a circle C that is contractible in M. Since F, G are incompressible, C bounds discs in both surfaces, see Lemma 3.3.5. Suppose that C is an innermost trivial circle with respect to G. Then it bounds a disc $D \subset G$ such that $D \cap F = \partial D$. Denote by D_1 the disc in F bounded by C. Since M is irreducible, the sphere $D \cup D_1$ bounds a 3-ball X. We use the ball for constructing an isotopy of G which annihilates C together with other circles in $D_1 \cap G$.

MOVE 2 (*Half-disc shift*). Assume that $F \cap G$ contains a proper arc l that is trivial in G, that is, cuts a half-disc D out of G. We may assume that l is an outermost trivial arc in G. Since F is boundary incompressible, l cuts a half-disc D_1 out of F as well. By boundary irreducibility of (M, Γ), the clean proper disc $D \cup D_1$ cuts a clean 3-ball X out of M. We use the ball for constructing an admissible isotopy of G which annihilates l together with all other circles and arcs in $D_1 \cap G$.

MOVE 3 (*Interior annulus shift*). Assume that the closure of a connected component of $G \setminus F$ is an incompressible annulus $G_1 \subset \text{Int } M$ such that it is parallel rel ∂G_1 to an annulus $F_1 \subset F$. This means that $F_1 \cup G_1$ bound a solid torus $X = D^2 \times S^1$ in M such that $F_1 = \alpha \times S^1, G_1 = \beta \times S^1$, where α, β are two complementary arcs in ∂D^2. One may assume that $F_1 \cap G = \partial F_1$. We use the torus $X = h(F_1 \times I)$ for constructing a shift of G_1 through F_1 which annihilates $\partial F_1 \subset F \cap G$.

MOVE 4 (*Boundary annulus shift*). This move is similar to Move 3. Assume that the closure of a connected component of $G \setminus F$ is an incompressible annulus $G_1 \in$ Int M such that one circle of ∂G_1 (say, C_1) is in ∂M while the other (say, C_2) is in Int M. We require also that G_1 is admissibly parallel to F_1. This means that there is a clean solid torus $X = \Delta^2 \times S^1$ in M such that $G_1 = \alpha \times S^1, F_1 = \beta \times S^1$, and $X \cap \partial M = \gamma \times S^1$, where Δ^2 is a triangle and α, β, γ are its edges. The shift moves G_1 through X to the other side of F, thus eliminating one circle of $F \cap G$.

Move 4 is applicable only when F, G are annuli with disjoint boundaries. For the case of annuli with intersecting boundaries we introduce another move.

MOVE 5 (*Band shift*). Assume that the closure G_1 of a connected component of $G \setminus F$ is a band (homeomorphic image of a rectangle) admissibly parallel to a connected component F_1 of $F \setminus G$ bounded by two radial segments of F. We assume that $F_1 \cap G = F \cap G_1$. The move consists in shifting G_1 through the ball X bounded by F_1, G_1 and ∂M.

All five moves are shown in Fig. 6.13.

Theorem 6.4.1. *Suppose F, G are two proper surfaces in a 3-manifold (M, Γ) with boundary pattern such that each is either an incompressible torus or a clean incompressible annulus and F, G are in general position. If the surfaces can be separated by an admissible isotopy, then they can be separated by Moves 1–5.*

Remark 6.4.2. One can easily see that application of elementary separating moves is an algorithmic procedure.

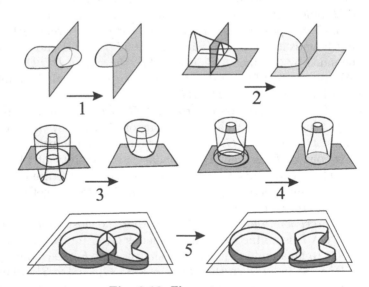

Fig. 6.13. Elementary moves

The proof of the theorem is based on three Lemmas 6.4.3–6.4.5. The first lemma is a partial case of the theorem when F, G can be separated by an admissible parallel shift of F.

Lemma 6.4.3. *Let F, G be two proper surfaces in a 3-manifold (M, Γ) with boundary pattern such that each is either an incompressible torus or a clean incompressible annulus and F, G are in general position. Additionally, we assume that there exists a surface $F' \subset M$ such that $F' \cap G = \emptyset$ and F' is admissibly parallel to F. Then F, G can be separated by Moves 1–5.*

Proof. Since F, F' are admissibly parallel, there is an embedding $h \colon F \times I \to M$ such that $h(F \times \{0\}) = F$, $h(F \times \{1\}) = F'$, and $h(\partial F \times I)$ is a clean subset of ∂M. The image of h is denoted by W. Clearly, W is homeomorphic either to $T^2 \times I$ or to $A^2 \times I$, where T^2 is a torus and A^2 is an annulus.

First we remove all trivial circles and arcs in $F \cap G$ by disc and half-disc shifts of G. F' remains fixed. We get a new surface (still denoted by G). Since $F \cap G$ contains now no trivial circles or arcs, it is either a collection of parallel nontrivial circles or a collection of radial arcs joining different boundary components of F and of G (the latter is possible only if F, G are annuli with intersecting boundaries).

CASE 1. $F \cap G$ consists of nontrivial circles. Since nontrivial circles decompose G into annuli, $G \cap W$ is the union of incompressible annuli. Let $A \subset G \cap W$ be one of them. Then either $\partial A \subset F$ or A has one boundary circle in F and the other in ∂M. Since W is either a thick torus or a thick annulus, in both cases A is parallel to an annulus $A_1 \subset F$. By an outermost annulus argument, we may assume that $G \cap \operatorname{Int} A_1 = \emptyset$. Therefore, one can apply Move 3 or Move 4 to destroy, respectively, two or one circle in $F \cap G$ without touching F'. Applying the moves for as long as possible, we annihilate all the circles in $F \cap G$.

CASE 2. F, G are annuli such that $F \cap G$ consists of radial segments (simple curves that join different boundary circles). In this case W is $A^2 \times I$ and $G \cap W$ consists of bands that have ends in F and lateral sides in different annuli of $W \cap \partial M$. Any such band B is parallel to a part of F. Starting from outermost bands, we apply Move 5 until annihilating all segments in $F \cap G$. □

The next lemma can be considered as an inductive step.

Lemma 6.4.4. *Let F, G be two proper surfaces in a 3-manifold (M, Γ) with boundary pattern such that each is either an incompressible torus or a clean incompressible annulus and F, G are in general position. Additionally, we assume that there exists a surface $F' \subset M$ such that F' is admissibly parallel to F and G can be separated from F' by Moves 1–5. Then G can also be separated from F by Moves 1–5.*

Proof. As in the proof of Lemma 6.4.3, we begin with removing all trivial circles and arcs in $G \cap (F \cup F')$ by Moves 1, 2. Then we apply to F', G the

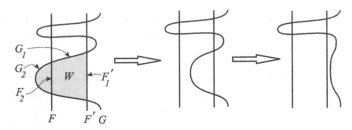

Fig. 6.14. Any shift of G through F' is a superposition of shifts of G through F and a shift of G through F' far from F

sequence of Moves 3–5 that shifts G from F'. Each of these moves shifts a part G_1 of G through a part F'_1 of F'. Before carrying out the move, we apply Moves 3–5 to G, F in order to annihilate all curves in $G_1 \cap F$. See Fig. 6.14. Each of these moves shifts a component of $G_2 = \mathrm{Cl}(G_1 \setminus (F \cap G_1))$ (an annulus or a band) through the parallel component of $F_2 = \mathrm{Cl}(F \setminus (F \cap G_1))$. Such parallel pairs do exist, since the manifold W bounded by F'_1, G_1 is either a thick annulus or a thick disc and hence has the same property as earlier: any proper annulus or band in W is parallel to the boundary.

After getting $F' \cap G = \emptyset$, we apply Lemma 6.4.3 to surfaces F, G for annihilating all curves in $F \cap G$.   □

The last of the three lemmas shows that any admissible isotopy that separates F and G can be replaced by a superposition of admissible parallel displacements of F.

Lemma 6.4.5. *Let F, F' be admissibly isotopic surfaces in a 3-manifold (M, Γ). Then there exists a sequence of surfaces $F_0, \ldots, F_n \subset M$ such that $F' = F_0$, $F = F_n$, and F_i is disjoint from and admissibly parallel to F_{i+1} for all $0 \leq i < n$.*

Proof. Let $h_t : F \to M, 0 \leq t \leq 1$, be an admissible isotopy such that $h_0(F) = F'$ and $h_1(F) = F$. Decompose the time interval $[0, 1]$ into so small subintervals $[\varepsilon_k, \varepsilon_{k+1}], 0 = \varepsilon_1 < \varepsilon_2 < \ldots < \varepsilon_m = 1$ that for $\varepsilon_k \leq t \leq \varepsilon_{k+1}$ the surface $h_t(F)$ remains strictly inside a regular neighborhood $N_k \approx F \times I$ of $h_{\varepsilon_k}(F)$. Put $F_{2k} = h_{\varepsilon_k}(F)$. Then we replace the isotopy $h_t, \varepsilon_k \leq t \leq \varepsilon_{k+1}$, by the superposition of two isotopic shifts. The first shift moves F_{2k} to a surface F_{2k+1} which is very close to a boundary component of N_k, the second one moves F_{2k+1} to F_{2k+2}, see Fig. 6.15. By construction, F_i is parallel to F_{i+1} for all $0 \leq i < n$, where $n = 2m$.   □

Proof of Theorem 6.4.1. Since G can be shifted from F, F is admissibly isotopic to a surface F' disjoint from F. By Lemma 6.4.5, there exists a sequence of surfaces $F_0, \ldots, F_n$ such that $F' = F_0$, $F = F_n$, and F_i is admissibly parallel to F_{i+1} for all $0 \leq i < n$. Then the conclusion of Theorem 6.4.1 can be easily obtained by induction on n. Lemmas 6.4.3 and 6.4.4 supply us with the base of the induction and the inductive step.   □

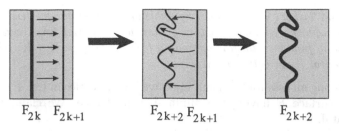

Fig. 6.15. F_{2k+1} is parallel to both F_{2k} and F_{2k+1}

Corollary 6.4.6. *Let $F_i, 1 \le i \le n$, be disjoint incompressible tori and clean incompressible boundary incompressible annuli in a 3-manifold (M, Γ). Suppose $G \subset M$ is another incompressible torus or clean incompressible, boundary incompressible annulus that can be shifted from each F_i by an admissible isotopy. Then one can construct an admissible isotopy $M \to M$ that shifts G from $\cup_{i=1}^{n} F_i$ and keeps fixed those surfaces F_i that were disjoint from G from the very beginning.*

Proof (By Induction on n). For $n = 1$ the conclusion coincides with the assumption. Assume that the assertion of the theorem has been proved for $n - 1$ surfaces F_i. Now consider n surfaces. By the inductive assumption, G can be shifted from $F_1, \ldots, F_{n-1}$, so we can assume that G is already disjoint from $F_i, 1 \le i \le n - 1$. Then we apply Theorem 6.4.1 to construct a sequence of elementary separating moves that shift G from F_n. Each elementary move shifts a part G_1 of G through X, where X is either a clean ball or a solid torus bounded by G_1 and the corresponding part of F_n. Since for $i, 1 \le i \le n - 1$, $G \cap F_i = F_n \cap F_i = \emptyset$, and since X contains no incompressible tori or incompressible boundary incompressible annuli, we can conclude that $X \cap F_i = \emptyset$. Therefore the shift of G_1 through X can be carried out so that G would remain disjoint from F_i. Doing so, we annihilate all intersections of G and F_n keeping G disjoint from other F_i. □

6.4.2 Does M^3 Contain Essential Tori and Annuli?

In this section, we describe algorithms for finding essential tori and annuli in 3-manifolds. We need them to prove an algorithmic version of the JSJ-decomposition Theorem. In case the algorithmic approach is irrelevant, it is possible to proceed directly to the statement and proof of the theorem in Sect. 6.4.4.

Recall that an incompressible torus T in a 3-manifold (M, Γ) is called essential, if it is not parallel to a torus in ∂M. An incompressible boundary incompressible clean annulus A in (M, Γ) is essential, if it is not parallel rel ∂ to an almost clean annulus in ∂M.

Lemma 6.4.7. *Let (M, Γ) be an irreducible orientable triangulated 3-manifold. If (M, Γ) contains an essential torus, then an essential torus can be found either among fundamental tori or among tori of the type $K + K$, where K is a fundamental Klein bottle in (M, Γ).*

Proof. Among all essential tori and injective Klein bottles in (M, Γ) we choose a normal surface F having the minimal edge degree. We claim that F is fundamental.

Suppose, on the contrary, that there is a nontrivial presentation $F = G_1 + G_2$. By Lemma 3.3.30 and Theorem 4.1.36, we can assume that G_1, G_2 are connected, incompressible, and different from S^2, RP^2. Since $\chi(F) = 0$, the additivity of the Euler characteristic tells us that each G_i is a torus parallel to the boundary (we cannot get a Klein bottle or a torus not parallel to the boundary because of minimality of F). It follows from Proposition 6.3.21 (case $m = 0, n = 1$) that F is not minimal, a contradiction.

To conclude the proof, we note that F is either an essential torus (then we are done) or an injective Klein bottle. In the second case the surface $K + K$ is an incompressible torus. It cannot be parallel to the boundary, since otherwise M would be homeomorphic to $K^2 \tilde{\times} I$, which contains no essential tori. Thus the torus $K + K$ is essential.   □

Lemma 6.4.8. *Let (M, Γ) be a triangulated Haken manifold and U, V two subsets of ∂M consisting of connected components of $\partial M \setminus \Gamma$ (the cases $U = \partial M \setminus \Gamma$ or $V = \emptyset$ are allowed). Suppose that (M, Γ) contains a clean essential annulus having at least one boundary circle in U and no boundary circles in V. Then such an annulus can be found either among clean fundamental annuli or among annuli of the type $B + B$, where B is a clean fundamental Möbius band in (M, Γ).*

Proof. Consider the set of all surfaces F in (M, Γ) possessing the following properties:

1. F is either a clean essential annulus or a clean injective Möbius band
2. $U \cap \partial F \neq \emptyset$ and $V \cap \partial F = \emptyset$

From among all surfaces of this set we choose a normal surface F_0 having the smallest edge degree. We claim that F_0 is fundamental.

Suppose, on the contrary, that there is a nontrivial presentation $F_0 = G_1 + G_2$. By Lemma 3.3.30 and Theorem 4.1.36, we can assume that G_1, G_2 are connected, incompressible, boundary incompressible, and different from S^2, RP^2, and clean D^2. To be definite, we suppose that $U \cap \partial G_1 \neq \emptyset$. Since $\chi(G_1) + \chi(G_2) = \chi(F) = 0$ and S^2, RP^2, D^2 are the only connected surfaces with positive Euler characteristics, $\chi(G_1) = \chi(G_2) = 0$. It follows that G_1 is either an annulus (essential or not), or a Möbius band. If G_1 is an essential annulus or a Möbius band, we get the direct contradiction with the choice of F_0. If G_1 is inessential annulus, i.e., if it is parallel rel ∂ to an almost clean annulus in ∂M, then F_0 is not minimal by Lemma 6.3.20, and we get a contradiction again.   □

Lemma 6.4.9. *There are algorithms to decide if a given torus or a clean proper annulus in an irreducible boundary irreducible 3-manifold (M, Γ) is parallel to a torus or, respectively, is admissibly parallel rel ∂ to an almost clean annulus in ∂M.*

Proof. Let A be a clean proper annulus in (M, Γ). We investigate whether A decomposes (M, Γ) into two parts and whether one of these parts is a solid torus for which the boundary circles of A are longitudes. The latter can be easily done by finding a nontrivial compressing disc D and calculating the intersection number of ∂D with a circle from ∂A (compare with Corollary 4.1.14). Obviously, A is boundary parallel if and only if both questions are answered positively.

Similarly, let T be a torus in (M, Γ). We investigate whether T decomposes (M, Γ) into two parts, and whether one of them (denote it by W) is homeomorphic to $S^1 \times S^1 \times I$. The latter can be done by constructing an essential annulus A in W. Indeed, W is $S^1 \times S^1 \times I$ if and only the 3-manifold W_A, obtained by cutting W along A, is a solid torus for which the traces of ∂A under cutting are longitudes. $\qquad\square$

Theorem 6.4.10. *There exist algorithms to decide whether an irreducible boundary irreducible 3-manifold (M, Γ) contains an essential torus and whether it contains a clean essential annulus. Moreover, if U, V are two subsets of ∂M consisting of connected components of $\partial M \setminus \Gamma$, then one can decide algorithmically whether (M, Γ) contains an essential annulus having at least one boundary circle in U and no boundary circles in V. In all these cases a surface of the required type can be constructed algorithmically (if it exists).*

Proof. To find an essential torus, we triangulate (M, Γ) and construct all fundamental surfaces. Then we construct the set $T_1, \ldots, T_n$ of all fundamental tori and tori of the type $K + K$, where K is a fundamental Klein bottle. Lemma 6.4.7 tells us that (M, Γ) contains an essential torus if and only if at least one of T_i is essential, i.e. incompressible and not parallel to a component of ∂M. It remains to test each T_i for possessing these properties. That can be done by Theorem 4.1.15 and Lemma 6.4.9.

Also, (M, Γ) contains a clean essential annulus if and only if one of the clean fundamental annuli and annuli of the type $B + B$, where B is a clean fundamental Möbius band, is essential (Lemma 6.4.8). We look for such annuli using Theorems 4.1.15, 4.1.19, and Lemma 6.4.9. If we are interested in essential annuli having a boundary circle in U and no boundary circles in V, we restrict our attention to fundamental annuli and annuli of the type $B + B$ which possess this property. $\qquad\square$

Corollary 6.4.11. *Let (M, Γ) be a Haken manifold and C a simple closed curve on $\partial M \setminus \Gamma$. Then one can algorithmically decide if (M, Γ) contains an essential annulus A such that C is a boundary circle of A.*

Proof. Let N be a regular neighborhood of C in $\partial M \setminus \Gamma$. We apply the algorithm described in Theorem 6.4.10 to the 3-manifold $(M, \Gamma \cup \partial N)$ and $U = \text{Int } N, V = \partial M \setminus (\Gamma \cup \partial N)$. Obviously, (M, Γ) contains an essential annulus with a boundary circle C if and only if $(M, \Gamma \cup \partial N)$ contains an essential annulus having a boundary circle in U.   □

6.4.3 Different Types of Essential Tori and Annuli

In this step we will classify tori and annuli according to how they intersect each other (Definitions 6.4.12 and 6.4.13). Not only will this classification be useful in progressing towards our final goal, it also happens to be algorithmic, which is what the next section is all about.

Consider an irreducible boundary irreducible 3-manifold (M, Γ).

Definition 6.4.12. *Let F be an incompressible torus or a clean incompressible boundary incompressible annulus in (M, Γ). Then F is called* rough *if any incompressible torus and any clean incompressible boundary incompressible annulus in M is admissibly isotopic to a surface disjoint from F.*

The terminology is derived from the observation that people tend to avoid contacts with a rough person. Let us introduce two types of essential annuli in (M, Γ).

Definition 6.4.13. *A clean incompressible boundary incompressible annulus $A \subset (M, \Gamma)$ is called* longitudinal *if any clean incompressible boundary incompressible annulus $A_1 \subset (M, \Gamma)$ is admissibly isotopic to an annulus A_1' such that $\partial A \cap \partial A_1' = \emptyset$. Otherwise A is called* transverse.

It follows from the definition that all rough annuli are longitudinal, while transverse annuli are never rough. Note that if A and A_1 are any clean incompressible boundary incompressible annuli in (M, Γ), then all trivial circles and trivial arcs in $A \cap A_1$ can be easily eliminated by an admissible isotopy of A_1. After that $A \cap A_1$ will either be empty, or consist of core circles of both annuli, or consist of their common radial segments. A is transverse if and only if for at least one annulus A_1 we can never get the first or the second option.

Example 6.4.14. Let us investigate the types of annuli and tori in $(K^2 \tilde{\times} I, \emptyset)$, the orientable twisted product of the Klein bottle K^2 and I. It is convenient to represent $K^2 \tilde{\times} I$ as the "thick" cylinder $S^1 \times I \times I$ with its base annuli $S^1 \times I \times \{0, 1\}$ identified. The identification map $\varphi \colon S^1 \times I \times \{0\} \to S^1 \times I \times \{1\}$ can be presented as the superposition of an involution $\alpha \colon S^1 \times I \times \{0\} \to S^1 \times I \times \{0\}$ having exactly two fixed points and the parallel translation $\tau \colon S^1 \times I \times \{0\} \to S^1 \times I \times \{1\}$ given by the rule $\tau(x, y, 0) = (x, y, 1)$. To be definite, we will assume that the restriction of α onto $S^1 \times \{1/2\} \times \{0\}$ is the reflection on a diameter and that α preserves the decomposition of $S^1 \times I \times \{0\}$ into radial segments. Thus the two fixed points p, q of α are the intersection points of $S^1 \times \{1/2\} \times \{0\}$ and the diameter. See Fig. 6.16.

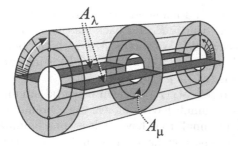

Fig. 6.16. Two essential annuli in the thick Klein bottle

Alternatively, one can present $K^2 \tilde{\times} I$ as the twisted product $M^2 \tilde{\times} S^1$ (the fibers have the form $S^1 \times \{*\} \times \{*\}$). It admits also a structure of a Seifert manifold fibered over D^2 with two exceptional fibers with parameters $(2,1), (2,-1)$ (see Remark A.1.1 for the definition of Seifert manifolds). Any regular fiber is obtained by gluing two segments of the type $\{x\} \times I$ and $\{\varphi(x)\} \times I$ while each of the segments $\{p, q\} \times I$ gives an exceptional fiber.

There is only one incompressible torus T in $K^2 \tilde{\times} I$, which is parallel to the boundary. Recall that, up to isotopy, K^2 contains exactly two different orientation preserving simple closed curves: the meridian μ, which does not decompose K^2, and the longitude λ, which decomposes K^2 into two Möbius bands (see Example 3.2.7). The annuli $A_\mu = \mu \tilde{\times} I$ and $A_\lambda = \lambda \tilde{\times} I$ are essential. Any other essential annulus in $K^2 \tilde{\times} I$ is isotopic to the one of them.

It is easy to see that A_μ, A_λ are transversal. If we forbid one of them by inserting an appropriate boundary pattern (for example, a circle parallel to boundary components of the other annulus), then the other annulus becomes longitudinal, but still not rough. If we forbid also the second annulus by taking a pattern that decomposes $\partial(K^2 \tilde{\times} I)$ into 2-cells, then T becomes rough.

This information helps us to answer the following question: which 3-manifolds can be obtained by attaching a solid torus to $K^2 \tilde{\times} I$? The answer is that they are all Seifert. Indeed, $K^2 \tilde{\times} I$ possesses two Seifert fibrations; A_μ is saturated (i.e., consists of fibers) with respect to the one of them, A_λ is saturated with respect to the other. If the meridian m of the attached torus is not isotopic to μ or λ, then both Seifert fibrations on $K^2 \tilde{\times} I$ can be extended to Seifert fibrations on the resulting 3-manifold. It is easy to see that in both exceptional cases, when m is isotopic to μ or λ, we get reducible 3-manifolds: $S^2 \times S^1$ (if m is isotopic to μ) and $RP^3 \# RP^3$ (if m is λ). $\qquad\square$

The aim of this section is to show how one can determine the type of a given torus and a given annulus algorithmically. This is done in Theorem 6.4.23 for tori and in Theorem 6.4.28 for annuli.

THE TORUS CASE. Let F be an incompressible torus in an orientable 3-manifold (M, Γ). We wish to know if it is rough. The answer depends on how incompressible annuli approach to F from different sides. It is convenient to cut M along F and get a 3-manifold M_F containing two copies F_+, F_- of F in the boundary. If we identify F_+, F_- back via the natural identification map $\varphi \colon F_- \to F_+$, we recover M. We supply M_F with the boundary pattern Γ_F, which is a copy of Γ in ∂M_F. Of course, if (M, Γ) is irreducible and boundary irreducible, then so is (M_F, Γ_F).

We introduce two sets $\mathcal{A}_\pm(F)$ of clean essential annuli in (M_F, Γ_F). The set $\mathcal{A}_-(F)$ consists of annuli having at least one boundary circle in F_-, the set $\mathcal{A}_+(F)$ consists of annuli that have at least one boundary circle in F_+. If $A \in \mathcal{A}_-(F)$ has only one boundary circle in F_-, then this circle is denoted by $\partial_- A$. If both boundary circles of A are in F_-, then $\partial_- A$ is one of them (the other is parallel to it). Similarly, if $A \in \mathcal{A}_+(F)$, then $\partial_+ A$ is a circle in $\partial A \cap F_+$.

Now we recall the definition of the intersection number of curves on a surface. Let C_1, C_2 be two transversal closed oriented 1-dimensional submanifolds of an oriented surface G. If p is a crossing point of C_1, C_2, then the orientations of C_1, C_2 determine a local orientation of G at p. We assign to p its sign $\varepsilon = \pm 1$ by setting $\varepsilon = +1$ if the local orientation at p agrees with the global orientation of G, and $\varepsilon = -1$, if not. The intersection number $\lambda(C_1, C_2)$ is the sum of the signs of all the crossing points.

It is well known that $\lambda(C_1, C_2)$ is invariant under isotopy (and even homotopy) of C_1, C_2 and hence is determined also for 1-dimensional submanifolds which are not in general position. If C_1, C_2 and G are not oriented, but C_1, C_2 are connected and G is orientable, then $\lambda(C_1, C_2)$ is also defined, but only up to sign. Evidently, if G is a torus and $C_1, C_2 \subset G$ are nontrivial simple closed curves, then C_1, C_2 are isotopic if and only if $\lambda(C_1, C_2) = 0$.

In our situation we will use the intersection number in a slightly more general sense. Recall that we have an incompressible torus F in an orientable 3-manifold (M, Γ) and two copies $F_\pm$ of F in the boundary of M_F. Let C_1, C_2 be two circles in $F_- \cup F_+$. Then their intersection number $\lambda'(C_1, C_2)$ is defined as the intersection number of the corresponding curves in F. The case when one circle is in F_-, the other in F_+ is also allowed. If both C_1, C_2 lie in the same surface $F_\pm$, then $\lambda'(C_1, C_2) = \lambda(C_1, C_2)$.

We are ready now to suggest a simple criterion for an incompressible torus to be rough.

Proposition 6.4.15. *Let F be an incompressible torus in an orientable Haken manifold (M, Γ) with boundary pattern. Then F is not rough if and only if there exist annuli $A \in \mathcal{A}_-(F), B \in \mathcal{A}_+(F)$ such that $\lambda'(\partial_- A, \partial_+ B) = 0$.*

Proof $(\Longrightarrow)$. Assume that F is not rough. Then there is a surface $G \subset (M, \Gamma)$ which is either an incompressible torus or a clean incompressible boundary incompressible annulus and which cannot be shifted from F by an admissible isotopy. Applying to F, G elementary separating moves for as long as possible, we get a new surface of the same type (denoted also by G) that still intersects F. Then the intersection of F and G consists of nontrivial circles. If G is a torus and $F \cap G$ consists of only one circle C, then F cuts G into an annulus

Fig. 6.17. (a) $F \cap G$ decomposes G into annuli; (b) A and two copies of B constitute G

$A \subset M_F$ such that $A \in \mathcal{A}_-(F) \cap \mathcal{A}_+(F)$ and $\lambda'(\partial_-A, \partial_+A) = \lambda(C, C) = 0$. If G is an annulus or $F \cap G$ is the union of ≥ 2 circles, then the circles $F \cap G$ decompose G into a collection $\{A_1, \ldots, A_k\}, k \geq 2$ of clean annuli such that each annulus has at least one boundary circle in $F_- \cup F_+$, see Fig. 6.17a.

The annuli are essential, since otherwise we could perform elementary separating moves. Obviously, $\mathcal{A}_-(F)$ contains at least $[k/2]$ of them, and the same is true for $\mathcal{A}_+(F)$. Since G is embedded, all the boundary circles of the annuli are parallel in F and thus have zero intersection number.

($\Longleftarrow$) Suppose that there exist annuli $A \in \mathcal{A}_-(F), B \in \mathcal{A}_+(F)$ such that $\lambda'(\partial_-A, \partial_+B) = 0$. To prove that F is not rough, we find an incompressible torus or a clean proper incompressible boundary incompressible annulus in (M, F) that cannot be shifted from F.

CASE 1. Suppose that there exists an annulus $R \in \mathcal{A}_-(F) \cap \mathcal{A}_+(F)$ such that $\lambda'(\partial_-R, \partial_+R) = 0$. We can assume that the identification map $\varphi \colon F_- \to F_+$ takes ∂_-R to ∂_+R. After the identification we get either a torus in M, which intersects F along one circle, or a Klein bottle in M, which also intersects F along one circle and which we transform into a torus by taking the boundary of its regular neighborhood. In both cases the torus G thus obtained and F admit no elementary separating moves. It follows from Theorem 6.4.1 that G cannot be shifted from F by an isotopy.

CASE 2. Suppose that there is no annulus $R \in \mathcal{A}_-(F) \cap \mathcal{A}_+(F)$ such that $\lambda'(\partial_-R, \partial_+R) = 0$. According to our assumption, one can find two clean essential annuli $A \in \mathcal{A}_-(F), B \in \mathcal{A}_+(F)$ such that $\lambda'(\partial_-A, \partial_+B) = 0$. Applying to them elementary separating moves for as long as possible, we get two new annuli (still denoted by A, B) such that their intersection either is empty or consists of core circles of both annuli. The second option is impossible, since otherwise it would be easy to construct an annulus $R \in \mathcal{A}_-(F) \cap \mathcal{A}_+(F)$ with $\lambda'(\partial_-R, \partial_+R) = 0$ as follows. Let C be a unique circle in $A \cap B$ such that the annulus $A_1 \subset A$ bounded by C and ∂_-A contains no other circles of $A \cap B$. Denote by B_1 the annulus in B bounded by C and ∂_+B. Then $R = A_1 \cup B_1$ is in $\mathcal{A}_-(F) \cap \mathcal{A}_+(F)$. Therefore we can conclude that A, B are disjoint.

CASE 2.1. $\partial A \subset F_-, \partial B \subset F_+$. Since $\lambda'(\partial_-A, \partial_+B) = 0$, we can assume that φ takes ∂A to ∂B. Then, just as in Case 1, the union of A and B determines a torus or a Klein bottle in M, which we convert into a torus by taking the boundary of its regular neighborhood. It follows from Theorem 6.4.1 that in both cases the resulting torus cannot be shifted from F.

CASE 2.2. $\partial A \subset F_-$ and only one circle of ∂B is in F_+. Then we construct a clean incompressible annulus $G \subset (M, \Gamma)$ from a copy of A in M and two parallel copies of B in M, see Fig. 6.17b. Again, Theorem 6.4.1 assures us that one cannot shift G from F. The case when $\partial B \subset F_+$ and only one circle of ∂A is in F_- is similar.

CASE 2.3. Each of the annuli A, B has only one boundary circle in $F_- \cup F_+$. This case is similar to Case 2.2: We construct $G \subset M$ as the union of a copy of A and a copy of B.

We can conclude that in all cases F is not rough.   □

This criterion is not satisfactory, since it is not algorithmic. Clearly, we can use Theorem 6.4.10 for finding two annuli $A \in \mathcal{A}_-(F), B \in \mathcal{A}_+(F)$. Suppose that they do exist. Then F is not rough provided that $\lambda'(\partial_- A, \partial_+ B) = 0$. But if $\lambda'(\partial_- A, \partial_+ B) \neq 0$, then the criterion tells us nothing. Nevertheless, it turns out that with exactly two types of exceptions the inequality $\lambda'(\partial_- A, \partial_+ B) \neq 0$ guarantees us that F is rough. To clarify the appearance of these exceptions, we recall the notion of Stallings manifold [118].

Definition 6.4.16. *Let S be a surface and $f \colon S \to S$ a homeomorphism. Then the* Stallings manifold *$M_f = S \times I / f$ with fiber S and monodromy homeomorphism f is obtained from $S \times I$ by identifying each point $(x, 1) \in S \times \{1\}$ with the point $(f(x), 0) \in S \times \{0\}$.*

If S is a torus with a fixed coordinate system, then f can be presented by the monodromy matrix A_f of order 2. The trace $\mathrm{Tr}(A_f)$ of A_f does not depend on the choice of the coordinate system on S.

Lemma 6.4.17. *Let (M, Γ) be an orientable Haken manifold with boundary pattern and F an essential torus in (M, Γ). Suppose that for $\varepsilon = +$ or for $\varepsilon = -$ there exist annuli $A, A_1 \in \mathcal{A}_\varepsilon(F)$ such that $\lambda(\partial_\varepsilon A, \partial_\varepsilon A_1) \neq 0$. Then either M is a Stallings manifold with fiber F or F bounds in M a thick Klein bottle $K^2 \tilde\times I$.*

Proof. By symmetry, we can assume that $\varepsilon = -$. Since A, A_1 are essential, we can eliminate all trivial circles and trivial arcs in $A \cap A_1$ by admissible isotopies of the annuli. It follows that the intersection of the new annuli (still denoted by A, A_1) consists of radial arcs of both annuli.

Denote by X the union of A and A_1. Then the natural fibrations of A and A_1 into radial segments determine an I-fibration of X.

CASE 1. Suppose that the I-fibration of X is trivial, i.e., $X = Y \times I$, where Y is a graph. Then $Y \times \partial I$ consists of two copies of Y such that one copy, say, Y_0, is in F_- while the other, Y_1, is in $\partial M_F \setminus F_-$. Obviously, Y_0 coincides with the union of $\partial_- A$ and $\partial_- A_1$, which decomposes F_- into 2-cells. See Fig. 6.18.

Denote by $N = N(X \cup F_-)$ a regular neighborhood of $X \cup F_-$ in (M_F, Γ_F). Then its *interior boundary* $\partial_{int}(N) = \mathrm{Cl}(\partial N \setminus \partial M_F)$ is the disjoint union of clean proper discs. Since (M_F, Γ_F) is Haken, these discs cut clean balls out

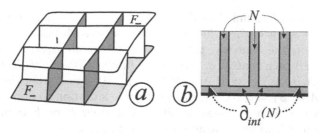

Fig. 6.18. (a) Y_0 decomposes F_- into 2-cells. (b) $\partial_{int}N$ consists of clean discs

of M_F. Adding these balls to N, we get a clean connected component of M_F homeomorphic to $F_- \times I$. One boundary torus of this component is F_-, the other must be F_+ (otherwise F would be parallel to a clean torus in ∂M, which contradicts our assumption that F is essential). It follows that M is a Stallings manifold with fiber F.

CASE 2. Suppose that the fibering of X is nontrivial, i.e., $X = Y \tilde{\times} I$, where Y is a graph. Then $Y \tilde{\times} \partial I$ is connected and coincides with $\partial A \cup \partial A_1 \subset F_-$. In this case the interior boundary $\partial_{int}N$ of $N = N(X \cup F_-)$ consists of 2-spheres. By irreducibility of M_F, they must be trivial. Filling the spheres by 3-balls, we get a connected component $(M_F)_0$ of M_F bounded by the torus F_-. As in Case 1, the I-bundle structure on X can be extended to an I-bundle structure on $(M_F)_0$. It remains to observe that the unique I-bundle whose boundary is a torus is $K^2 \tilde{\times} I$. □

Remark 6.4.18. Suppose that under the assumptions of Lemma 6.4.17, M turns out to be a Stallings 3-manifold with fiber F. It follows from the proof of the lemma that a Stallings structure on M, i.e., a homeomorphism $M_F \to F \times [0,1]$ and the corresponding monodromy homeomorphism $F \to F$ can be constructed algorithmically. In particular, we can calculate the trace of the monodromy matrix.

Similarly, if a connected component $(M_F)_0$ of M_F is homeomorphic to $K^2 \tilde{\times} I$, then a homeomorphism $(M_F)_0 \to K^2 \tilde{\times} I$ can also be constructed algorithmically.

The following proposition supplements Proposition 6.4.15.

Proposition 6.4.19. *Let F be an incompressible torus in an orientable Haken manifold (M, Γ) with boundary pattern. Suppose that M is not a Stallings manifold with fiber F and that F does not bound a thick Klein bottle in M. Then F is rough if and only if either at least one of the sets $\mathcal{A}_-(F), \mathcal{A}_+(F)$ is empty or there exist annuli $A \in \mathcal{A}_-(F), B \in \mathcal{A}_+(F)$ such that $\lambda'(\partial_- A, \partial_+ B) \neq 0$.*

Proof. It follows from Proposition 6.4.15 that F is rough if and only if either at least one of the sets $\mathcal{A}_-(F), \mathcal{A}_+(F)$ is empty or $\lambda'(\partial_- A, \partial_+ B) \neq 0$ for all pairs of annuli $A \in \mathcal{A}_-(F), B \in \mathcal{A}_+(F)$. It remains to prove that if for at least one pair of annuli $A \in \mathcal{A}_-(F), B \in \mathcal{A}_+(F)$ the intersection number $\lambda'(\partial_- A, \partial_+ B)$ is not 0, then the same holds for all such pairs.

Suppose, on the contrary, that there exist annuli $A, A_1 \in \mathcal{A}_-(F)$ and $B, B_1 \in \mathcal{A}_+(F)$ such that $\lambda'(\partial_- A, \partial_+ B) \neq 0$, but $\lambda'(\partial_- A_1, \partial_+ B_1) = 0$. Then either $\lambda(\partial_- A, \partial_- A_1) \neq 0$ or $\lambda(\partial_+ B, \partial_+ B_1) \neq 0$, since otherwise we would have $\lambda'(\partial_- A, \partial_+ B) = \lambda'(\partial_- A_1, \partial_+ B_1)$. In this situation Lemma 6.4.17 tells us that either M is a Stallings manifold with fiber F or F bounds a thick Klein bottle in M, which contradicts our assumption. $\square$

Before describing the algorithm to decide whether a given incompressible torus F in a given 3-manifold (M, Γ) is rough, we investigate the aforementioned exceptional cases:

1. M is a Stalling 3-manifold with fiber F
2. F bounds $K^2 \widetilde{\times} I$ in M

To be more precise, we explain how to recognize exceptions and how to determine the type of F, if we encounter one of them. The following three lemmas solve these problems.

Lemma 6.4.20. *There are algorithms to decide if a given 3-manifold M is homeomorphic to $T^2 \times I$ or to $K^2 \widetilde{\times} I$.*

Proof. Let M be a given 3-manifold. The desired algorithm can be described as follows (compare with Lemma 6.4.9 and Example 6.4.14).

1. We check whether M is orientable and Haken. If not, then M is neither $T^2 \times I$ nor $K^2 \widetilde{\times} I$.
2. We look for an incompressible boundary incompressible annulus $A \subset M$, see Theorem 6.4.10. If A does not exist, then $M \neq T^2 \times I, K^2 \widetilde{\times} I$. Otherwise, we insert A into M.
3. Suppose that A decomposes M into two solid tori such that A runs twice along their longitudes. Then $M = K^2 \widetilde{\times} I$.
4. Suppose that after cutting M along A we get a solid torus Q such that each of the two copies of A in ∂Q runs once along a longitude of Q. Then $M = K^2 \widetilde{\times} I$, if ∂M is a torus, and $M = T^2 \times I$, if ∂M consists of two tori.
5. If neither 3 nor 4 happen, then $M \neq T^2 \times I, K^2 \widetilde{\times} I$. $\square$

Now that we know how to recognize the exceptional cases, let us see what we do when the first of them (a Stallings 3-manifold) comes our way.

Lemma 6.4.21. *Let $M_f = T^2 \times I/f$ be an orientable Stallings manifold, where T^2 is a torus and $f: T^2 \to T^2$ an orientation preserving homeomorphism. Then the fiber $T^2 \times \{0\}$ of M_f is not rough if and only if for some (and thus for every) coordinate system on T^2 the trace of the monodromy matrix A_f of f is equal to ± 2.*

Proof. Suppose that the trace of A_f equals ± 2. Then the characteristic polynomial of A_f is $\lambda^2 \pm 2\lambda + 1$ and A_f has an eigenvector (p, q) with coprime integer coordinates whose eigenvalue is ± 1. This means that f can be changed by an isotopy such that afterwards it would take a simple closed curve $C \subset T^2$ of type (p, q) to itself. Then $G = C \times I/f \subset T^2 \times I/f$ is either an incompressible torus or a Klein bottle in M_f such that the boundary of its regular neighborhood is an incompressible torus. In both cases the torus cannot be shifted from $T^2 \times \{0\}$ by an isotopy. This means that $T^2 \times \{0\}$ is not rough.

Conversely, suppose that an incompressible torus $G \subset M_f$ cannot be shifted from $T^2 \times \{0\}$ by an isotopy. We can assume that the pair $(G, T^2 \times \{0\})$ does not admit elementary separating moves. Then all circles in $G \cap (T^2 \times \{0\})$ are nontrivial and decompose G into annuli, say $A_1, \ldots, A_n$. These annuli, considered as annuli in $T^2 \times I$, are essential and hence run from $T^2 \times \{0\}$ to $T^2 \times \{1\}$. If there is only one such annulus, i.e., if $G \cap (T^2 \times \{0\})$ consists of only one circle C, then C is invariant with respect to f and thus determines an eigenvector of A_f with eigenvalue ± 1. If there are more, then f takes any circle $C \subset G \cap (T^2 \times \{0\})$ to a circle which is disjoint from C and thus parallel to it. Therefore C again determines an eigenvector of A_f with eigenvalue ± 1. Evidently, ± 1 is an eigenvalue of A_f if and only if the trace is ± 2. $\square$

Suppose now that we have stumbled upon the second of our problematic cases, which is the thick Klein bottle. Recall that $K^2 \tilde{\times} I$ contains two essential annuli A_μ, A_λ (the inverse images of a meridian μ and a longitude λ of K^2 under the projection $K^2 \tilde{\times} I \to K^2$, see Example 6.4.14). By $(\tilde{\mu}, \tilde{\lambda})$ we will denote a meridian-longitude pair on $\partial(K^2 \tilde{\times} I)$ composed from a circle $\tilde{\mu} \subset \partial A_\mu$ and a circle $\tilde{\lambda} \subset \partial A_\lambda$. It does not matter, which circle of ∂A_μ (or of ∂A_λ) is taken, since they are parallel.

Lemma 6.4.22. *Let F be an incompressible torus in an orientable Haken manifold (M, Γ). Suppose that F decomposes M into two components $M_F^- \supset F_-$ and $M_F^+ \supset F_+$ such that one of them (say, M_F^-) is $K^2 \tilde{\times} I$. Then F is not rough if and only if there is a clean annulus $B \subset M_F^+$ such that ∂B contains a circle isotopic to either $\tilde{\mu}$ or $\tilde{\lambda}$.*

Proof. Follows from Proposition 6.4.15, since any essential annulus in $K^2 \tilde{\times} I$ is isotopic to either A_μ or A_λ. $\square$

Having learnt how to handle the exceptional cases, we are ready to establish the desired conclusion.

Theorem 6.4.23. *There is an algorithm to decide, whether or not a given incompressible torus F in an orientable Haken manifold (M, Γ) is rough.*

Proof. First we decide whether or not we have one of the two exceptional cases. It can be done algorithmically by Lemma 6.4.20. If the answer is positive, i.e., if we are in an exceptional situation, then the type of F can be determined by Lemma 6.4.21 or Lemma 6.4.22. Suppose that the answer is negative. Then for $\varepsilon = \pm$ we apply Theorem 6.4.10 to the 3-manifold (M_F, Γ_F) and $U = F_\varepsilon, V = \emptyset$ for finding clean essential annuli $A \in \mathcal{A}_-(F), B \in \mathcal{A}_+(F)$. If at least one of these annuli does not exist, then F is rough by Proposition 6.4.19. Otherwise we construct them and calculate $\lambda'(\partial_- A, \partial_+ B)$. Propositions 6.4.15 and 6.4.19 tell us the type of F in this case. $\qquad\square$

ANNULUS CASE. According to Definition 6.4.13, there are three types of clean incompressible boundary incompressible annuli: longitudinal which are rough, longitudinal which are not rough, and transverse. We now need to learn how to determine algorithmically the type of a given annulus.

Let F be a clean proper annulus in an orientable 3-manifold (M, Γ). Cutting M along F, we get a 3-manifold M_F containing two copies F_+, F_- of F in the boundary. If we identify F_+, F_- back via the natural identification map $\varphi: F_- \to F_+$, we recover M. We supply M_F with the boundary pattern $\Gamma_F = \Gamma' \cup \partial F_- \cup \partial F_+$, where Γ' is the copy of Γ in ∂M_F. Clearly, if (M, Γ) is Haken, then so is (M_F, Γ_F).

Just as in the torus case, we introduce two sets $\mathcal{A}_\pm(F)$ of clean annuli in (M_F, Γ_F). Assume that either $\varepsilon = -$ or $\varepsilon = +$. Then we say that a proper annulus $A \subset (M_F, \Gamma_F)$ belongs to $\mathcal{A}_\varepsilon(F)$ if and only if the following conditions hold:

1. A is clean, incompressible, and boundary incompressible
2. At least one boundary circle of A is in F_ε
3. A is not parallel rel ∂ to an annulus $A' \subset \partial M_F$ such that $A' \cap \Gamma_F$ is one of the boundary circles of F_ε

Note that $\mathcal{A}_\varepsilon(F)$ can contain inessential annuli. Each of them is parallel rel ∂ to an annulus $A' \subset \partial M_F$ such that $A' \cap \Gamma_F$ consists of more than one circles of Γ_F.

Below we show how the sets $\mathcal{A}_\pm(F)$ help us to recognize longitudinal annuli. Two other sets $\mathcal{B}_\varepsilon(F), \varepsilon = \pm$, are responsible for F to be transverse. They consist of semi-clean bands and are defined below.

Definition 6.4.24. *Let $F \subset (M, \Gamma)$ be a clean incompressible annulus. We say that a proper disc $B \subset (M_F, \Gamma_F)$ is a semi-clean band for $F_- \cup F_+$, if $B \cap \Gamma' = \emptyset$ and $\partial B \cap (F_- \cup F_+)$ consists of two radial arcs r_1, r_2 of the annuli F_-, F_+. The arcs are called the* ends *of the band. B is trivial, if ∂B bounds a disc $B' \subset \partial M_F$ such that $B' \cap \Gamma_F = \partial r_1 \cup \partial r_2$, see Fig. 6.19.*

Note that any semi-clean band having one end in F_-, the other in F_+ is always nontrivial.

Remark 6.4.25. Denote by Γ'_F the boundary pattern in ∂M_F obtained from Γ_F by removing $\partial F_- \cup \partial F_+$ and inserting a core circle C_- of F_- and a core circle C_+ of F_+. Then any semi-clean band B for $F_- \cup F_+$ determines a semi-clean disc D in (M_F, Γ'_F) (see Sect. 3.3.3) having both points of $\partial D \cap \Gamma'_F$

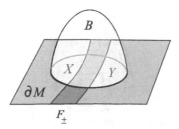

Fig. 6.19. B is trivial if and only if X, Y are clean discs

in $C_- \cup C_+$. Vice versa, any semi-clean disc D in M_F, Γ'_F whose boundary crosses $C_- \cup C_+$ in two points determines a semi-clean band B for $F_- \cup F_+$. Moreover, B is nontrivial if and only if so is D. It follows that there is an algorithm to decide whether F_- admits a nontrivial semi-clean band. We simply use Lemma 4.1.18 for finding a nontrivial semi-clean disc for $C_- \cup C_+$.

Now we introduce two sets $\mathcal{B}_\varepsilon(F)$, $\varepsilon = \pm 1$, of bands in (M_F, Γ_F) as follows: $\mathcal{B}_\varepsilon(F)$ consist of nontrivial semi-clean bands having at least one end in F_ε. As we have mentioned above, the sets $\mathcal{A}_\pm(F), \mathcal{B}_\pm(F)$ help us to determine the type of a given annulus F. They also help us to recognize the exceptional cases, of which, just as in the torus case, there are two.

Lemma 6.4.26. *Let F be a clean proper incompressible annulus in an orientable Haken manifold (M, Γ) with boundary pattern such that all four sets $\mathcal{A}_-(F)$, $\mathcal{A}_+(F)$, $\mathcal{B}_-(F)$, $\mathcal{B}_+(F)$ are nonempty. Then $M = T^2 \times I$ or $M = K^2 \tilde{\times} I$, and $\Gamma = \emptyset$.*

Proof. CASE 1. Suppose that there is an annulus $A \in \mathcal{A}_-(F) \cap \mathcal{A}_+(F)$, which has one boundary circle in F_-, the other in F_+. Let $B_- \in \mathcal{B}_-(F)$ be a band. We can assume that each end of B_- contained in F_- crosses the circle $\partial_- A = \partial A \cap F_-$ at exactly one point. Further, since A is incompressible and boundary incompressible and B is a disc, we can assume that $A \cap B_-$ contains no circles or arcs having both endpoints in $\partial_- A$. Then the intersection of A and B consists of an arc joining F_- and F_+. Denote by N a regular neighborhood of $A \cup B \cup F_- \cup F_+$ in M_F. The interior boundary $\partial_{int} N = \mathrm{Cl}(\partial N \setminus \partial M_F)$ is the union of two disjoint clean proper discs in (M_F, Γ_F). Since (M_F, Γ_F) is Haken, these discs are parallel rel ∂ to clean discs in ∂M. It follows that (M_F, Γ_F) is homeomorphic to $(A \times I, \partial A \times \{0, 1\})$.

Recall that (M, Γ) can be reconstructed from (M_F, Γ_F) by identifying F_-, F_+ via $\varphi \colon F_- \to F_+$ and removing $\partial F_- \cup \partial F_+$ from the pattern. Therefore, (M, Γ) can be obtained from $(A \times I, \partial A \times \{0, 1\})$ by identifying $A \times \{0\}$ with $A \times \{1\}$. There are only two (isotopy classes of) homeomorphisms $A \times \{0\} \to A \times \{1\}$ which give an orientable 3-manifold. One of them can be obtained from the other by postcomposing with an orientation preserving involution $A \times \{1\} \to A \times \{1\}$ having two fixed points. One of these homeomorphisms gives us $(T^2 \times I, \emptyset)$, the other $(K^2 \tilde{\times} I, \emptyset)$.

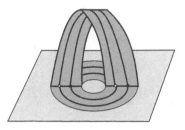

Fig. 6.20. The arcs $A_- \cap B_-$ join different ends of B_- and different circles of ∂A_- (only the boundary of A_- is shown)

CASE 2. Suppose that $\mathcal{A}_-(F) \cap \mathcal{A}_+(F) = \emptyset$. Choose an annulus $A_- \in \mathcal{A}_-(F)$ and a band $B_- \in \mathcal{B}_-(F)$. As above, we can assume that each end of B_- in F_- crosses each circle of $\partial A_- \cap F_-$ at exactly one point and that $A_- \cap B_-$ contains no circles and no arcs having both endpoints in the same end of B_- or in the same circle of ∂A_-. It follows that both ends of B_- are in F_- and that $A_- \cap B_-$ consists of two arcs such that the endpoints of each arc lie in different ends of B_- and different circles of ∂A_-, see Fig. 6.20.

It is not hard to prove that in this situation M_F consists of two components and the component containing A_-, B_- is a solid torus with F_- running twice along the longitude. Since the same is true for the other component of M_F, (M_F, Γ_F) is $K^2 \tilde{\times} I$. Nevertheless, we prefer to save ourselves some effort by reducing Case 2 to Case 1 as follows.

Construct an annulus $A_+ \in \mathcal{A}_+(F)$ and a band $B_+ \in \mathcal{B}_+(F)$ having the same properties as A_- and B_-. We may assume that the identification map $\varphi: F_- \to F_+$ takes $(A_- \cup B_-) \cap F_-$ to $(A_+ \cup B_+) \cap F_+$. Then, reconstructing M, i.e., identifying F_- with F_+ via φ, we get an incompressible annulus $F' = B_- \cup_\varphi B_+ \subset M$ and a torus $T = A_- \cup_\varphi A_+ \subset M$. Cutting M along F', we obtain the situation of Case 1. Indeed, the cut transforms T into even two incompressible annuli joining F'_- with F'_+, and it transforms F into even two semi-clean bands with ends in F'_-, F'_+. It follows that (M, Γ) is $(K^2 \tilde{\times} I, \emptyset)$.   □

Now we relate the type of F with the sets $\mathcal{A}_\pm(F), \mathcal{B}_\pm(F)$. By doing so, we will provide a key to the sought-after algorithmic recognition of types of annuli.

Proposition 6.4.27. *Let F be a clean proper incompressible boundary incompressible annulus in an orientable Haken manifold (M, Γ) with boundary pattern. Suppose that (M, Γ) is not $(T^2 \times I, \emptyset)$ or $(K^2 \tilde{\times} I, \emptyset)$. Then the following implications are true:*

1. *F is longitudinal but not rough $\overset{1}{\Longleftrightarrow} \mathcal{A}_-(F) \neq \emptyset, \mathcal{A}_+(F) \neq \emptyset$;*
2. *F is transverse $\overset{2}{\Longleftrightarrow} \mathcal{B}_-(F) \neq \emptyset, \mathcal{B}_+(F) \neq \emptyset$;*
3. *F is rough $\overset{3}{\Longleftrightarrow}$ at least one of the sets $\mathcal{A}_-(F), \mathcal{A}_+(F)$ is empty and at least one of the sets $\mathcal{B}_-(F), \mathcal{B}_+(F)$ is empty.*

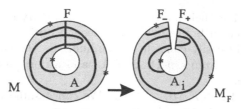

Fig. 6.21. $\mathcal{A}_-(F), \mathcal{A}_+(F)$ consist of, respectively, 3 and 2 annuli. The boundary patterns of M and M_F are shown by *stars*

Proof. ($\overset{1}{\Longrightarrow}$) Let F be longitudinal but not rough. Then there is a surface $G \subset (M, \Gamma)$ which is either an incompressible torus or a clean incompressible boundary incompressible annulus such that $\partial G \cap \partial F = \emptyset$, but G cannot be shifted from F by an admissible isotopy. Applying to F, G elementary separating moves for as long as possible, we get a new surface (denoted also by G), which still intersects F. Then the intersection of F with G consists of nontrivial circles. If $F \cap G$ consists of one circle, then G is a torus, which determines an annulus $G_F \subset M_F$ in $\mathcal{A}_-(F) \cap \mathcal{A}_+(F)$. If $F \cap G$ is the union of more than one circles, then these circles decompose G into a collection $\{A_1, \ldots, A_k\}$ of $k \geq 2$ clean annuli in (M_F, Γ_F) such that each annulus has at least one boundary circle in $F_- \cup F_+$. The annuli are in $\mathcal{A}_-(F) \cup \mathcal{A}_+(F)$, since otherwise we could perform elementary separating moves further. Obviously, $\mathcal{A}_-(F)$ contains at least $[k/2]$ of them, and the same is true for $\mathcal{A}_+(F) \neq \emptyset$, see Fig. 6.21.

($\overset{2}{\Longrightarrow}$) Let F be transverse. Then there is a clean incompressible boundary incompressible annulus $G \subset (M, \Gamma)$ which intersects F along radial segments and cannot be shifted from F by an admissible isotopy. It follows that the segments decompose G into bands in (M_F, Γ_F) such that at least one is in $\mathcal{B}_-(F)$ and at least one in $\mathcal{B}_+(F)$.

($\overset{3}{\Longrightarrow}$) Let F be rough. We wish to prove that at least one of the sets $\mathcal{A}_-(F), \mathcal{A}_+(F)$ is empty and at least one of the sets $\mathcal{B}_-(F), \mathcal{B}_+(F)$ is empty. Suppose, on the contrary, that either $\mathcal{A}_-(F)$ and $\mathcal{A}_+(F)$ are nonempty or $\mathcal{B}_-(F)$ and $\mathcal{B}_+(F)$ are nonempty. To obtain a contradiction, we construct an incompressible torus or a clean proper incompressible boundary incompressible annulus in (M, F) that cannot be shifted from F.

CASE A1. There exists an annulus $A \in \mathcal{A}_-(F) \cap \mathcal{A}_+(F)$. We can assume that the identification map $\varphi \colon F_- \to F_+$ identifies the boundary circles of A. Therefore, after the identification, we get either a torus in M, which intersects F along one circle, or a Klein bottle in M, which also intersects F along one circle and which we transform into a torus by taking the boundary of its regular neighborhood. It follows from Theorem 6.4.1 that in both cases the obtained torus G cannot be shifted from F by an isotopy.

CASE A2. There exist two clean essential annuli $A_- \in \mathcal{A}_-(F)$, $A_+ \in \mathcal{A}_+(F)$, but $\mathcal{A}_-(F) \cap \mathcal{A}_+(F) = \emptyset$. Applying to the annuli elementary separating moves for as long as possible, we get two new annuli (still denoted by A_-, A_+) such that their intersection either is empty or consists of circles which are nontrivial in both annuli. The second option is impossible, since otherwise one could easily construct an annulus in M_F connecting the core circles of F_- and F_+ from a portion of A_- and a portion of A_+. Therefore, A_-, A_+ are disjoint.

CASE A2.1. $\partial A_- \subset F_-, \partial A_+ \subset F_+$. We can assume that φ takes ∂A_- to ∂A_+. Then, just as in Case A1, their union determines a torus or a Klein bottle in M, which we convert into a torus by taking the boundary of its regular neighborhood. It follows from Theorem 6.4.1 that in both cases the resulting torus cannot be shifted from F.

CASE A2.2. $\partial A_- \subset F_-$ and only one circle of ∂A_+ is in F_+. Then we construct a clean incompressible annulus $G \subset (M, \Gamma)$ from an annulus $A'_- \subset M$ corresponding to A_- and two parallel copies of the annulus $A'_+ \subset M$ corresponding to A_+. Again, Theorem 6.4.1 assures us that one cannot shift G from F. The case when $\partial A_+ \subset F_+$ and only one circle of ∂A_- is in F_- is similar.

CASE A2.3. Each of the annuli A_-, A_+ has only one boundary circle in $F_- \cup F_+$. This case is similar to Case A2.2: we construct $G \subset M$ as the union of a copy of A_- and a copy of A_+.

CASE B1. There exists a band $B \in \mathcal{B}_-(F) \cap \mathcal{B}_+(F)$. This case is similar to Case A1. We can assume that the identification map $\varphi: F_- \to F_+$ identifies the ends of B, and we get either a clean annulus in M which intersects F along one segment, or a clean Möbius band in M which also intersects F along one segment and which can be replaced by a clean annulus by taking the interior boundary of its regular neighborhood. It follows from Theorem 6.4.1 that in both cases the annulus G thus obtained cannot be shifted from F by an admissible isotopy.

CASE B2. There exist two bands $B_- \in \mathcal{B}_-(F)$, $B_+ \in \mathcal{B}_+(F)$, but $\mathcal{B}_-(F) \cap \mathcal{B}_+(F) = \emptyset$. Just as in Case A2.1, we may assume that they are disjoint and that φ takes the ends of B_- to the ends of B_+. Then their union determines in (M_F, Γ_F) either an annulus, or a Möbius band, which we double to an annulus. In both cases the annulus thus obtained cannot be shifted from F by an admissible isotopy.

$(\overset{1,2,3}{\Longleftarrow})$. Let us divide the set X of all clean incompressible boundary incompressible annuli in (M, Γ) into three disjoint subsets X_1, X_2, X_3, where X_1 consists of longitudinal annuli which are not rough, X_2 is the set of all transverse annuli in X, and X_3 is the set of all rough annuli. Consider also three other subsets of X: $Y_1 = \{F \in X : \mathcal{A}_-(F) \neq \emptyset, \mathcal{A}_+(F) \neq \emptyset\}$, $Y_2 = \{F \in X : \mathcal{B}_-(F) \neq \emptyset, \mathcal{B}_+(F) \neq \emptyset\}$, and $Y_3 = X \setminus (Y_1 \cup Y_2)$.

Implications $(\overset{1,2,3}{\Longrightarrow})$, which we have already proved, tell us that $X_i \subset Y_i, i = 1, 2, 3$. Obviously, $Y_1 \cap Y_3 = Y_2 \cap Y_3 = \emptyset$. If there is an annulus $F \in Y_1 \cap Y_2$, then we get a contradiction between Lemma 6.4.26 and our assumption that (M, Γ) is not $(T^2 \times I, \emptyset), (K^2 \tilde{\times} I, \emptyset)$. Therefore Y_1 and Y_2 are also disjoint. Since $Y_1 \cup Y_2 \cup Y_3 = X$, it follows that $X_i = Y_i, i = 1, 2, 3$, and that implications $(\overset{1,2,3}{\Longleftarrow})$ are also true. □

Theorem 6.4.28. *There is an algorithm to determine the type of a given clean incompressible boundary incompressible annulus F in an orientable Haken manifold (M, Γ).*

Proof. First we use Lemma 6.4.20 to decide whether or not (M, Γ) is homeomorphic to one of the exceptional 3-manifolds $(T^2 \times I, \emptyset)$, $(K^2 \tilde{\times} I, \emptyset)$. If it is, then F is transverse, because so are all incompressible boundary incompressible annuli in $(T^2 \times I, \emptyset)$ and $(K^2 \tilde{\times} I, \emptyset)$.

Suppose that (M, Γ) is not exceptional. Then we take $\varepsilon = \pm$ and look for a clean annulus $A_\varepsilon \in \mathcal{A}_\varepsilon(F)$ and for a band $B_\varepsilon \in \mathcal{B}_\varepsilon(F)$ as follows. First we use Lemma 6.4.8 for finding an essential annulus having at least one boundary circle in F_ε. If it does not exist, we look for an annulus in $\mathcal{A}_\varepsilon(F)$ by considering all the annuli which are parallel to almost clean annuli in ∂M_F (there are only finitely many of them). The search for bands can be performed by using Lemma 4.1.18 and Remark 6.4.25. Proposition 6.4.27 tells us how the information thus obtained allows us to determine the type of F.   □

6.4.4 JSJ-Decomposition Exists and is Unique

Definition 6.4.29. *Let (M, Γ) be a 3-manifold. A JSJ-system $\mathcal{S}$ for (M, Γ) is a collection of disjoint clean connected surfaces in M such that:*

1. *Every surface in $\mathcal{S}$ is either a rough torus not parallel to a clean torus in ∂M, or a rough annulus not parallel to an almost clean annulus in ∂M.*
2. *No two surfaces from $\mathcal{S}$ are admissibly parallel.*
3. *$\mathcal{S}$ is maximal with respect to 1, 2. It means that any other system $\mathcal{S}'$ satisfying 1, 2 and containing $\mathcal{S}$ must be isotopic to $\mathcal{S}$.*

It is worth noticing that all annuli in $\mathcal{S}$ are essential, while some tori are not. Indeed, $\mathcal{S}$ can include tori that are parallel to nonclean tori in ∂M. Also, since all the surfaces in $\mathcal{S}$ are rough, any incompressible torus and any clean incompressible boundary incompressible annulus in (M, Γ) is admissibly isotopic to a surface in the complement of $\mathcal{S}$.

Example 6.4.30. If M is a graph manifold (see Sect. 2.4), then any canonical system for M is a JSJ-system for M.   □

Theorem 6.4.31. *For every irreducible boundary irreducible orientable 3-manifold (M, Γ) with boundary pattern a JSJ-system exists, is unique up to admissible isotopy, and can be constructed algorithmically.*

Proof. STEP 1. We begin with constructing a set $\mathcal{F}$ of disjoint tori and clean annuli in (M, Γ) such that $\mathcal{F}$ possesses the following properties:

(I) All tori in $\mathcal{F}$ are incompressible and nonparallel to each other or to clean tori of ∂M.
(II) All annuli in $\mathcal{F}$ are essential and no two of them are admissibly parallel.
(III) $\mathcal{F}$ cannot be included into a larger system satisfying I, II.

Starting from $\mathcal{F}_0 = \emptyset$ and arguing inductively, we assume that we have already obtained a system $\mathcal{F}_n = \{F_1, \ldots, F_n\}$ satisfying I, II. Denote by W the 3-manifold obtained from M by cutting along all the surfaces from $\mathcal{F}_n$. The boundary of W contains two copies F_i^-, F_i^+ of each surface F_i. We equip W with the boundary pattern Δ consisting of the copy of Γ and of the boundaries of all surfaces $F_i^\pm, 1 \leq i \leq n$.

To replace $\mathcal{F}_n$ by a larger system, we use Theorem 6.4.10 for finding an essential torus or a clean essential annulus in W that has no boundary circles in $\cup_{i=1}^n F_i^\pm$. If we find one, we add it to $\mathcal{F}_n$. If not, we look for a clean annulus A in W such that:

1. A is disjoint from $\cup_{i=1}^n F_i^\pm$ and parallel rel ∂ to an almost clean annulus $A' \subset \partial W$.
2. A, considered as an annulus in (M, Γ), is essential and not admissibly parallel to an annulus from $\mathcal{F}_n$.

If we find one, we add it to $\mathcal{F}_n$. If not, we look for a torus A in W which is parallel to a nonclean torus of ∂W but not parallel to a clean torus of ∂W. Again, having found such a torus, we add it to $\mathcal{F}_n$.

Now, let us perform this process for as long as possible. The procedure is finite by Theorem 6.3.10. By construction, we end up with a system $\mathcal{F}$ satisfying properties I–III above.

STEP 2. Using Theorems 6.4.23 and 6.4.28, we select from $\mathcal{F}$ rough tori and annuli. Denote the surfaces thus obtained by $S_1, \ldots, S_k$. We claim that they form a JSJ-system $\mathcal{S}$ for (M, Γ).

By construction, all these surfaces are rough and pairwise nonparallel. Let us prove that $\mathcal{S}$ is maximal. Suppose, on the contrary, that there is a larger system $\mathcal{S}'$ satisfying conditions 1, 2 of Definition 6.4.29. Choose a surface $S \in \mathcal{S}' \setminus \mathcal{S}$. Then S is a rough torus or a rough annulus disjoint with $S_1, \ldots, S_k$. By Corollary 6.4.6, we can assume that S is also disjoint with all the other surfaces from $\mathcal{F}$. It follows from construction of $\mathcal{S}$ and property III of $\mathcal{F}$ that S is admissibly parallel to a surface from $\mathcal{F}$, which is also rough and thus is one of S_i, a contradiction. Therefore, the system $S_1, \ldots, S_k$ is a JSJ-system.

STEP 3. Let us prove that the JSJ-system is unique. Suppose that $R = \{F_1, \ldots, F_k\}$ and $R' = \{G_1, \ldots, G_m\}$ are two JSJ-systems of rough tori and annuli. By Corollary 6.4.6, one can assume that the surfaces of both systems are disjoint. Since the system R is maximal, each surface of R' is parallel to a surface of R, and no two surfaces of R' are parallel to the same surface of R. We may conclude that R' is isotopic to a part of R. Since R' is also maximal, this part coincides with R.   □

Let $\mathcal{S} = \{S_1, \ldots, S_k\}$ be a JSJ-system in an irreducible boundary irreducible orientable 3-manifold (M, Γ). Denote by $M_{\mathcal{S}}$ the 3-manifold obtained from M by cutting along all the surfaces from $\mathcal{S}$. The boundary of $M_{\mathcal{S}}$ contains two copies S_i^-, S_i^+ of each surface $S_i, 1 \leq i \leq k$. We equip $M_{\mathcal{S}}$ with the boundary pattern $\Gamma_{\mathcal{S}}$ consisting of the copy of Γ in $\partial M_{\mathcal{S}}$ and all the boundary circles of $\cup_{i=1}^k (S_i^- \cup S_i^+)$. Since all surfaces S_i are incompressible and boundary incompressible, the manifold $(M_{\mathcal{S}}, \Gamma_{\mathcal{S}})$ is irreducible and boundary irreducible. In general, it consists of several connected components (Q_i, Δ_i), where Q_i are connected components of $M_{\mathcal{S}}$ and $\Delta_i = \Gamma_{\mathcal{S}} \cap \partial Q_i$. These connected components of $(M_{\mathcal{S}}, \Gamma_{\mathcal{S}})$ are called *JSJ-chambers*. It turns out that JSJ-chambers contain no essential rough tori or annuli anymore. Moreover, their JSJ-systems are empty.

Lemma 6.4.32. *Let* $\mathcal{S} = \{S_1, \ldots, S_k\}$ *be a JSJ-system for an irreducible boundary irreducible orientable 3-manifold* (M, Γ), *and* (Q, Δ) *a JSJ-chamber of* (M, Γ). *Then the JSJ-system for* (Q, Δ) *is empty.*

Proof. It suffices to prove the following assertion:

$(*)$ Any rough torus F in (Q, Δ) is parallel to a clean torus in ∂Q, and any rough annulus F in (Q, Δ) is parallel to an almost clean annulus in ∂Q.

Let us do that. First we suppose that F is an annulus and that a circle $C \subset \partial F$ is contained in some surface $S = S_i^{\varepsilon} \in \mathcal{S}, \varepsilon = \pm$. Denote by S' a proper surface in Q which is close and parallel to S. Since F is rough in (Q, Δ), it can be shifted from S' by an admissible isotopy. Therefore, we can assume that F is contained in the region $S \times I$ between S and S'. It cannot happen that both boundary circles of F lie in S, since then F would be boundary compressible, see Fig. 6.22. This contradicts our assumption that F is rough. If only one circle of ∂F lies in S, then F is parallel rel ∂ to an almost clean annulus in ∂Q. Hence in this case the assertion $(*)$ is true.

Now we suppose that F is either a torus or an annulus such that ∂F does not intersect surfaces $S_i^{\pm}, 1 \leq i \leq k$, and hence lies in ∂M. We claim that then F, considered as a proper surface in (M, Γ), is also rough.

Indeed, let G be an arbitrary incompressible torus or an arbitrary clean incompressible boundary incompressible annulus in (M, Γ). By Corollary 6.4.6, we can shift G from all surfaces S_i by an admissible isotopy. Then the new surface G' thus obtained is contained in only one JSJ-chamber. If this chamber is different from (Q, Δ), then $G' \cap F = \emptyset$. If the chamber coincides with (Q, Δ), then G' can be shifted from F since F is rough in (Q, Δ). In both

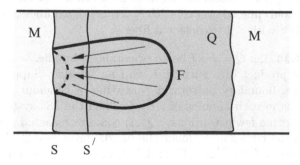

Fig. 6.22. F can be shifted from S' into $S \times [0, 1)$

cases, we can conclude that any G as above can be shifted from F by an admissible isotopy of (M, Γ). It follows that F is rough in (M, Γ).

We can conclude that F, being rough in (M, Γ), is admissibly parallel either to:

(1) A surface S_i from $\mathcal{S}$, or
(2) To a clean torus component of ∂M.

In both cases the assertion (*) is also true.   □

6.4.5 Seifert and I-Bundle Chambers

Let $\mathcal{S} = \{S_1, \ldots, S_k\}$ be a JSJ-system for an irreducible boundary irreducible orientable 3-manifold (M, Γ) and (Q, Δ) a JSJ-chamber of (M, Γ). Then the JSJ-system for (Q, Δ) is empty (by Lemma 6.4.32). It follows that (Q, Δ) contains no tori or annuli which are simultaneously essential and rough. It may happen that (Q, Δ) contains no essential tori and annuli at all. Then (Q, Δ) is simple, and this is fine, since any simple 3-manifold contains only finitely many surfaces of bounded complexity, see Theorem 6.3.17. On the other hand, (Q, Δ) may contain essential tori and annuli, which are not rough. What can we say about such nonsimple JSJ-chambers? They turned out to be fibered either into circles, or into segments. Let us present two basic examples.

Example 6.4.33. Let Q be an orientable Seifert manifold fibered over a surface F (see Remark A.1.1 for the definition of Seifert manifolds). Denote by X the finite set in F consisting of the images of all exceptional fibers under the fibration projection $p: Q \to F$. Suppose that Q is equipped with a boundary pattern Δ composed of a finite number of fibers. Then the JSJ-system for (Q, Δ) is empty, although, with a few exceptions, (Q, Δ) contains essential tori or annuli (which are necessarily not rough). Such tori and annuli can be found among surfaces of the type $p^{-1}(C)$, where C is an orientation preserving essential curve on $F \setminus X$. (Recall that a circle in $F \setminus X$ is *essential*, if it is not parallel to ∂F and does not bound a disc in F containing ≤ 1 points of X. A proper arc in $F \setminus X$ is *essential*, if it is not parallel in $F \setminus X$ to an arc in ∂F). The exceptional manifolds without essential tori or annuli are fibered solid tori, manifolds fibered over S^2 with ≤ 3 exceptional fibers, and manifolds fibered over RP^2 with ≤ 1 exceptional fibers.

Example 6.4.34. Let $Q = F \tilde{\times} I$ be an orientable I-bundle, i.e., an orientable direct or skew product of a surface F and an interval. Suppose that Q is equipped with a boundary pattern Δ consisting of nonempty collections of nontrivial circles in each annulus of $\partial F \tilde{\times} I$. Then the JSJ-system for (Q, Δ) is empty and, with a few exceptions, (Q, Δ) contains essential annuli (inverse images of essential circles in F under the bundle projection). The exceptions are I-bundles over S^2, RP^2, D^2, A^2, and M^2, where A^2 is an annulus and M^2 is a Möbius band.   □

Later we show (see Propositions 6.4.35 and 6.4.41) that these examples exhaust all types of orientable Haken manifolds that contain essential tori or annuli, but have empty JSJ-systems.

Proposition 6.4.35. *Suppose that an orientable Haken manifold (Q, Δ) with boundary pattern has the empty JSJ-system, but contains either an essential torus or an essential longitudinal annulus. Then Q admits a structure of a Seifert manifold with Δ consisting of fibers, and such a structure can be constructed algorithmically.*

Remark 6.4.36. Saying that a Seifert structure can be constructed algorithmically, we mean the following. One can construct a triangulation of (Q, Δ) and a simplicial map $p \colon Q \to S$ of Q to a triangulated surface S such that p is a Seifert fibration.

The proof of Proposition 6.4.35 requires a certain amount of preparatory work, which is carried out in Lemmas 6.4.38, 6.4.39 and 6.4.40. First go a few notations and definitions. Let us introduce a class $\mathcal{Z}_Q$ of 3-submanifolds of Q. We assume that the boundary ∂Z of each manifold $Z \in \mathcal{Z}_Q$ is the union of the *exterior boundary* $\partial_{ext} Z = \partial Z \cap \partial Q$ and the *interior boundary* $\mathrm{Cl}(\partial Z \setminus \partial_{int} Z) \subset Q$. The intersection $\partial_{ext} Z \cap \partial_{int} Z$ must be disjoint from Δ and consist of circles in ∂Q. We equip Z with the boundary pattern $\Delta_Z = (\partial_{int} Z \cap \partial_{ext} Z) \cup (\Delta \cap \partial_{ext} Z)$. For convenience of terminology, we call a subset of a Seifert fibered manifold *saturated*, if it the union of whole fibers.

Definition 6.4.37. *Let (Q, Δ) be an orientable Haken manifold. A connected submanifold $Z \subset Q$ with the boundary pattern Δ_Z belongs to the class $\mathcal{Z}_Q$, if Z possesses a Seifert fibration which is faithful in the following sense:*

1. *Δ_Z is saturated*
2. *Z contains a surface F_Z which is also saturated and, being considered as a surface in (Q, Δ), is either an essential torus or a clean essential longitudinal annulus.*

For example, any regular neighborhood N of an essential torus or a clean essential annulus in (Q, Δ) belongs to $\mathcal{Z}_Q$. We will use N as a starting point for constructing a Seifert structure on (Q, Δ).

Lemma 6.4.38. *Let (Q, Δ) be an orientable Haken manifold with boundary pattern, and let a submanifold Z of Q belong to the class $\mathcal{Z}_Q$. Then for any collection of disjoint incompressible boundary incompressible clean annuli $A_i \subset (Z, \Delta_Z), 1 \leq i \leq k$, there exists a faithful Seifert fibration of Z such that all A_i are saturated.*

Proof. It suffices to prove the lemma for $k = 1$; the general case can be easily obtained by induction. Let $A = A_1$ be given.

CASE 1. Suppose that A has a boundary circle C either in an annular component of $\partial_{int} Z$ or $\partial_{ext} Z$, or in a torus of $\partial_{ext} Z$ containing a circle of Δ_Z.

Then, since Δ_Z and $\partial_{int}Z \cap \partial_{ext}Z$ consist of fibers of a faithful fibration of (Z, Δ_Z) and A is clean, C is admissibly isotopic to a fiber of the same fibration. It follows that there exists an admissible isotopy $h_t: Q \to Q, 0 \le t \le 1$, such that $h_0 = 1$ and the annulus $A' = h_1(A)$ is saturated. Then the inverse isotopy $h_{1-t}h_1^{-1}: Q \to Q, 0 \le t \le 1$, deforms the faithful fibration of (Z, Δ_Z) to a new faithful fibration so that A is saturated.

CASE 2. Suppose that both circles of ∂A are in clean torus components of ∂Z and that Z is not homeomorphic to a thick torus $T \times I$ or a thick Klein bottle $K \tilde{\times} I$. Then A is boundary incompressible and hence essential in Z considered as a 3-manifold without boundary pattern. It is well known that any essential annulus in any Seifert manifold is saturated with respect to some Seifert fibration. Therefore, A A is saturated with respect to a Seifert fibration of Z. This new fibration is isotopic to the original one, since $T \times I$ and $K \tilde{\times} I$ are the only 3-manifolds with nonempty boundary that contain essential annuli and admit more than one isotopy class of Seifert fibrations. Certainly, this isotopy can be made admissible.

CASE 3. In this remaining case, when (Z, Δ_Z) is either $(T \times I, \emptyset)$ or $(K \tilde{\times} I, \emptyset)$, we simply change a given Seifert fibration of Z in order to make A saturated. The new fibration remains faithful, since $\Delta_Z = \emptyset$.   □

Let a submanifold $Z \subset Q$ belong to the class $\mathcal{Z}_Q$. Our intention is to extend a faithful fibration of Z to a faithful fibration of a larger submanifold $Z' \in \mathcal{Z}_Q$, with the final goal to obtain (after a finite number of steps) a faithful fibration of Q. We consider five cases when such an extension is always possible. In all of them the extension consists in "filling a gap". Namely, we replace Z by $Z' = Z \cup U$, where the "hole" U is the closure of a connected component of $Q \setminus Z$. Certainly, we should care on extension of the fibration. We point out that all the nonsingular fibers of Z are homotopic. It follows that all of them are nontrivial in Q, since otherwise all fibered tori and annuli in Z would be compressible, in contradiction with condition 2 of Definition 6.4.37. Let us describe these five types of extensions:

1. (*Filling up a solid torus gap*). Suppose that U is a solid torus in the interior of Q. Since the fibers of ∂U are nontrivial in Q, they are not meridians of U. It follows that any faithful fibration of Z can be extended to a faithful fibration of $Z' = Z \cup U$.
2. (*Filling up a thick torus gap*). Let $U \approx T \times I$ be the region between to parallel boundary tori T, T_1 of ∂Z. Assume that there is a proper annulus $A \subset U$ whose boundary circles lie in distinct boundary tori of U and are fibers of a faithful fibration of Z. This assumption guarantees us that the faithful fibration of Z can be extended to a faithful fibration of $Z' = Z \cup U$.
3. (*Adding a torus collar*). Let $U \approx T \times I$ be a collar of a boundary torus $T \subset \partial Q$ such that the other end T_1 of U is a torus of $\partial_{int}Z$. Suppose that there is an annulus $A \subset U$ joining a clean circle in T with a fiber $C \subset T_1$ of a faithful fibration of Z. Then the fibration of Z extends to a faithful fibration of $Z' = Z \cup U$ in an evident manner.

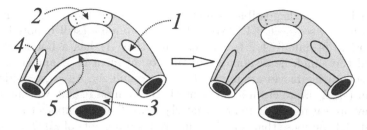

Fig. 6.23. The behavior of base surfaces of Seifert manifolds under five completion moves

4. (*Filling up an annular tunnel*). Suppose that U is a tunnel, i.e., a region between a clean incompressible annulus $A \subset \partial_{int} Z$ which is inessential in (Q, Δ), and the corresponding almost clean annulus A' in ∂Q. Then any faithful fibration of Z extends easily to a faithful fibration of $Z' = Z \cup U$.

5. (*Filling up a thick annulus gap*). Suppose that $U \approx A \times I$ is a region between two parallel annuli $A_0 = A \times \{0\}, A_1 = A \times \{1\}$ of $\partial_{int} Z$ such that $\partial_{ext} U = \partial A \times I$ consists of two almost clean annuli in ∂Q. Then any faithful fibration of Z can be extended to a faithful fibration of $Z' = Z \cup U$.

By construction, in all these cases Z' also belongs to the class $\mathcal{Z}_Q$. We will refer to the transitions from Z to Z' described above as to *completion moves*. All five completion moves are shown in Fig. 6.23. If neither of them is possible, Z is called *complete*.

If Z is a submanifold of a 3-manifold Q, then the total number of connected components of $\partial_{int} Z$ is denoted by $\#(\partial_{int} Z)$.

Lemma 6.4.39. *Let* (Q, Δ) *be an orientable Haken manifold such that its JSJ-system is empty. Suppose that a submanifold* $Z \in \mathcal{Z}_Q$ *is complete. Then all tori and annuli in* $\partial_{int} Z$ *are essential in* (Q, Δ).

Proof. Suppose, on the contrary, that a component F of $\partial_{int} Z$ is an inessential torus or a clean inessential annulus.

CASE 1. Let F be a compressible torus. Then, since Q is irreducible, F bounds a solid torus $U \subset Q$. If $Z \cap \text{Int } U \neq \emptyset$, then $Z = U$, in contradiction with the requirement that Z contains an essential surface F_Z. If $Z \cap \text{Int } U = \emptyset$, then U is a solid torus gap for Z, and we get a contradiction with our assumption that Z is complete.

CASE 2. Let F be parallel to a torus $T \subset \partial Q$. Then F cuts off from Q a collar $U \approx T \times I$ of T. If $Z = U$ or if T is clean, we get the same contradictions as above.

CASE 3. Let F be parallel to a torus $T \subset \partial Q$ and cut off from Q a collar $U \approx T \times I$ such that $Z \neq U$ and T is not clean. Since the JSJ-system for (Q, Δ) is empty, F is not rough. It follows that there exists an incompressible boundary incompressible clean annulus $A' \subset (Q, \Delta)$ such that A' cannot be shifted from F by an admissible isotopy. Let us apply to A' and all the annuli and tori in $\partial_{int}Z$ all possible elementary separating moves (A' is moving, $\partial_{int}Z$ is fixed), see Sect. 6.4.1. We get a new annulus (still denoted by A') that consists of annuli $A_1, A_2, \ldots, A_m$ such that A_1 has a boundary circle in T and the annuli lie alternately in the complement of Z and in Z. Since further separating moves are impossible, the annuli $A_2, A_4, \ldots$, considered as annuli in (Z, Δ_Z), are clean and essential. By Lemma 6.4.38, we can assume that they are saturated. It follows that A_1 joins T with a fiber of F.

CASE 3.1. Suppose that not all connected components of $\Delta \cup T$ are circles. Then $\Delta \cup T$ contains a *pattern strip*, i.e., a connected component of Δ which is not a circle and which is contained in an incompressible annulus in $A' \subset T$ with the clean boundary. Denote by $\bar{A}$ a proper annulus in Q obtained from A' by shifting it rel ∂ inward. Then $\bar{A}$ is rough and essential, which contradicts our assumption that the JSJ-system for (Q, Δ) is empty. This contradiction shows that $\Delta \cap T$ consists of disjoint circles.

CASE 3.2. Suppose that $\Delta \cup T$ is a collection of disjoint circles. Then we can extend Z by adding the torus collar U. This contradicts our assumption that Z is complete.

Since in all above cases we get contradictions, we may conclude that if F is a torus, then it is essential.

CASE 4. Suppose F is an annulus. Then it cannot be compressible, since each circle of ∂F is a fiber of Z and hence is nontrivial in Q. Moreover, it cannot be boundary compressible, since then it would be parallel to a clean annulus in ∂Q. This contradicts our assumption that Z is complete and, in particular, admits no annular tunnels. By the same reason F cannot be parallel to an almost clean annulus in ∂Q. It follows that F is essential. $\square$

Lemma 6.4.40. *Let (Q, Δ) be an orientable Haken manifold such that the JSJ-system for (Q, Δ) is empty. Suppose a manifold $Z \subset Q, Z \neq Q$, belongs to the class $\mathcal{Z}_Q$. Then one can construct a complete manifold $Z' \in \mathcal{Z}_Q$ so that $Z \subset Z'$ and either a connected component of $\partial_{int}Z'$ is not admissibly parallel in (Q, Δ) to a surface in Z, or $\#(\partial_{int}Z') < \#(\partial_{int}Z)$.*

Proof. Evidently, each completion move decreases $\#(\partial_{int}Z)$, so if Z is not complete, then we can easily get Z' by applying to Z all possible completion moves. Assume that Z is complete. Consider a connected component F of $\partial_{int}Z$, which by Lemma 6.4.39 is either an essential torus or a clean essential annulus. Since the JSJ-system for (Q, Δ) is empty, F is not rough in Q. Therefore, there is a clean surface $G \subset Q$ such that G is either a clean incompressible boundary incompressible annulus or an incompressible torus, and G cannot be shifted from F by an admissible isotopy. Applying to G and all the components of $\partial_{int}Z$ elementary separating moves as long as possible (G is moving, $\partial_{int}Z$ remains fixed), we may assume that G admits no further separating moves. Then either the intersection of G with $\partial_{int}Z$ consists of nontrivial circles, or $G \cap \partial_{int}Z$ consists of radial arcs. In both cases $G \cap F \neq \emptyset$.

Let us prove that the second case, when $G \cap \partial_{int}Z$ consists of radial arcs and hence F is a transverse annulus, is impossible. Assume the contrary. Then the arcs $G \cap \partial_{int} Z$ decompose G into quadrilaterals, which lie alternately in Z and $Cl(Q \setminus Z)$. Since G admits no further elementary separating moves, each quadrilateral disc $D = I \times I$ from $G \cap Z$ is a nontrivial compressing disc for ∂Z. The only Seifert 3-manifold with compressible boundary is a solid torus, hence Z can be identified with $S^1 \times I \times I$ so that $D = \{*\} \times I \times I$, $\partial_{ext} Z = S^1 \times \partial I \times I$, and $\partial_{int} Z = S^1 \times I \times \partial I$. It follows that the annulus F_Z, which is essential in (Q, Γ) and is contained in Z by Definition 6.4.37, is admissibly isotopic to $F = S^1 \times I \times \{*\}$. All these annuli are transverse in (Q, Δ), since G intersects the annuli $\partial_{int} Z = S^1 \times I \times \partial I$ along radial segments and this intersections cannot be removed by an isotopy of G. This contradicts the assumption that F_Z is longitudinal.

Therefore $G \cap \partial_{int} Z$ consists of nontrivial circles. They decompose G into annuli, which lie alternately in Z and $Cl(Q \setminus Z)$ and which are essential in (Z, Δ_Z). Denote by $A_i, 1 \leq i \leq k$, the annuli which are in Z. By Lemma 6.4.38, we can assume that they are saturated with respect to a faithful fibration of Z. Then a regular neighborhood $N = N(Z \cup G)$ of $Z \cup G$ in Q also belongs to the class $\mathcal{Z}_Q$. We extend $N(Z \cup G)$ to a complete submanifold $Z' \in \mathcal{Z}_Q$.

In order to prove the conclusion of the lemma, let us suppose that each connected component F' of $\partial_{int} Z'$ is admissibly parallel to a surface in Z. Then F' is admissibly parallel to a connected component of $\partial_{int} Z$ which lies between F' and the aforementioned surface in Z. Since distinct components of $\partial_{int} Z'$ are not admissibly parallel to each other (otherwise we could fill the thick torus or annulus gap between them), the corresponding components of $\partial_{int} Z$ are distinct. This implies $\#(\partial_{int} Z') \leq \#(\partial_{int} Z)$. The equality $\#(\partial_{int} Z) = \#(\partial_{int} Z')$ cannot hold, because otherwise the admissible parallelism between $\partial_{int} Z'$ and $\partial_{int} Z$ would ensure the existence of an admissible isotopic deformation of Z' into the interior of Z. This would contradict to the assumption that G cannot be shifted from F. □

Proof of Proposition 6.4.35. Let F be an essential torus or an essential longitudinal annulus in Q. Then a regular neighborhood $N(F)$ of F in Q belongs to the class $\mathcal{Z}_Q$. Applying to $Z = N(F)$ and to each of the subsequent submanifolds Lemma 6.4.40 as long as possible, we construct a sequence $Z = Z_1 \subset Z_2 \subset \ldots$ of manifolds in $\mathcal{Z}_Q$. Each time either $\#(\partial_{int} Z_i)$ decreases or a new torus or annulus appears in $\partial_{int} Z_{i+1}$ that is not parallel to a surface in Z_i. It follows from the First Finiteness Property (Theorem 6.3.10) that the sequence $Z = Z_1 \subset Z_2 \subset \ldots$ has finite length. The last manifold Z_n coincides with Q, therefore Q is a Seifert manifold possessing a faithful fibration. □

The following proposition extends Proposition 6.4.35 to the case of transverse annuli.

Proposition 6.4.41. *Suppose that an orientable Haken manifold (Q, Δ) with boundary pattern has the empty JSJ-system, but contains a transverse annulus. Then Q admits a structure of an I-bundle $F \tilde{\times} I$ with Δ consisting of a nonempty collection of nontrivial circles in each annulus of $\partial F \tilde{\times} I$. The bundle structure on (Q, Δ) can be constructed algorithmically.*

Proof. This is very similar to the proof of Proposition 6.4.35, so we restrict ourselves to sketching details. First we define a class $\mathcal{I}_Q$ of submanifolds of Q as follows. A submanifold $Z \subset Q$ belongs to the class $\mathcal{I}_Q$ if:

1. Z is an I-bundle $F_1 \tilde{\times} I$ such that $F_1 \tilde{\times} \partial I$ is a clean surface in ∂Q and each annulus from $\partial F_1 \tilde{\times} I$ is either an annulus in $\partial_{int} Z$ or an almost clean annulus in ∂Q containing at least one circle of Δ.
2. Z contains a clean proper annulus A_Z which is transverse in (Q, Δ).

Next we describe three types of *completion moves*, which extend the I-bundle structure of Z to an I-bundle structure of a larger submanifold Z'. Each such move consists in filling up a gap , i.e, in adding to Z a connected component U of $\mathrm{Cl}(Q \setminus Z)$:

1. Suppose that U is a 3-ball $D^2 \times I$ such that $D^2 \times \partial I$ are two clean discs in ∂Q and $\partial D^2 \times I$ is an annulus in $\partial_{int} Z$. Then we extend the I-fibration of Z to an I-fibration of $Z' = Z \cup U$.
2. Suppose that $U = S^1 \times I \times I$ is the region between two admissibly parallel annuli $A_0 = S^1 \times \{0\} \times I, A_0 = S^1 \times \{1\} \times I$ of $\partial_{int} Z$. Then the I-fibration of Z can be easily extended to an I-fibration of $Z' = Z \cup U$.
3. Suppose that U is a tunnel, i.e., a region between a clean incompressible boundary incompressible annulus $A \subset \partial_{int} Z$ which is inessential in (Q, Δ) and the corresponding almost clean annulus A' in ∂Q. Then the I-bundle structure of Z extends easily to an I-fibration of $Z' = Z \cup U$.

Note that $\mathcal{I}_Q$ is nonempty, since it contains a regular neighborhood $N(A_Z)$ of the transverse annulus $A_Z \subset (Q, \Delta)$, which exists by the assumption. Starting with $Z_1 = N(A_Z)$, we will extend Z by completion Moves 1–3 as long as possible. Suppose that some Z_i obtained in this way is complete, that is, it does not admit further completion moves. Then each annulus in $\partial_{int} Z_{i+1}$ is essential and thus is not rough in (Q, Δ). Therefore, we can replace Z_i by a regular neighborhood $Z_{i+1} = N(Z_i \cup G)$ of $Z_i \cup G$, where G is a clean annulus in (Q, Δ) that intersects $\partial_{int} Z_i$ along radial segments and cannot be shifted from $\partial_{int} Z_i$ by an admissible isotopy. Then we apply completion Moves 1–3 again, and so on. The same arguments as in the proof of Proposition 6.4.35 ensure us that the process ends up with a manifold $Z_n \in \mathcal{Z}_Q$ coinciding with Q. $\square$

Finally, we are summing up.

Theorem 6.4.42. *Any JSJ-system for an irreducible boundary irreducible orientable 3-manifold (M, Γ) decomposes M into JSJ-chambers $\{(Q_i, \Delta_i)\}$ of the following three types: simple 3-manifolds, Seifert manifolds (as in Example 6.4.33), and I-bundles (as in Example 6.4.34). The types of the JSJ-chambers as well as Seifert or I-bundle structures on them (if exist) can be found algorithmically.*

Proof. Follows from Propositions 6.4.35 and 6.4.41. $\square$

Remark 6.4.43. There are exactly two types of JSJ-chambers which are simultaneously Seifert manifolds and I-bundles: $S^1 \times S^1 \times I$ and $K^2 \tilde{\times} I = M^2 \tilde{\times} S^1$ with the empty boundary pattern, where M^2 is the Möbius band and K^2 is the Klein bottle. They contain no essential tori or longitudinal annuli, so we prefer to consider them as I-bundles.

6.4.6 Third Finiteness Property

Now we are ready to establish the main result of Sect. 6.4 (the Third Finiteness Property, see Theorem 6.4.44). Recall that two surfaces F, F' in a 3-manifold (M, Γ) are equivalent, if they are related by an admissible homeomorphism $h \colon (M, \Gamma) \to (M, \Gamma)$, and strongly equivalent, if h can be chosen so that its restriction onto ∂M is admissibly isotopic to the identity, see Definitions 6.3.11 and 6.3.12.

Theorem 6.4.44 (Third Finiteness Property). *Any irreducible boundary irreducible orientable 3-manifold (M, Γ) contains only finitely many essential tori and longitudinal annuli up to strong equivalence, and only finitely many transverse annuli up to equivalence. Representatives of all equivalence classes of tori and annuli can be constructed algorithmically.*

We would like to emphasize an important difference between longitudinal and transverse annuli: homeomorphisms $(M, \Gamma) \to (M, \Gamma)$ that relate transverse annuli can be nontrivial (i.e., not admissibly isotopic to the identity) on ∂M. In other words, we consider tori and longitudinal annuli up to strong equivalence while transverse annuli only up to admissible homeomorphisms. If M is a chamber of a larger manifold (M', Γ'), then such homeomorphisms are usually not extendible to homeomorphisms of (M', Γ'). On the contrary, strong equivalence can be always realized by homeomorphisms $(M, \Gamma) \to (M, \Gamma)$ which are extendible to (M', Γ').

Before proving the theorem, we need to prove its 1-dimensional analogue. Let F be a compact surface with a boundary pattern $\gamma \subset \partial F$ consisting of a finite number of points. We say that two connected simple proper curves $C_1, C_2 \subset F$ disjoint from γ are *strongly equivalent*, if there is a homeomorphism $h \colon F \to F$ such that $h(C_1) = C_2$ and $h_{|\partial F}$ is isotopic to the identity via an isotopy fixed on γ. The set of strong equivalence classes of essential curves in F is denoted by $\mathcal{C}(F, \gamma)$.

Lemma 6.4.45. *For any compact surface F with boundary pattern γ the set $\mathcal{C}(F, \gamma)$ is finite. Representatives of all the equivalence classes from $\mathcal{C}(F, \gamma)$ can be constructed algorithmically.*

Proof. We can assume that F is connected. Denote by $\mathcal{C}(F)$ the set of simple proper curves in F modulo homeomorphisms $F \to F$ which induce homeomorphisms $\partial F \to \partial F$ isotopic to the identity. Let $\mathcal{C}'(F)$ be the set of the same curves considered modulo arbitrary homeomorphisms $F \to F$. Finally let $\mathcal{F}(F)$ be the set of one- or two-component surfaces (considered modulo homeomorphisms) whose Euler characteristic is greater than or equal to $\chi(F)$.

There are natural finite-to-one maps $p_1: \mathcal{C}(F, \gamma) \rightarrow \mathcal{C}(F)$, $p_2: \mathcal{C}(F) \rightarrow \mathcal{C}'(F)$, and $p_3: \mathcal{C}'(F) \rightarrow \mathcal{F}(F)$. To get p_1, we simply forget about γ, to get p_2, we forget about the requirement that the induced homeomorphism $F \rightarrow F$ must be isotopic to the identity. The last map assigns to every simple proper curve $C \subset F$ the surface F_C obtained from F by cutting along C. Note that $\chi(F_C) = \chi(F)$, if C is a circle, and $\chi(F_C) = \chi(F) + 1$, if C is a proper arc. It follows from the classification theorem for compact surfaces that $\mathcal{F}(F)$ is finite. Therefore $\mathcal{C}(F, \gamma)$ is also finite.

Analyzing this proof, one can notice that two proper curves C_1, C_2 determine the same element of $\mathcal{C}(F, \gamma)$ if and only if one of the following conditions holds:

1. Both curves are closed, do not separate F, and either both reverse or both preserve orientation
2. Both curves are closed and decompose F into two surfaces that have the same topological type and contain the same components of ∂F
3. Both curves are nonseparating arcs with endpoints contained in the same components of ∂F and in the same intervals of $\partial F \setminus \gamma$
4. Both curves are separating arcs that have endpoints in the same component of ∂F and decompose F into parts which contain the same remaining components of ∂F and the same points of $\partial \gamma$

Representing each situation described in 1–4 by a curve (if possible), we get an algorithmically constructible list of all elements of $\mathcal{C}(F, \gamma)$.   □

Proof of Theorem 6.4.44. Let $\mathcal{S} = \{S_1, \ldots, S_k\}$ be a JSJ-system for (M, Γ) and F an incompressible torus or a clean incompressible boundary incompressible annulus in (M, Γ). Since all tori and annuli of $\mathcal{S}$ are rough, F is admissibly isotopic to a surface which is disjoint with $\mathcal{S}$. So we can assume that from the very beginning F is contained in a JSJ-chamber (Q, Δ).

Suppose that (Q, Δ) is a simple JSJ-chamber. Then F is either a torus parallel to a torus of ∂Q or an annulus parallel rel ∂ to an almost clean annulus in ∂Q. All such surfaces can be easily enumerated.

Suppose that (Q, Δ) is a Seifert JSJ-chamber. Then S is either a torus isotopic to a saturated torus in (Q, Δ) or a longitudinal annulus admissibly isotopic to a saturated annulus in (Q, Δ). Therefore it can be represented as the inverse image (with respect to the fibration projection) of a curve on the base surface of the chamber. It follows from Lemma 6.4.45 that the number of strong equivalence classes of such surfaces is finite and that their representatives can be constructed algorithmically. It remains to select essential tori and longitudinal annuli among them.

Finally, we suppose that (Q, Δ) is an I-bundle JSJ-chamber with $Q = G \tilde{\times} I$ and Δ consisting of nontrivial circles in $\partial G \tilde{\times} I$. Then S is a transverse annulus admissibly isotopic to an annulus of the form $C \tilde{\times} I$, where C is an essential curve on F. Therefore, transverse annuli correspond to simple closed curves in the base surface. Then we apply Lemma 6.4.45 again.   □

Remark 6.4.46. Let F be a clean proper surface in (M, Γ). Recall that *a twist* of M along F is a homeomorphism $(M, \Gamma) \rightarrow (M, \Gamma)$ which is fixed outside a small regular neighborhood of F in M. For example, if F is an annulus, then a twist along F can be described as follows: we cut M along F, rotate one of the two annuli arising in this way by 2π, and glue them again via the identity homeomorphism.

Theorem 6.4.44 can be sharpened by stating that any irreducible boundary irreducible 3-manifold M with boundary pattern Γ contains only finitely many incompressible tori and longitudinal annuli up to twists along incompressible tori, and only finitely many transverse annuli up to twists along transverse annuli. The proof is the same; all what we need is to replace Lemma 6.4.45 by a version of Mennike–Lickorish result [68] that any self-homeomorphism of a compact surface can be presented as a product of Dehn twists along simple closed curves. In the case of a Seifert manifold M the twists along curves in the base surface correspond to the twists along their inverse images, which are tori in M. In the case of an I-bundle the inverse images are proper annuli.

6.5 Extension Moves

Now we are ready to begin the realization of the program described in Sect. 6.3.2: we define extension moves and prove that they possess properties C_1–C_4.

6.5.1 Description of General Extension Moves

We describe here the first five extension moves. They are called *general* since they fit into the general scheme of constructing admissible subpolyhedra described in Sect. 6.3.1: we extend a given admissible subpolyhedron by inserting a proper surface. We prove that those moves possess properties C_1–C_3. Yet, in general, they do not possess property C_4: in many cases we end up with an admissible polyhedron having not only 3-ball chambers. Special extension moves, which help us to overcome this difficulty and prove the recognition theorem for Haken 3-manifolds, will be described later.

Let P be an admissible subpolyhedron of an orientable Haken manifold (M, Γ) and (Q, Δ) a chamber of P. Recall that Q is a compact 3-manifold such that $Q \cap P = \partial Q$ and $\Delta = (SP \cup \Gamma) \cap \partial Q$. In other words, Δ consists of those parts of the pattern Γ and those of the singular graph SP of P that are in ∂Q. It follows from the definition of an admissible subpolyhedron that (Q, Δ) is Haken. Each of the following extension moves E_1–E_5 consists in inserting a proper surface into (Q, Δ). The surface either is connected or consists of two parallel copies of a connected surface (we use such double surfaces for

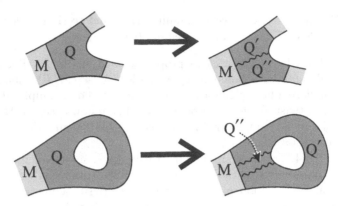

Fig. 6.24. Inserting surfaces into chambers

avoiding nonseparating 2-components, see condition 2 of Definition 6.3.6). In both cases, we decompose Q into two new chambers Q', Q'', see Fig. 6.24.

E_1: Inserting an Essential Torus

Suppose that Q contains an essential torus T (recall that the property of a torus to be essential does not depend on the boundary pattern: T is essential, if it is incompressible and not parallel to a torus of ∂Q). Then, we add to P either T or two parallel copies of T depending on whether or not T separates Q.

E_2: Inserting a Longitudinal Annulus Which is not Parallel rel ∂ to an Annulus in ∂Q

Suppose that (Q, Δ) contains a longitudinal annulus A which, considered as an annulus in Q, is not parallel rel ∂ to an annulus in ∂Q. Then we add to P either A or two parallel copies of A depending on whether or not A separates Q.

E_3: Inserting a Clean Essential Annulus Which is Parallel to an Annulus in ∂Q

Suppose that (Q, Δ) contains a clean essential annulus A which, considered as an annulus in Q, is parallel to an annulus A' in ∂Q. Of course, $A' \cap \Delta$ must contain noncircle components, since otherwise A would be inessential. Then we add A to P.

E_4: Inserting a p-minimal proper disc

. Suppose that (Q, Δ) is a simple 3-manifold, which contains no essential tori and no clean essential annuli. Note that Q, being boundary irreducible as a 3-manifold with the boundary pattern Δ, can be boundary reducible as a 3-manifold with the empty boundary pattern. Suppose that:

(a) Q is boundary reducible, i.e., ∂Q is compressible
(b) Q is not a solid torus having a clean longitude

Among all nontrivial compressing discs for ∂Q we choose a disc D having the smallest p-complexity. Then we add to P either D or two parallel copies of D, depending on whether or not D separates Q.

E_5: Inserting a p-Minimal Surface with Nonempty Boundary

Suppose that:

(a) (Q, Δ) is a simple 3-manifold and Q is boundary irreducible
(b) Q contains a connected proper incompressible boundary incompressible surface which is not S^2 or D^2 and, if $\partial Q \neq \emptyset$, has nonempty boundary

Among all surfaces described in (b) we choose a surface F having the smallest p-complexity. Then we add to P either F or two parallel copies of F depending on whether or not F separates Q.

Remark 6.5.1. It follows from Lemma 6.2.8 that if $\partial Q \neq \emptyset$ and Q is not a 3-ball, then (a) implies (b). Suppose that Q is closed. Then $M = Q$ (i.e., we are applying the very first move), and the existence of an incompressible surface in Q follows from our assumption that M is sufficiently large. This is the only place when this assumption is used. At all subsequent moves we will have nonclosed chambers, which are either handlebodies or are automatically sufficiently large.

Our next goal is to prove that moves E_1–E_5 satisfy properties C_1–C_3 (see Sect. 6.3.2). Recall that two admissible subpolyhedra of a 3-manifold (M, Γ) with boundary pattern are called equivalent, if there is a homeomorphism $(M, \Gamma) \to (M, \Gamma)$ transforming one subpolyhedron into the other.

Proposition 6.5.2. *Extension moves E_1–E_5 satisfy properties C_1, C_2.*

In a more detailed form Proposition 6.5.2 can be formulated as follows. Let P be an admissible subpolyhedron of an orientable Haken 3-manifold (M, Γ). Then there exist only finitely many nonequivalent admissible subpolyhedra that can be obtained from P by performing exactly one of moves $E_i, 1 \leq i \leq 5$. Moreover, representatives of the equivalence classes can be constructed algorithmically.

Proof. For moves E_1, E_2, E_3 the conclusion follows from the Third Finiteness Property (Theorem 6.4.44). Indeed, applying this theorem to each chamber (Q_i, Δ_i), we obtain an algorithmically constructible set of essential tori and longitudinal annuli which represent all the strong equivalence classes of such tori and annuli. Then we select among them annuli which are essential in

(Q_i, Δ_i). Each of these tori and annuli determines one of moves E_1, E_2, E_3 and hence an extension of P. By construction, we obtain all such extensions up to strong equivalence, i.e., up to homeomorphisms $(Q_i, \Delta_i) \to (Q_i, \Delta_i)$ of the corresponding chambers that induce homeomorphisms $\partial Q_i \to \partial Q_i$ admissibly isotopic to the identity. Any such homeomorphism can be extended to a homeomorphism $(M, \Gamma) \to (M, \Gamma)$.

Let us prove the proposition for moves E_4, E_5. Let (Q_i, Δ_i) be a simple chamber. It follows from the second finiteness property (Theorem 6.3.17) that for any k there exists a finite algorithmically constructible set $\mathcal{F}_k$ of proper surfaces in (Q_i, Δ_i) such that any proper incompressible boundary incompressible surface F of p-complexity $\leq k$ is strongly equivalent to a surface from $\mathcal{F}_k$. As above, strong equivalence of surfaces F_1, F_2 implies equivalence of the corresponding extensions $P \cup F_1, P \cup F_2$. This means that if k is given, then the number of different extensions of P obtained by adding a surface of p-complexity $\leq k$ is finite.

Suppose now that (Q_i, Δ_i) contains a proper disc D which is nontrivial in Q considered as a 3-manifold without boundary pattern. Its p-complexity we denote by k. Clearly, the number of different discs of p-complexity $\leq k$ is finite. Therefore, the number of p-minimal nontrivial discs and corresponding moves E_4 is also finite. Certainly, the discs and the moves can be constructed algorithmically.

Suppose that Q_i is boundary irreducible and contains an incompressible boundary incompressible surface F such that $F \neq S^2, D^2$ and $\partial F \neq \emptyset$. Denote by k the p-complexity of F. Using that the set $\mathcal{F}_k$ is finite at the same way as above, we can conclude that the number of moves E_5 is also finite.   □

For proving property C_3 we need a preparation. To any Haken 3-manifold (Q, Δ) we associate seven numerical characteristics:

(1) The maximal number $t(Q)$ of disjoint pairwise nonparallel essential tori in Q

(2–4) The extended complexity $\bar{c}(Q) = (c(Q), c_1(Q), c_2(Q))$ of Q, see Definition 4.2.7

(5) The maximal number $a(Q, \Delta)$ of disjoint longitudinal annuli in Q which, considered as annuli in Q, are not parallel to each other and to annuli in ∂Q

(6) The maximal number $d(Q)$ of disjoint nontrivial proper discs in (Q, Γ) not parallel to each other

(7) The number $s(Q, \Delta)$ of *pattern strips* in ∂Q, where by a pattern strip in ∂Q we mean a connected component of Δ which is not a circle and which is contained in an incompressible annulus in ∂Q with the clean boundary. Two examples of pattern strips are shown in Fig. 6.25.

Note that $t(Q), \bar{c}(Q), d(Q)$ depend only on Q; the boundary pattern is irrelevant. In contrast to this, $a(Q, \Delta)$ and $s(Q, \Delta)$ depend on Q as well as on Δ. All these numbers are nonnegative and finite. For $t(Q)$ and $a(Q, \Delta)$ it

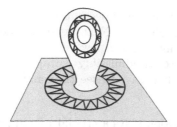

Fig. 6.25. Two pattern strips

follows from the first finiteness property (Theorem 6.3.10), for $\bar{c}(Q, \Delta)$ from Definition 4.2.7, and for $s(Q, \Delta)$ it is evident. $d(Q)$ is finite since it coincides with the length of the longest sequence of nontrivial cuts that transforms Q into its core (see Definition 4.1.21 and Proposition 4.1.25).

Definition 6.5.3. *Let P be an admissible subpolyhedron of an orientable Haken manifold (M, Γ), and let $(Q_i, \Delta_i), 1 \leq i \leq n,$ be all the chambers of P different from a solid torus with a clean longitude. Then the* chamber complexity $C(P)$ *of P is the tuple of seven numbers*

$$\left(\sum_{i=1}^{n} t(Q_i), \sum_{i=1}^{n} \bar{c}(Q_i), \sum_{i=1}^{n} a(Q_i, \Delta_i), \sum_{i=1}^{n} d(Q_i), \sum_{i=1}^{n} s(Q_i, \Delta_i) \right),$$

where the sum $\sum_{i=1}^{n} \bar{c}(Q_i)$ of the extended complexities of the chambers consists of three numbers $\sum_{i=1}^{n} c(Q_i), \sum_{i=1}^{n} c_1(Q_i),$ and $\sum_{i=1}^{n} c_2(Q_i)$. The tuples are considered in lexicographic ordering.

Remark 6.5.4. We point out that if Q is a solid torus, then $t(Q), \bar{c}(Q),$ and $a(Q, \Delta)$ are zeros. Therefore, taking the sums above over only those chambers which are not homeomorphic to a solid torus with a clean longitude, we ignore the meridional discs of such tori and pattern strips in their boundaries.

Lemma 6.5.5. *Let F be a compact connected surface and G a proper incompressible subsurface of $F \times I$ such that $G \cap (F \times \partial I) = \emptyset$ and G is either a torus, or an annulus not parallel rel ∂ to an annulus in $\partial F \times I$, or a disc not parallel rel ∂ to a disc in $\partial F \times I$. Then there exists an isotopy $F \times I \to F \times I$ which keeps $F \times \partial I$ fixed and takes G to a surface of the form $F \times \{*\}$.*

Proof. Suppose that G is a disc not parallel to a disc in $\partial F \times I$. Then ∂G is isotopic to a core circle of the annulus of $\partial F \times I$ it is contained in. This can happen only if F is a disc and G has the form $F \times \{*\}$.

Suppose that G is an annulus not parallel to an annulus in $\partial F \times I$. Then it is boundary incompressible in the manifold $(F \times I, \partial F \times \partial I)$ with boundary pattern. If G is a torus, then it is also boundary incompressible because it has no boundary. Choose a triangulation T of $(F \times I, \partial F \times \partial I)$ such that all its vertices are in $F \times \partial I$. Such a triangulation can be easily constructed as follows. First we triangulate F and decompose $F \times I$ into prisms of the type $\Delta \times I$, where Δ is a triangle in F. Next we subdivide coherently each prism into tetrahedra without introducing new vertices.

By Corollary 3.3.25, G is admissibly isotopic to a clean normal surface $G' \subset (F \times I, \partial F \times \partial I)$. It is easy to see that any such surface has the form $F \times \{*\}$.   □

The following proposition plays a crucial role in proving property C_3.

Proposition 6.5.6. *Let (M, Γ) be an orientable Haken manifold. Suppose that an admissible polyhedron $P' \subset (M, \Gamma)$ is obtained from an admissible polyhedron $P \subset (M, \Gamma)$ by one move E_i, $1 \le i \le 5$. Then the chamber complexity of P' is strictly less then the one of P.*

Proof. Let E_i consists in inserting a surface G into a chamber (Q, Δ) of P. Recall that G either is connected and separates Q, or consists of two parallel copies of a nonseparating surface. In both cases G decomposes Q into two chambers $(Q', \Delta'), (Q'', \Delta'')$, see Fig. 6.24.

In the following five steps we obtain information on the behavior of $t(Q), \bar{c}(Q), a(Q, \Delta), d(Q)$, and $s(Q, \Delta)$ under the move. The information is summarized in the table shown in Fig. 6.26 (to the left). Each column of the table corresponds to one step and tells us whether the contribution of (Q', Δ') and (Q'', Δ'') to the corresponding numerical characteristic of P' is less, equal, or less or equal than the contribution of (Q, Δ) to the same characteristic of P.

By performing move E_i, i.e., inserting G into Q, the topological types of other chambers remain the same, but ∂G can contribute new circles and arcs to the boundary patterns of the neighboring chambers. Step 6 describes the behavior of the numerical characteristics of such neighboring chambers. The results are summarized in the right-handside table of Fig. 6.26. It follows from the tables that any move E_i, $1 \le i \le 5$, strictly decreases $C(P)$. Since $C(P)$ is bounded from below by the zero tuple, for proving the proposition it remains to describe Steps 1–6.

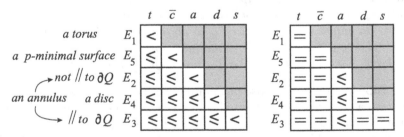

		t	$\bar{c}$	a	d	s		t	$\bar{c}$	a	d	s
a torus	E_1	<					E_1	=				
a p-minimal surface	E_5	≤	<				E_5	=	=			
→ not ∥ to ∂Q	E_2	≤	≤	<			E_2	=	=	≤		
an annulus a disc	E_4	≤	≤	≤	<		E_4	=	=	≤	=	
↳ ∥ to ∂Q	E_3	≤	≤	≤	≤	<	E_3	=	=	≤	=	=

Fig. 6.26. The behavior of $t(Q), \bar{c}(Q), a(Q, \Delta), d(Q), s(Q, \Delta)$ under application of moves E_1–E_5 to a chamber (Q, Δ) *(left)* and to a neighboring chamber *(right)*. For convenience we include nick-names of surfaces which are inserted under the corresponding moves

STEP 1. Let us prove that $t(Q') + t(Q'') < t(Q)$ for $i = 1$, and $t(Q') + t(Q'') \leq t(Q)$ for $2 \leq i \leq 5$. Consider maximal collections $T' = \{T_1, \ldots, T_k\}$ and $T'' = \{T_{k+1}, \ldots, T_{k+m}\}$ of disjoint pairwise nonparallel essential tori in Q' and Q'', respectively. Then $t(Q') + t(Q'') = k + m$.

Claim. $T_1, \ldots, T_{k+m}$, considered as tori in Q, are pairwise nonparallel and essential.

Suppose, on the contrary, that there is a direct product $T \times I \subset Q$ such that $T \times \{0\}$ is a torus $T_j, 1 \leq j \leq k + m$, and $T \times \{1\}$ is either another torus $T_{j'}, j' \neq j$, or a torus in ∂Q. Since T' consists of pairwise nonparallel essential tori and the same holds for T'', at least one connected component $G_0 \subset \partial Q' \cap \partial Q''$ of G must lie in $T \times I$. By Lemma 6.5.5, it is parallel to T_j, which contradicts our assumption that T_j is not parallel to the boundary of Q' or Q''.

Recall that $t(Q)$ is the maximal number of disjoint pairwise nonparallel essential tori in Q. It follows from the claim that $t(Q) \geq k+m$, which implies that $t(Q') + t(Q'') \leq t(Q)$. Moreover, if we are performing move E_1, i.e., if a connected component G_0 of G is a torus, then the system $G_0, T_1, \ldots, T_{k+m}$ consists of $k+m+1$ pairwise nonparallel essential tori in Q. Therefore, $t(Q) \geq k + m + 1$ and $t(Q') + t(Q'') < t(Q)$.

STEP 2. It follows from Theorem 4.2.15 that if $i \neq 1, 5$ (when E_i inserts into Q an incompressible surface with nonempty boundary), then $\bar{c}(Q') + \bar{c}(Q'') \leq \bar{c}(Q)$. Let $i = 5$. If Q is closed then $c(Q') + c(Q'') < c(Q)$ and thus $\bar{c}(Q') + \bar{c}(Q'') < \bar{c}(Q)$ by Theorem 4.2.14. If $\partial Q \neq \emptyset$, then the inequality $\bar{c}(Q') + \bar{c}(Q'') < \bar{c}(Q)$ follows from Theorem 4.2.15.

STEP 3. Let us prove that $a(Q', \Delta') + a(Q'', \Delta'') < a(Q, \Delta)$ for $i = 2$ and $a(Q', \Delta') + a(Q'', \Delta'') \leq a(Q, \Delta)$ for $i = 3, 4$. If $i = 3$, then one of the chambers Q', Q'' is a solid torus while the other is homeomorphic to Q, so the inequality $a(Q', \Delta') + a(Q'', \Delta'') \leq a(Q, \Delta)$ (even the equality $a(Q', \Delta') + a(Q'', \Delta'') = a(Q, \Delta)$) is evident. Suppose that $i = 2, 4$. Then we do the same as in Step 1. Consider maximal collections $\mathcal{A}' = \{A_1, \ldots, A_k\}$ and $\mathcal{A}'' = \{A_{k+1}, \ldots, A_{k+m}\}$ of disjoint clean annuli in (Q', Δ') and, respectively, in (Q'', Δ''), which are nonparallel to each other and to ∂Q.

We can assume that $\partial A_j \cap G = \emptyset$ for all $j, 1 \leq j \leq k + m$. Indeed, suppose that G consists of one or two annuli, i.e., that we are considering E_2. Then we can eliminate all the circles in $\partial A_j \cap G$ by letting them jump over a boundary circle of ∂G to the outside of G in $\partial Q'$ or $\partial Q''$. Evidently, such jump can be realized by an isotopy of A_j, see Fig. 6.27. If we are considering move E_4, then G consists of one or two discs. In this case $\partial A_j \cap G$ is automatically empty for all j, since all A_j are clean and ∂G is included into Δ' and Δ''.

We claim that $A_1, \ldots, A_{k+m}$, considered as annuli in Q, are nonparallel to each other and to annuli in ∂Q. Indeed, suppose that there is a direct product $A \times I \subset Q$ such that $A \times \{0\}$ is an annulus $A_j, 1 \leq j \leq k + m$, and $A \times \{1\}$ is either another annulus $A_{j'}, j' \neq j$ or an annulus in ∂Q. Then at least one connected component G_0 of G must lie in $A \times I$. By Lemma 6.5.5,

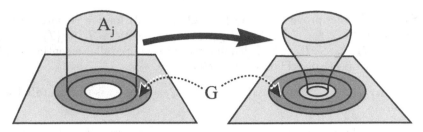

Fig. 6.27. Moving ∂A_j away from G

it is parallel to A_j, which contradicts the assumption that A_j is not parallel to the boundary.

It follows from the claim that $a(Q, \Delta) \geq k + m$ and thus $a(Q', \Delta') + a(Q'', \Delta'') \leq a(Q, \Delta)$. Moreover, if we are performing E_2, i.e., if a component G_0 of G is a clean annulus in (Q, Δ), then the system $G_0, A_1, \ldots, A_{k+m}$ consists of $k + m + 1$ pairwise nonparallel essential annuli in Q. Therefore, $a(Q, \Delta) > k + m$ and $a(Q', \Delta') + a(Q'', \Delta'') < a(Q, \Delta)$.

STEP 4. Let us investigate the behavior of $d(Q)$ with respect to E_4. This move inserts into Q a surface G consisting of a nontrivial separating disc or of two parallel copies of a nonseparating disc. To prove the inequality $d(Q') + d(Q'') < d(Q)$, we do the same as in Steps 1, 2. Consider maximal collections $\mathcal{D}' = \{D_1, \ldots, D_k\}$ and $\mathcal{D}'' = \{D_{k+1}, \ldots, D_{k+m}\}$ of disjoint discs in (Q', Δ') and, respectively, in (Q'', Δ''), which are nonparallel to each other and to disc in ∂Q. Shifting each D_j from G, we can assume that $\partial D_j \cap G = \emptyset$ for all $j, 1 \leq j \leq k + m$.

We claim that $D_1, \ldots, D_{k+m}$, considered as discs in Q, are nonparallel to each other and to discs in ∂Q. Indeed, suppose that there is a direct product $D \times I \subset Q$ such that $D \times \{0\}$ is a disc $D_j, 1 \leq j \leq k + m$, and $D \times \{1\}$ is either another disc $D_{j'}, j' \neq j$ or a disc in ∂Q. Then at least one connected component G_0 of G must lie in $D \times I$. By Lemma 6.5.5, it is parallel to D_j, which contradicts the assumption that D_j is not parallel to the boundary.

It follows from the claim that the system $G_0, D_1, \ldots, D_{k+m}$ consists of $k + m + 1$ pairwise nonparallel essential discs in Q. Therefore, $d(Q) > k + m$ and $d(Q') + d(Q'') < d(Q)$.

Suppose that $i = 3$, i.e., that move E_i consists in inserting in Q a clean boundary parallel annulus. Let us prove that the contribution of Q', Q'' to $\sum_j d(Q_j)$ coincides with the one of Q. Indeed, one of the manifolds Q', Q'' (let Q') is homeomorphic with Q while the other is a solid torus having a clean longitude (for example, a core circle of the inserted annulus). Since by calculating $C(P)$ we ignore solid tori with clean longitudes, Q'' contributes nothing. Actually, this is the first reason for ignoring such tori (the second one is explained in the next step).

STEP 5. Let us investigate the behavior of $s(Q, \Delta)$ with respect to E_3. This move consists in inserting a boundary parallel clean annulus A which is essential in (Q, Δ). As we have seen above, A decomposes Q into a solid torus Q'' and a chamber Q' homeomorphic to Q. Since A is essential, Q'' must contain at least one pattern strip of Δ, so (Q', Δ') has a fewer number of pattern strips than (Q, Δ). It follows that $s(Q', \Delta') < s(Q, \Delta)$ (since Q'' is a solid torus with a clean longitude, we do not count the strips in $\partial Q''$. This is the second reason for ignoring such tori).

STEP 6. Performing move E_i, i.e., inserting G into Q, we do not change the topological types of other chambers. Therefore, their characteristics $t, \bar{c}, d$ remain the same. On the other hand, if $2 \leq i \leq 5$, then E_i creates new circles or arcs in the boundary pattern Δ_j of any neighboring chamber Q_j and thus can change $a(Q_j, \Delta_j)$ and $s(Q_j, \Delta_j)$.

Let us show that E_i never increases $a(Q_j, \Delta_j)$ (we will need that fact only for $2 \leq i \leq 4$). Indeed, denote by $\tilde{\Delta}_j \supset \Delta_j$ the boundary pattern of Q_j after the move. Let $A_1, \ldots, A_k, k = a(Q_j, \tilde{\Delta}_j)$ be the maximal set of disjoint clean annuli in $(Q_j, \tilde{\Delta}_j)$ which are not parallel in Q_j to each other and to annuli in ∂Q_j. Then the same annuli form a set of disjoint clean annuli in (Q_j, Δ_j), which are also not parallel in Q_j to each other and to annuli in ∂Q_j. It follows that $a(Q_j, \tilde{\Delta}_j) = k \leq a(Q_j, \Delta_j)$.

Concerning the number of pattern strips, we will need only the evident fact that if we apply E_3 to a chamber (Q, Δ), then for any neighboring chamber (Q_j, Δ_j) the number of its pattern strips $s(Q_j, \Delta_j)$ remains the same.   □

Corollary 6.5.7. *Extension moves E_1–E_5 satisfy property C_3. In other words, any sequence $P_1 \subset P_2 \subset \ldots$ of admissible subpolyhedra of any orientable Haken manifold (M, Γ), where each P_{i+1} is obtained from P_i by one of moves E_1–E_5, is finite.*

Proof. It follows from Proposition 6.5.6 (see also Fig. 6.26) that each of the moves E_1–E_5 strictly decreases $C(P_i)$. On the other hand, any 7-tuple representing the chamber complexity of an admissible subpolyhedron is bounded from below by the zero 7-tuple. Thus the process of applying moves E_1–E_5 must be finite.   □

Remark 6.5.8. Figure 6.28 shows, why we have forbidden inserting discs into solid tori with clean longitudes (see the description of move E_4). Indeed, if such insertions were allowed, we could get an infinite sequence of moves $E_3, E_4, E_3, \ldots$.   □

6.5.2 Structure of Chambers

Suppose that (M, Γ) is an orientable Haken manifold such that $\partial M \neq \emptyset$. Let us apply to (M, Γ) general extension moves for as long as possible. By Corollary 6.5.7, after a finite number of moves we shell stop. Let as describe the structure of chambers of the resulting admissible polyhedron $P \subset M$.

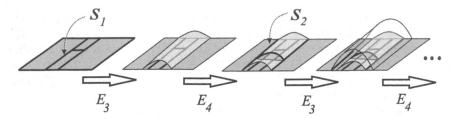

Fig. 6.28. Each E_3 disables a pattern strip s_i, but each next E_4 creates a new strip s_{i+1}

Definition 6.5.9. *A chamber* (Q, Δ) *of an admissible subpolyhedron* $P \subset (M, \Gamma)$ *is called an* I-bundle chamber, *if it is a twisted or an untwisted* I-*bundle* $F \tilde{\times} I$ *such that* $\chi(F) < 0$ *and* Δ *consists of nontrivial circles in each annulus of* $\partial F \tilde{\times} I$ *(see Example 6.4.34).*

It is convenient to decompose the boundary ∂Q of any I-bundle chamber $Q = F \tilde{\times} I$ into two parts: the *lateral boundary* $\partial F \tilde{\times} I$ and the *base boundary* $F \tilde{\times} \partial I$. The lateral boundary consists of annuli, each containing at least one circle of Δ. In the case of a direct product $F \times I$ the base boundary of Q consists of two copies of F, the orientable base surface of the fibration $F \tilde{\times} I \to F$. If the product $F \tilde{\times} I$ is twisted, then F is nonorientable and the base boundary of Q is an orientable double covering of F. Since $\chi(F) < 0$, the base boundary of any I-bundle chamber is neither a disc, nor an annulus, nor the union of two discs or two annuli. Möbius strips are also impossible. In contrast, the boundary pattern of any 3-ball or solid torus chamber Q of P decomposes ∂Q into discs and/or annuli.

Lemma 6.5.10. *Let* P *be an admissible subpolyhedron of an orientable Haken manifold* (M, Γ). *If* P *admits no extension moves* E_1–E_5, *then for any chamber* (Q, Δ) *of* P *exactly one of the following holds:*

1. *Q is a 3-ball*
2. *Q is a solid torus having a clean longitude*
3. *(Q, Δ) is an I-bundle chamber*

Proof. First, (Q, Δ) does not contain essential tori and essential longitudinal annuli (otherwise, we could apply one of moves E_1–E_3). Since any rough annulus is longitudinal, (Q, Δ) does not contain essential rough annuli either. It follows from Proposition 6.4.41 that if (Q, Δ) contains a transverse annulus, then (Q, Δ) is an I-bundle of the desired type.

It remains to consider the case when (Q, Δ) does not contain clean essential annuli and tori at all, i.e., it is simple. Suppose that Q is boundary reducible. Then Q is a solid torus with a clean longitude, since otherwise we could apply move E_4. It cannot happen that Q is boundary irreducible and Q is not a 3-ball, since then we could apply move E_5 (see Remark 6.5.1). We can conclude that Q is a 3-ball. $\square$

Recall that our goal is to construct extension moves so that we always end up with a simple skeleton, which has only 3-ball chambers. A natural idea to proceed further would be to subdivide solid torus chambers by inserting meridional discs. However, doing so we take chances to run into an infinite process. Indeed, inserting meridional discs can produce pattern strips in neighboring chambers. Therefore, we should apply move E_3 again, thus producing new solid torus chambers, and so on, see Remark 6.5.8. Thus we prefer to postpone subdividing solid torus chambers until Sect. 6.5.8 and turn our attention to I-bundle chambers.

Let F_0 be a connected component of the base boundary of an I-bundle chamber. Since $\chi(F_0) < 0$, it cannot lie in the boundary of a 3-ball or solid torus chamber. Therefore, F_0 either lies in ∂M or separates two different I-bundle chambers. It follows that all the I-bundle chambers are organized into closed or nonclosed chains. Nonclosed chains of chambers can end up at ∂M as well as inside M.

Remark 6.5.11. Recall that the *lateral boundary* $\partial F \tilde{\times} I$ of any I-bundle chamber $Q = F \tilde{\times} I$ consists of annuli, each intersecting the boundary pattern Δ in a nonempty collection of disjoint circles. To be definite, we will always assume that Δ contains $\partial F \tilde{\times} \partial I$ and that Q has no common lateral annuli with other I-bundle chambers. This can be easily achieved (at the expense of there appearing a few new solid torus chambers with clean longitudes) by inserting into Q additional clean annuli which are parallel to annuli from $\partial F \tilde{\times} I$, see Fig. 6.29. The new chamber Q' (a slightly squeezed copy of Q) possesses the required property and thus can be presented as $F \tilde{\times} I$ with $\Delta' \supset \partial F \tilde{\times} \partial I$.

As we observed above, I-bundle chambers form closed or nonclosed chains. To give a rigorous formulation of this observation, we describe the so-called *quasi-Stallings 3-manifolds*.

Definition 6.5.12. Let F be an orientable surface and $\alpha, \beta \colon F \to F$ two orientation reversing free involutions. Then the quasi-Stallings 3-manifold $M_{(\alpha,\beta)}$ with fiber F is obtained from $F \times I$ by identifying each point $(x,0) \in F \times \{0\}$ with the point $(\alpha(x),0)$, and each point $(x,1) \in F \times \{1\}$ with the point $(\beta(x),1)$.

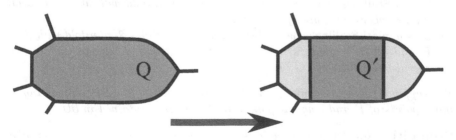

Fig. 6.29. Improving I-bundles

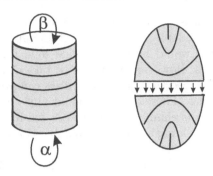

Fig. 6.30. Two ways of looking at a quasi-Stallings manifold

Recall that any Stallings manifold fibers over S^1. Similarly, any quasi-Stallings manifold $M_{(\alpha,\beta)}$ admits a fibration $p: M_{(\alpha,\beta)} \to I$. All but two fibers of this fibration are homeomorphic to the fiber F of $M_{(\alpha,\beta)}$. The two exceptional fibers $G_0 = p^{-1}(0), G_1 = p^{-1}(1)$ are homeomorphic nonorientable surfaces and admit two-sheeted coverings by F. It follows that the inverse images $p^{-1}([0,1/2])$ and $p^{-1}([1/2,1])$ are homeomorphic to the twisted I-bundles $G_0 \tilde{\times} I$, respectively, $G_1 \tilde{\times} I$. This provides us another way of looking at quasi-Stalling 3-manifolds. See Fig. 6.30. According to our definition, all quasi-Stallings manifolds are orientable.

Now we introduce four types of chains of I-bundle chambers.

Definition 6.5.13. *Let an admissible subpolyhedron P of an orientable Haken manifold (M, Γ) decompose M into 3-balls, solid tori, and I-bundle chambers without common lateral annuli. Consider a connected component U of the union of all I-bundle chambers. Then we say that:*

1. *U is a* direct chain *of I-bundle chambers, if U is a direct product $F \times I$ of an orientable surface F and I such that $F \times \partial I \subset \partial M$. In this case all the I-bundle chambers of U have the form $F \times I$*
2. *U is a* twisted chain*, if U is a twisted product $G \tilde{\times} I$ of a nonorientable surface G and I such that the surface $F = G \tilde{\times} \partial I$ lies in ∂M. In this case all the I-bundle chambers in U have the form $F \times I$, except exactly one chamber of the form $G \tilde{\times} I$*
3. *U is a* Stallings chain*, if U is a Stallings 3-manifold with fiber F and all the chambers in U are of the form $F \times I$*
4. *U is a* quasi-Stallings chain*, if U is a quasi-Stallings 3-manifold with fiber F and all the chambers in U are of the form $F \times I$, except exactly two chambers of the form $G \tilde{\times} I$*

We supply U with the boundary pattern $\Delta_U = \partial U \cap (\Gamma \cup SP)$ consisting of those points of Γ and singular points of P that are contained in ∂U.

Proposition 6.5.14. *Let an admissible subpolyhedron P of an orientable Haken manifold (M, Γ) decompose M into 3-balls, solid tori, and I-bundle* *chambers without common lateral annuli. Then any connected component of the union of all I-bundle chambers is a chain having one of the above four types, i.e., it is either direct, or twisted, or Stallings, or quasi-Stallings.*

Proof. To describe the mutual position of the I-bundle chambers of P, we construct a graph $\gamma(P)$ as follows. The vertices of $\gamma(P)$ are of two types: *black* and *white*. Black vertices correspond bijectively to the I-bundle chambers of P, white vertices represent connected components of the base boundaries of the I-bundle chambers. Two vertices are joined by an edge if and only if they correspond to an I-bundle chamber and a connected component of its base boundary.

Recall that any I-bundle chamber has either one base boundary component (if it is twisted) or two base boundary components (if it is a direct product). Therefore any black vertex has valence 1 or 2. Similarly, any component of the base boundary of any I-bundle chamber either separates two different chambers or lies in ∂M. Thus any white vertex has also valence 1 or 2. It follows that $\gamma(P)$ is a 1-dimensional manifold, i.e., a collection of disjoint circles and arcs. The white ends of the arcs correspond to surfaces in ∂M, the black ends stand for quasi-Stallings chambers in M.

Consider now a connected component U of the union of all I-bundle chambers and the corresponding component γ_U of $\gamma(P)$, see Fig. 6.31.

CASE 1. γ_U is an arc with white endpoints. Then U consists of trivial I-bundle chambers such that two connected components of their base surfaces lie in ∂M. Therefore, U is a direct chain.

CASE 2. γ_U is an arc having one black endpoint and one white one. Then U consists of a twisted I-bundle chamber Q_0 and a chain of trivial I-bundles, whose addition does not change the topological type of Q_0. It follows that U is a twisted chain.

CASE 3. γ_U is a circle. Then U is a closed chain of trivial I-bundle chambers and hence a Stallings chain.

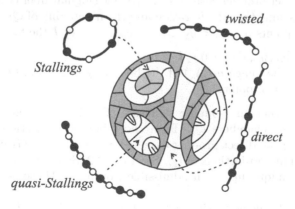

Fig. 6.31. I-bundle chambers form chains of four types: direct, twisted, Stallings, and quasi-Stallings

CASE 4. γ_U is an arc with black endpoints. Then U is a chain of I-chambers which begins and ends up with twisted bundles. Therefore, U is a quasi-Stalling chain. We have considered all the possibilities.   □

6.5.3 Special Extension Moves: Easy Case

Our next goal consists in subdividing I-bundle chambers by inserting proper annuli and discs. However, to preserve properties C_1, C_2 (see Sect. 6.3.2) we will insert such surfaces into all the chambers of a given chain U simultaneously. In other words, we will insert into U finite collections of discs and annuli which are organized in a regular manner. There are three types of such collections: *perturbed strips, annuli, and tori*. Therefore, there are three types of corresponding *special extension moves*: E_6, E_7, E_8. Move E_6 completely solves the problem of subdividing direct and twisted chains. Applying E_6 together with E_3, one can decompose into solid tori with clean longitudes all the I-bundle chambers of any direct or twisted chain. In contrast to that, moves E_7, E_8 only improve the structure of Stallings and quasi-Stallings chains by transforming them into chains whose underlying manifolds are simple.

First we describe E_6. Let an admissible subpolyhedron P of an orientable Haken manifold (M, Γ) decompose M into 3-balls, solid tori, and I-bundle chambers without common lateral annuli. Let (Q, Δ) be an I-bundle chamber. Then (Q, Δ) admits a decomposition into segments, i.e., a presentation as a direct product $F \times I$ or as a twisted product $G \tilde{\times} I$ We always assume that Δ consists of nontrivial circles of the type $C \times \{*\}$, where C is a boundary circle of F, respectively, of G. Any two such decompositions are equivalent in the following sense: there exists an admissible isotopy $(Q, \Delta) \to (Q, \Delta)$ taking one decomposition to the other. Certainly, this isotopy can be extended to an admissible isotopy of M.

Now, we consider a direct or twisted chain $U = Q_0 \cup \ldots \cup Q_m$ of I-bundle chambers. There is a fibration $p_1 : U \to I$ with fiber F, which has no singular fibers in the direct product case and exactly one singular fiber G in the twisted one. We can assume that the fibration and the numbering of Q_k are chosen so that for some points $0 = y_0 < y_1 < \ldots < y_{m+1} = 1$ of I the following holds:

1. $Q_k = p_1^{-1}([y_k, y_{k+1}])$ for $0 \le k \le m$;
2. For each k the image $p_1(\Delta_k)$ of the boundary pattern of Q_k consists of a finite number of points.

Let us construct an I-bundle structure of U by concatenating I-bundle structures of the chambers Q_i. Then the I-bundle projection $p_2 : U \to S$, where $S = F$ in the direct product case and $S = G$ in the twisted one, agrees with the I-bundle projections of Q_i. See Fig. 6.32. Since the structures of the chambers are unique up to an admissible isotopy of M, so is the structure of U.

Denote by P_U the admissible subpolyhedron $P \cap U$ of U. It consists of ∂Q and the fibers $F_i = p_1^{-1}(y_i), 1 \le i \le m$, which decompose U into chambers Q_i.

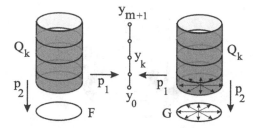

Fig. 6.32. Any direct or twisted chain admits two fibrations

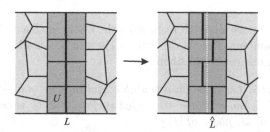

Fig. 6.33. Perturbing a strip

Our idea is to subdivide U into simpler chambers by inserting discs, which are called *nontrivial strips*. To realize it, we choose a nontrivial proper arc $l \subset S$ and insert into U the strip $L = p_2^{-1}(l)$. Obviously, L cuts the chambers Q_i of U into simpler pieces. However, the polyhedron $P_U \cup L$ is not admissible, since it contains fourfold lines $L \cap F_i, 1 \leq i \leq m$. The lines decompose L into shorter strips $L_i = L \cap Q_i, 0 \leq i \leq m$. We convert all the four-fold lines into pairs of parallel triple lines by perturbing L. The perturbation procedure consists in replacing each strip $L_i \subset Q_i, 0 \leq i \leq m$, by a parallel strip $L_i' \subset Q_i$ such that all strips thus obtained are disjoint. We will refer to the collection $\hat{L}$ of strips L_i' as a *perturbed strip* corresponding to L, see Fig. 6.33.

E_6: Subdividing a Direct or a Twisted Chain

Suppose that U is a direct or twisted chain of I-bundle chambers of an admissible subpolyhedron $P \subset (M, \Gamma)$ and L a nontrivial strip in U. Then we add to P two parallel copies of L and perturb them.

Let us comment on this move. The inserted perturbed strips decompose the I-bundle chambers of U into solid tori and smaller I-bundle chambers. We take two copies of L in order to preserve the condition that different I-bundle chambers have no common lateral annuli.

Now we turn our attention to moves E_7, E_8. Basically, they consist in inserting essential tori and annuli into a given Stallings or quasi-Stallings chain U. If such tori and annuli exist, then they can be found algorithmically. However, the intersection of such tori and annuli with fibers of U can be very nasty. Therefore, in order to get the finiteness property C_1 one should take

Fig. 6.34. An example of a vertical torus

into account their position with respect to the fibers of U which divide U into the I-bundle chambers. Let us describe a class of tori and annuli which are placed well with respect to all fibers.

Definition 6.5.15. *Let U be an orientable Stallings or quasi-Stallings manifold with fiber F. Then a proper two-sided surface $G \subset U$ is called* vertical*, if it is transversal to all fibers of U.*

For any vertical surface G in U and any fiber F the intersection $G \cap F$ is a proper 1-dimensional submanifold of F, i.e., a collection of disjoint circles and proper arcs. If we replace F by a close neighboring fiber, then the topological type of the intersection remains the same. It follows that any vertical surface in U consists of disjoint tori and proper annuli. So any connected vertical surface is either a torus or a proper annulus. See Fig. 6.34.

Let an admissible subpolyhedron P of an orientable 3-manifold (M, Γ) decompose M into 3-balls, solid tori, and I-bundle chambers without common lateral annuli. Let $U \subset M$ be a Stallings or quasi-Stallings chain with fiber F. Denote by P_U the admissible subpolyhedron $P \cap U$ of U, which consists of ∂U and several fibers of U. The fibers decompose U into I-bundle chambers Q_i. We wish to subdivide U into simpler chambers by inserting vertical tori and annuli. Let T be a vertical torus in U. Then $T \cap P_U$ contains four-fold circles. We convert them into pairs of parallel triple circles by perturbing T. The perturbation procedure consists in replacing each annulus in $T \cap Q_i$ by a parallel annulus so that all resulting annuli are disjoint. We will call the collection $\hat{T}$ of these new annuli a *perturbed torus* corresponding to T.

The description of a *perturbed annulus* $\hat{A}$ corresponding to a vertical annulus $A \subset U$ is similar: It consists of parallel copies of discs in the intersection of A with chambers Q_i.

E_7: Inserting a Perturbed Pair of Vertical Parallel Tori

Suppose that a direct or a twisted chain U of I-bundle chambers for an admissible subpolyhedron $P \subset (M, \Gamma)$ contains a vertical torus T which is essential in U. Then we add to P two parallel copies of T and perturb them.

E_8: Inserting a Perturbed Pair of Vertical Parallel Annuli

Suppose that a direct or a twisted chain U of I-bundle chambers for an admissible subpolyhedron $P \subset (M, \Gamma)$ contains a vertical annulus A which is essential in U. Then we add to P two parallel copies of A and perturb them.

Our next goal is to show that extension moves E_1–E_8 satisfy properties C_1, C_2, C_3. For this we need a few facts about vertical tori and annuli in Stallings and quasi-Stallings 3-manifolds.

Lemma 6.5.16. *Let M be an orientable Stallings or quasi-Stallings manifold with fiber F such that $\chi(F) < 0$. Then for any essential torus and any essential annulus G in M one can construct an isotopic vertical surface.*

Proof. Choose a fiber $F \subset M$ transversal to G. Since G and F are incompressible and boundary incompressible, one can easily eliminate all trivial circles and arcs in the intersection of G with F. Nontrivial circles and arcs in $G \cap F$ remain fixed. They decompose G either into annuli or into quadrilaterals, depending on the type of G.

Denote by M_F the 3-manifold obtained from M by cutting along F. If M is a Stallings manifold, then M_F is $F \times I$. If M is a quasi-Stallings manifold, then M_F is the disjoint union of two twisted I-bundles $S_i \widetilde{\times} I$ over the singular fibers S_1, S_2 of M. We supply M_F with the boundary pattern $\Gamma_F = \partial F_- \cup \partial F_+$, where F_-, F_+ are two copies of F in ∂M_F. Denote by $G_1, \ldots, G_n$ the pieces of G in (M_F, Γ_F). Since $G \cap F$ contains no trivial circles and arcs, there are two cases:

1. G is a torus and all G_i are annuli
2. G is an annulus and all G_i are quadrilaterals, each having two opposite sides in $F_- \cup F_+$, the other two in $\partial M_F \setminus (F_- \cup F_+)$

It cannot happen that G and G_i are annuli, since otherwise either G would be boundary compressible or F would be an annulus.

Consider the first case. If an annulus G_i is boundary compressible in (M_F, Γ_F), then G_i is parallel to an annulus A contained in $F_+ \cup F_-$. It follows that we can deform the annulus in M corresponding to G_i to the other side of F by an isotopy of M, thus diminishing the total number of circles in $G \cap F$. Performing such deformations for as long as possible, we get a new torus (still denoted by G) such that all annuli $G_i \subset M_F$ are incompressible and boundary incompressible. It follows that they are isotopic rel ∂ to annuli which are transversal to all the fibers of M_F. Therefore, we obtain an isotopy of G to a transversal torus.

The case when G is an annulus and G_i are quadrilaterals is similar. If a quadrilateral G_i is boundary compressible in (M_F, Γ_F), then the corresponding quadrilateral in M can deformed by an isotopy to the other side of F. The number of arcs in $G \cap F$ becomes smaller. Performing such deformations as long as possible, we get a new annulus (still denoted by G) such that all quadrilaterals $G_i \subset M_F$ are boundary incompressible. Then there is an isotopy

$M_F \to M_F$ which keeps $F_- \cup F_+$ fixed and takes all G_i to discs transversal to all fibers. This isotopy determines an isotopy of G to an annulus transversal to fibers.   □

Let M be a Stallings manifold with fiber $F \subset M$ and the monodromy map $f \colon F \to F$. Our next goal is to show that the vertical position of any essential torus or annulus $G \subset M$ is unique up to fiber-preserving isotopy of M. The corresponding statement for quasi-Stalling manifolds will be obtained later as a corollary. We suppose that G is transversal to a fixed fiber $F \subset M$. It turns out that the vertical position of G is completely determined by any essential curve in $G \cap F$. Let us discuss some properties of such curves. They can be extracted from or are related to the proof of Lemma 6.5.16.

Let C be either an essential circle or a proper essential arc in F. Then the set $O_f(C) = \{f^k(C), -\infty < k < \infty\}$ of curves in F is called the *orbit* of C. We will consider the curves up to isotopy, so any orbit determines a set of isotopy classes of curves in F. If two orbits determine the same set, we do not distinguish them.

Lemma 6.5.17. *Let G be an essential torus or a proper essential annulus in an orientable Stallings 3-manifold M with the monodromy map f. Suppose that the fibers of M have negative Euler characteristics. Let us choose a fiber $F \subset M$ intersecting G transversally. Then the following holds:*

1. *Any two essential curves in $G \cap F$ have isotopic orbits.*
2. *If a surface $G' \subset M$ is isotopic to G, then the orbits of essential curves in $G' \cap F$ are isotopic to the orbits of essential curves in $G' \cap F$.*
3. *If G is vertical, then the set $\{C_0, \ldots, C_{n-1}\}$ of all the curves in $G \cap F$ is isotopic to the orbit of any of them.*
4. *Let G be vertical. Suppose that G' is another essential vertical surfaces in M such that a curve in $G \cap F$ is isotopic to a curve in $G' \cap F$. Then G, G' are fiberwise isotopic.*

Proof. 1. We may assume that all curves in $G \cap F$ are essential (inessential curves can be eliminated by an isotopy of G without touching essential curves). They decompose G into a finite number of annular or quadrilateral pieces G_i. It is convenient to think of G_i as being contained in the 3-manifold M_F obtained by cutting M along F. As before, we supply M_F with the boundary pattern $\Gamma_F = \partial F_- \cup \partial F_+$, where F_-, F_+ are two copies of F in ∂M_F.

Suppose that a piece G_i joins a curve on F_- with a curve in F_+. Then the corresponding curves in F are in the same orbit, since one of them is isotopic to the image under f of the other. If G_i joins two curves contained both in F_- or both in F_+, then the corresponding curves in F are isotopic. Evidently, any curve $C \subset G \cap F$ can be obtained from any other curve $C' \subset G \cap F$ by such jumps along annular or quadrilateral pieces of G. Therefore, the orbits of C and C' are isotopic.

2. Let us orient $G, F,$ and M. After doing that it makes sense to speak about positive and negative normal vectors for G, F and about positive and

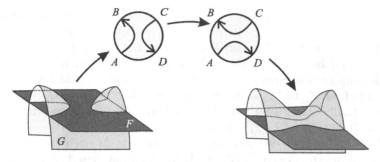

Fig. 6.35. One of the two modifications of the Morse type

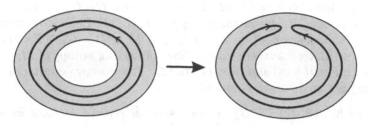

Fig. 6.36. Two essential circles on a torus fuse to an inessential circle

negative ordered triples of linearly independent vectors. Next, we orient every component of $G \cap F$ according to the following rule: at each point $x \in G \cap F$ the triple $(\bar{n}_G(x), \bar{n}_F(x), \bar{\tau}(x))$ must be positive, where $\bar{n}_G, \bar{n}_F$ are positive normal vectors for G, F and the tangent vector $\bar{\tau}(x)$ for $G \cap F$ shows its orientation at x. By the general position argument, any isotopy between G, G' can be replaced by a finite sequence of local moves so that each move either preserves the isotopy classes of $G \cap F$ in F and in G or subjects $G \cap F$ to one of the following two Morse modifications:

(a) Emerging a circle or a arc which is trivial in F and hence in G, or eliminating such circle or arc
(b) Replacing two disjoint proper oriented arcs $AB, CD \subset G \cap F$ contained in a small disc in F by two disjoint arcs AD, CB, see Fig. 6.35.

Recall that any two disjoint essential curves in a torus or an annulus are parallel. It follows that both modifications preserve the set of essential curves in $G \cap F$ except the case when two essential circles fuse together to an inessential circle or, vice-versa, appear after a fusion of an inessential circle with itself (see Fig. 6.36). In both exceptional situations at least one of the remaining curves in $G \cap F$ (denote it by C) is essential. Otherwise G would be isotopic to an essential surface $G' \subset M$ such that $G' \cap F = \emptyset$. This is impossible. Indeed, any essential torus or a clean essential annulus in $(M_F, \partial F_- \cup \partial F_+) = (F \times I, \partial F \times \partial I)$ has the form $F \times \{*\}$ by Lemma 6.5.5, in contradiction with our assumption that $\chi(F) < 0$. Therefore, the orbits of curves in $G \cap F$ before and after the modification can be determined by the same curve C. It follows from item 1 that then they coincide.

3. Let G be vertical. We can assume that the curves $\{C_0, \ldots, C_{n-1}\}$ in $G \cap F$ are numbered in a cyclic order with respect to their position in G. They decompose G into annuli or quadrilaterals which, considered as surfaces in M_F, run from F_+ to F_-. It follows that for each i the curve C_{i+1} is isotopic to $f^\varepsilon(C_i)$ (indices are taken modulo n), where ε is ± 1 and does not depend on the i. Thus $\{C_0, \ldots, C_{n-1}\}$ is the orbit of any C_i.

4. Let G, G' be two essential vertical surfaces in M such that a curve on $G \cap F$ is isotopic to a curve in $G' \cap F$. It follows from item 3 that then $G \cap F$ and $G' \cap F$ are isotopic. After an appropriate fiber-preserving isotopy of G we can assume that $G \cap F = G' \cap F$. Let the surfaces $G_F, G'_F \subset M_F = F \times I$ be obtained from G, G' by cutting along $G \cap F, G' \cap F$. Both are vertical, i.e., transversal to all fibers $F \times \{*\}$. Moreover, $G_F \cap (F \times \partial I) = G'_F \cap (F \times \partial I)$. Since $\chi(F) < 0$, it follows that there exists a fiber-preserving isotopy of $M_F = F \times I$ which keeps $F \times \partial I$ fixed and takes G_F to G'_F. This isotopy determines a fiber-preserving isotopy between G and G'.   □

Corollary 6.5.18. *Let G_1, G_2 be two essential vertical surfaces in an orientable Stallings manifold M with fiber F such that $\chi(F) < 0$. Suppose that they are isotopic. Then they are isotopic by a fiber-preserving isotopy.*

Proof. We can assume that G_1, G_2 are connected. It follows from conclusions 2, 3 of Lemma 6.5.17 that $G_1 \cap F$ and $G_2 \cap F$ are isotopic. Then G_1, G_2 are fiberwise isotopic by conclusion 4 of the same lemma.   □

A similar statement is true for quasi-Stallings manifolds, but for proving it we need to relate quasi-Stallings manifolds to Stallings ones.

Lemma 6.5.19. *Let $M_{(\alpha, \beta)}$ be the quasi-Stallings manifold defined by involutions $\alpha, \beta \colon F \to F$ and let $M_{\alpha\beta}$ be the Stallings manifold having the monodromy $\alpha\beta \colon F \to F$. Then there is a 2-sheeted covering $p \colon M_{\alpha\beta} \to M_{(\alpha, \beta)}$ such that the restriction of p onto each fiber of $M_{\alpha\beta}$ is either a homeomorphism onto a nonsingular fiber or a 2-sheeted covering map onto a singular fiber of $M_{(\alpha, \beta)}$.*

Proof. Consider two copies $(F \times I)_1, (F \times I)_2$ of $F \times I$ and identify their bases $(F \times \{0, 1\})_1, (F \times \{0, 1\})_2$ by the following rules: $(x, 0)_1 = (\alpha(x), 0)_2, (x, 1)_1 = (\beta(x), 1)_2$. See Fig. 6.37. The manifold $M_{\alpha\beta}$ thus obtained is a Stallings manifold with the monodromy $\alpha\beta$.

On the other hand, the rule $(x, t)_1 \leftrightarrow (x, t)_2$ determines a free involution $i \colon M_{\alpha\beta} \to M_{\alpha\beta}$ such that the quotient map $p \colon M_{\alpha\beta} \to M_{(\alpha, \beta)}$ possesses the required properties.   □

Corollary 6.5.20. *Let $M_{(\alpha, \beta)}$ be a quasi-Stallings manifold with fiber F such that $\chi(F) < 0$. Then the following holds: if two essential vertical surfaces in $M_{(\alpha, \beta)}$ are isotopic, then they are isotopic by a fiber-preserving isotopy.*

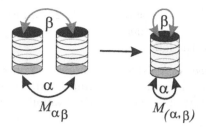

Fig. 6.37. $M_{\alpha\beta}$ is a 2-sheeted covering of $M_{(\alpha,\beta)}$

Proof. Let G_1, G_2 be two isotopic essential vertical surfaces in $M_{(\alpha,\beta)}$. Denote by $\tilde{G}_1, \tilde{G}_2$ their inverse images under the covering map $M_{\alpha\beta} \to M_{(\alpha,\beta)}$ constructed in Lemma 6.5.19. Note that any isotopy $M_{(\alpha,\beta)}$ can be replaced by a superposition of isotopic local moves and that any local move can be lifted to a pair of isotopic local moves of $M_{\alpha\beta}$. It follows that $\tilde{G}_1, \tilde{G}_2$ are isotopic in $M_{\alpha\beta}$.

Let F be a nonsingular fiber of $M_{(\alpha,\beta)}$. Then $p^{-1}(F)$ consists of two homeomorphic copies of F. Denote by F_0 one of them. Now we apply Corollary 6.5.18: since $\tilde{G}_1, \tilde{G}_2$ are isotopic in $M_{\alpha\beta}$, they are fiberwise isotopic. The restriction of any such isotopy onto F_0 takes $\tilde{G}_1 \cap F_0$ to $\tilde{G}_2 \cap F_0$. Projecting it to F, we get an isotopy of F taking $G_1 \cap F$ to $G_2 \cap F$.

The remaining part of the proof is similar to the proof of conclusion 4 of Lemma 6.5.17. Indeed, since $G_1 \cap F$ and $G_2 \cap F$ are isotopic in F, we can assume that $G_1 \cap F = G_2 \cap F$. Let us cut $M_{(\alpha,\beta)}$ along F. We get two twisted I-bundles, each contained a copy $F_\pm$ of F in the boundary. All the pieces of G_1, G_2 in them are transversal to fibers, incompressible and boundary incompressible. Since $\chi(F) < 0$, one can construct fiber-preserving isotopies φ_1, φ_2 of the bundles that keep F_-, F_+ fixed and take the pieces of G_1 to the pieces of G_2. Reconstructing $M_{(\alpha,\beta)}$ and concatenating φ_1, φ_2, we get a fiber-preserving isotopy of $M_{(\alpha,\beta)}$ that takes G_1 to G_2. $\square$

Proposition 6.5.21. *Extension move E_1–E_8 satisfy properties C_1, C_2, C_3.*

In a more detailed form Proposition 6.5.21 can be formulated as follows. Let P be an admissible subpolyhedron of an orientable Haken manifold (M, Γ). Then

1. There exist only finitely many admissible subpolyhedra that can be obtained from P by performing exactly one of moves $E_i, 1 \leq i \leq 8$ (the polyhedra are considered modulo the equivalence relation generated by homeomorphisms $(M, \Gamma) \to (M, \Gamma)$).
2. Representatives of the equivalence classes of such polyhedra can be constructed algorithmically.
3. Any sequence $P_1 \subset P_2 \subset \ldots$ of admissible subpolyhedra of (M, Γ), where each P_{i+1} is obtained from P_i by one of moves E_1–E_8, is finite.

Proof. For moves $E_i, i \leq 5$ the statements 1, 2 are already known (Proposition 6.5.2), so it suffices to consider moves E_6–E_8. To apply one of them to a chain U, we should do the following:

1. If U is either direct or twisted, we choose a nontrivial strip in U. Let U be a Stallings or a quasi-Stallings manifold. Suppose that it contains essential vertical tori or essential vertical annuli. Then we choose one of these surfaces.
2. Insert two parallel copies of the surface thus obtained into U and perturb them.

Let us show that both steps are algorithmic and that at each step we have only a finite choice. Indeed, if U is a direct chain $S \times I$ or a twisted chain $S \tilde\times I$, then L is determined by a proper arc $l \subset S$. By Lemma 6.4.45, we can choose l in a finite number of ways up to homeomorphisms $S \to S$ which are isotopic to the identity on ∂S. Any such homeomorphism can be extended to a homeomorphism $(M, \Gamma) \to (M, \Gamma)$ taking P to P. This means that we have only a finite choice for L.

Consider the case of a Stallings chain U. By Theorem 6.4.10, we can determine if U contains an essential torus or an essential annulus. If the answer is affirmative, we construct one. Note that any essential annulus in a Stallings or quasi-Stallings manifold with fiber $F, \chi(F) < 0$, is longitudinal. By Theorem 6.4.44, the number of such tori and annuli is finite up to strong equivalence. Then we use Lemma 6.5.16 to replace the torus or the annulus constructed above by an isotopic vertical surface, which is determined uniquely by Corollaries 6.5.18, 6.5.20. It remains to note that each of these surfaces can be perturbed in a finite number of ways.

For proving property C_3, we use the chamber complexity $C(P)$ of P. Moves E_6, E_8 can be presented as superpositions of moves E_4, move E_7 is a superposition of moves E_2. By Proposition 6.5.6, each of them decreases $C(P)$, which guarantees us that, performing moves E_1–E_8, we inevitably stop. $\square$

6.5.4 Difficult Case

In this section, we describe move E_9 for subdividing simple Stallings and quasi-Stallings chains. Let $U = (F \times I)/f$ be a Stallings chain of direct I-bundle chambers with fiber F in an orientable Haken manifold (M, Γ) with an admissible simple subpolyhedron P. Why cannot we proceed at the same manner as in the case of, say, a direct chain $F \times I$: choose a nontrivial proper arc $l \subset F$, insert the strip $L = l \times I$ into $F \times I$, and perturb it in the quotient space $(F \times I)/f$? Well, we can, but doing so we run into the following crucial obstacle: the set of strips obtained in this way is infinite, even if we consider them up to homeomorphisms of (M, Γ). The reason is that not every homeomorphism $F \to F$ of the fiber of U can be extended to a homeomorphism of U and hence to a homeomorphism of M. As we will see later, the set of extendible homeomorphisms is very small, even finite, if we consider them up to multiplication by a power of the monodromy map f.

Definition 6.5.22. *A 1-dimensional subpolyhedron X of a compact connected surface F with nonempty boundary is called a* simple skeleton *of F, if the following conditions hold:*

1. *X is the union of ∂F and disjoint simple proper arcs $C_1, \dots, C_n$*
2. *The arcs decompose F into discs*
3. *Each arc C_i separates different discs*

Definition 6.5.23. *Let (Q, Δ) be an I-bundle chamber. Then a simple skeleton P_Q of Q is called* vertical, *if there are a presentation $F \tilde\times I$ of Q and a simple skeleton X of F such that $P_Q = (X \tilde\times I) \cup (F \tilde\times \partial I)$.*

In other words, P_Q is obtained from ∂Q by inserting quadrilaterals $p^{-1}(C_i)$, where $p\colon Q \to F$ is the bundle projection and $C_1, \dots, C_n$ are the arcs of X which are not in ∂Q.

Definition 6.5.24. *Let U be an orientable Stallings or quasi-Stallings manifold presented as a chain of I-bundle chambers $Q_0, \dots, Q_{m-1}$. Then a simple skeleton S of U is called* vertical *, if $S \cap Q_i$ is a vertical skeleton of $Q_i, 0 \le i \le m-1$.*

Note that any vertical skeleton S of U contains all fibers that decompose U into the union of I-bundle chambers. These fibers contain all the true vertices of S. The remaining part of S consists of quadrilaterals. As usual, we will measure the complexity of S by the total number $c(S)$ of its true vertices. The next move allows us to subdivide Stallings and quasi-Stallings chains of I-bundle chambers.

E_9: Inserting a Minimal Vertical Skeleton of a Simple Stallings or Quasi-Stallings Chain

Suppose that a Stallings or quasi-Stallings chain U of I-bundle chambers for an admissible subpolyhedron $P \subset (M, \Gamma)$ contains no essential annuli or tori, i.e., is simple. Then we add to P a vertical skeleton of U having the minimal number of true vertices. See Fig. 6.38.

Our goal (which we achieve at the end of Sect. 6.5.7) is to show that any simple Stallings or quasi-Stallings chain U has only finitely many minimal vertical skeletons (up to homeomorphisms $(M, \Gamma) \to (M, \Gamma)$ taking P to P), and that all of them can be constructed algorithmically. One can easily find an upper bound for the complexity of minimal vertical skeletons of U: it suffices to construct a particular vertical skeleton of U and take the number k of its true vertices. It is also easy to show that the number of vertical skeletons of complexity $\le k$ and hence the number of minimal vertical skeletons of U is finite. But how can one enumerate all minimal skeletons? One successful strategy for doing that consists in forgetting for a while about 3-manifolds and concentrating on skeletons considered as abstract special polyhedra.

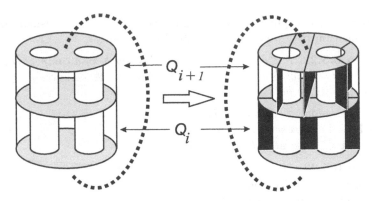

Fig. 6.38. Inserting a vertical skeleton

Definition 6.5.25. *An abstract special polyhedron S is vertical, if it is home-omorphic to a vertical skeleton of an orientable Stallings or quasi-Stallings manifold.*

Let k be an integer and F an orientable surface having a negative Euler characteristic. Denote by $\mathcal{S}(F,k)$ and $\mathcal{SQ}(F,k)$ the sets of all abstract special polyhedra such that each polyhedron $S \in \mathcal{S}(F,k) \cup \mathcal{SQ}(F,k)$ has $\leq k$ true vertices and is homeomorphic to a vertical skeleton of an orientable Stallings or, respectively, quasi-Stallings chain with fiber F. Of course, we consider the polyhedra up to homeomorphisms.

Lemma 6.5.26. *For any surface F and for any integer k the sets $\mathcal{S}(F,k)$ and $SQ(F,k)$ are finite and can be constructed algorithmically.*

Proof. By Theorem 2.1.1, one can construct a finite list of polyhedra containing all special spines with $1, 2, \ldots, k$ vertices. It remains to select among them vertical skeletons of Stallings and quasi-Stallings manifolds with fiber F. Let us describe in detail the selection procedure.

Consider a special polyhedron S with $\leq k$ vertices. First we use Theorem 1.1.20 to decide if S is a special spine of an orientable 3-manifold W. If it is, then W is unique up to homeomorphism (see Theorem 1.1.17). To decide if S is a vertical skeleton of a Stallings manifold, we subject it to several tests.

TEST 1. Does ∂W consist only of spheres and tori? If not, we reject S and take the next spine. If yes, then we denote by W' the 3-manifold obtained by attaching 3-balls to all spherical components of ∂W.

TEST 2. Does S contain a finite collection of disjoint tori such that they cut off a collar of $\partial W'$? If not, we reject S again. If yes, we denote by W'' the 3-manifold obtained from W' by cutting off the collar of $\partial W'$. The boundary $\partial W'' \subset S$ of W'' consists of the tori constituting the collection.

TEST 3. Does S contain the union $G = F_0 \cup \ldots \cup F_{m-1} \subset S$ of $m \geq 2$ disjoint copies of F such that the following holds:

1. G is proper in W'', and G contains all true vertices of S.
2. G decomposes W'' into m connected components $Q_i, 0 \le i \le m-1$, such that $Q_i \cap Q_{i-1} = F_i$ for all i (indices are taken modulo m).
3. The closure $\mathrm{Cl}(\alpha)$ of any 2-component α of S contained in Q_i and not contained in $F_i \cup F_{i+1}$ is a quadrilateral having one side in F_i and the opposite side in F_{i+1}.

All three questions are of algorithmic nature, i.e., they can be answered algorithmically. It is easy to show that $S \in \mathcal{S}(F, k)$ if and only if all three answers are affirmative. Indeed, if they are, then every pair $(Q_i, F_i \cup F_{i+1})$ is homeomorphic to the pair $(F \times I, F \times \{0,1\})$. It follows that W'' is a Stallings manifold and S is a vertical skeleton of W''.

The quasi-Stallings case is similar. To decide whether or not S is in $\mathcal{SQ}(F, k)$, we subject it to the same Tests 1 and 2. Test 3 should be modified as follows:

(A) In item 2 we require that G decompose W'' into $m+1$ connected components $Q_i, 0 \le i \le m$, such that $Q_i \cap Q_{i+1} = F_i$ for $0 \le i \le m-1$ and $Q_m \cap Q_0 = \emptyset$;

(B) In item 3 we preserve the constraint that the closure $\mathrm{Cl}(\alpha)$ of any 2-component $\alpha \subset S$ contained in $Q_i \backslash (F_i \cup F_{i-1})$ is a quadrilateral. However, two opposite sides of $\mathrm{Cl}(\alpha)$ must lie in distinct copies of F only for $0 < i < m$. If $i = 0$ or $i = m$, then $\mathrm{Cl}(\alpha)$ must have two opposite sides in the unique copy $F' = F_0$ or $F' = F_{m-1}$ of F contained in Q_i. Moreover, $\mathrm{Cl}(\alpha)$ must be boundary incompressible in Q_i considered as a 3-manifold with the boundary pattern $\partial F'$.

These modifications of Test 3 reflect the distinction between the structure of Stallings and quasi-Stallings manifolds. Evidently, all three answers to the modified tests are affirmative if and only if W'' is quasi-Stallings and S is a vertical skeleton of W''. Components Q_0 and Q_m correspond to the twisted I-bundle chambers of W'', which contain singular fibers. $\square$

At first glance Lemma 6.5.26 is sufficient for proving that E_9 possesses properties C_1, C_2 of extension moves. Indeed, let U be a Stallings or quasi-Stallings chain of I-bundle chambers. To subdivide it into balls, i.e., to perform E_9, one can choose among the polyhedra in $\mathcal{S}(F, k)$ or $\mathcal{SQ}(F, k)$ an abstract vertical polyhedron homeomorphic to a minimal vertical spine of U and insert it into U. However, we have to overcome two obstacles.

First $\mathcal{S}(F, k)$ and $\mathcal{SQ}(F, k)$ contain abstract polyhedra homeomorphic to vertical skeletons of many different simple Stallings or quasi-Stallings manifolds. Therefore, a recognition algorithm for Stallings and quasi-Stallings manifolds is needed to select skeletons of 3-manifolds homeomorphic to U.

Second any abstract polyhedron homeomorphic to a skeleton of U can be inserted into U in many different ways. We have to show that the number of such insertions is essentially finite. Sections 6.5.5–6.5.7 are devoted to solving these problems.

6.5.5 Recognition of Simple Stallings Manifolds with Periodic Monodromy

Suppose that (M, Γ) is an orientable Haken manifold. Let us apply to (M, Γ) moves E_1–E_8 as long as possible. We get an admissible polyhedron $P \subset M$, which decomposes M into 3-balls, solid tori, and I-bundle chambers such that every chain U of I-bundle chambers is a simple Stallings or quasi-Stallings manifold. As we mentioned earlier, Haken's idea of controlled decomposition of the 3-manifold into smaller pieces does not work anymore. One can sharpen the situation by considering the case when from the very beginning M is a simple Stallings manifold (say, closed). Then our construction of simple skeletons for M stops just after the first step (inserting two parallel copies of the fiber).

As we will see later, any simple orientable Stallings manifold M_f belongs to one of the following two types:

1. M_f is a Seifert manifolds fibered over S^2 with 3 exceptional fibers. Its monodromy map f is periodic up to isotopy. This means that all curves on F are periodic.

2. M_f is not a Seifert manifold and f admits no essential periodic curves at all.

Let us give an exact definition of an *essential periodic curve*. By a *singular curve* in a surface F we mean an arbitrary map $c: S^1 \to F$ or $c: (I, \partial I) \to (F, \partial F)$. A singular curve is *essential*, if it cannot be contracted to a point by a proper homotopy.

Definition 6.5.27. *Let $f: F \to F$ be a homeomorphism of a surface F onto itself. Then a singular curve c in F is called* periodic *(with respect to f), if for some nonnegative integer number n the curves c and $f^n c$ are properly homotopic.*

Clearly, any inessential singular closed curve is periodic with period 1.

The condition that $f: F \to F$ have no essential periodic curves admits a very convenient geometric interpretation. Recall that a singular torus $\varphi: T \to M$ in an orientable Haken 3-manifold M is *essential*, if the induced homomorphism $\varphi_*: \pi_1(T) \to \pi_1(M)$ is injective and its image is not conjugate to a subgroup of $\pi_1(\partial M)$. A proper singular annulus $\varphi: (A, \partial A) \to (M, \partial M)$ is *essential*, if $\varphi_*: \pi_1(A) \to \pi_1(M)$ is injective and the restriction of φ onto a proper arc $l \subset A$ joining different components of ∂A is not properly homotopic to a map into ∂M.

Lemma 6.5.28. *Let M_f be a Stallings manifold with fiber F and monodromy map $f: F \to F$ such that $\chi(F) < 0$. Then f has an essential periodic curve if and only if M_f contains either an essential singular torus or an essential singular annulus.*

Proof. Suppose that f admits a closed essential singular curve c of period n. Then c determines a singular torus T in the finite covering M_{f^n} of M_f, which is obtained by cyclic gluing n exemplars of $F \times I$ via f. Indeed, T is composed from n copies of $c \times I \subset F \times I$. Of course, T is essential and projects into an essential singular torus in M_f.

Vice versa, by the same arguments as in the proof of Lemma 6.5.16, any essential singular torus T can be shifted to vertical position, where each circle in $F \cap T$ is a periodic essential curve.

The proof for singular annuli and arcs is similar.   □

This lemma shows that if the monodromy map $f \colon F \to F, \chi(F) < 0$, has no periodic curves, then M_f is simple. The converse is also true, with the exception of Seifert manifolds fibered over S^2 with three exceptional fibers.

Proposition 6.5.29. *Let M_f be an irreducible Stallings or quasi-Stallings manifold. Suppose there exists an essential singular annulus in M. Then there exists an embedded essential annulus in M_f.*

Proposition 6.5.30. *Let M_f be an irreducible Stallings or quasi-Stallings manifold. Suppose there exists an essential singular torus in M. Then at least one of the following assertions is true:*

1. There exists an embedded essential torus in M_f;
2. M_f is a Seifert manifold.

Both propositions are partial cases of the famous torus-annulus theorem [52, 55]. In [57] Johannson proved them by using the machinery of Nielsen [98], which, however, can be replaced by a more efficient elementary technique of Jaco and Shalen [53].

Corollary 6.5.31. *Let M_f be an orientable simple Stallings manifold with fiber $F, \chi(F) < 0$, and monodromy map $f \colon F \to F$. Then exactly one of the following holds:*

1. f has no essential periodic curves;
2. M_f is a Seifert manifold fibered over S^2 with three exceptional fibers.

Proof. Suppose that conclusion 1 does not hold, i.e., that f has essential periodic curves. Since M_f is simple, it follows from Propositions 6.5.29 and 6.5.30 that M_f is a Seifert manifold. All simple orientable Seifert manifolds are known: they are fibered either:

(a) Over S^2 with ≤ 3 exceptional fibers, or
(b) Over RP^2 with ≤ 1 exceptional fiber

In case (a) we cannot have less than three exceptional fibers, since otherwise M_f would be either a lens space or $S^2 \times S^1$. This is impossible, since a Stallings manifold with $\chi(F) < 0$ cannot have a finite fundamental group or be reducible. It remains to note that every orientable Seifert manifold as in (b) admits another fibration as in (a). The only exception is the manifold $RP^2 \tilde{\times} S^1$, which is reducible and thus cannot be Stallings.   □

Denote by $\mathcal{S}(3)$ the class of all Seifert manifolds fibered over S^2 with three exceptional fibers. Let us consider the recognition problem for Stallings manifolds which are in $\mathcal{S}(3)$. First we recall a general way for constructing such manifolds. Suppose that $(\alpha_i, \beta_i), 1 \le i \le 3$, are three pairs of coprime integers, where $\alpha_i > 1$. Let us cut two solid tori $D_1 \times S^1, D_2 \times S^1$ out of the solid torus $D^2 \times S^1$. We get the manifold $N^2 \times S^1$, where $N^2 = D^2 \backslash \mathrm{Int}\,(D_1 \cup D_2)$ is a disc with two holes. N^2 and S^1 are considered as subsets of the standard plane R^2. They inherit the standard orientation of R^2 and thus determine orientations of $N^2, \partial N^2, N^2 \times S^1$. We equip the tori $\partial N^2 \times S^1$ with coordinate systems composed from oriented meridians $c_i \times \{*\}$ and oriented longitudes $\{*\} \times S^1$, where $c_i, 1 \le i \le 3$, are the boundary circles of N^2. Then we attach each torus $D_i \times S^1$ back to $N^2 \times S^1$ via a homeomorphism taking the meridian $\partial D_i \times \{*\}$ to a curve of the type (α_i, β_i). The oriented Seifert manifold M thus obtained is denoted by $M(S^2, (\alpha_1, \beta_1)(\alpha_2, \beta_2)(\alpha_3, \beta_3))$. Pairs (α_i, β_i) are called *nonnormalized parameters* of its exceptional fibers. The sum $e(M) = \sum_i \beta_i/\alpha_i$ is called the *Euler number* of M.

The classification of such manifolds is well known [96, 99]: Two manifolds $M = M(S^2, (\alpha_1, \beta_1)(\alpha_2, \beta_2)(\alpha_3, \beta_3))$ and $M' = M(S^2, (\alpha'_1, \beta'_1)(\alpha'_2, \beta'_2)(\alpha'_3, \beta'_3))$ are homeomorphic via an orientation-preserving homeomorphism if and only if after an appropriate renumbering of (α_i, β_i) we have $\alpha_i = \alpha'_i, \beta_i = \beta'_i \bmod \alpha_i, 1 \le i \le 3$, and $e(M) = e(M')$. Changing signs of all β_i corresponds to reversing orientation of the manifold.

We are interested in fiber-preserving classification of Stallings manifolds from $\mathcal{S}(3)$ (saying "fiber-preserving," we mean fibers of Stallings fibrations). To begin with, we describe an example of a surface in a Seifert manifold.

Example 6.5.32. Let $M = M(S^2, (\alpha_1, \beta_1)(\alpha_2, \beta_2)(\alpha_3, \beta_3))$ be a Seifert manifold presented as $N^2 \times S^1$ whose boundary components are filled with three solid tori. Suppose that $e(M) = 0$. Denote by n the smallest integer $n > 0$ divisible by each α_i. Let $x_0, \ldots, x_{n-1} \in S^1$ be the cyclically ordered vertices of a regular n-gon with vertices in S^1. Denote by γ_1, γ_2 two disjoint segments in N^2 joining the first and the second boundary circles of N^2 with the third one. Let $N_k = N^2 \times \{x_k\}, 0 \le k \le n-1$ be n copies of N^2 in $N^2 \times S^1$. Then we transform them into a surface $G_0 \subset N^2 \times S^1$ as follows. For $i = 1, 2$, we cut $N^2 \times S^1$ along the annulus $A_i = \gamma_i \times S^1$, rotate one of the annuli $A_i^{\pm}$ arising in this way to the right by $2\pi\beta_i/\alpha_i$, and glue the annuli back. See Fig. 6.39 for $\beta_1 = 2$ and $\beta_2 = -1$.

One can easily show that the boundary circles of G_0 on the first two tori $\partial N^2 \times S^1$ have types $(\alpha_i, \beta_i), i = 1, 2$. The assumption $e(M) = \beta_1/\alpha_1 + \beta_2/\alpha_2 + \beta_3/\alpha_3 = 0$ says that the boundary circles of G_0 on the third boundary torus have type (α_3, β_3). Adding to G_0 meridional discs of the solid tori attached to $N^2 \times S^1$, we get a closed surface $G \subset M$.

Lemma 6.5.33. *The surface G constructed in Example 6.5.32 possesses the following properties:*

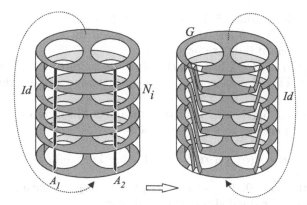

Fig. 6.39. A Stallings fiber of a Seifert manifold over S^2 with three exceptional fibers of types (5,2),(5,-1),(5,-1)

1. G *is a connected orientable nonseparating surface in* M.
2. G *is a fiber of a Stallings fibration* $p: M \to S^1$ *whose monodromy map* $g: G \to G$ *has period* n.
3. $\chi(G) = n(-1 + 1/\alpha_1 + 1/\alpha_2 + 1/\alpha_3)$.

Proof. Since n is the smallest number divisible by $\alpha_1, \alpha_2, \alpha_3$, G is connected. Obviously, it is orientable and nonseparating. Indeed, G decomposes each circle $\{*\} \times S^1 \subset N^2 \times S^1$ into arcs joining different sides of G. The same arcs determine also a direct product structure $G \times I$ on the manifold M_G obtained from M by cutting along G. Since G is composed from n sheets, the monodromy map has period n. By construction, G is obtained by attaching $n/\alpha_1 + n/\alpha_2 + n/\alpha_3$ discs to G_0 having the Euler characteristic $-n$. Therefore, $\chi(G) = n(-1 + 1/\alpha_1 + 1/\alpha_2 + 1/\alpha_3)$.

Proposition 6.5.34. *A 3-manifold* $M = M(S^2, (\alpha_1, \beta_1)(\alpha_2, \beta_2)(\alpha_3, \beta_3))$ *contains a connected closed incompressible surface* $F \neq S^2$ *if and only if* $e(M) = 0$. *Such surface is unique up to isotopy and is a fiber of a Stallings fibration on* M *with periodic monodromy. Moreover, if* $\chi(F) < 0$, *then the period* n *of the monodromy map does not exceed* $-42\chi(F)$.

Proof. Let us present M as $N^2 \times S^1$ with three attached solid tori. Consider an incompressible surface $F \subset M$ and deform it so that afterwards it crosses the solid tori at the minimal number of meridional discs. Since $F \neq S^2$, it crosses each torus along at least one meridional disc. Let us investigate how the surface $F_0 = F \cap (N^2 \times S^1)$ intersects the annuli $A_i = \gamma_i \times S^1$, see Example 6.5.32 and Fig. 6.39. Deforming F_0 by an isotopy, we can ensure that the intersections $F_0 \cap A_i, i = 1, 2$, consist of radial segments of the annuli. The remaining part of F_0 in the solid torus obtained from $N^2 \times S^1$ by cutting along A_i is the union of meridional discs of this solid torus. It follows that F is isotopic to the surface G constructed in Example 6.5.32, with some n

divisible by each α_i. Since F is connected, n is as small as possible. We can conclude that $e(M) = 0$ and that any two closed incompressible surfaces in M are isotopic (since they are isotopic to G).

Conversely, if $e(M) = 0$, then $H_1(M; Z)$ is infinite and M is Haken by Lemma 4.1.29. Therefore, it contains a closed incompressible surface different from S^2.

By Lemma 6.5.33, F is a fiber of a Stallings fibration of M with monodromy map of period $n = -\chi(F)/\kappa$, where

$$\kappa = \kappa(\alpha_1, \alpha_2, \alpha_3) = 1 - 1/\alpha_1 - 1/\alpha_2 - 1/\alpha_3.$$

If $\chi(F) < 0$, then $\kappa > 0$. The set of all triples $(\alpha_1, \alpha_2, \alpha_3)$ with $2 \leq \alpha_1 \leq \alpha_2 \leq \alpha_3$ and $\kappa > 0$ can be divided into three subsets:

1. $\alpha_1 = 2, \alpha_2 = 3, \alpha_3 \geq 7$
2. $\alpha_1 = 2, \alpha_2 \geq 4, \alpha_3 \geq 5$
3. $\alpha_1 \geq 3, \alpha_2 \geq 3, \alpha_3 \geq 4$

The minimal value of κ for all such triples is $1 - 1/2 - 1/3 - 1/7 = 1/42$. Therefore, $n = -\chi(F)/\kappa \geq -42\chi(F)$.   □

Corollary 6.5.35. *There is an algorithm to decide if a given simple Stallings manifold M with fiber $F, \chi(F) < 0$, is a Seifert manifold fibered over S^2 with three exceptional fibers. In case M is Seifert, a Seifert structure on M can be constructed algorithmically.*

Proof. Let f be the monodromy map for M. It follows from Propositions 6.5.31 and 6.5.34 that $M \in \mathcal{S}(3)$ if and only if f is periodic with period $n \leq -42\chi(F)$. Therefore, in order to decide if $M \in \mathcal{S}(3)$, it suffices to test all powers $f^n, 1 \leq n \leq -42\chi(F)$, for being homotopic to the identity map $F \to F$.

Suppose that $M \in \mathcal{S}(3)$. Then a Seifert structure on M can be constructed as follows. Let T be a triangulation of M. For any k, there are finitely many star subdivisions of T with $\leq k$ tetrahedra. We can find them all and test each triangulation for existence of a simplicial map p of M onto a triangulated 2-sphere such that p is the projection of a Seifert fibration with three exceptional fibers. Eventually we must find such p. This follows from the assumption that $M \in \mathcal{S}(3)$ and the Alexander theorem [1] that any two triangulation of the same 3-manifold have combinatorially isomorphic star subdivisions.   □

A more economic way of constructing a Seifert structure can be extracted from [53], see also [73, 115].

Proposition 6.5.36. *There is an algorithm to decide if two given simple Stallings manifolds with fiber $F, \chi(F) < 0$, from $\mathcal{S}(3)$ are homeomorphic via a homeomorphism taking Stallings fibers onto Stallings fibers.*

Proof. Let M, M' be two given simple Stallings manifolds. By Corollary 6.5.35, we can construct Seifert fibrations of M, M' and hence calculate parameters

(α_i, β_i) and $(\alpha'_i, \beta'_i), 1 \leq i \leq 3$, of their exceptional fibers. Recall that any two homeomorphic manifolds from $\mathcal{S}(3)$ are fiber-preserving homeomorphic. Taking into account that $e(M) = e(M')$, we can conclude that M, M' are homeomorphic via an orientation preserving homeomorphism if and only if, after an appropriate renumbering of (α'_i, β'_i), we have $\alpha_i = \alpha'_i$ and $\beta_i = \beta'_i$ mod α_i for $1 \leq i \leq 3$. It remains to note that every homeomorphism $M \to M'$ is isotopic to a fiber-preserving one. This follows from the fact that, up to isotopy, M' contains a unique closed incompressible surface different from S^2, namely, the fiber. See Proposition 6.5.34.   □

6.5.6 Recognition of Simple Stallings Manifolds with Nonperiodic Monodromy

In this section we consider the recognition problem for simple Stallings manifolds whose monodromy maps are nonperiodic and hence have no essential periodic curves. To continue the construction of admissible skeletons, we must solve this problem. Numerous attempts to overcome this obstacle had no success. Mathematicians came to the conclusion that Haken's approach was insufficient and that the recognition problem for such Stallings manifolds (which can be easily reduced to the conjugacy problem for their monodromy maps) should be solved by independent methods. This indeed was done by Hemion [41, 42]. Clearly, the recognition problem for quasi-Stallings manifolds also requires an independent solution. Surprisingly, Hemion's solution for the Stallings case is insufficient for the quasi-Stallings one; this problem turned out to be much more difficult. We tackle it in Sect. 6.5.7.

Note that at the present time an alternative solution of the algorithmic recognition problem for simple Stallings manifolds with nonperiodic monodromy and for simple quasi-Stallings manifolds can be obtained by using the fact that they are hyperbolic [62, 100, 120, 121]. Nevertheless, I prefer to keep the exposition within the limits of elementary combinatorial approach by using the Hemion solution for the Stallings case (Proposition 6.5.42). For the quasi-Stallings case I borrow the notion of stretching factor from Thurston's theory of surface homeomorphisms (see the proof of Proposition 6.5.49).

First we reduce the recognition problem for Stallings manifolds to the conjugacy problem for their monodromy maps. Let $f, g: F \to F$ be two homeomorphisms of a surface F onto itself. Suppose that there exists a homeomorphism $h: F \to F$ such that hfh^{-1} is isotopic to g. In this case, we say that h *conjugates* f to g or that f and g are *conjugate*. The set of all isotopy classes of homeomorphisms $F \to F$ that conjugate f to g is denoted by $\mathrm{Conj}(f, g)$. Let M_f, M_g be two Stallings manifolds with fiber F. It is convenient to represent them in the form $M_f = (F \times I)/\sim, M_g = (F \times I)/\sim$, where the equivalence relation $\sim$ is generated by identifications $(x, 1) = (f(x), 0)$ in the first case and identifications $(x, 1) = (g(x), 0)$ in the second. We will think of F as being contained in M_f and M_g as $F \times \{0\}$.

Recall that a homeomorphism $M_f \to M_g$ is fiber-preserving, if it takes each fiber $F \times \{t\}$ of M_f onto the corresponding fiber $F \times \{t\}$ of M_g. Denote by Homeo(M_f, M_g) the set of all fiber-preserving homeomorphisms $M_f \to M_g$ considered modulo fiber-preserving isotopy. It is evident that the restriction of any $H \in$ Homeo(M_f, M_g) onto F conjugates f to g, i.e., belongs to Conj(f, g). Indeed, rewrite $H(x, t)$ as $(h_t(x), t)$. Then, recalling that H is a homeomorphism between the Stallings manifolds, not just between direct products $F \times I$, we obtain $g h_1 = h_0 f$. Since h_0 and h_1 are isotopic, f and g are conjugate. Any h_t can be taken as a conjugating homeomorphism.

Let us assign now to every homeomorphism $H \in$ Homeo(M_f, M_g) its restriction onto F. This assignment induces a map $\psi \colon$ Homeo$(M_f, M_g) \to$ Conj(f, g). If $f = g$ and hence $M_f = M_g$, then Homeo(M_f, M_f) and Conj(f, f) are groups and ψ is a homomorphism.

The following lemma is easy.

Lemma 6.5.37. *For any two Stallings manifolds M_f, M_g the map ψ defined above is a bijection between Homeo(M_f, M_g) and Conj(f, g).*

Proof. Let a homeomorphism $h \colon F \to F$ conjugate f to g, that is, $h f h^{-1}$ be isotopic to g. Then h and $g h f^{-1}$ are also isotopic. Choose an isotopy $h_t \colon F \to F$ such that $h_0 = g h f^{-1}$ and $h_1 = h$. Then the map $H' \colon F \times I \to F \times I$ given by the rule $H'(x, t) = (h_t(x), t)$ determines a fiber-preserving homeomorphism $H \colon M_f \to M_g$. It is easy to verify that assigning $h \to H$ determines the inverse map $\psi^{-1} \colon$ Conj$(f, g) \to$ Homeo(M_f, M_g). $\Box$

To formulate Hemion's theorem, we need a preparation. Let F be a compact surface with nonempty boundary. Then any disc Δ obtained from F by cutting it along nonseparating disjoint proper arcs is called a *fundamental disc* of F. All the liftings of these arcs to the universal covering $\tilde{F}$ of F cut $\tilde{F}$ into copies of Δ. Any such copy is called a *fundamental region* of $\tilde{F}$. Recall that any homeomorphism $f \colon F \to F$ can be lifted to a homeomorphism $\tilde{f} \colon \tilde{F} \to \tilde{F}$.

Definition 6.5.38. *Let Δ be a fundamental disc for a surface F, let $\delta \subset \tilde{F}$ be a fundamental region corresponding to Δ, and let $f \colon F \to F$ be a homeomorphism. Then the Δ-size $d(f)_\Delta$ of f is the smallest number N such that $\tilde{f}(\delta)$ is contained in N fundamental regions of $\tilde{F}$, see Fig. 6.40.*

Evidently, $d(f)_\Delta$ does not depend on the choice of δ. The following property of the Δ-size is easy, see [41].

Lemma 6.5.39. *Let F be a compact surface and Δ a fundamental disc for F. Then for any number N there exists an algorithmically constructible finite set of homeomorphisms $F \to F$ such that every homeomorphism $h \colon F \to F$ of size $d(f)_\Delta \leq N$ is isotopic to one of them.*

Theorem 6.5.40 (Hemion [41, 42]). *Suppose that homeomorphisms f, g of a surface F onto itself admit no essential periodic curves. Let Δ be a fundamental disc for F. Then there exists an integer number $N = N(f, g)$ having the following properties:*

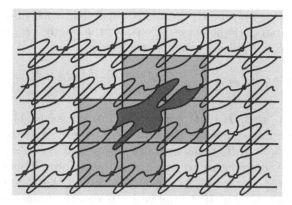

Fig. 6.40. Seven fundamental regions are required in order to cover the image of one fundamental region

1. N can be calculated algorithmically.
2. Any homeomorphism $h: F \to F$ that conjugates f to g is isotopic to a homeomorphism of the form $h'f^n$, where n is an integer and $h': F \to F$ is of size $d(h')_\Delta \le N$.

Corollary 6.5.41. *Suppose that homeomorphisms $f, g: F \to F$ of a surface F onto itself admit no essential periodic curves. Then one can algorithmically decide whether or not f, g are conjugate. Moreover, if they are, then one can construct a finite set $h_1, \ldots, h_k$ of homeomorphisms $F \to F$ such that the following holds:*

1. *Each h_i conjugates f to g*
2. *Any homeomorphism $h: F \to F$ conjugating f to g is isotopic to a homeomorphism of the form $h_i f^n$, where $1 \le i \le k$ and $-\infty < n < \infty$.*

Proof. Let Δ and N be as in Theorem 6.5.40. Using Lemma 6.5.39, we construct a finite set of homeomorphisms $F \to F$ representing all homeomorphisms of size $\le N$. Then we select among them homeomorphisms which conjugate f to g (this can be done algorithmically). If the set thus obtained is empty, then f, g are not conjugate. Suppose that it is nonempty, i.e., consists of homeomorphisms $h_1, \ldots, h_k$. Let $h: F \to F$ be any other homeomorphism that conjugates f to g. Applying Theorem 6.5.40 again, we can replace h by an isotopic homeomorphism of the form $h'f^n$ such that $d(h')_\Delta \le N$. Since h' also conjugates f to g, it is isotopic to one of h_i. $\qquad \square$

It may be illuminating to reformulate Corollary 6.5.41 as follows. Recall that $\mathrm{Conj}(f, g)$ denotes the set of isotopy classes of homeomorphisms $F \to F$ that conjugate f to g, and that f acts on $\mathrm{Conj}(f, g)$ by the rule $h \to hf$. Suppose that f, g admit no essential periodic curves. Then the orbit set $\mathrm{Conj}(f, g)/f$ of this action is finite and representatives of the orbits can be constructed algorithmically (for example, one can take $h_1, \ldots, h_k$, see above).

Let $M_f = (F \times I)/f$ be the Stallings 3-manifold with fiber F such that f have no essential periodic curves. Consider the family $\{H_s, 0 \leq s \leq 1\}$ of maps $F \times I \to F \times I$ given by the following rule:

$$H_s(x,t) = \begin{cases} (x, t+s), & \text{if } t+s \leq 1; \\ (f(x), t+s-1), & \text{if } t+s \geq 1. \end{cases}$$

If $t + s = 1$, then $H_s(x,t)$ consists of two points $(x,1)$ and $(f(x),0)$. Since they determine the same point of M_f, we get an isotopy $H_s: M_f \to M_f$ such that $H_0 = 1$, H_s takes fibers to fibers, and H_1 takes each fiber to itself. It may be illuminating to think of H_s as follows: it moves each fiber F around M_f and returns to its initial position so that the resulting map $F \to F$ is f. If we descend H_s to the base circle of the fibration $p: M_f \to S^1$, we get a full rotation of S^1. Having that in mind, we redenote H_1 by R_f. Clearly, $R_f = \psi^{-1}(f)$, where the map ψ^{-1} is inverse to the isomorphism $\psi: \text{Homeo}(M_f, M_f) \to \text{Conj}(f, f)$, see Lemma 6.5.37. We emphasize that R_f is isotopic to the identity, but there is no such isotopy taking each fiber onto itself.

The next proposition is actually a geometric reformulation of Corollary 6.5.41. It solves the algorithmic recognition problem for Stallings manifolds whose monodromy maps have no essential periodic curves.

Proposition 6.5.42. *Suppose that the monodromy maps $f, g: F \to F$ of Stallings manifolds M_f, M_g with fiber F have no essential periodic curves. Then one can algorithmically decide whether or not M_f and M_g are fiberwise homeomorphic. Moreover, if they are, then one can construct a finite set $H_1, \ldots, H_k$ of fiber-preserving homeomorphisms $M_f \to M_g$ such that any fiber-preserving homeomorphism $H: M_f \to M_g$ is fiberwise isotopic to a homeomorphism of the form $H_i R_f^n$, where $1 \leq i \leq k$ and $-\infty < n < \infty$.*

Proof. Follows from Lemma 6.5.37, Corollary 6.5.41, and the fact that f corresponds to R_f.   □

Propositions 6.5.36 and 6.5.42 solve the recognition problem for simple Stallings manifolds and thus eliminate the first obstacle for subdividing Stallings chains, which was mentioned at the end of Sect. 6.5.4. To overcome the second obstacle, i.e., to prove that each abstract vertical skeleton can be inserted into a Stallings chain U in essentially a finite number of ways, let us analyze fiber-preserving homeomorphisms between Stalling manifolds. Our goal is to prove that the set of their restrictions onto the boundaries of the manifolds is finite up to fiber-preserving isotopy.

Definition 6.5.43. *Let M_f, M_g be Stallings manifolds with fiber F. We say that two fiber-preserving homeomorphisms $h, h' \in \text{Homeo}(M_f, M_g)$ are boundary equivalent, if their restrictions onto ∂M_f are fiberwise isotopic (with respect to the induced fiberings of $\partial M_f, \partial M_g$ into the boundary circles of the fibers of M_f, M_g).*

We emphasize that the isotopy must be fiber-preserving in the strong sense: the image of each boundary circle must remain fixed during the isotopy. The set of all boundary equivalence classes of homeomorphisms $M_f \to M_g$ will be denoted by $\mathrm{Homeo}_\partial(M_f, M_g)$. There is a natural surjective map $\varphi \colon \mathrm{Homeo}(M_f, M_g) \to \mathrm{Homeo}_\partial(M_f, M_g)$, which assigns to each homeomorphism $h \in \mathrm{Homeo}(M_f, M_g)$ its boundary equivalence class. The proof of the following proposition is based on the observation that any element of $\mathrm{Homeo}_\partial(M_f, M_g)$ is essentially determined by the induced bijection between the boundary tori of M_f and M_g.

Proposition 6.5.44. *For any simple orientable Stallings manifolds M_f, M_g the set $\mathrm{Homeo}_\partial(M_f, M_g)$ is finite. Homeomorphisms $M_f \to M_g$ representing all its elements can be constructed algorithmically.*

Proof. It is sufficient to prove the proposition for the case $f = g$, when $\mathrm{Homeo}(M_f, M_f)$ and $\mathrm{Homeo}_\partial(M_f, M_f)$ are groups and φ is a homomorphism. Let us define a subgroup $\mathrm{Homeo}_0(M_f, M_f) \subset \mathrm{Homeo}(M_f, M_f)$ by setting that $h \in \mathrm{Homeo}(M_f, M_f)$ belongs to $\mathrm{Homeo}_0(M_f, M_f)$ if and only if h takes each torus of ∂M_f onto itself. We claim that $\varphi(\mathrm{Homeo}_0(M_f, M_f)) = 1$ or, equivalently, that the restriction of any $h \in \mathrm{Homeo}_0(M_f, M_f)$ onto ∂M_f is fiberwise isotopic to the identity. Since the permutation group of the boundary tori of M_f is finite, the proposition follows from the claim.

Let us prove the claim. Consider a torus T of ∂M_f. Since h is fiber-preserving, the restriction $h_{|T}$ is fiberwise isotopic to a power τ_C^k of the Dehn twist τ_C along a circle $C \subset F \cap T$ (all such circles are parallel in T). It follows that we get a homomorphism from $\mathrm{Homeo}_0(M_f, M_f)$ to the infinite cyclic group $\langle \tau_C \rangle$ generated by τ_C. Note that the image of the rotation R_f under this homomorphism is trivial, since $R_f \colon M_f \to M_f$ is isotopic to the identity. On the other hand, Proposition 6.5.42 tells us that, up to multiplication by powers of R_f, the number of homeomorphisms in $\mathrm{Homeo}(M_f, M_f)$ is finite. We can conclude that the image of $\mathrm{Homeo}_0(M_f, M_f)$ in $\langle \tau_C \rangle$ is finite and hence trivial. □

Remark 6.5.45. Proposition 6.5.44 can be extracted from the fact that any simple Stallings manifold whose monodromy map has no essential periodic curves is hyperbolic [101] and thus has a finite self-homeomorphism group.

6.5.7 Recognition of Quasi-Stallings Manifolds

Our last business before we are done with subdividing I-bundle chambers is to construct a recognition algorithm for quasi-Stallings manifolds. As we have mentioned in the introduction to Sect. 6.5.6, Hemion's theorem is insufficient for this. This was discovered a long time after the classification theorem for Haken manifolds (see Theorem 6.1.1 and its corollary) was announced. For so many years the correctness of the theorem remained unsettled, not only because no proof was written, but also because some crucial ideas had not been discovered yet. The text you are reading now contains the first complete resolution of the problem. We follow the main lines of Sect. 6.5.5. The crucial difficulty consists in solving the recognition problem for quasi-Stallings manifolds, i.e., in obtaining an analog of Proposition 6.5.42. Let $M_{(\alpha,\beta)}$ be a quasi-Stallings manifold with fiber F. Similar to the Stallings case, it is convenient to represent it in the form $M_{(\alpha,\beta)} = (F \times I)/ \sim$, where the equivalence relation $\sim$ is generated by the equalities $(x,0) = (\alpha(x),0), (x,1) = (\beta(x),1)$. We assume that F is contained $M_{(\alpha,\beta)}$ as $F \times \{1/2\}$. Let $M_{(\alpha',\beta')}$ be another quasi-Stallings manifold with fiber F. We assume that it is also presented in the form $(F \times I)/ \sim$. In particular, $F \subset M_{(\alpha',\beta')}$. Then a homeomorphism $M_{(\alpha,\beta)} \to M_{(\alpha',\beta')}$ is called *fiber-preserving*, if it takes each fiber $F \times \{t\}$ of $M_{(\alpha,\beta)}$ onto the corresponding fiber $F \times \{t\}$ of $M_{(\alpha',\beta')}$.

Denote by $\mathrm{Homeo}(M_{(\alpha,\beta)}, M_{(\alpha',\beta')})$ the set of all fiber-preserving homeomorphisms $M_{(\alpha,\beta)} \to M_{(\alpha',\beta')}$ considered modulo fiber-preserving isotopy. Let $\mathrm{Conj}(\alpha, \beta, \alpha', \beta')$ be the set of all homeomorphisms $F \to F$ that conjugate α to α' and β to β'. Just as in the Stallings case, restricting homeomorphisms $M_{(\alpha,\beta)} \to M_{(\alpha',\beta')}$ onto F, we get a map $\psi \colon \mathrm{Homeo}(M_{(\alpha,\beta)}, M_{(\alpha',\beta')}) \to \mathrm{Conj}(\alpha, \beta, \alpha', \beta')$. The following lemma is evident.

Lemma 6.5.46. *For any two quasi-Stallings manifolds $M_{(\alpha,\beta)}, M_{(\alpha',\beta')}$ the map $\psi \colon \mathrm{Homeo}(M_{(\alpha,\beta)}, M_{(\alpha',\beta')}) \to \mathrm{Conj}((\alpha, \beta); (\alpha', \beta'))$ defined above is bijective.*

This lemma tells us that in order to decide if $M_{(\alpha,\beta)}, M_{(\alpha,'\beta')}$ are fiberwise homeomorphic, it suffices either to find h which conjugates α to α' and β to β', or to prove that such h does not exist. However, Hemion's solution of the conjugacy problem cannot be directly used here, since α, β are periodic and we wish to find a common conjugating homeomorphism. On the other hand, if h does exist, then it conjugates $\alpha\beta$ to $\alpha'\beta'$ and this property can be used for reducing the set of potential candidates for h to a finite number of one-parameter series. However, we should be sure that $\alpha\beta$ to $\alpha'\beta'$ have no essential periodic curves.

Lemma 6.5.47. *If a quasi-Stallings manifold $M = M_{(\alpha,\beta)}$ with fiber F, where $\chi(F) < 0$, is simple, then the homeomorphism $\alpha\beta \colon F \to F$ has no essential periodic curves.*

Proof. Suppose, on the contrary, that $\alpha\beta$ has at least one essential periodic curve. Then the Stallings manifold $M_{\alpha\beta}$ contains an essential singular surface (a torus or an annulus) by Lemma 6.5.28. Projecting this surface to M (which is a quotient of $M_{\alpha\beta}$), we get an essential singular surface in M. Since M is simple, then it is a Seifert manifold by Propositions 6.5.29 and 6.5.30. Clearly, its base surface is S^2 and it has three exceptional fibers (see the proof of Corollary 6.5.31). By Proposition 6.5.34, any such manifold can contain only one incompressible connected surface different from S^2, and this surface is nonseparating. On the other hand, M is a quasi-Stallings manifold and thus its fiber F is a separating surface of the above type. This is a contradiction.

$\square$

Lemma 6.5.48. *Let* $\alpha, \beta, \alpha', \beta' : F \to F$ *be orientation reversing free involutions of an orientable surface* F *such that* $\chi(F) < 0$ *and the homeomorphisms* $\alpha\beta, \alpha'\beta'$ *admit no essential periodic curves. Then one can construct a finite set* $X = X(\alpha, \beta, \alpha', \beta')$ *of homeomorphisms* $F \to F$ *such that a homeomorphism* $h' : F \to F$ *is in* $Conj(\alpha, \beta, \alpha', \beta')$ *if and only if* h' *has the form* $h' = h(\alpha\beta)^n$, *where* $h \in X$ *and* n *is an integer solution of the equation* $(\alpha\beta)^{2n} = h^{-1}\alpha'h\alpha$.

Proof. Using Corollary 6.5.41, we construct a finite set X of homeomorphisms $F \to F$ so that the following holds:

1. Each $h \in X$ conjugates $\alpha\beta$ to $\alpha'\beta'$
2. Any homeomorphism $h' : F \to F$ conjugating $\alpha\beta$ to $\alpha'\beta'$ is isotopic to a homeomorphism of the form $h(\alpha\beta)^n$, where $h \in X$ and $-\infty < n < \infty$.

Let us prove that for this X the conclusion of the lemma is true. We begin with the observation that a homeomorphism $h' : F \to F$ of the form $h' = h(\alpha\beta)^n$ conjugates α to α' if and only if $(\alpha\beta)^{2n} = h^{-1}\alpha'h\alpha$. Indeed, since α and β are involutions, we have

$$h'\alpha(h')^{-1} = h(\alpha\beta)^n \alpha (\alpha\beta)^{-n} h^{-1} = h(\alpha\beta)^n \alpha (\beta\alpha)^n h^{-1} = h(\alpha\beta)^{2n}\alpha h^{-1}.$$

It follows that $h'\alpha(h')^{-1} = \alpha'$ if and only if $h(\alpha\beta)^{2n}\alpha h^{-1} = \alpha'$, i.e., $(\alpha\beta)^{2n} = h^{-1}\alpha'h\alpha$.

Now suppose h' conjugates α to α' and β to β'. Then it conjugates $\alpha\beta$ to $\alpha'\beta'$ and hence by item 2 has the form $h' = h(\alpha\beta)^n$, where $h \in X$. The above observation tells us that n is satisfies the equation $(\alpha\beta)^{2n} = h^{-1}\alpha'h\alpha$.

Vice versa, if h' has the form $h' = h(\alpha\beta)^n$ such that $h \in X$ and $(\alpha\beta)^{2n} = h^{-1}\alpha'h\alpha$, then it conjugates $\alpha\beta$ to $\alpha'\beta'$ and α to α'. It follows that h' conjugates β to β' and hence $h' \in Conj(\alpha, \beta, \alpha', \beta')$. $\square$

At first glance, Lemma 6.5.48 provides a solution of the recognition problem for simple quasi-Stallings manifolds. Given $M_{(\alpha,\beta)}$ and $M_{(\alpha',\beta')}$, we construct $X = X(\alpha, \beta, \alpha', \beta')$ and for every $h \in X$ look for an integer number n such that $(\alpha\beta)^{2n} = h^{-1}\alpha'h\alpha$. It follows from Lemma 6.5.46 that the manifolds are fiberwise homeomorphic if and only if we find one. However, we cannot test all n, therefore this procedure is potentially infinite and thus not algorithmic. This crucial fact was overlooked by Waldhausen [131], Johannson [58,59] and Hemion [42].

Considering this situation from the geometric point of view, we come to the same conclusion. Indeed, if $M_{(\alpha,\beta)}, M_{(\alpha,\beta)}$ are fiberwise homeomorphic, then so are their 2-sheeted coverings $M_{\alpha\beta}$ and $M_{\alpha'\beta'}$. In general, the converse is not true, since not every fiber-preserving homeomorphism $h : M_{\alpha\beta} \to M_{\alpha\beta'}$ is fiberwise isotopic to an equivariant one. Not only h, but also all homeomorphisms of the form $hR^n_{\alpha\beta}$ should be tested for this property.

Let us return to the equation $(\alpha\beta)^{2n} = h^{-1}\alpha'h\alpha$. Denoting $\alpha\beta$ by f and $h^{-1}\alpha'h\alpha$ by g, we come naturally to the following problem.

Problem. *Can we decide algorithmically whether or not g is isotopic to an integer power of f?*

If the answer is affirmative, even only for homeomorphisms admitting no periodic curves, we get a recognition algorithm for quasi-Stallings manifolds. It turns out that the answer is indeed affirmative.

Proposition 6.5.49. *Let two homeomorphisms $f, g: F \to F$ of a surface F onto itself admit no essential periodic curves. Then one can algorithmically decide whether or not g is isotopic to an integer power f^n of f. Moreover, if it is, then n is uniquely defined and can be calculated algorithmically.*

The proof is based on the Thurston theory of surface homeomorphisms [124] and references therein. We derive from it a few facts needed for the proof. The main fact is that if a homeomorphism $f: F \to F$ has no periodic curves and $\chi(F) < 0$, then f is isotopic to a pseudo-Anosov homeomorphism. Therefore, one can assign to f a number $\lambda(f)$ called a *stretching factor*. It possesses the following properties (see [21, 124]):

1. $\lambda(f)$ is an algebraic number greater than 1
2. $\lambda(f^n) = \lambda(f)^{|n|}$ for any integer $n \neq 0$
3. Isotopic homeomorphisms have the same stretching factor

Moreover, it is proved in [9] that the stretching factor can be calculated algorithmically. One can write a computer program that assigns to any f a matrix with non-negative integer elements such that $\lambda(f)$ is the maximal eigenvalue of that matrix.

Proof of Proposition 6.5.49. Let f, g be given, say, as compositions of Dehn twists. We calculate $\lambda(f)$ and $\lambda(g)$. Since $\lambda(f) > 1$ by property 1, one can find an integer number N such that $\lambda^N(f) > \lambda(g)$. It follows from properties 2, 3 that if an integer power f^n of f is isotopic to g, then $\mid n \mid < N$. So to answer the question whether or not g is isotopic to an integer power of f it suffices to test all integer numbers n between $-N$ and N for possessing the required property. Note that if such n does exist, then it is unique. Indeed, if $g = f^n$ and $g = f^m$, then $n = m$, since f is not periodic. $\square$

Proposition 6.5.50. *Suppose that quasi-Stallings manifolds $M_{(\alpha,\beta)}, M_{(\alpha',\beta')}$ with fiber F are simple. Then one can algorithmically decide whether or not they are fiberwise homeomorphic. Moreover, if they are, then the set of fiber-preserving homeomorphisms $M_{(\alpha,\beta)} \to M_{(\alpha',\beta')}$ is finite (up to fiber-preserving isotopy) and all the homeomorphisms can be constructed algorithmically.*

Proof. Since $M_{(\alpha,\beta)}, M_{(\alpha',\beta')}$ are simple, then $\alpha\beta$ and $\alpha'\beta'$ admit no essential periodic curves by Lemma 6.5.47. Therefore we can apply Lemma 6.5.48 and construct a finite set $X = X(\alpha, \beta, \alpha', \beta')$ of homeomorphisms $F \to F$

such that $h' \in \mathrm{Conj}(\alpha, \beta, \alpha', \beta')$ if and only if h' has the form $h' = h(\alpha\beta)^n$, where $h \in X$ and n is an integer solution of the equation $(\alpha\beta)^{2n} = h^{-1}\alpha' h\alpha$. Then for each h we use Proposition 6.5.49 to find n (if it does exist). The corresponding homeomorphisms $h(\alpha\beta)n$ are in $\mathrm{Conj}(\alpha, \beta, \alpha', \beta')$. By Lemma 6.5.46, they determine a finite set of fiber-preserving homeomorphisms $M_{(\alpha,\beta)} \to M_{(\alpha',\beta')}$ which by construction contains all fiber-preserving homeomorphisms $M_{(\alpha,\beta)} \to M_{(\alpha',\beta')}$. The manifolds $M_{(\alpha,\beta)}, M_{(\alpha',\beta')}$ are fiberwise homeomorphic if and only if this set is nonempty. □

We are ready to deliver the final blow to the problem of I-bundle chambers. Let an admissible subpolyhedron P of an orientable 3-manifold (M, Γ) decompose M into 3-balls, solid tori, and I-bundle chambers without common lateral annuli. Suppose that $U \subset M$ is a Stallings or quasi-Stallings chain of I-bundle chambers with fiber F. We will say that two vertical skeletons of U are equivalent, if there exist a homeomorphism $(M, \Gamma) \to (M, \Gamma)$ which is invariant on P. The following proposition tells us that E_9 satisfies properties C_1, C_2 of extension moves (property C_3 is evident, since there are only finitely many Stallings and quasi-Stallings chains, and each move E_9 decomposes one of these chains into 3-balls).

Proposition 6.5.51. *For any simple Stallings or quasi-Stallings chain $U \subset M$ the set of equivalence classes of minimal vertical skeletons of U is finite. Representatives of the equivalence classes can be constructed algorithmically.*

Proof. Let $F_0, \ldots, F_{m-1}$ be the fibers of U that decompose it into the I-bundle chambers. All fibers are homeomorphic to a fixed surface F. We divide the proof into several steps.

STEP 1. Construct a vertical skeleton of U and denote by k the number of its true vertices.

STEP 2. Let $\mathcal{S}(F, k)$ and $\mathcal{SQ}(F, k)$ be the set of all vertical skeletons with $\leq k$ true vertices of all Stallings and quasi-Stallings manifolds with fiber F. We consider the skeletons up to homeomorphisms. By Lemma 6.5.26, these sets are finite and can be constructed algorithmically.

STEP 3. We choose in $\mathcal{S}(F, k)$ or in $\mathcal{SQ}(F, k)$ (depending on the type of U) all skeletons $S_1, \ldots, S_n$ which satisfy the following conditions:

1. Each S_i contains exactly m disjoint exemplars $F'_0, \ldots, F'_{m-1}$ of the fiber F such that any true vertex of S_i is contained in one of them.
2. The Stalling or quasi-Stallings manifold U_i determined by S_i is fiberwise homeomorphic to U.
3. S_i is minimal, i.e., its complexity (the number of its true vertices) does not exceed the complexity of any other skeleton satisfying 1 and 2.

It follows from Propositions 6.5.42 and 6.5.50 that the selection can be performed algorithmically.

STEP 4. For each U_i we choose a fibration $p_i : U_i \to I$ so that $F'_0, \ldots, F'_{m-1}$ are fibers. Consider the set $\mathcal{H}_i$ of all homeomorphisms $U_i \to U$ which take fibers to fibers and $\cup_{i=1}^{m-1}F_i'$ to $\cup_{i=1}^{m-1}F_i$. The homeomorphisms are considered up to the equivalence relation generated by postcomposing with fiber-preserving homeomorphisms $U \to U$ whose restrictions onto ∂U are fiberwise isotopic to the identity. It follows from Proposition 6.5.44 in the non-Seifert Stallings case and Proposition 6.5.50 in the quasi-Stallings one that $\mathcal{H}_i$ consists of finitely many equivalence classes of homeomorphisms. If U is a Seifert manifold fibered over S^2 with three exceptional fibers, then $\mathcal{H}_i$ is finite automatically. Let $h_1^{(i)}, h_2^{(i)}, \ldots, h_N^{(i)}$ be representatives of all the elements of $\mathcal{H}_i$ (of course, they can be constructed algorithmically). We use all these homeomorphisms to insert S_i into U and get vertical skeletons $h_j^{(i)}(S_i), 1 \le j \le N$, of U.

STEP 5. Let us prove that vertical skeletons $h_j^{(i)}(S_i), 1 \le i \le k, 1 \le j \le N$, represent all the equivalence classes of minimal vertical skeletons of U. Indeed, let S be a minimal vertical skeleton of U. Then it is homeomorphic to a skeleton $S_i, 1 \le i \le n$ and hence is obtained by an insertion of S_i into U. Since the set $h_1^{(i)}, h_2^{(i)}, \ldots, h_N^{(i)}$ contains representatives of all elements of $\mathcal{H}_i$, the homeomorphism $U_i \to U$ used for the insertion differs from one of $h_j^{(i)}, 1 \le j \le N$ by a fiber-preserving isotopy. It follows that S equivalent to one of the skeletons $h_j^{(i)}(S_i)$. $\square$

6.5.8 Subdivision of Solid Tori

In this final section we deal with the last remaining case, the chambers that are solid tori with clean longitudes. Let P be an admissible subpolyhedron of a Haken 3-manifold (M, Γ) such that P does not admit moves E_1–E_9. Then any chamber (Q, Δ) of P is either a 3-ball or a solid torus having a clean longitude on the boundary. Δ decomposes ∂Q either into discs (if Q is a ball) or into discs and annuli (if Q is a solid torus). Those discs and annuli will be called *disc* and *annular patches* of P. The only difference between patches and 2-components of P is that Γ decomposes some 2-components of P contained in ∂M into smaller patches. If (Q, Δ) is a solid torus chamber, then any annular patch of ∂Q contains a longitude of Q. Otherwise we could apply either E_4 or E_3, depending on whether or not (Q, Δ) contains essential annuli. By the same reason, Δ consists of disjoint circles and not more than one pattern strip.

To extend P to a simple skeleton of (M, Γ), we need to subdivide the solid torus chambers. Here is an idea how to do that. Let U be a connected component of the union of all the solid torus chambers of P. We supply U with the boundary pattern $\Delta_U = \partial U \cap (\Gamma \cup SP)$ consisting of those points of Γ and singular points of P that are contained in ∂U. Since any annulus and any solid torus can be fibered into circles, we have good chances to represent U as a circle bundle over a surface. Any such bundle admits a section. The idea consists in inserting into U a minimal section and perturbing it.

It turns out, however, that sometimes U admits no circle fibration. For example, it can happen that the intersection of two neighboring annular patches consists not of circles, but of disjoint arcs. This situation occurs when a pattern strip in the boundary ∂Q of a chamber (Q, Δ) degenerates to a necklace placed between two annular patches of ∂Q (by a *necklace* in ∂Q we mean a pattern strip that can be covered by discs joined by arcs into a closed chain which goes once along a longitude). Also, two solid torus chambers can intersect each other along several disjoint discs.

If at least one of these situations occurs, then the circle bundle structures on annuli and tori cannot be extended to a circle bundle structure on U. The explanation is simple: in both cases one can find fibers of neighboring solid torus chamber which have common points but do not coincide. So we begin the realization of the above idea with introducing a new class of admissible subpolyhedra, for which a circle bundle structure on U always exists.

Definition 6.5.52. *We say that an admissible subpolyhedron P of (M, Γ) is faithful, if it decomposes (M, Γ) into 3-balls and solid torus chambers with clean longitudes such that the following holds:*

1. *Each solid torus chamber (Q, Δ) is simple. This means that Δ consists of several disjoint longitudes and at most one pattern strip.*
2. *The intersection of any two annular patches of P either is empty or consists of one or two disjoint circles of their boundaries.*
3. *No two different solid torus chambers have a common disc patch.*

Lemma 6.5.53. *Let P be an admissible subpolyhedron of an orientable Haken 3-manifold (M, Γ) such that all its chambers are 3-balls and solid tori having clean longitudes. Suppose that P is faithful. Then any connected component (U, Δ_U) of the union of all solid torus chambers can be presented as a circle bundle over a surface so that all the annular patches and solid torus chambers of (U, Δ_U) are saturated (i.e., consist of fibers).*

Proof. We construct a fibration of (U, Δ_U) as follows. First, we decompose into circles all the annular patches of P. The remaining part of the boundary ∂Q of every solid torus chamber (Q, Δ) consists of annuli decomposed into disc patches. Here, we have used the first property of a faithful polyhedron. Indeed, the union of disc patches of ∂Q cannot have a connected component homeomorphic to a disc, since otherwise the intersection of the two annuli adjacent to this disc would consist of arcs.

It follows that the circle fibration of the annular patches can be extended to circle fibrations of the boundaries of all solid torus chambers. Since no disc patch lies in the boundary of two different solid torus chambers, this extension is well-defined. It remains to extend it to the interiors of all the solid torus chambers. One can do that without introducing exceptional fibers, since all fibers in the boundary of every solid torus chamber are longitudes.   □

The fibration of U constructed above possesses the property that all the annular patches and solid torus chambers of $P_U = P \cap U$ are saturated. We will call such fibrations *faithful* too. Evidently, U admits many faithful fibrations. Any two of them are isotopic by a *faithful isotopy* of U which is invariant on the solid torus chambers and annular patches. We point out that in general a faithful isotopy is not admissible: it can move disc patches of ∂U together with their boundaries, which are in Δ_U).

In view of Lemma 6.5.53, our next goal is quite natural: we wish to extend any P to a faithful subpolyhedron. For doing that we introduce two auxiliary extension moves. They are very similar to E_3 and E_4, so we denote them by E_3' and E_4'.

E_3': Inserting a Clean Annulus Which Embraces a Necklace, Followed by Adding a Double p-Minimal Meridional Disc

Suppose that the boundary pattern Δ of a solid torus chamber (Q, Δ) of P contains a necklace N. Then we insert into Q a clean annulus that cuts off a solid torus V such that $V \cap \Delta = N$. After that we choose a p-minimal meridional disc D of V and insert into V two parallel copies of D, see Fig. 6.41.

E_4': Doubling a Disc Patch that Separates two Solid Torus Chambers

Suppose that a 2-cell D of P separates two solid torus chambers. Then we add to P a parallel copy of D, see Fig. 6.42.

It is easy to see that each move E_3' decreases the number of necklaces and does not increase the number of disc patches that separate solid torus chambers. Also, each E_4' decreases the number of such disc patches and does not create new necklaces. It follows that after several moves E_3', E_4' we get a

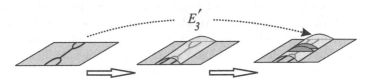

Fig. 6.41. Cutting off a necklace

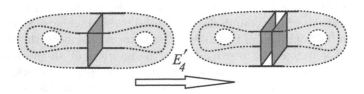

Fig. 6.42. Doubling a disc patch

faithful subpolyhedron. Evidently, moves E_3', E_4' possess properties C_1–C_3 of extension moves.

Further on we assume that P is a faithful subpolyhedron of an orientable Haken manifold (M, Γ). Thus every connected component U of the union of all solid torus chambers of P admits a faithful fibration. For a while the behavior of P and M outside U is irrelevant for us. So for now we are considering U as a 3-manifold with the boundary pattern $\Delta_U = \partial \cap (\Gamma \cup SP)$ and the faithful subpolyhedron $P_U = P \cap U$.

Definition 6.5.54. *A surface $F \subset U$ is called* a section *of (U, Δ_U), if there is a faithful fibration of U such that F intersects every fiber at exactly one point.*

Our next extension move consists in subdividing U into balls by inserting a section. Such a section always exists unless U is closed and hence $M = U$ is a circle bundle over a closed surface. Any such bundle is completely determined by its base surface S and the Euler number, which takes values in Z or Z_2, depending on whether or not S is orientable. Therefore, we could easily solve the recognition problem for such circle bundles separately, without using hierarchies and extension moves.

Nevertheless, we prefer stay within the general scheme at the expense of introducing another auxiliary move E_4'' (compare with E_4 from Sect. 6.5.1).

E_4'': Inserting a Pair of Parallel Meridional Discs of a Solid Torus Chamber

Suppose that a faithful subpolyhedron P of a closed orientable Haken manifold (M, Γ) decomposes it into chambers such that all of them are solid tori. Choose a simple chamber (Q, Δ) of P and a meridional disc $D \subset Q$ having the smallest p-complexity. Then we add to P two parallel copies of D.

The meaning of E_4'' is that the union of all solid torus chambers for the resulting faithful polyhedron $P' \subset M$ is not closed anymore. Therefore, any connected component of the union admits a section. By definition, E_4'' possesses properties C_1, C_2 of extension moves. Since we apply it not more than once, it possesses also property C_3.

We are ready now to describe the last special extension move. Let P be a faithful subpolyhedron of an orientable Haken 3-manifold (M, Γ) and $U \neq M$ a connected component of the union of all solid torus chambers of P. As above, we endow U with the boundary pattern $\Delta_U = \partial U \cap (\Gamma \cup SP)$. Let C be a 1-dimensional submanifold of ∂U. We assume that C is in general position with respect to Δ_U, that is, it does not contain vertices of Δ_U and intersects edges transversally. By the *length* $\ell(C)$ we mean the number of points in $\Delta_U \cap C$.

Definition 6.5.55. *A section F of U is called* minimal, *if for any other section F' of U we have $\ell(\partial F) \leq \ell(\partial F')$.*

Fig. 6.43. Perturbing sections

One should point out that we do not require F, F' above be sections of the same faithful fibration. So any minimal section of U is minimal globally, i.e., $\ell(F)$ takes the minimal value among all sections of all faithful fibrations.

Extension Move E_{10}: Inserting a Perturbed Minimal Double Section

Let P be a faithful subpolyhedron of a Haken 3-manifold (M, Γ) and U a connected component of the union of all solid torus chambers such that $U \neq M$. Let $\Delta_U = \partial U \cap (\Gamma \cup SP)$. Suppose that U admits a section. Then we insert into (U, Δ_U) two parallel copies of a minimal section and perturb them, see Fig. 6.43.

Our next goal is to prove that E_{10} satisfies properties C_1–C_2 of extension moves (property C_3 is evident). We begin with recalling a few well-known facts about sections of circle bundles. Let W be a circle bundle over a surface S with nonempty boundary (we do not assume that W is equipped with a boundary pattern or contains an admissible subpolyhedron). Then the following is true:

(i) *W admits a section F.* For example, if W is the direct product bundle $S \times S^1$, then one can take $F = S \times \{*\}$, where $*$ is a point of S^1. If W is a twisted S^1-bundle, then the existence of a section can be proved by using elementary obstruction theory [45]. Indeed, the obstruction to constructing a section vanishes, since $\partial S \neq \emptyset$ and thus the second homology group of S is trivial. It may be illuminating to consider an example of the twisted product $M^2 \tilde{\times} S^1$ of a Möbius strip and the circle, which is simply a thick Klein bottle. The annulus A_λ (see Example 6.4.14 and Fig. 6.16) decomposes $M^2 \tilde{\times} S^1$ into two twisted I-bundles over M^2. Each of them has a section, which simultaneously is a section of $M^2 \tilde{\times} S^1$.

(ii) *Fiber-preserving twists of W along saturated annuli and tori take sections to sections. Vice versa, any two sections F_1, F_2 of W are related by fiber-preserving isotopies and twists along saturated annuli and tori. Moreover, if $\partial F_1 = \partial F_2$, then one can take F_1 to F_2 by fiber-preserving isotopies and twists along saturated tori such that ∂W remains fixed.*

(iii) *A section C of ∂W (i.e., a 1-dimensional submanifold of ∂W intersecting each fiber of ∂W exactly once) is extendible to a section of W if and only* *if the intersection number $\lambda(C, \partial F)$ of C with the boundary of some (and hence of any) section F of W is zero. Indeed, if $\lambda(C, \partial F) = 0$, then we can take ∂F to C (and hence F to a section extending C) by fiber-preserving isotopies and twists along proper saturated annuli. Conversely, suppose that there is a section F' of W such that $C = \partial F'$. We can assume that F and F' are transversal. Then $\lambda(\partial F', \partial F) = 0$, since the endpoints of any proper arc in $F' \cap F$ have opposite signs.*

Remark 6.5.56. One should point out that the intersection number $\lambda(C, \partial F)$ used above is defined correctly, although we do not suppose that the circles of C and ∂F are oriented. We simply orient them such that the orientations be induced by the same orientation of ∂F_0, where F_0 is the base surface of the fibration. The value of $\lambda(C, \partial F)$ thus obtained does not depend on the chosen orientation of ∂F_0.

Let us return now to the union U of all solid torus chambers. We consider it as a 3-manifold with the boundary pattern Δ_U and the faithful subpolyhedron P_U.

Definition 6.5.57. *Two 1-dimensional submanifolds of ∂U are equivalent, if they are related by admissible isotopies of (U, Δ_U) and twist along clean circles.*

Obviously, equivalent 1-dimensional submanifolds have the same length.

Lemma 6.5.58. *For any number k the boundary of (U, Δ_U) contains only finitely many equivalence classes of 1-dimensional submanifolds of length $\leq k$. Representatives of the equivalence classes can be constructed algorithmically.*

Proof. To construct a 1-dimensional submanifold $C \subset (\partial U, \Delta_U)$ of length $l \leq k$, one should choose l points in Δ_U and join them into a collection of disjoint circles by arcs that intersect Δ_U only at their endpoints. The first choice is finite up to isotopy of P_U. Since all patches of $(\partial U, P_U)$ are discs and annuli, the second choice is also finite up to isotopy of $(\partial U, P_U)$ and twists along core circles of the annular patches. $\square$

Lemma 6.5.59. *Let C be a 1-dimensional submanifold of ∂U having one circle in each component of ∂U. Then it can be extended to a section F_C of a faithful fibration of U if and only if the following conditions hold:*

1. *If a boundary circle S of an annular patch is contained in ∂U, then C crosses S at exactly one point.*
2. *$\lambda(C, \partial F) = 0$, where F is a section of any faithful fibration of U.*

Proof. The "only if" part is easy. Indeed, suppose that C is the boundary of a section F_C. Denote by $S_1, S_2, \ldots, S_n$ those boundary circles of annular patches of P_U that are contained in Δ_U. Since they are fibers, we have item 1. Property (iii) above implies that $\lambda(C, \partial F) = 0$ for any section F of the same faithful fibration. If F and F_C are sections of different faithful fibrations of U, the equality $\lambda(C, \partial F) = 0$ remains true, since it is preserved under isotopy.

Let us prove the "if" part. Denote by $A_j, 1 \leq j \leq n$, the annuli in ∂U into which the circles S_i decompose ∂U. Since C crosses each S_i at exactly one point, it intersect each annulus A_j along an arc which is radial, i.e., joins different boundary circles of A_j. Therefore, one can construct a fibration of A_j into circles such that the arc $C \cap A_j$ crosses each fiber exactly once. Doing so for all A_j, we construct a fibration of ∂U into circles and then extend it to a faithful fibration of U. By construction, C is a section of ∂U. Since $\lambda(C, \partial F) = 0$, this section can be extended to a section of U, see property (iii) above. $\square$

Let (U, Δ_U) be a 3-manifold with a faithful subpolyhedron P_U and let F_1, F_2 be two sections of U. We say that they are *equivalent*, if there exists a homeomorphism $h: (U, \Delta_U) \to (U, \Delta_U)$ such that $h(F_1) = F_2$, $h(P_U) = P_U$, and the restriction of h onto Δ_U is isotopic to the identity.

Proposition 6.5.60. *Let P be an admissible subpolyhedron of an orientable Haken 3-manifold (M, Γ) such that all its chambers are 3-balls and solid tori having clean longitudes. Suppose that P is faithful. Then for any connected component (U, Δ_U) of the union of all solid torus chambers the number of equivalence classes of minimal sections of U is finite. Representatives of the equivalence classes can be constructed algorithmically.*

Proof. First we construct a section F of U. The length $\ell(\partial F)$ is denoted by k. Next we apply Lemma 6.5.58 to construct representatives of all equivalence classes of 1-submanifolds of ∂U having length $\leq k$.

CASE 1. Suppose ∂U contains no annular patches. Using Lemma 6.5.59, we select among the representatives constructed above all submanifolds which can be extended to sections of U. Then we construct those sections and choose among them sections of the minimal length.

CASE 2. Suppose that ∂U contains at least one annular patch A. Then we select among the representatives submanifolds $C_1, \ldots, C_m$ which satisfy condition 1 of Lemma 6.5.59. Twists along a core circle of A helps us to modify $C_1, \ldots, C_m$ so that afterwards condition 2 of Lemma 6.5.59 holds. This is possible, since each such twist change $\lambda(C_i, \partial F)$ by ± 1. Then we do the same as in Case 1: we extend the selected submanifolds to sections of U and choose minimal ones.

In both cases we get a finite list $F_1, \ldots, F_n$ of sections of U. This list turns out to consist of representatives of all the minimal sections. To prove that, consider an arbitrary minimal section F' of U. It follows from the construction of $F_1, \ldots, F_n$ that $\partial F'$ can be transformed into the boundary ∂F_i of one of them by an admissible isotopy and twists along clean circles in ∂U. Obviously, twists can appear only in the second case.

Let us fix a clean circle S_0 in ∂U and replace each twist of $\partial F'$ along any clean circle $S \subset (\partial U, \Delta_U)$ by the twist of F' along a saturated annulus

$A \subset U$ with $\partial A = S \cup S_0$. We obtain a section F'' such that $\partial F''$ is admissibly isotopic to ∂F_i inside all the patches of ∂U, except possibly the annular patch A_0 containing S_0. Property $\lambda(\partial F_i, \partial F'') = 0$ (which is true for any pair of sections) tells us that $\partial F''$ is isotopic to ∂F_i also inside A_0. Therefore, we can assume that $\partial F'' = \partial F_i$. It remains to conclude that by property (ii) for sections of fiber bundles (see above), F'' can be transformed into F_i by faithful isotopy and twists along saturated tori. It follows that F' is equivalent to F_i. □

Corollary 6.5.61. *Extension move E_{10} satisfies properties C_1–C_3.*

Proof. In a more detailed form this corollary can be formulated as follows. Let P be an admissible subpolyhedron of a Haken 3-manifold (M, Γ) such that any chamber is either a 3-ball or a solid torus and every connected component U of the union of all the solid torus chambers admits a faithful fibration. Then:

1. There exist only finitely many admissible subpolyhedra that can be obtained from P by performing exactly one move E_{10} (as usual, we consider the polyhedra up to homeomorphisms $(M, \Gamma) \to (M, \Gamma)$).
2. Representatives of the equivalence classes of such polyhedra can be constructed algorithmically.

First, we note that for every U all the annular patches of ∂U are contained in ∂M. Indeed, if A is an annular patch of ∂U, then A is contained neither in a ball chamber (since A is an annulus) nor in a solid torus chamber (otherwise U would be only a part of a connected component of the union of all solid torus chambers). Therefore, $U \cap \mathrm{Cl}(M \setminus \mathrm{Int}\, U)$ consists of disc patches. It follows that any homeomorphism $h: (U, \Delta_U) \to (U, \Delta_U)$ such that $h(P_U) = P_U$ and the restriction of h onto Δ_U is isotopic to the identity can be extended to an admissible homeomorphism $(M, \Gamma) \to (M, \Gamma)$. Therefore, equivalent sections of U differ by admissible homeomorphisms $(M, \Gamma) \to (M, \Gamma)$. By Proposition 6.5.60, there exist only finitely many inequivalent minimal sections. Each of them can be perturbed only in finitely many ways. So the number of admissible subpolyhedra of (M, Γ) that can be obtained from P by applying E_{10} is finite. Clearly, representatives can be constructed algorithmically. □

Theorem 6.5.62. *Extension moves $E_1 - E_{10}$ (together with auxiliary moves E_3', E_4', E_4'') satisfy properties C_1–C_4.*

Proof. Properties C_1–C_3 follow from Proposition 6.5.2 (C_1, C_2 for E_1–E_5), Corollary 6.5.7 (C_3 for E_1–E_5), Proposition 6.5.21 (for E_1–E_8), Proposition 6.5.51 (for E_9), and Corollary 6.5.61 (for E_{10}). At each step we get chambers of more and more restricted types: balls, solid tori and I-bundles after moves E_1–E_5, balls, solid tori and I-bundles whose union consists of simple Stallings and quasi-Stallings manifolds after E_7, E_8, only balls and solid tori after E_9. Auxiliary moves E_3', E_4' transform the union of solid torus chambers into faithful circle bundles. Finally E_{10} decomposes the bundles into balls. This means that we get a simple skeleton of (M, Γ). □

6.5.9 Proof of the Recognition Theorem

One can offer a short informal proof, which consists of two references: to
Theorem 6.5.62 and to the proof of Theorem 6.3.3 modulo the existence of
extension moves, see the end of Sect. 6.3.2. Indeed, the proof of Theorem 6.3.3
tells us that the Recognition Theorem is true provided that there exist extension moves with properties C_1–C_4. On the other hand, moves E_1–E_{10} together
with three auxiliary moves E_3', E_4', E_4'' possess the required properties by Theorem 6.5.62.

Nevertheless, for the sake of completeness, we present a formal proof of
Theorem 6.1.6 (which can be considered as the Recognition Theorem for
Haken manifolds with boundary pattern). In the case of the empty boundary
pattern Theorem 6.1.6 is equivalent to Theorem 6.1.1.

Proof of Theorem 6.1.6. Let (M, Γ) be a Haken 3-manifold with boundary
pattern. Starting with ∂M and arguing inductively, we construct a sequence
$\{\partial M\} = \mathcal{P}_0(M, \Gamma), \mathcal{P}_1(M, \Gamma), \ldots$ of finite sets of admissible subpolyhedra. We
think of each polyhedron $P \in \mathcal{P}_i(M, \Gamma)$ as being embedded into its own copy
of (M, Γ). Since P contains ∂M, it contains Γ. To carry out the inductive
step, we describe a multi-valued version $\tilde{E}_k$ of move $E_k, 1 \leq k \leq 10$. Unlike
the move E_k, which to a given polyhedron P assigns only one polyhedron, $\tilde{E}_k$
creates a finite set of polyhedra, each obtained from P by E_k. Clearly, multi-
valued versions $\tilde{E}_3'$, $\tilde{E}_4'$, $\tilde{E}_4''$, of auxiliary moves E_3', E_4', E_4'' are also needed.

Let k be fixed and let $P \in \mathcal{P}_i(M, \Gamma)$. By properties C_1, C_2 of extension
moves, we can construct a set $\tilde{E}_k(P)$ of admissible subpolyhedra of (M, Γ)
such that

1. $\tilde{E}_k(P)$ is finite
2. For any admissible subpolyhedron P' of (M, Γ) obtained by applying E_k
   to P, there exists a homeomorphism $(M, \Gamma) \to (M, \Gamma)$ taking P' to a
   subpolyhedron from $\tilde{E}_k(P)$

If P has no E_k-extensions, i.e., if E_k is not applicable, then we set $\tilde{E}_k(P) =
\{P\}$. Then we define $\mathcal{P}_{i+1}(M, \Gamma)$ to be the disjoint union $\cup_P \tilde{E}_k(P)$ over all
$P \in \mathcal{P}_i(M, \Gamma)$ of the sets $\tilde{E}_k(P)$, write $\mathcal{P}_{i+1}(M, \Gamma) = \tilde{E}_k(\mathcal{P}_i(M, \Gamma))$, and say
that $\mathcal{P}_{i+1}(M, \Gamma)$ is obtained from $\mathcal{P}_i(M, \Gamma)$ by applying $\tilde{E}_k$. See Fig. 6.44.

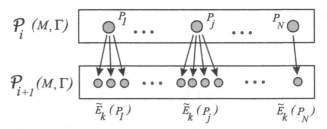

Fig. 6.44. The multi-valued move $\tilde{E}_k$

Let us now apply to $\mathcal{P}_0(M, \Gamma)$ multi-valued moves $\tilde{E}_k, 1 \leq k \leq 10$, as long as possible. The order of moves is not very important; the only requirement is that it should be the same for all manifolds (M, Γ). To be definite, we accept the following rule: we apply $\tilde{E}_k$ to $\mathcal{P}_i(M, \Gamma)$ only if $\tilde{E}_m(\mathcal{P}_i(M, \Gamma)) = \mathcal{P}_i(M, \Gamma)$ for all $m < k$, i.e., if no polyhedron $P \in \mathcal{P}_i(M, \Gamma)$ admits an extension move E_m with $m < k$. In other words, we must first always try to apply moves with smaller numbers.

Property C_3 of extension moves guarantees that the process stops: after a finite number of steps we get a finite set $\mathcal{P}_N(M, \Gamma)$ such that $\tilde{E}_k(\mathcal{P}_N(M, \Gamma)) = \mathcal{P}_N(M, \Gamma)$ for all k. By property C_4, $\mathcal{P}_N(M, \Gamma)$ consists of simple skeletons of M. Obviously, $\mathcal{P}_N(M, \Gamma)$ can contain many homeomorphic copies (duplicates) of the same skeleton. Saying "homeomorphic copies," we take also into account the pattern Γ, which is contained in $\partial M \subset P$. In other words, we consider the skeletons as pairs (P, Γ). Removing aforementioned duplicates, we get a new set $\mathcal{P}(M, \Gamma)$.

Let us show that $\mathcal{P}(M, \Gamma)$ is characteristic, i.e., that $\mathcal{P}(M_1, \Gamma_1), \mathcal{P}(M_2, \Gamma_2)$ consist of the same number of pairwise homeomorphic polyhedra if and only if (M_1, Γ_1) and (M_2, Γ_2) are homeomorphic. By construction, $\mathcal{P}(M, \Gamma)$ depends only on the homeomorphism type of (M, Γ). Therefore, $(M_1, \Gamma_1) = (M_2, \Gamma_2)$ implies $\mathcal{P}(M_1, \Gamma_1) = \mathcal{P}(M_2, \Gamma_2)$. The inverse implication follows from Lemma 6.3.2.

To prove Theorem 6.1.6, it remains to note that the recognition problem for 2-dimensional polyhedra with pattern is algorithmically solvable in an evident manner. □

This mathematically rigorous proof guarantees that the recognition algorithm for Haken manifolds does exist. However, to facilitate the understanding how the algorithm works, we describe below once more its main procedures. We are far from anticipating a practical realization of the algorithm by a computer program, since many its steps are at least exponential. Nevertheless, to emphasize that theoretically such realization is possible, we employ the terminology from computer programming and call the procedures subprograms.

Subprogram (A): Constructing Fundamental Surfaces

Let (Q, Δ) be a 3-manifold with boundary pattern. Then we do the following:

1. Triangulate Q so that Δ consists of edges.
2. Write down the matching system of normal equations.
3. Construct the set of all fundamental solutions.
4. Select admissible fundamental solutions and realize them by surfaces.

Subprogram (B): Constructing a JSJ-Decomposition

Let (Q, Δ) be an irreducible boundary irreducible 3-manifold with boundary pattern. We construct a JSJ-decomposition of (Q, Δ) and determine the types of its JSJ-chambers as follows.

1. Using (A), triangulate Q and construct the set of all fundamental surfaces.
2. Test all fundamental tori and clean annuli as well as doubles of all fundamental Klein bottles and clean Möbius bands for being essential. Having found such a surface F, we insert it into (Q, Γ). See Sect. 6.4.2.
3. Then we cut Q along F and apply to the manifold thus obtained Steps 1, 2 again. We do that as long as possible until the procedure stops. Denote by $\mathcal{F}$ the resulting set of disjoint tori and annuli.
4. Using Theorems 6.4.23 and 6.4.28, we select from $\mathcal{F}$ rough tori and annuli. By Theorem 6.4.31, they form a JSJ-system for (M, Γ).
5. Using Theorem 6.4.42, we determine the types of the JSJ-chambers and construct Seifert and I-bundle structures on Seifert and, respectively, I-bundle JSJ-chambers.

Our next task is to realize by subprograms general multi-valued moves $\tilde{E}_1$–$\tilde{E}_{10}$. Let P be an admissible subpolyhedron of a Haken 3-manifold (M, Γ) with boundary pattern and $(Q_i, \Delta_i), 1 \leq i \leq m$, the chambers of P. Using (B), we construct a JSJ-decomposition of each (Q_i, Δ_i) and determine the types of the JSJ-chambers.

Subprogram $(\tilde{E}_1)$: Inserting Essential Tori

Let (S_{ij}, δ_{ij}) and (F_{ij}, γ_{ij}) be all Seifert JSJ-chambers of (Q_i, Δ_i) and their base surfaces, respectively. Denote by C_{ijk} representatives of all strong equivalence classes of simple closed curves in (F_{ij}, γ_{ij}). Let T_{ijk} be the corresponding tori in Q_i, i.e., the inverse images of C_{ijk} under the fibration projections. We select among them tori $T_1, \ldots, T_N$ which are essential in the JSJ-chambers they are contained in. Then we take N copies of P and add each $T_s, 1 \leq s \leq N$, or two exemplars of T_s to the corresponding copy of P inside the corresponding copy of (M, Γ). See Sect. 6.4.6 and the description of E_1 in Sect. 6.5.1 for details. The set of polyhedra thus obtained is $\tilde{E}_1(P)$.

Remark 6.5.63. According to our rule, the construction of the characteristic set $\mathcal{P}(M, \Gamma)$ begins with applying $\tilde{E}_1$ as long as possible. The chambers of all resulting polyhedra contain no essential tori. A pleasant (although not needed for the proof) observation is that this property is preserved under all other extension moves. This means that we will never apply move $\tilde{E}_1$ again.

Subprogram $(\tilde{E}_2)$: Inserting Longitudinal Annuli not Parallel to the Boundary

We do the same as above, but this time $\{C_{ijk}\}$ is the set of proper arcs in F_{ij}, whose inverse images A_{ijk} are annuli. We select among them annuli $A_1, \ldots, A_N$ which are not parallel to the boundary of the corresponding JSJ-chambers. Then we take N copies of P and add each $A_s, 1 \leq s \leq N$, or two exemplars of A_s to the corresponding copy of P inside the corresponding copy of (M, Γ).

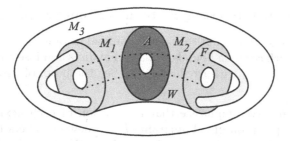

Fig. 6.45. F divides the simple manifold $W \cup M_3$ into the non-simple manifold W and its simple complement M_3

Remark 6.5.64. In contrast to the previous move, the property of P to not admit E_2-moves is not preserved under other moves. For example, move E_5 can create new nontrivial longitudinal annuli. To show that, consider three orientable simple 3-manifolds M_1, M_2, M_3 such that $\partial M_1, \partial M_2$ are genus two surfaces, while a connected component F of ∂M_3 is a surface of genus three. We suppose that manifolds are placed in some R^N so that $M_1 \cap M_2 = \partial M_1 \cap \partial M_2$ is a nonseparating annulus and $W \cap M_3 = \partial W \cup \partial M_3 = F$, where $W = M_1 \cup_A M_2$. Then the union $W \cup M_3$ is a simple manifold. Nevertheless, F cuts off from it the manifold W, which admit an essential annulus A. See Fig. 6.45.

Subprogram $(\tilde{E}_3)$: Inserting Clean Essential Annuli Which are Parallel to the Boundary

Each Q_i contains only finitely many clean incompressible annuli A_{ij} such that each A_{ij} is parallel rel ∂ to an annulus $A'_{ij} \subset \partial Q$ containing at least one pattern strip, see the definition of E_3 in Sect. 6.5.1 and Fig. 6.25. Let N be the total number of such annuli in all chambers (Q_i, Δ_i). Then we take N exemplars of (M, Γ) and add each A_{ij} to the copy of P inside the corresponding exemplar of (M, Γ).

Subprogram $(\tilde{E}_4)$: Inserting p-Minimal Proper Discs

This time we consider simple chambers which, considered as manifolds without boundary pattern, are boundary reducible. Let (Q_i, Δ_i) be one of them and let $D_i \subset Q_i$ be a nontrivial boundary compressing disc. By Theorem 4.1.13, such disc can be constructed algorithmically. Denote by k_i its p-complexity. By Theorem 6.3.17, there exists an algorithmically constructible finite set of surfaces in (Q_i, Δ_i) containing all two-sided proper incompressible boundary incompressible connected surfaces of complexity $\leq k_i$. Since all nontrivial p-minimal proper discs in Q_i are contained in this set, they can be constructed algorithmically. We take all such discs in all boundary reducible chambers and add them to P, each inside the corresponding copy of (M, Γ).

Subprogram ($\tilde{E}_5$): Inserting p-Minimal Proper Surfaces

We do the same as above, but consider p-minimal incompressible boundary incompressible surfaces with boundary instead of discs and insert them into simple chambers (Q_i, Δ_i) with irreducible Q_i. See the proof of Proposition 6.5.2 for details.

Remark 6.5.65. We emphasize that application of each next move $\tilde{E}_i$ requires often application of subprogram (B) to new chambers that appeared on the previous step as well as to their neighbors. We determine if those chambers are simple, and if not, construct their JSJ-decompositions. We need that for testing whether they admit some move E_m with $m < k$. So the subprogram (B) is repeatedly used many times. The subprogram (A) is used even more, since it is needed not only for (B), but also for testing if Q_i is boundary reducible and if a given surface $F \subset Q_i$ is incompressible and boundary incompressible.

Let us now turn our attention to special extension moves. Let P be an admissible subpolyhedron of a Haken 3-manifold (M, Γ) with boundary pattern and $(Q_i, \Delta_i), 1 \leq i \leq m$, the chambers of P. According to our rule, we assume that P admits no moves $\tilde{E}_i, 1 \leq i \leq 5$. We wish to subdivide all chains of I-bundle chambers of P. Since I-bundle structures of these chambers can be constructed algorithmically, we can assume that each component U of the union of all I-bundle chambers of P is already presented as a direct, twisted, Stallings, or quasi-Stallings chain.

Subprogram ($\tilde{E}_6$): Subdividing Direct and Twisted Chains

This move is easy, since nontrivial strips in U correspond to nontrivial arcs in the base surface of U. The number of such strips is finite up to strong equivalence. We insert the strips into U, double them, and perturb in all possible ways, placing each perturbed double strip into its own copy of (M, Γ). See the description of E_6 in Sect. 6.5.3.

Subprograms ($\tilde{E}_7$), ($\tilde{E}_8$): Inserting Perturbed Pairs of Vertical Annuli and Tori

These moves are also easy. Just as in the case of moves $\tilde{E}_1, \tilde{E}_2$, we construct all essential tori and annuli in U. Then we bring them into vertical position (see Lemma 6.5.16), take two parallel copies of each such surface, perturb them in all possible ways, and insert into different copies of M.

Subprogram ($\tilde{E}_9$): Inserting Minimal Vertical Skeletons into Simple Stallings and Quasi-Stallings Chains

To perform $\tilde{E}_9$, we need several other subprograms. First we should be able to find out if the monodromy map f of a given simple Stallings manifold U

is periodic. Since the upper bound for the conjectural period is known (see Proposition 6.5.34), it suffices to test only finitely many powers of f for being isotopic to the identity. For that we need a separate subprogram. Such subprogram can be easily written, since one can recognize isotopic curves by eliminating all biangles bounded by their union and testing if the resulting curves are disjoint and parallel. If f happens to be periodic, then M coincides with U and is a Seifert manifold fibered over S^2 with three exceptional fibers. Since a Seifert structure on M can be constructed algorithmically (Corollary 6.5.35), recognition of such manifolds is easy.

Suppose that f is not periodic. Then it has no essential periodic curves and we can apply Hemion's theorem. Creating a subprogram for constructing all homeomorphisms conjugating f to another given homeomorphism g seems to be the most difficult part of the whole recognition algorithm. It is not known yet whether the upper bound $N(f, g)$ of the size of conjugating homeomorphisms (see Theorem 6.5.40) is exponential, super-exponential, or even more. Nevertheless, Theorem 6.5.40 says that $N(f, g)$ can be calculated algorithmically, and this is enough for recognition of simple Stallings manifolds.

We must have a subprogram for constructing all minimal vertical skeletons of U. Such subprogram is described in the proof of Proposition 6.5.26. Adding to P those minimal vertical skeletons of all Stallings chains, we get the desired set of values of $\tilde{E}_9$ for the Stallings case.

The case of simple quasi-Stallings manifolds is similar. Here we need a subprogram for calculating the stretching factor. Such subprogram can be easily written on the base of the original algorithm [9].

Subprogram $(\tilde{E}_{10})$: Inserting Perturbed Minimal Double Section

This subprogram can be written on the base of Sect. 6.5.8. We construct a section of the union U of all solid torus chambers and calculate its length k. Then we construct all 1-dimensional submanifolds of ∂U of length $\le k$ (Lemma 6.5.58), select among them boundaries of sections (Lemma 6.5.59), and construct representatives of all equivalence classes of minimal sections (Proposition 6.5.60). It remains to double these sections, perturb them, and insert into different copies of (M, Γ).

3-Manifold Recognizer

In preceding chapters we described a number of important algorithms, which make heavy use of the Haken method of normal surfaces. As a rule, algorithms based on that method have exponential complexity and hence are impractical. In particular, although the recognition problem for Haken manifolds has an algorithmic solution, there is no chance of it being be realized by a computer program, at least in the foreseeable future. On the other hand, quite often experienced topologists recognize 3-manifolds rather quickly. They use other algorithms which, whether based on rigorous mathematics or intuition, are much more efficient. However, this gain does not come for free. The price is that one has to allow an algorithm to be only partial (i.e., not an algorithm at all in the formal meaning of this term). The problem of finding an efficient partial algorithm for answering a particular class of geometric questions is in itself a well-stated mathematical problem. Trying to solve it, we inevitably discover new structural properties of geometric objects.

In this chapter we describe the theoretical foundations and applications of the *3-Manifold Recognizer*, which is a computer program for working with 3-manifolds written by Tarkaev according to an algorithm elaborated by Matveev and other members of the Chelyabinsk Topology Group. The program passed several extensive tests. It recognized successfully all closed irreducible orientable 3-manifold up to complexity 12 (altogether more than 30,000 manifolds). It recognized also several monster-sized examples prepared especially for that purpose. See Sects. 7.5 and 7.6 for the detailed survey of the obtained results. We begin with a description of how one can present 3-manifolds in computer memory and how one can manipulate them.

7.1 Computer Presentation of 3-Manifolds

What we need now is an economic way of presenting 3-manifolds in a form that computer would understand. The idea consists in replacing 3-manifolds

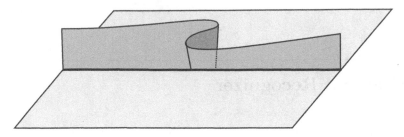

Fig. 7.1. Returns inside edges are forbidden

by cell complexes (their special spines) and encoding the spines by strings of integers.

7.1.1 Cell Complexes

In topology of manifolds the notion of a cell complex is usually considered in a more general sense than in computer topology (see [65, 67]). So we briefly recall it, restricting ourselves to the 2-dimensional case. We prefer an inductive definition:

1. A 0-dimensional cell complex $X^{(0)}$ is a finite collection of points called *vertices*.
2. A 1-dimensional cell complex $X^{(1)}$ is obtained from a 0-dimensional complex $X^{(0)}$ by attaching several 1-dimensional cells (i.e., arcs). The endpoints of the arcs are attached to the vertices of $X^{(0)}$, and the arcs must have no other common points. In other words, a 1-dimensional cell complex is simply a graph (loops and multiple edges are allowed).
3. A 2-dimensional cell complex $X^{(2)}$ is obtained from a 1-dimensional complex $X^{(1)}$ by attaching several 2-dimensional cells. In other words, we take a collection $\{D_1, \ldots, D_n\}$ of disjoint 2-dimensional discs and attach each disc D_i to $X^{(1)}$ via an *attaching map* $\varphi_i \colon \partial D_i \to X^{(1)}$. It is convenient to assume that the inverse image $\varphi^{-1}(e)$ of each open 1-cell e of $X^{(1)}$ consists of open connected subarcs of ∂D_i such that each of them is mapped onto e homeomorphically. In other words, we require that the boundary curve $l_i = \varphi_i(\partial D_i)$ of every 2-cell passes along the edges monotonically, without returns inside them. For example, the situation shown in Fig. 7.1 is forbidden. $X^{(1)}$ is called the *1-dimensional skeleton* of $X^{(2)}$.

In general, the incidence relation between cells does not determine $X^{(2)}$. It is also necessary to know in which direction the boundary curve l of each 2-cell passes along each edge as well as the cyclic order in which l_i passes along different edges. For example, consider 2-cell complexes obtained from a rectangle by identification of its vertices to one point and by different identifications of its sides (see Fig. 7.2). Each of them has one vertex, two loop edges, and one 2-cell attached to each edge exactly twice. Nevertheless, the

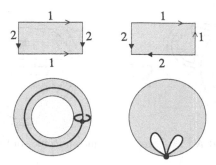

Fig. 7.2. Two complexes with the same incidence relation

complexes are different, since one of them is a torus while the other is a 2-sphere with three identified points.

On the other hand, information on the graph $X^{(1)}$, the orientations of edges, and the order in which the boundary curves of the 2-cells of $X^{(2)}$ pass along the oriented edges is quite sufficient for reconstructing $X^{(2)}$. The corresponding data can be expressed numerically as follows. First we number the vertices and the edges, and write down all the edges as a sequence of pairs $(i_1, j_1), \ldots, (i_m, j_m)$. Here i_k, j_k are the numbers assigned to the vertices joined by k-th edge. Simultaneously, we orient each edge (i_k, j_k) by an arrow directed from the vertex i_k to the vertex j_k. Second, for each 2-cell we write down a string $(p_1\ p_2 \ldots p_n)$ of nonzero integers ($\pm$ edges' numbers) which show how the boundary curve of the 2-cell passes along edges. The signs of p_i show the directions. For example, the complexes in Fig. 7.2 can be described by the following data:

1. Edges: $(1, 1), (1, 1)$ (in both cases)
2. 2-Cells: $(1\ 2\ -1\ -2)$ in the first case and $(1\ -1\ 2\ -2)$ in the second

Obviously, the total number of vertices, the sequence of pairs, and the collection of strings determine $X^{(2)}$. It is less obvious that if the link of each vertex is connected (i.e., if each vertex of $X^{(2)}$ has a cone neighborhood with a connected base), then $X^{(2)}$ is completely determined by the strings only. Indeed, to recover $X^{(2)}$, it suffices to do the following:

1. Realize the strings by boundaries of disjoint polygons with oriented and numbered edges.
2. Identify the edges that have the same numbers via orientation preserving homeomorphisms.

Then the 2-dimensional cell complex thus obtained is homeomorphic to $X^{(2)}$.

We do not describe any specific realization of the above data in the computer memory. One can use arrays of integers; when an object-oriented programming language is used, the concepts of the cell complex, the spine, and others lend themselves to be used as classes. In that case, this chapter serves toward developing these classes; indeed, in the upcoming sections we describe

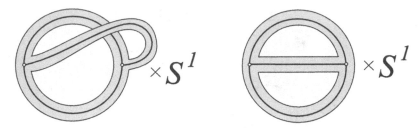

Fig. 7.3. $T_0^2 \times S^1$ and $N^2 \times S^1$ are not homeomorphic, but have the same simple spine $\Theta \times S^1$

the main algorithms we will apply to spines, and this description will help us to determine which methods should be incorporated into that class, as well as which operations would be applied to its objects.

7.1.2 3-Manifolds as Thickened Spines

Recall that by Theorem 1.1.17, any 3-manifold is determined by its special spine. Therefore, numerical representation of cell complexes is sufficient for presenting 3-manifolds: We simply use representation of their special spines. However, while manipulating manifolds, it is convenient to have more freedom by allowing spines which are not special (almost simple would be more than enough). The problem is that an almost simple and even simple spine does not determine the manifold.

For example, direct and twisted I-bundles over a closed surface F are not homeomorphic, but have the same simple spine (a copy of F). In this example one of the manifolds is orientable, while the other is not. An example of closed orientable 3-manifolds having homeomorphic almost simple spines can be obtained by considering connected sums $M_1 \# M_2$ and $M_1 \# (-M_2)$, where M_1, M_2 are oriented. If neither M_1 nor M_2 admit an orientation reversing homeomorphism, then $M_1 \# M_2$ and $M_1 \# (-M_2)$ are distinct, although they have homeomorphic almost simple spines obtained from simple spines of M_1, M_2 by joining them by an arc. Finally, an example of two orientable 3-manifolds with boundary having the same simple spine is shown in Fig. 7.3. The direct products of a punctured torus T_0^2 and a twice punctured disc N^2 by S^1 are distinct, but have the same simple spine $\Theta \times S^1$, where Θ is a theta-curve.

It is clear now that an almost simple spine P of a 3-manifold M determines M only if we supply P with additional information. Let us describe an example of what information that should be, for the case when P is simple. We assume that P is decomposed into cells, i.e., is presented as a cell complex. Since $P \subset M$, each 2-cell of P has two *sides*. It is easy to show that P determines M, if we know how these sides are glued together in a neighborhood of each edge of P. This information can be written into a matrix which encodes the gluing of the sides.

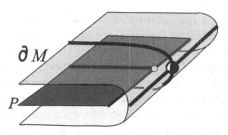

Fig. 7.4. The behavior of a DS-diagram near ∂P

However, this approach does not seem very natural. We are forced to decode the information written into the matrix each time we need it, and then to encode it again. On the other hand, the sides of 2-cells have a very clear geometric interpretation, which can be used for specifying the thickening. Let us identify M with a close regular neighborhood of P. There is a natural retraction $M \to P$ such that its restriction $p \colon \partial M \to P$ possesses the following properties:

1. The inverse image $p^{-1}(C)$ of every open k-dimensional cell C of P consists of two open 2-cells if $k = 2$, three or two open arcs if $k = 1$, and two, three, or four points if $k = 0$.
2. The restriction of p onto each of these 2-cells and arcs is a homeomorphism onto the corresponding 2-cell or edge of P.

The 2-cells, arcs and points in ∂M described above form a cell decomposition of ∂M. So the sides of 2-cells of P are just the 2-cells of ∂M. Under the name "DS-diagrams" such decompositions had been introduced and investigated by Ikeda and Inoue [50]. An example of a DS-diagram for the hyperbolic 3-manifold Q_1 can be found on Fig. 2.22. There, we show a special spine P_1 decomposed into two vertices, four edges, and two 2-cells. The induced cell decomposition of ∂Q_1 (which is a torus) consists of eight vertices, 12 edges, and four 2-cells.

Let us consider now almost simple spines containing no points with disconnected links, in particular, no 1-dimensional part. They too determine DS-diagrams on ∂M. The only difference is that each *free* (i.e., having valence 1) edge of P has only one counterpart in ∂M, see Fig. 7.4.

It is evident that P and ∂M, represented by cell complexes as described above (together with the correspondence relation between cells of ∂M and P induced by $p \colon \partial M \to P$), determine M in a unique way.

The case when P contains a 1-dimensional part is slightly more complicated. The same relates to all almost simple spines having points with disconnected links, like the cone over several disjoint segments. The problem is that we loose the correspondence between cells of ∂M and P. For example, if e is a principal edge, i.e., an edge of the 1-dimensional part of P, then $p^{-1}(e)$ is a tube (an annulus) in ∂M.

There are several ways to work with such spines. First, one can allow annular or more complicated components of the corresponding DS-diagrams. However, this is not convenient, since one of the advantages of DS-diagrams is that they are composed form 2-cells. Another way consists in forgetting principal edges. This is equivalent to cutting M along proper discs (meridional discs of the solid tubes that correspond to the edges). This is also not very convenient, since one would have to store information on which components of the boundary of the resulting 3-manifold should be joined by tubes. We prefer the third way: Spines having points with disconnected links are not considered at all.

7.2 Simplifying Manifolds and Spines

Let M be a 3-manifold and P its almost simple spine without points having disconnected links. We will simplify M and P by different transformations (moves), which can be divided into four groups. Moves of the first group reduce the cell structures of P and ∂M, but preserve their topological types. Moves of the second and fourth group also preserve M, but change P. The third group consists of surgeries, which change M and P, but only in a controlled manner.

For example, one of the surgery moves cuts off $N^2 \times S^1$ from M, where N^2 is a disc with two holes. Clearly, $N^2 \times S^1$ and the remaining piece M_1 of M do not determine M. In order to reconstruct M, it suffices to know the following:

1. Which tori of ∂M_1 and $\partial(N^2 \times S^1)$ should be glued together.
2. Coordinate systems on them.
3. Matrices that determine the gluing homeomorphisms.

This observation motivates our further strategy. On the one hand, we reduce ourselves to considering orientable 3-manifolds whose boundaries consist only of spheres and tori. Some tori must be equipped with coordinate systems. On the other hand, we organize such manifolds into special structures just as chemical atoms are organized into molecules. These structures are called *labeled molecules*. The exact definition will be given later; for now we describe a convenient way for representing coordinate systems on the boundary tori.

7.2.1 Coordinate Systems on Tori

Let T be a torus presented as a cell complex. We suppose that T and all its edges are oriented. By a coordinate system on T we mean an ordered pair (μ, λ) of oriented simple closed curves on T with their intersection number ± 1. We are going to use coordinate systems for describing homeomorphisms of tori by integer matrices. So only the homology classes of μ and λ are essential. Those can be represented by 1-dimensional chains, i.e., by linear combinations

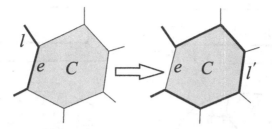

Fig. 7.5. Instead of passing along e, we go around the remaining part of ∂C

$\mu = \sum_{i=1}^{n} m_i e_i$, $\lambda = \sum_{i=1}^{n} l_i e_i$, where $e_i, 1 \leq i \leq n$, are the edges of T and m_i, l_i are integers.

Further on under a presentation of an oriented 3-manifold M by its cellular spine we mean a set of the following data:

1. An oriented cell complex P whose body is an almost simple spine of M having no points with disconnected links.
2. An oriented nonempty surface (the boundary of M) which consists of spheres and tori and is also presented as an oriented cell complex.
3. A cellular map $p: \partial M \to P$ such that p preserves orientations of all cells and the restriction of p onto each open cell is a homeomorphism.
4. Coordinate systems on some boundary tori presented by ordered pairs of 1-dimensional chains.

We assume that the orientation of ∂M agrees with the orientation of M. Before considering simplification moves, let us describe an auxiliary move E_1. We need it for shifting the coordinate curves aside and clear space for other moves.

Definition 7.2.1. *Let a 3-manifold M with boundary be presented by its cellular spine P. Then an edge $e \subset \partial M$ is called* clean, *if either:*

1. *e is contained in a spherical component of ∂M.*
2. *e lies on a torus $T \subset \partial M$ such that either T is not endowed with a coordinate system or the coefficients at e of the coordinate chains μ, λ of T are 0.*

An edge a of P is clean, *if so are all its preimages on ∂M.*

E_1: *Cleaning edges.* Let $\mu = \sum_{i=1}^{n} m_i e_i$, $\lambda = \sum_{i=1}^{n} l_i e_i$ be coordinate chains on a torus $T \subset \partial M$. Suppose that $\partial C = \sum_{i=1}^{n} \varepsilon_i e_i$ is the boundary chain of a 2-cell $C \subset T$ such that it passes through an edge e_k only once, i.e., $\varepsilon_k = \pm 1$. Then we clean e_k by replacing μ and λ by the new chains $\mu' = \mu - m_k \varepsilon_k \partial C$ and $\lambda' = \lambda - l_k \varepsilon_k \partial C$, whose coefficients at e_k are 0. Obviously, the new chains determine the same homology classes. A visual interpretation of this move is shown in Fig. 7.5, where l stands either for μ or for λ.

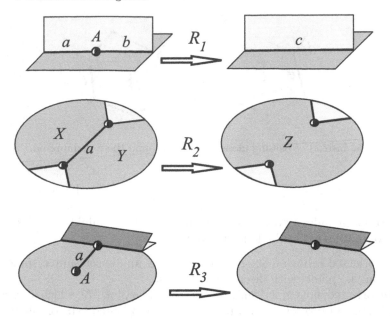

Fig. 7.6. Reducing cell decomposition

7.2.2 Reduction of Cell Structures

Now we begin describing the promised moves. Let M be a 3-manifold presented by its cellular spine P. It can happen that the decomposition of P into cells is unnecessarily complicated. Below we describe several transformations that simplify it. The topological types of M and P remain the same.

R_1: *Removing vertices of valence 2.* Suppose that P contains a vertex A where exactly two different edges a, b meet together. Suppose also that all boundary curves that pass through A do that without returns. This means that a boundary curve is allowed to proceed from a to b or from b to a. Situations when it enters A and goes at once back are forbidden. Then we remove this vertex and amalgamate a, b into a new edge c.

R_2: *Fusing 2-cells.* Let an edge a of P be a common edge of exactly two different 2-cells X, Y of P. Suppose that a is clean (if not, we clean it by applying E_1). Then we remove a and amalgamate X, Y to a new 2-cell Z.

R_3: *Removing needles.* Let a be a *needle* in P, i.e., an edge of valence 2 having a free vertex A. We assume that the needle is not degenerate, i.e., its other vertex is not free. Then we remove a and A.

All three moves are shown in Fig. 7.6.

Remark 7.2.2. There is only one case when P contains a degenerate needle: A connected component of P is a 2-sphere decomposed into two vertices, one edge, and one 2-cell. In this case we do not collapse the needle, since after doing that we would get a 2-sphere without edges. This would be confusing,

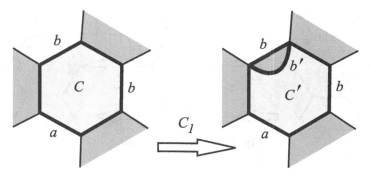

Fig. 7.7. Doubling an edge

since the attaching map of the unique 2-cell (which should be presented by a string of integers) would be presented by the empty set.

7.2.3 Collapses

Let M be a 3-manifold presented by its cellular spine P. Suppose that P contains a free edge a. This means that a lies in the boundary of exactly one 2-cell C of P. We wish to collapse P onto $P_1 = P \setminus (\text{Int } a \cup \text{Int } C)$. At first glance, this operation is trivial: we simply remove a and C. However, one should be careful, since P_1 thus obtained can contain points with disconnected links, and we have decided to consider spines without such points. Indeed, there can appear principal edges as well as vertices with disconnected links. To avoid that, we replace the collapse by several more elementary moves that either realize it or reduce to cutting move S_1 (see Sect. 4.2.8). The first move prevents appearance of principal edges.

C_1: *Doubling an edge.* Let a 2-cell C of P have a free edge a. Suppose that the boundary curve of C passes two or three times through another edge b such that b is not incident to any other 2-cells of P. Then we insert into C a new edge b' which joins the endpoints of b, see Fig. 7.7. The new edge decomposes C into a biangle and a new 2-cell C'.

This move improves the cell decomposition of P, since the collapse of C' through a does not convert b into a free edge while the collapse of C does so.

The second move collapses a 2-cell through its free edge without creating vertices with disconnected links. Let C be a 2-cell of P and l an arc in C joining interior points of two consecutive edges of ∂C. Then l cuts off a triangle from C. We will call this triangle a *vertex wing*. Clearly, creating a vertex wing requires introducing not only a new edge l, but also two new vertices (its endpoints).

C_2: *Winged collapsing.* Suppose that the boundary curve of a 2-cell C of P contains a free edge a and that every other edge of ∂C is incident to at least one 2-cell of P different from C. Let us clean a by move E_1 and insert

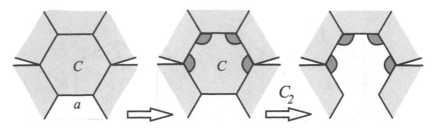

Fig. 7.8. Winged collapsing

vertex wings into all corners of C except the ones at the endpoints of a. Then we collapse the remaining part of C through a, see Fig. 7.8.

This move eliminates C completely, only the wings remain. Thanks to introducing the wings, the links of the vertices are not changed and thus remain connected. Quite often the wings are useless, i.e., the links are still connected without them. The next move allows us to eliminate such superfluous wings.

C_3: *Collapsing nonseparating wings.* Let P contain an embedded triangle 2-cell W having exactly one free edge a (such 2-cells appear under winged collapses). Let A be the vertex of W opposite to a. Suppose that the collapse of W through a does not produce vertices with disconnected links (it suffices to require that the link of A remains connected). Then we clean a and perform this collapse.

If a triangle wing cannot be removed without creating disconnected links, then we convert it into a quadrilateral convenient for performing move S_1 (see Sect. 7.2.4).

C_4: *Transforming a triangle wing into a quadrilateral with two opposite free sides.* Let P contain an embedded triangle 2-cell W having exactly one free edge a. Suppose that the collapse of W through a converts the link of the vertex A opposite to a into a disconnected graph $G = G' \cup G''$. Then we replace W by a quadrilateral W' having two opposite free edges. In other words, we stretch up A into a new free edge a' that joins the cones over G', G''. See Fig. 7.9, where we illustrate C_3 and C_4.

7.2.4 Surgeries

Let us describe surgery moves. In contrast to reductions and collapses, they can change the manifold. Each time when this happens, we specify which information should be stored in order to enable the reconstruction of the original manifold. Let M be a 3-manifold presented by its cellular spine P.

S_1: *Cutting 2-cells.* Suppose that a 2-cell C of P contains two free edges a, b such that they cannot be joined by a chain of consecutive free edges of C. Let us join a middle point of a with a middle point of b by a proper arc l in C. Then we cut P along l, see Fig. 7.10.

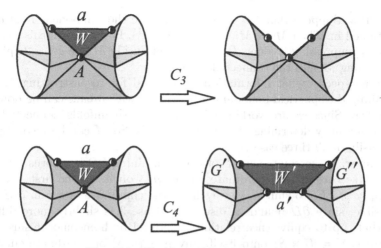

Fig. 7.9. Working with triangle wings

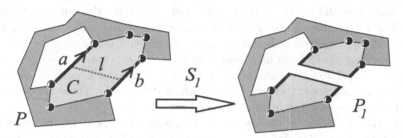

Fig. 7.10. Cutting a 2-cell

Obviously, the resulting polyhedron P_1 is a spine of a 3-manifold M_1 obtained by cutting M along a proper disc $D \subset M$ such that $D \cap P = l$. Let us investigate the relation between M and M_1 (a similar investigation was carried out in the proof of Corollary 2.4.4). There are three cases. Figure 2.17 illustrates them quite adequately, although it was designed for a slightly different purpose.

CASE 1. Suppose D separates M into two components M' and M''. Then $M = (M' \cup D^3) \# M''$, where the notation $M' \cup D^3$ means that we fill up a boundary sphere of M' with a 3-ball. To get a sphere S that separates M into the connected sum, it suffices to take $\partial M'$ and slightly push it inside M.

CASE 2. Suppose ∂D separates the component F of ∂M it is contained in, but D does not separate M. Then $M = (M_1 \cup D^3) \# (S^2 \times S^1)$. Indeed, let F', F'' be the two components of ∂M_1 obtained from F. Recall that F is either a sphere or a torus. Therefore, at least one of the surfaces F', F'' is a 2-sphere. If we push it inside M, we get a nonseparating sphere S in M. The existence of such sphere explains the appearance of the summand $S^2 \times S^1$.

CASE 3. Suppose that ∂D does not separate the component F of ∂M it is contained in. Then $M = (M_1 \cup D^3)\#(D \times S^1)$. Evidently, in this case F is a torus. Compressing it along D and pushing inside M, we get a 2-sphere S decomposing M into the connected sum.

Let us specify what information is required for reconstructing M from M_1. Filling the spherical boundary components with 3-balls can be done in a unique way. Since we are working with oriented 3-manifolds, connected sums are also uniquely determined by the summands. So M can be reconstructed quite easily in all three cases.

However, if the component T of ∂M containing ∂D is a torus, then one should take care of the coordinate curves μ, λ on it. In the first two cases we simply use E_1 to shift μ, λ away from the copies $\tilde{a}, \tilde{b}$ of a, b in ∂M. This is possible, since ∂D bounds a disc in T. Case 3 is slightly more difficult. Note that, up to equivalence relation generated by homeomorphisms of the solid torus $V = D \times S^1$ onto itself, any pair (μ, λ) of coordinate curves on ∂V is completely determined by their intersection numbers w_μ, w_λ with the meridian $m = \partial D \times \{*\}$ (pairs w_μ, w_λ and $-w_\mu, -w_\lambda$ determine equivalent coordinate systems). These numbers can be calculated quite easily. Indeed, denote by m_a, m_b the coefficients of μ at the edges $\tilde{a}, \tilde{b} \subset T$, and by l_a, l_b the corresponding coefficients of λ. Suppose that $\tilde{a}, \tilde{b}$ are oriented coherently as shown in Fig. 7.10. Then $w_\mu = m_a + m_b, w_\lambda = l_a + l_b$. Clearly, if $\tilde{a}, \tilde{b}$ have opposite orientations, then $w_\mu = m_a - m_b, w_\lambda = l_a - l_b$. These two numbers should be stored for the future reconstruction of M.

In order to describe the geometric meaning of w_μ, w_λ, let us present m in the form $m = \mu^p \lambda^q$. It follows from properties of the intersection number that $p = -w_\lambda \varepsilon, q = w_\mu \varepsilon$, where $\varepsilon = \pm 1$ is the intersection number of μ and λ. Therefore, $\pm(p, q)$ are the coordinates of m with respect to the coordinate system μ, λ.

S_2: *Delicate piercing.* Suppose that ∂M consists of at least two components, and at least one of them is a sphere. Then we can find a 2-cell C of P that separates a spherical component S of ∂M from another one. This means that one of the two copies of C in ∂M lies in S while the other one is contained in a different component of ∂M. Choose a disc $D \subset \text{Int } C$ and join it with a vertex of C by a simple arc. Then the piercing consists in removing Int D from P, see Fig. 7.11.

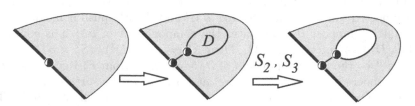

Fig. 7.11. Piercing

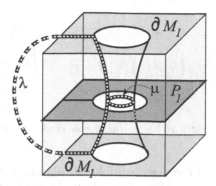

Fig. 7.12. Coordinate curves on the boundary torus

We have called this piercing delicate since it induces a very mild modification of M. Indeed, the new manifold M_1 is homeomorphic to a 3-manifold obtained from M by filling up a spherical boundary component with a 3-ball. Clearly, all points of the new spine P_1 have connected links (for this reason we do not remove the whole cell C but only make a small hole).

S_3: *Rough piercing.* We apply this move only in case when $\partial M = S^2$ and all opportunities for other simplification moves have already been exhausted. In particular, P must be special. Let C be a 2-component of P. To apply S_3, we do the same as in the case of S_2. We choose a disc $D \subset \mathrm{Int}\, C$, join it with a vertex of C by a simple arc, and remove $\mathrm{Int}\, D$ from P.

Rough piercing transforms M into a new 3-manifold M_1 whose boundary T is a torus. According to our rule, we should introduce coordinate curves on T. We do that as follows. The first curve $\mu \subset T$ corresponds to ∂D, the second curve λ is an arbitrary closed curve on T which crosses μ exactly once. See Fig. 7.12.

S_4: *Cutting along a proper annulus with nontrivial boundary components.* Suppose that a 2-component α of P contains a nontrivial orientation-preserving simple closed curve l. We assume that l intersects the edges of P transversally. Let A be a proper annulus in M such that $A \cap P = l$. We assume that the boundary circles of A are nontrivial in ∂M (see move U_1 in Sect. 7.2.5 for the case when at least one of them is trivial). Then we cut P along l and M along A, see Fig. 7.13 (only a half of A is shown).

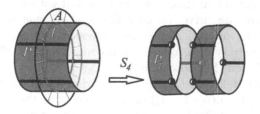

Fig. 7.13. Cutting along a circle

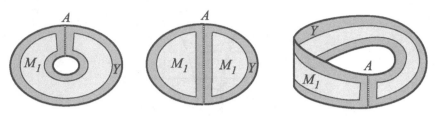

Fig. 7.14. $Y = \text{Cl}(M \setminus M_1)$ is an orientable circle bundle over N^2 or M_0^2

The new spine P_1 of the new manifold M_1 obtained by the cut contains n pairs of new vertices and n pairs of new free edges, where n is the number of intersection points of l with the edges.

Let us investigate what information should be stored to enable the subsequent reconstruction of M from M_1. Denote by L the connected component of $A \cup \partial M$ that contains A. It consists of A and one or two boundary tori. Let Y be a regular neighborhood of L in M such that P_1 is contained in the manifold $\text{Cl}(M \setminus Y)$, which can be identified with M_1. Then P_1 is a spine of M_1, $M = M_1 \cup Y$, and $M_1 \cap Y$ consists of one or two common boundary tori. Since the boundary circles of A are nontrivial in ∂M, L fibers onto circles. It follows that Y is a circle bundle over a surface F. It is easy to see that $\chi(F) = -1$ and ∂F consists of three or two circles. Therefore, F is homeomorphic either to a twice punctured disc N^2 or to a once punctured Möbius band M_0^2. The second option takes place when the coherently oriented boundary circles of A lie on the same boundary torus of M and have opposite orientations there. See Fig. 7.14, where we show 2-dimensional pictures. We may conclude that $Y = N^2 \times S^1$ or $Y = M_0^2 \tilde{\times} S^1$.

A coordinate system on every torus $T \subset Y \cap M_1$ can be chosen arbitrarily, but the most natural choice for λ is an arbitrarily oriented fiber of Y contained in T. Of course, the fiber should be replaced by a curve composed of edges, i.e., by a homologous 1-chain. The choice of μ does not really matter; one can take any curve on T crossing λ once. Nevertheless, in the case when $T \subset Y \cap M_1$ consists of two tori it is convenient to take coherent orientations for their longitudes and to orient the meridians so as to have equal intersection numbers with the longitudes.

However, we face the following problem. Since M_1 is presented by its spine P_1, the coordinate systems on the tori of ∂M_1 can be presented as usual, i.e., by 1-chains. In contrast to this, Y is not endowed with a spine. How can we describe the coordinate curves on ∂Y? We must do that for the new curves on the tori $Y \cap M_1$ as well as for the curves on the tori $Y \cap \partial M$, which determine coordinate systems on the toral components of ∂M. This problem has a simple solution. Suppose that $Y = N^2 \times S^1$. Let $T_i, 1 \leq i \leq 3$, be the boundary tori of Y and (μ_i, λ_i) the coordinate curves on them. Choose canonical coordinate curves (s_i, f_i) on T_i, where s_i is the oriented boundary circle of $N^2 \times \{*\}$ contained in T_i and f_i is an oriented fiber, $1 \leq i \leq 3$. Of course, the orientations of s_i and f_i must be induced by the same orientations of N^2 and S^1, respectively. Then μ_i, λ_i can be written in the form $\mu_i = s_i^{a_i} f_i^{b_i}, \lambda_i = s_i^{c_i} f_i^{d_i}$, i.e., they can be represented by (2×2)-matrices $C_i = \begin{pmatrix} a_i & c_i \\ b_i & d_i \end{pmatrix}$ with determinants ± 1.

If $Y = M_0^2 \tilde{\times} S^1$, then we do the same. Choose a section $F \subset Y$, i.e., a surface intersecting each fiber at exactly one point. F is of course homeomorphic to M_0^2. The canonical coordinate systems (s_i, f_i) on the boundary tori $T_i, i = 1, 2$, of Y are composed of the boundary circles of F and fibers. The matrices C_1, C_2 representing the systems have the same meaning: they express μ_i, λ_i through (s_i, f_i).

To be definite, we will always assume that for if $T_i \subset Y \cap M_1$, then C_i is the unit matrix. This corresponds to the "must natural choice" described above.

S_5: *Cutting along a theta-curve.* Suppose that M contains a proper punctured torus T_0 such that $\Theta = T_0 \cap P$ is a theta-curve which does not separate T_0, and $C = \partial T_0$ is contained in a spherical component S of ∂M. We supply T_0 with a coordinate system (μ, λ) composed of two distinct circles contained in Θ. Then we cut M along T_0 and P along Θ. The manifold M_1 obtained in this way contains two new toral boundary components $T_\pm$, each consisting of a copy of T_0 and a disc of S. We supply them with coordinate systems $(\mu_\pm, \lambda_\pm)$ composed of the copies of μ and λ.

Let us describe a method of finding a torus T_0 which satisfies the above conditions. For simplicity, we consider only the case when the spine P is special and its cell structure is reduced, i.e., coincides with the intrinsic decomposition of P into true vertices, triple edges, and 2-components. Suppose that there exist two edges a, b and three 2-cells C_1, C_2, C_3 of P such that their boundary curves, being properly oriented, pass through a in the same direction and pass through b also in the same direction. Let us choose two points $A \in a, B \in b$ and join them by three arcs $l_i \subset C_i, 1 \le i \le 3$. We get a theta-curve $\Theta \subset P$ so that its regular neighborhood $N(\Theta)$ in P is $\Theta \times I$ and thus consists of three strips joining a subarc of a with a subarc of b. Suppose that each of these strips approaches the union of the other two from different sides, see Fig. 7.15 to the left. Then the inverse image $C = p^{-1}(\Theta)$ under the projection map $p: \partial M \to P$ is a circle in ∂M bounding a proper punctured torus $T_0 \subset M$ such that $T_0 \cap P = \Theta$, see Fig. 7.15 to the right. If C is contained in a spherical component of ∂M, then move S_5 is applicable.

S_6: *Cutting along a theta-curve and adding a torus.* This move is a twin of S_5. Suppose M contains a proper punctured torus T_0 such that $\Theta = T_0 \cap P$ is a theta-curve, $C = \partial T_0$ is contained in a toral component of ∂M, and C bounds a disc $D \subset \partial M$. Such punctured tori, if they exist, can be found by the same method as above. We supply T_0 with a coordinate system (μ, λ) composed of two distinct circles contained in Θ. This time we cannot cut M along T_0, since then we would obtain a manifold with a genus 2 surface in the boundary. The procedure is slightly more complicated. First, we push the

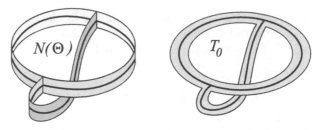

Fig. 7.15. A regular neighborhood of Θ in P and the transversal punctured torus

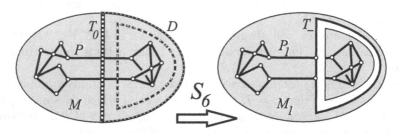

Fig. 7.16. Cutting M along a torus intersecting P at a theta-curve

torus $T_0 \cup D$ inward and get a torus $T \subset \text{Int } M$ parallel to $T_0 \cup D$. Next we cut M along T and supply both boundary tori created in this way by the coordinate systems inherited from the system (μ, λ) on T_0. A spine P_1 of the new manifold M_1 obtained in this way can be constructed as follows. We replace P by a polyhedron $P \cup T_-$, where T_- is a parallel copy of T placed between T and $T_0 \cup D$. Finally we cut P along the theta-curve $T \cap P$, see Fig. 7.16.

Remark 7.2.3. Move S_5 considered above produces a spine P_1 which has free edges and thus can be collapsed onto a smaller spine. In fact, the combined move (cutting along a theta-curve and collapsing) is very powerful, since it entirely eliminates the cells C_1, C_2, C_3 of P that intersect T_0.

The situation with the move S_6 is more delicate. On the one hand, adding T_- to P creates two new vertices, whose links are even more complicated than the link of the standard true vertex. On the other hand, cutting and collapsing destroys three half-cells and all true vertices contained in their boundaries. Therefore, in order to have a real simplification, we should take care that the number k of such vertices be at least 3. One can easily show that, since C_1, C_2, C_3 are distinct, k is always greater than one. Moreover, $k = 2$ if and only if T cuts off from M a piece which contains exactly two true vertices of P. This piece is always homeomorphic either to a thick torus $T^2 \times S^1$ or to a solid torus $D^2 \times S^1$. In all other cases we have $k \geq 3$.

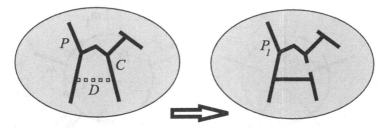

Fig. 7.17. A disc replacement move

7.2.5 Disc Replacement Moves

We are now going to describe the last portion of transformations. Let M be an oriented 3-manifold presented by its simple cellular spine P. Any disc replacement can be considered as 2-cell trading: we attach to P a new 2-cell $D \subset M$ and remove another one. The manifold remains the same. We define here a general disc replacement move and then describe some specific versions which quite often simplify P.

Let us identify M with a close regular neighborhood of P. Suppose $\tilde{\ell}$ is a trivial circle on ∂M. Denote by ℓ the closed curve $p(\tilde{\ell}) \subset P$, where $p \colon \partial M \to P$ is the standard projection. We always assume that ℓ is in general position, i.e., intersects the edges of P and itself transversally. The total number of points where l crosses SP and itself is called the *weight* of ℓ. Since $\tilde{\ell}$ is trivial in ∂M, there is a disc D in M with the embedded interior (but maybe with self-intersecting boundary) such that $\partial D = \ell$ and D has no other common points with P. Then D cuts off from $M \setminus P$ a 3-ball B. Let $C \neq D$ be a 2-component of the simple polyhedron $P' = P \cup D$ which separates B from another component of $M \setminus P'$. Removing from C an open disc, i.e., performing a delicate piercing (move S_2), we get another simple spine P_1 of M. We say that P_1 is obtained from P by a *disc replacement move*, see Fig. 7.17.

Remark 7.2.4. Suppose that the component $T \subset \partial M$ containing $\tilde{\ell}$ is a torus with a coordinate system (μ, λ). What happens to μ, λ (represented by integer 1-dimensional chains) under the disc replacement move described above? Let $\tilde{\ell}$ cross edges $e_0, e_1, \ldots, e_{n-1}$ of T in a cyclic order. Denote by D_1 the disc on T bounded by $\tilde{\ell}$. Then the cell decomposition of the new torus T' corresponding to T can be constructed in two steps. First, we insert into T new coherently oriented edges whose union is $\tilde{\ell}$. Then we replace the union of all 2-cells contained in D_1 by a new 2-cell D_1', see Fig. 7.18.

Denote by m_i the coefficient of μ at $e_i, 0 \leq i \leq n-1$. Then the new 1-dimensional chain μ' can be described as follows:

1. The coefficients of μ' at the initial segments e_i' of e_i are $m_i, 0 \leq i \leq n-1$.
2. For each k the coefficient of μ' at the edge of $\tilde{\ell} = \partial D_1'$ between the endpoints of e_{k-1}', e_k' (indices are taken modulo n) is $c_k = c_0 + \sum_{i=0}^{k-1} m_i$, where c_0 is a fixed constant. For instance, one can take $c_0 = 0$.

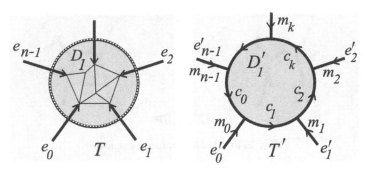

Fig. 7.18. The behavior of μ under the disc replacement move

Since $\sum_{i=0}^{n-1} m_i = 0$, the chain μ' is defined correctly. The second coordinate curve λ' can be described in the same way.

Disc replacements had been used in Sect. 2.2.1 for converting almost simple spines into special ones. Note that addition of D creates new true vertices at self-crossings of ℓ and at points where ℓ crosses the triple edges of P. The total number of such vertices is called the *weight* of the move (it is equal to the weight of ℓ). We are mainly interested in cases when, after collapsing P_1, we get an almost simple polyhedron P_2 which is simpler than P (recall that we measure the complexity of an almost simple polyhedron X by the number $c(X)$ of its true vertices).

Definition 7.2.5. *A disc replacement move is called* admissible *if the following conditions are satisfied:*

1. *The weight of the move does not exceed 4. It means that the addition of D creates not more than four new true vertices, i.e., $c(P_1) - c(P) \leq 4$.*
2. *The combined move (attaching D, piercing, and collapsing) does not increase $c(P)$, i.e., $c(P_2) \leq c(P)$.*

An admissible disc replacement move is called monotone *if $c(P_2) < c(P)$, and* horizontal *if $c(P_2) = c(P)$.*

We describe several admissible disc replacement moves. The first move has weight 0 and is actually a form of Move 3 from the proof of Theorem 2.2.4.

U_1: ℓ *has weight 0.* Let D be a disc in Int M such that $D \cap P = \partial D$ is a nontrivial embedded circle in a 2-component C of P. Then we attach D to P and perform a delicate piercing. See Fig. 2.6 in Chap. 2.

It is easy to see that this move is admissible. We do not create new true vertices while some true vertices and triple lines may disappear under collapsing. In any case, C becomes simpler, since its Euler characteristic increases.

We want a simple method of finding a disc D and a circle ℓ satisfying the required conditions. Here is one. Let $\tilde{C}$ be the inverse image of a 2-component C of P under the projection $p \colon \partial M \to P$. Then $\tilde{C}$ is either the union of two

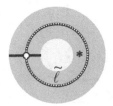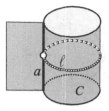

Fig. 7.19. ℓ is of weight 1

copies of C or a 2-sheeted covering of C, depending on whether or not C is two-sided. In both cases $\tilde{C}$ contains a circle trivial in ∂M and nontrivial in C if and only if such a circle can be found among boundary components of $\tilde{C}$. It follows that we need to consider only those circles in $\tilde{C}$ which are parallel to $\partial \tilde{C}$. In our situation, when ∂M consists of spheres and tori, all boundary circles of $\tilde{C}$ are nontrivial only if $\tilde{C}$ is either a nontrivial annulus in a toral component of ∂M or the union of two such annuli. Evidently, C is a Möbius strip in the first case and an annulus in the second.

Further on we consider only the case when P is special and its cell structure is reduced, i.e., coincides with the intrinsic decomposition of P into true vertices, triple edges, and 2-components. Recall that then M possesses the induced cell decomposition (DS-diagram). We describe all possible types of curves of weight ≤ 4 which bound discs in the complement of P and thus can be used for performing disc replacement moves. It turns out that in all these cases a place for piercing can be chosen so that the resulting move be admissible, i.e., is either monotone or horizontal. Altogether, there are six nontrivial moves.

The moves are presented by pictures, each consisting of two parts. The left part shows a circle $\tilde{\ell} \subset \partial M$ contained in a subdisc of ∂M. The places where we can potentially puncture $P \cup D$ are marked by asterisks. The right part shows a regular neighborhood of ℓ in P. Obviously, the collapse destroys all true vertices at the boundary of a chosen asterisked region.

U_2: ℓ *is of weight 1.* Denote by C the 2-component of P whose closure contains ℓ and by a the edge of P intersecting ℓ. Then the boundary curve of C has a counterpass on ℓ. So the move is monotone, since after puncturing and collapsing there disappears not only the new true vertex, but also at least one old vertex. See Fig. 2.9 and 7.19.

U_3: ℓ *is of weight two.* In this case the combined move (U_3 plus collapsing) destroys both new vertices, see Fig. 7.20. It is always monotone, except the situation when ℓ crosses the same edge twice and the move produces a homeomorphic spine. The inverse of the lune move (see Definition 1.2.9) is a partial case of U_3.

U_4: ℓ *is of weight 3 and has no self-intersections.* The corresponding combined move destroys two new vertices and at least one old vertex. Moreover, if ℓ is not contained in a regular neighborhood of a true vertex, then one of

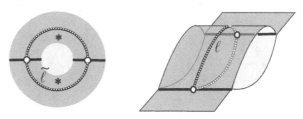

Fig. 7.20. ℓ is an embedded circle of weight 2

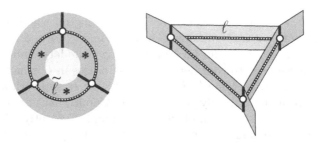

Fig. 7.21. ℓ is of weight 3 and has no self-intersections

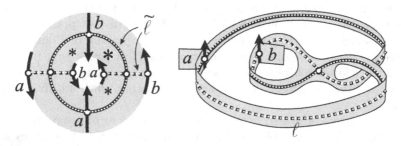

Fig. 7.22. ℓ is of weight 3 and has one self-intersection

the three asterisked 2-components contains at least two old vertices. So the move can be made monotone by a right choice of a 2-component for piercing. See Fig. 7.21.

U_5: ℓ *is of weight 3 and has one self-intersection.*

In this case the removal of the 2-component marked by the large asterisk destroys all three new true vertices and at least one old true vertex. So this move is always monotone. See Fig. 7.22.

U_6: ℓ *is of weight 4 and has no self-intersections.*

In this case the boundary curve of at least one of the asterisked 2-components contains two new and at least two old vertices. Therefore, U_6 is at least horizontal. See Fig. 7.23.

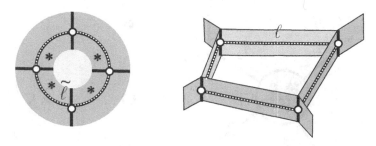

Fig. 7.23. ℓ is an embedded circle of weight 4

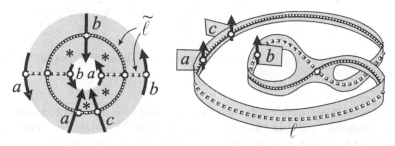

Fig. 7.24. ℓ is of weight 4 and has one self-intersection

U_7: ℓ *is of weight 4 and has one self-intersection.*

In this case the removal of at least one of the asterisked components destroys all four or more true vertices. So this move is either horizontal or monotone. See Fig. 7.24.

One can easily show that moves U_2–U_6 exhaust all types of disc replacement moves on special spines such that addition of the new disc creates ≤ 4 true vertices.

7.3 Labeled Molecules

In this section we use a convenient terminology introduced by Fomenko for investigation of integrable Hamiltonian systems [11, 30].

7.3.1 What is a Labeled Molecule?

A *labeled molecule* is an arbitrary graph G, some of whose vertices and edges are equipped with labels of different kind, see Fig. 7.25.

There are five types of vertices and two types of edges. Labeled vertices are called *atoms*. Let us describe them:

1. *Solid torus vertices.* They have valence 1 and are shown as fat black dots.
2. *Möbius vertices.* They are of valence 2 and labeled by white circles.
3. *Triple vertices.* They have valence 3 and no label.

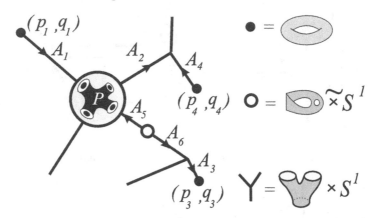

Fig. 7.25. Edges labeled by matrices join atoms of different types into a labeled molecule

4. *Exceptional vertices.* Each such vertex stands for an oriented 3-manifold X presented by its simple spine P and a map $\partial X \to P$. The boundary of X consists of spheres and tori. We assume that there is a fixed bijection between the boundary tori of X and the edges of G having an endpoint at the vertex (if an edge has both endpoints at the vertex, then it corresponds to two boundary tori). The boundary tori which are joined by edges with vertices of types 1–4 are equipped with coordinate systems.

5. *Virtual vertices.* They are of valence 1 and not labeled.

These labels have a clear meaning. Vertices of the first three types correspond to standard 3-manifolds and are considered as known atoms. Each Möbius atom is an oriented S^1-bundle $M_0^2 \tilde{\times} S^1$ over a once punctured Möbius strip M_0^2. Since M_0^2 is homeomorphic to the twice punctured projective plane, the boundary tori are symmetric in the sense that they are related by an orientation preserving involution of the bundle. We equip the tori with coordinate systems so that the meridians are the boundary components of a section of the bundle while the longitudes are fibers. The meridians and longitudes are oriented so that the intersection numbers of the meridian with the corresponding longitudes are equal. There are eight choices of such orientations. Any two of them are related by a homeomorphism of $M_0^2 \tilde{\times} S^1$.

Triple vertices correspond to $N^2 \times S^1$, where N^2 is the twice punctured disc. We introduce coordinate systems on the boundary tori of $N^2 \times S^1$ as follows: The meridians are the boundary circles of an oriented section endowed with the induced orientations. The longitudes are oriented fibers, and the intersection numbers of the meridians with the corresponding longitudes are equal. Any two such triples of coordinate systems differ by a homeomorphism of $N^2 \times S^1$.

Clearly, solid torus vertices correspond to solid tori. Each such torus $V = D^2 \times S^1$ must be labeled by an ordered pair (p, q) of coprime integers. These integers are the coordinates of the meridian $m = \partial D \times \{*\}$ of V in a coordinate system (μ, λ) on ∂V. An explicit presentation of (μ, λ) is not needed. We only require that the matrix label (see below) for the incoming or outgoing edge must be written with respect to the same system.

Exceptional atoms can correspond to known 3-manifolds, for example, to hyperbolic 3-manifolds contained in known tables [46] or to complements of tabulated knots [44, 109]. They can also correspond to 3-manifolds which are unknown in the sense that we only know their simple spines.

Virtual vertices correspond to empty atoms. They are introduced to indicate the valences of other atoms and stand for their toral boundary components.

Let us describe the edges of G. They can be divided into two groups: *true edges* with vertices of types 1–4 and *virtual edges* that have a virtual vertex. Each true edge e is oriented and labeled by order 2 matrix A with determinant ± 1. We can reverse the orientation of e, but then A must be replaced by A^{-1}. Virtual edges have no labels.

Finally we assume that any labeled molecule is equipped with three global labels (g_1, g_2, g_3), which are non-negative integers. They tell us how many 3-balls have been added and how many solid tori and copies of $S^2 \times S^1$ have been cut out during the construction the molecule.

It turns out that any labeled molecule G determines a connected 3-manifold $M(G)$ whose boundary consists of spheres and tori. To reconstruct $M(G)$, we take into account the meaning of atoms and edge labels. This information allows us to compose a 3-manifold $M'(G)$ as follows. We replace each vertex v of type 1–4 by a copy of the corresponding manifold. The boundary tori of this manifold are parameterized by the edges that have an endpoint in v. If e is a true edge, then the corresponding tori are equipped with coordinate systems, and e is labeled with a matrix. We glue together the manifolds that correspond to the endpoints of e via the homeomorphism described by the matrix. Of course, if G is disconnected, then so is the obtained 3-manifold $M'(G)$. In this case we replace it by the connected sum of all its components (since we are considering only oriented manifolds, the connected sum is well defined). We denote the connected sum also by $M'(G)$.

What is the role of the global labels g_1, g_2, g_3? To get $M(G)$, we add to $M'(G)$ exactly g_1 copies of the standard 3-ball. This is equivalent to puncturing $M'(G)$ exactly g_1 times. Then we add g_2 copies of the solid torus and g_3 copies of the manifold $S^2 \times S^1$. In other words,

$$M(G) = M'(G) \# g_1 B^3 \# g_2 (D^2 \times S^1) \# g_3 (S^2 \times S^1).$$

7.3.2 Creating a Labeled Molecule

What we have said in Sect. 7.3.1 implies that each labeled molecule determines a manifold. Now we go in the opposite direction. Let M be an oriented 3-manifold whose boundary consists of spheres and tori. We describe a procedure

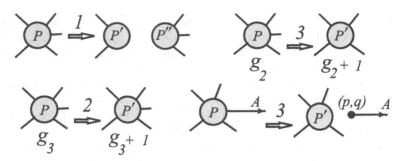

Fig. 7.26. Action of S_1 onto labeled molecules

that assigns to M a labeled molecule G such that $M = M(G)$. Clearly, we wish the atoms be as simple as possible.

We begin with constructing a special spine P of M and creating a molecule G_0 consisting of only one exceptional atom M with label P. The atom has only virtual outgoing edges. We then apply to G_0 and all labeled molecules arising in this way all moves described in Sect. 7.2 for as long as possible.

Moves E_1 (cleaning edges) and R_1–R_3 (reductions of cells) do not affect the molecule. Moves C_1–C_4 (collapses) as well as disc replacement moves U_1–U_7 modify the spines of exceptional vertices. The molecules remain the same. In contrast with this, surgeries S_1–S_6 do change the labeled molecule. Let us describe what happens with the molecule when we apply move S_i to an exceptional atom X of G represented as a thickening of its simple spine P.

MOVE S_1. Recall that this move consists in cutting X and P along a proper disc $D \subset X$ and along the arc $l = D \cap P$, respectively. It either divides X into two new atoms (Case 1), or replaces (X, P) by a new pair (X', P') and increases the global label g_3 by 1 (Case 2), or tears off from X a solid torus (Case 3). See Fig. 7.26, where we show the first two possibilities and two versions of the third one. Which version takes place depends on the behavior of D, see the description of move S_1. If ∂D is a nonseparating curve on a torus $T \subset \partial X$, then we take into account the type of the edge e of G corresponding to T. If e is virtual, then we remove e and add 1 to g_2. If e is true, then we cut off a solid torus atom V together with e. Then we label V with two integers (p, q), which are the coordinates of the meridian of V. The matrix label of e remains unchanged.

MOVES S_2, S_3. Delicate piercing S_2 replaces the pair (X, P) by a new pair (X', P') and increases the global label g_1 by 1. Rough piercing S_3 transforms an exceptional atom of valence 0 into a new atom joined with a unique solid torus atom by an edge. It follows from the description of S_3 that the labels for the atom and the edge are (1,0) and the unit matrix $E = \begin{pmatrix} 1 & 0 \\ 0 & 1 \end{pmatrix}$, see Fig. 7.27 to the left.

MOVE S_4. This move consists in cutting X and P along a proper annulus $A \subset X$ and the circle $l = A \cap P$, respectively. There appears a new atom X'

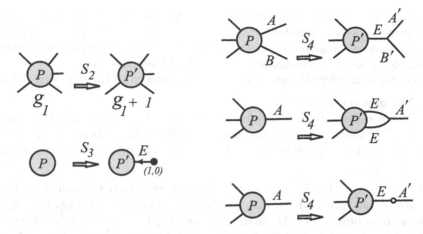

Fig. 7.27. How S_2, S_3, S_4 transform molecules

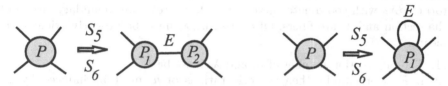

Fig. 7.28. How S_5, S_6 transform molecules

with its simple spine P' such that $X = X' \cup Y$, where Y homeomorphic to either $N^2 \times S^1$ or $M_0^2 \tilde{\times} S^1$. The corresponding transformation of the molecule is shown in Fig. 7.27 to the right. There are three cases, depending on the type of Y and the number of edges that join Y with the remaining part X' of X. The new matrices A', B' determine the same gluing maps, but are written for a standard coordinate system on Y.

MOVES S_5, S_6. These moves consist in cutting X along a torus intersecting P along a theta-curve Θ. In the case of move S_5 the torus is punctured and its boundary is contained in a spherical component of ∂X. In the case of move S_6 the torus is closed and lies in the interior of X. In both cases we cut P along Θ. The corresponding transformation of the molecule depends on whether or not Θ splits P into two parts. We replace X either by two new atoms joined by a new edge, or by one new atom with a new loop, see Fig. 7.28. We label the new edge by the matrix E, since the coordinate systems on the corresponding boundary tori are composed of oriented copies of the same circles in Θ.

7.3.3 Assembling Seifert Atoms

Let G be a labeled molecule obtained for a 3-manifold M by the procedure described above. Note that the atoms of the types $M_0^2 \tilde{\times} S^1$, $N^2 \times S^1$ are Seifert manifolds. We will denote them by squares. Now we begin the reverse process of assembling them into larger Seifert manifolds. Combined Seifert atoms thus obtained will also be denoted by squares. We consider each square as a closed box containing complete information about the corresponding oriented Seifert manifold X and the coordinate systems on its boundary tori.

Let us show why this information can be reduced to the following data:

1. The base surface F of X.
2. The non-normalized Seifert invariants $(p_1, q_1), \ldots, (p_m, q_m)$ of m distinguished fibers such that all exceptional fibers (i.e., fibers with $p \neq \pm 1$) are included. We can always add a new fiber $(1, 0)$ and thus assume that $m \geq 1$.

Let F and $(p_i, q_i), 1 \leq i \leq m$, be given. Then we reconstruct X as follows. Let a surface F_0 be obtained from F by cutting out m disjoint discs $D_i, 1 \leq i \leq m$. Consider an oriented circle bundle X_0 over F_0, i.e., an orientable direct or twisted product of F_0 and S^1. If F_0 is orientable, then the product is direct, otherwise twisted. In both cases it admits a section $S \subset X_0$. We endow the tori of ∂X_0 with coordinate curves (μ_i, λ_i), where μ_i are boundary circles of the section and λ_i are fibers. Of course, μ_i, λ_i must be properly oriented. It means:

1. The intersection number of μ_i and λ_i must be 1.
2. If F is orientable, then the orientations of μ_i must be induced by an orientation of F.

Let $T_i, 1 \leq i \leq n$, be the boundary tori of ∂X_0 such that $T_i = \partial D_i \times S^1$ for $1 \leq i \leq m$ and $T_i \subset \partial F \times S^1$ for $m + 1 \leq i \leq n$. In order to get X, it remains to attach solid tori $V_i, 1 \leq i \leq m$, so that the meridian of each torus V_i is mapped to a circle of the type $\mu_i^{p_i} \lambda_i^{q_i}$. We will call the coordinate systems $(\mu_i, \lambda_i), m + 1 \leq i \leq n$, on the remaining tori of ∂X_0 (i.e., on the tori of ∂X) *canonical*.

The difference between normalized and non-normalized Seifert invariants of exceptional fibers is that in the first case we know q_i only modulo p_i. If the manifold is closed, then the normalized invariants do not determine it; we must know one additional integer parameter. In the case of non-normalized parameters each q_i can be an arbitrary number coprime to p_i, and no additional parameters are needed.

Remark 7.3.1. Consider an oriented Seifert atom X with the base surface F, distinguished fibers $(p_i, q_i), 1 \leq i \leq m$, and coordinate curves $(\mu_i, \lambda_i), m + 1 \leq i \leq n$, for ∂X. Let us reverse the orientation of X, the signs of all q_i, and the orientations of all μ_i. Then X and the new atom $\bar{X}$ thus obtained are homeomorphic via an orientation preserving homeomorphism $h: X \to \bar{X}$ which takes coordinate curves to the corresponding coordinate curves and preserves their orientations. The same remains true, if we reverse orientations of all λ_i. In the first case h is induced by a map $F \to F$ which reverses orientations of all circles of ∂F. In the second case it is induced by an orientation reversing map $S^1 \to S^1$.

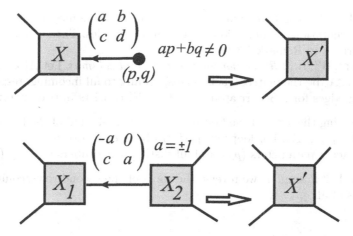

Fig. 7.29. Assembling moves W_1, W_2

Let us describe three assembling moves.

MOVE W_1. *Adjoining a solid torus atom.* Let a molecule contain a solid torus atom V with parameters (p, q) joined by an edge e with a Seifert atom X. If e is directed from X to V, then we reverse it, and replace its matrix label by the inverse matrix. Suppose that the label of e has the form $\begin{pmatrix} a & b \\ c & d \end{pmatrix}$ such that $ap + bq \neq 0$. This means that the corresponding attaching map $\varphi: \partial V \to \partial X$ takes the meridian of V to a curve which is not isotopic to a fiber. We remove V and replace X by the combined Seifert atom $X' = X \cup_\varphi V$. The base surface F' of X' is obtained from the base surface F of X by attaching a disc to the corresponding circle of ∂F. Also, X' obtains an additional distinguished fiber with Seifert invariants $(ap + bq, cp + dq)$.

MOVE W_2. *Uniting two Seifert atoms.* Suppose that two Seifert atoms X_1, X_2 are joined by an edge e with the label of the form $\begin{pmatrix} -1 & 0 \\ -c & 1 \end{pmatrix}$. This means that the gluing map φ of the corresponding tori takes oriented longitudes to oriented longitudes. We replace X_1, X_2 by the combined Seifert atom $X' = X_1 \cup_\varphi X_2$. The base surface F' of X' is obtained by the induced gluing of the base surfaces of X_1, X_2. The set of distinguished fibers of X' consists of those of X_1 and X_2 with the same Seifert invariants, and of one additional regular fiber with parameters $(1, c)$. See Fig. 7.29. We can apply this move also in the case $X_1 = X_2$, when e is a loop.

MOVE W_3. *Removing loops.* Suppose that a Seifert atom X with the base surface F has a loop edge e with the label of the form $\begin{pmatrix} 1 & 0 \\ c & -1 \end{pmatrix}$. We replace X by a new Seifert atom X' whose base surface F' is obtained by the induced gluing of two boundary circles of F. The set of distinguished fibers of X' consists of those of X and of one additional fiber with parameters $(1, c)$.

We describe also two auxiliary moves induced by homeomorphisms which reverse orientations of all meridians or of all longitudes. The homeomorphisms are described in Remark 7.3.1.

MOVES W_4', W_4''. *Reversing signs of rows and columns.* Let $A_1, \ldots, A_k$ and $A_{k+1}, \ldots, A_s$ be the matrix labels corresponding to all incoming, respectively, outgoing edges for a Seifert atom X. Then W_4' consists in two operations:

1. Reversing the signs of the first rows of $A_1, \ldots, A_k$ and of the first columns of $A_{k+1}, \ldots, A_s$. For loops we perform both operations.
2. Replacing parameters (p_i, q_i) of all distinguished fibers of X by $(p_i, -q_i)$.

Move W_4'' is similar: we reverse the signs of the second rows/columns and the signs of all q_i.

7.4 The Algorithm

Let M be a 3-manifold whose boundary is either empty or consists of tori. Let G be a labeled molecule obtained for M by the disassembling process described above. Suppose that G contains only atoms of the first three types (solid tori, $M_0^2 \tilde{\times} S^1$, $N^2 \times S^1$), which are Seifert manifolds. Let us begin the reverse process of applying to G assembling Moves W_1–W_3. Suppose that at some step we obtain two different atoms joined by an edge with the label of the form $\begin{pmatrix} a & 0 \\ c & d \end{pmatrix}$. We use moves W_4', W_4'' to get the label of the form $\begin{pmatrix} -1 & 0 \\ \pm c & +1 \end{pmatrix}$. Then we unite the atoms by W_2.

Let us perform moves W_1–W_3 for as long as possible. Of course, one must keep in mind the possibility of replacing the manifold $(D^2; (2,1), (2,-1))$ by the manifold $M^2 \tilde{\times} S^1$ and conversely. Suppose that the process stops at a connected molecule such that $g_1 = g_2 = g_3 = 0$. The following cases are possible:

1. If all the atoms of the molecule turn out to be Seifert, then we obtain a graph manifold with known parameters. In particular, if the molecule has only one atom and no loops, then we obtain a Seifert manifold. If there remains a single atom $T^2 \times I$ with one loop edge, then the type of the manifold (Seifert or Sol) is determined by the monodromy matrix.
2. If the molecule consists of a known hyperbolic atom of type Q_i and an atom $D^2 \times S^1$ that are joined by a single edge, then the original manifold is of the form $(Q_i)_{p,q}$, where the parameters p and q are computed by using the gluing matrix.
3. The molecule can contain both Seifert and hyperbolic atoms. In this case we get the JSJ decomposition of M presented in the form of geometric chambers of M and gluing matrices.
4. The molecule may contain unknown atoms (that is, atoms that are not yet recognized). In this case the solution of the recognition problem is conditional, up to recognition of the unknown atoms in the molecule.

Finally, suppose that the obtained molecule is disconnected or g_1, g_2, g_3 are not zeros. Then M is the connected sum of 3-manifolds that correspond to the connected components of the molecule plus g_1 balls, g_2 solid tori, and g_3 copies of $S^2 \times S^1$.

It turned out that the software program based on the above principles recognizes manifolds rather well. The program (called 3-Manifold Recognizer) was written by V. Tarkaev and is available from

$< http : //www.topology.kb.csu.ru/ \sim recognizer >$.

Why is the program useful? Suppose that, doing mathematics, mechanics, physics, you got a 3-manifold and would like to get info on it. Then you can put it into the Recognizer and get a lot of information, including values of different invariants, JSJ decomposition, type of geometry, and, in best cases, its name.

7.5 Tabulation

Theorem 7.5.1. *The numbers of closed orientable irreducible 3-manifolds up to complexity 12 are given in Table 7.1:*

As we described in Sect. 2.3, the first three columns of the table had been obtained by Matveev, Savvateev, and Ovchinnikov (see [80, 91, 102]). Later Martelli wrote a computer program which tabulates 3-manifolds in two steps. First, it enumerates some special building blocks (bricks), and only then assembles bricks into 3-manifolds. An interesting relative version of the complexity theory (see [74]) serves as a theoretical background for the program. We describe it in Sect. 7.7. Using this method, Martelli composed tables of closed orientable manifolds up to complexity 9 and corrected an error of Ovchinnikov, who had not noticed that two of the manifolds of complexity 7 were homeomorphic. Elements of manual recognition were still present. The manifolds of complexity 10 were tabulated by Martelli, who used his earlier approach, and (slightly earlier) by Matveev, who recognized manifolds by a computer program written by Tarkaev. A comparison of intermediate results helped to

Table 7.1. Closed 3-manifolds up to complexity 12

Type\c	≤ 5	6	7	8	9	10	11	12	Total
S^3	61	61	117	214	414	798	1,582	3,118	6,365
E^3	0	6	0	0	0	0	0	0	6
Nil	0	7	10	14	15	15	15	15	91
$H^2 \times R$	0	0	0	2	0	8	4	24	38
$\widetilde{SL_2 R}$	0	0	39	162	513	1,416	3,696	9,324	15,150
Sol	0	0	5	9	23	39	83	149	308
H^3	0	0	0	0	4	25	120	459	608
Composite	0	0	4	35	185	777	2,921	10,345	14,267
Total	61	74	175	436	1,154	3,078	8,421	23,434	36,833

get rid of some deficiencies of both the Martelli method and the Matveev algorithm. At last, the results coincided for all complexities ≤ 10. The last step was taken by Matveev and Tarkaev, who tabulated all closed irreducible orientable 3-manifolds up to complexity 12 by using an improved tabulation program [86, 87].

In Fig. 7.30 the growth of the number of manifolds of diverse types is shown graphically. Note that the scale for the ordinate axis is logarithmic. Since all

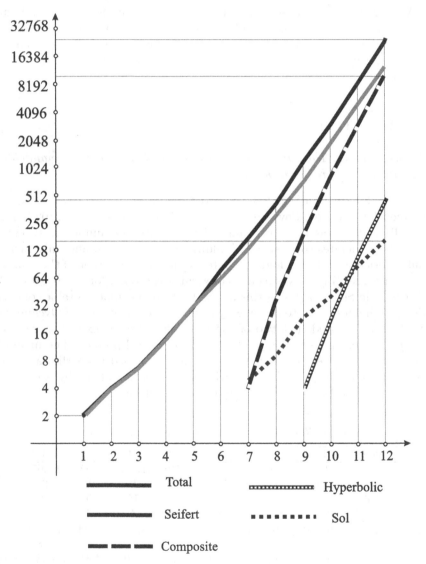

Fig. 7.30. Growth of the number of manifolds in dependence on the growth of their complexity

the manifolds possessing geometries S^3, E^3, Nil, $H^2 \times R, \widetilde{SL_2R}$ are Seifert manifolds, the data on them are presented on a single graph. Both manifolds $S^2 \times S^1$ and $RP^3 \# RP^3$ possessing the geometry $S^2 \times R$ are reducible and thus have no influence on the table and graphs.

A few words on enumeration of nonorientable 3-manifolds. First results in this direction had been obtained by Casali with the help of the theory of crystallizations, see Sect. 7.6.2. She found all seven nonorientable 3-manifolds admitting crystallizations with up to 26 vertices [17], among them five P^2-irreducible manifolds. Amendola and Martelli [2] proved that there are no closed nonorientable 3-manifolds of complexity ≤ 5 and only 5-manifolds of complexity 6 (they coincide with P^2-irreducible manifolds from Casali's list). Later Amendola and Martelli showed that there are three closed nonorientable 3-manifolds of complexity 8 [3]. See also [18]. The same result has been proved independently by Burton [14] using his computer program *Regina* [13]. More than that, he classified all minimal singular triangulations with at most seven and later with eight tetrahedra [16]. Note that *Regina* turned out to be very useful for the orientable case too [15].

7.5.1 Comments on the Table

The first conclusion which one can make after looking at the table and some information about specific manifolds is that the table contains nothing unexpected. Everything is just as is predicted by the theory. Namely, every manifold is either geometric or composite, that is, can be assembled from geometric manifolds. Moreover, the complexity of the manifolds increases as they become more complicated in the informal meaning of the word.

As we have mentioned in Sect. 2.3.3, all closed orientable irreducible 3-manifolds up through complexity 5 are elliptic. One of the most interesting elliptic manifolds, the *dodecahedron space* (the Poincaré homology sphere), is of complexity 5. All the six *flat* (that is, having geometry E^3) orientable 3-manifolds have complexity 6. The first seven Nil-manifolds are also at this level.

First manifolds $(S^2, (2, 1), (2, 1), (3, 1), (3, -4))$ and $(RP^2, (3, 1), (3, -1))$ with the geometry $H^2 \times R$ have complexity 8. Note that all closed 3-manifolds of complexity ≤ 8 are graph-manifolds (this was theoretically proved in [80]). The structure of nontrivial graph-manifolds of complexity ≤ 11 is very simple. There are only three types of manifolds of this kind:

1. Two Seifert manifolds glued together. Each of them is fibered over a disc and has two or three exceptional fibers.
2. Three Seifert manifolds glued together. One of them is $(S^1 \times I; (\alpha, \beta))$, the other two are fibered over the disc and have two exceptional fibers.
3. Seifert manifold $(S^1 \times I; (\alpha, \beta))$, with boundary tori glued together.

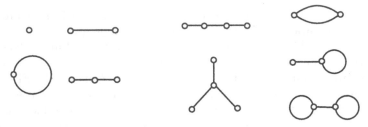

Fig. 7.31. Graphic structure of manifolds of complexity ≤ 12

The corresponding reduced molecules are as follows: an isolated vertex, a circle with single vertex, a segment, and a wedge of two segments (see Fig. 7.31 to the left). At complexity level 12, five new types of molecules arise. Of course, the four types encountered at complexity levels ≤ 11 are preserved.

The first manifolds that have hyberbolic structure and therefore are not graph-manifolds occur at complexity level 9. We discuss hyperbolic manifolds in detail in Sect. 7.5.2. For now, we mention only that all 3-manifolds up through complexity 11 whose JSJ-decomposition is nontrivial and contains a hyperbolic chamber have the same structure, namely, they are of the form $(M^2 \tilde{\times} S^1) \cup Q_i$, where $i = 1$. Other 24-manifolds of the same type appear at complexity 12 for $i = 1, 4, 9$, see Proposition 2.4.11 for the description of Q_i.

7.5.2 Hyperbolic Manifolds up to Complexity 12

The complete table of closed hyperbolic manifolds up to complexity 12 contains 608 manifolds. We restrict ourselves to a small part of the table, the manifolds of complexity 9 and 10, and order these manifolds by their volumes. In Table 7.2 we present the volumes with three digits after the decimal point.

Theorem 7.5.2. *All closed orientable hyperbolic 3-manifolds of complexity ≤ 10 are given in Table 7.2.*

We comment on some interesting points which can be noted by studying both this table and the complete electronic table of hyperbolic manifolds.

(1) It turns out that every closed hyperbolic manifold up through complexity 12 can be presented in the form $Q_{p,q}$, where Q is a finite volume hyperbolic manifolds with a cusp that have complexity from 2 to 7. All these manifolds can be found in the table of manifolds with cusps in SnapPea software by Weeks [132]. See also [32] and Proposition 2.4.11 for the description of all eleven cusped hyperbolic manifolds Q_i, $1 \leq i \leq 11$ of complexities 2 and 3. Most of the closed hyperbolic manifolds have several such representations. For instance, the Thurston manifold $(Q_2)_{5,1}$ is homeomorphic to the manifolds

Table 7.2. Closed hyperbolic manifolds of complexity ≤ 10

no.	H_1	volume	no.	H_1	volume
1	$Z_5 \oplus Z_5$	0.942	16	$Z_3 \oplus Z_9$	1.583
2	Z_5	0.981	17	Z_{30}	1.588
3	$Z_3 \oplus Z_6$	1.014	18	Z_{30}	1.588
4	$Z_5 \oplus Z_5$	1.263	19	Z_5	1.610
5	Z_6	1.284	20	Z_7	1.649
6	0	1.398	21	Z_{15}	1.649
7	Z_6	1.414	22	Z_7	1.757
8	Z_{10}	1.414	23	$Z_3 \oplus Z_3$	1.824
9	Z_{35}	1.423	24	$Z_2 \oplus Z_{12}$	1.831
10	Z_3	1.440	25	$Z_7 \oplus Z_7$	1.885
11	Z_5	1.529	26	Z_{39}	1.885
12	Z_{21}	1.543	27	Z_{40}	1.885
13	Z_{35}	1.543	28	Z_{30}	1.910
14	Z_{40}	1.583	29	Z_{35}	1.953
15	Z_{21}	1.583			

$(Q_1)_{1,2}, (Q_7)_{1,1}$ and $(Q_{11})_{2,-1}$. One can reduce the number of needed hyperbolic bricks Q_i by means of such homeomorphisms. On studying the results of the computer experiment it turned out that, to represent four closed hyperbolic manifolds of complexity 9, it suffices to have two bricks Q_1 and Q_2. For complexity 10 one needs all bricks $Q_i, 1 \leq i \leq 11$. For complexity 11 we need 12 additional bricks from the cusped census of Weeks, while at the complexity level 12 Dehn fillings of Q_i represent only about a half of all closed hyperbolic manifolds.

(2) Three manifolds in the table (with numbers $14, 19, 20$) are missed in the Weeks table of closed hyberbolic manifolds which is contained in the accompanying package of the "SnapPea" software. Moreover, the volume 1.610 of manifold number 19 is also absent in the monotone list of volumes presented in this package. Among 120 closed hyperbolic manifolds of complexity 11 the number of missed manifolds is 52. The reason is that Weeks did not pose for himself the problem of listing the closed hyperbolic manifolds and volumes without gaps. He intentionally neglected some manifolds since their geometries were too close to the geometry of the corresponding cusped manifolds. Therefore, it is not surprising that a systematic and exhaustive search in order of increasing complexity has led to new manifolds and new volumes.

(3) The volume of the manifold $(Q_2)_{7,-9}$ of complexity 10 is equal to 1.649, whereas the manifold $(Q_2)_{7,-8}$ of complexity 11 has smaller volume 1.463. This disproves the conjecture on correlation between the complexity of 3-manifolds and its hyperbolic volume (within a single series of manifolds of the form Q_{pq}). See [32].

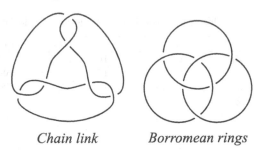

Chain link   Borromean rings

Fig. 7.32. Useful links

(4) Tarkaev observed that among 608 closed orientable hyperbolic manifolds of complexity ≤ 12 almost all have Heegaard genus 2. To be more precise,
he proved that 600 manifolds have genus 2 and 4 manifolds have genus greater
than 2. Most probably that the remaining 4-manifolds also have genus greater
than 2. The first closed hyperbolic manifold of genus 3 has complexity 11. It
has volume 2.468 and $H_1 = Z_3 \oplus Z_3 \oplus Z_3$.

Tarkaev has also shown that the majority (577 from 608) of closed orientable hyperbolic manifolds of complexity ≤ 12 can be obtained by Dehn
surgery of S^3 along either the 3-component *Chain link* or *Borromean rings*
(see Fig. 7.32).

7.5.3 Why the Table Contains no Duplicates?

Inaccurate results have been obtained (and published) several times in the
course of creating tables of manifolds, namely, some tables have contained
duplicates. For this reason, we carried out an independent check of the table
of manifolds up to complexity 12 by comparing their first homology groups
and the values of the Turaev–Viro invariants (see [126] and Chap. 8) of orders
≤ 14 and, in some cases, of orders 15 and 16. Note that the invariants have
been computed from spines while the manifolds have been tabulated after
recognition, according to their names. So the absence of duplicates can be
considered as an indirect confirmation that the recognition program works
correctly.

The invariants were computed by software developed by Pervova and
Tarkaev. Here the lens spaces were set aside, because the problem which lens
spaces have the same Turaev–Viro invariants has been completely solved [117].
To avoid the problem of the approximate nature of computer calculations, we
used the representations of the invariants by values of polynomials with integer coefficients at the corresponding roots of unity rather than the numerical
values of the invariants. This joint use of homology and Turaev–Viro invariants turned out to be extremely effective: We succeeded in distinguishing all
manifolds of complexity ≤ 11 excepts for a few pairs. The first pair among
manifolds of complexity 11 is obtained from the Seifert manifold with base

an annulus and with one exceptional fiber of type $(2,1)$ by gluing together its boundary tori along the homeomorphisms given by the matrices $\begin{pmatrix} 1 & 4 \\ 0 & -1 \end{pmatrix}$ and $\begin{pmatrix} -1 & -4 \\ 0 & 1 \end{pmatrix}$. The second pair differs only in the type of the exceptional fiber, which is given by the parameters $(3,2)$ and in the matrices, which are now $\begin{pmatrix} 1 & 3 \\ 0 & -1 \end{pmatrix}$ and $\begin{pmatrix} -1 & -3 \\ 0 & 1 \end{pmatrix}$.

The coincidence of all the Turaev–Viro invariants of these manifolds is explained by the fact that in both cases one of the matrices differs from the other by changing the signs of all its elements. This is sufficient for the coincidence of invariants of all orders (and not just of the computed invariants), see Sect. 8.2.

The manifolds of the all other pairs are Seifert manifolds fibered over the sphere with three exceptional fibers. Here is an example of such a pair: the manifolds $(S^2, (2,1), (3,1), (6,1))$ and $(S^2, (2,1), (3,1), (6,-11))$ have the same homology groups and the same values of Turaev–Viro invariants up through order 14 (including the homologically trivial summands correspond-ing to integral colorings). We note that the first parameters of the excep-tional fibers of these manifolds are the same. This observation holds also for all the other pairs of complexity 11: If the calculated Turaev–Viro in-variants do not distinguish Seifert manifolds of the indicated type, then the homology and the first invariants of the exceptional fibers of these manifolds coincide. However, Seifert manifolds $(S^2, (2,1), (7,3), (13,-5))$ and $(S^2, (3,1), (3,2), (13,-2))$ of complexities 10 and 11 also have the same homol-ogy group Z_{99} and the same values of Turaev–Viro invariants up through order 14 while manifolds $(S^2, (2,1), (5,2), (5,-8))$ and $(S^2, (2,1), (5,1), (5,-7))$ of complexities 7 and 8 having $H_1 = Z_{35}$ can be distinguished by invariants of order 5.

Let us describe the results of the corresponding computer experiment for manifolds of complexity 12. Here we found eight new pairs of different graph manifolds having the same homology groups and the same values of all Turaev–Viro invariants. First four pairs have the same structure as above. The only difference is that they are obtained by gluing boundary tori of Seifert manifolds fibered not over annuli, but over punctured Möbius bands. Each of the manifolds of the next two pairs is composed from a Seifert manifold fibered over the disc with two exceptional fibers and the direct product $N^2 \times S^1$, where N^2 is a disc with two holes. The reason why all Turaev–Viro invari-ants of paired manifolds coincide, it the same: The matrix labels of the loop edges of the corresponding reduced molecules differ by reversing signs of all elements.

We found also two pairs of manifolds with isomorphic homology groups such that they have the same values of Turaev–Viro invariants up to order 16, but this fact cannot be explained by reversing signs of gluing matrices. Each

Table 7.3. Comparison of Turaev–Viro invariants

k	3	4	5	6	7	8	9	10–12
%	9	22	59	4	5	0.2	0.06	<0.005

of those four manifolds is composed from two Seifert manifolds fibered over a disc with two exceptional fibers. Their types are the following:
$(D^2, (2,1), (3,-8)) \cup (D^2, (2,1), (3,-2))$, $(D^2, (2,1), (3,1)) \cup (D^2, (2,1), (3,7))$
for the first pair and
$(D^2, (2,1), (3,-5)) \cup (D^2, (2,1), (3,-5))$, $(D^2, (2,1), (3,4)) \cup (D^2, (2,1), (3,4))$
for the second one.

The gluing matrix for all 4-manifolds is $\begin{pmatrix} 1 & 3 \\ 0 & -1 \end{pmatrix}$, but the homology groups are different: Z_{123} in the first case and Z_{321} in the second. We proved that the manifolds are distinct, by using the fact that the JSJ-decomposition is unique up to isotopy (Theorem 6.4.31). This is the only case when the manual work was necessary.

Below we present statistical information about utility of Turaev–Viro invariants of various orders. Tarkaev made an interesting experimental observation: If 3-manifolds of complexity $c \leq 12$ can be distinguished by Turaev–Viro invariants of order 14 or less, then they can be distinguished by invariants of order $\leq c$.

From 10^8 pairs only 34,8602 cannot be distinguished by homology. Among them 32% are distinguished only by Turaev–Viro invariants, 8% only by the ε-invariant, and 60% by both of them. Table 7.3 shows the percent of pairs of the tabulated manifolds, which have isomorphic homology groups and can be distinguished by invariants of order k, $3 \leq k \leq 12$, while all their invariants of orders $< k$ coincide.

Five pairs of manifolds have been distinguished by invariants of order 10, ten pairs by invariants of order 11 and only two pairs required invariants of order 12.

7.6 Other Applications of the 3-Manifold Recognizer

7.6.1 Enumeration of Heegaard Diagrams of Genus 2

Let $(F; u_1, u_2; v_1, v_2)$ be a genus 2 Heegaard diagram of a closed orientable 3-manifold M. Here F is a closed surface which decomposes M into two handlebodies H_1, H_2 of genus 2, u_1, u_2 are meridians of H_1, and v_1, v_2 are meridians of H_2. We will always assume that all crossing points of the meridians are transversal and that the diagram is normalized (the latter means that among the regions into which the meridians split F there are no biangles). The total number of those crossing points is called the *Heegaard complexity* of the diagram.

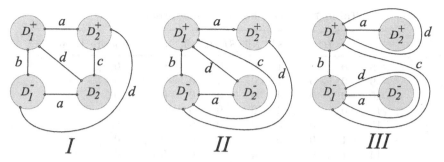

Fig. 7.33. Three types of genus 2 Heegaard diagrams

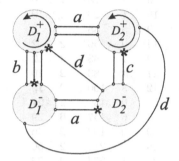

Fig. 7.34. 6-Tuple $(2, 3, 2, 1, 2, 4)$ represents a Heegaard diagram of Seifert manifold $(S^2, (2,1), (3,1), (3,-1))$. The gluing maps φ_1, φ_2 take stars to stars.

Let us cut F along u_1, u_2. We obtain a sphere with four holes $D_1^{\pm}, D_2^{\pm}$ which are conveniently interpreted as distinguished discs on the sphere. The meridians v_1, v_2 will then be cut into arcs which join the holes. We agree to depict k parallel arcs as one arc marked by the number $k \geq 0$.

It is well known (see [29, 31, 48]) that the set of all genus 2 Heegaard diagrams can be decomposed into three types shown in Fig. 7.33.

Since type III diagrams are definitely nonminimal (they admit waves [29]) and type II diagrams appear only episodically, we will mainly consider diagrams of type I. Each such diagram can be determined by a 6-tuple (a, b, c, d, e, f), where a, b, c, d are as above and e, f determine the gluing maps $\varphi_i \colon \partial D_i^- \to \partial D_i^+, i = 1, 2$. In order to give exact descriptions of φ_i, we introduce topological symmetries $s_i \colon \partial D_i^- \to \partial D_i^+$ and topological rotations $R_i \colon \partial D_i^+ \to \partial D_i^+$ by the following rules:

1. s_i takes $\partial D_i^- \cap (v_1 \cup v_2)$ to $\partial D_i^+ \cap (v_1 \cup v_2)$ such that the endpoint of each b-arc (respectively, c-arc) is taken to the other endpoint of the same arc.
2. R_i shifts each point of $D_i^- \cap (v_1 \cup v_2)$ to the next point of $D_i^+ \cap (v_1 \cup v_2)$ (here we assume that the orientation of ∂D_i^+ is chosen so that R_i takes the last endpoint of d-arcs to the first endpoint of a-arcs). See Fig. 7.34.

Now we define φ_1 and φ_2 as follows: $\varphi_1 = R_1^e s_1$ and $\varphi_2 = R_2^f s_2$.

Table 7.4. Manifolds of Heegaard complexity ≤ 32

c	E^3	$H^2 \times R$	S^3	Nil	$\widetilde{SL_2R}$	Sol	H^3	Composite	Total
8	0	0	1	0	0	0	0	0	1
9	0	0	1	0	0	0	0	0	1
10	0	0	4	0	0	0	0	0	4
11	0	0	4	0	0	0	0	0	4
12	2	0	5	0	0	0	0	0	7
13	1	0	9	1	1	0	0	0	12
14	1	0	8	4	4	0	0	0	17
15	0	0	11	2	12	0	0	0	25
16	0	0	6	4	18	1	0	1	30
17	0	0	16	1	26	2	0	1	46
18	0	1	9	5	38	2	2	4	61
19	0	0	14	0	56	2	2	4	78
20	0	1	12	1	63	4	2	9	92
21	0	1	21	1	83	2	10	14	132
22	0	0	12	3	100	2	15	16	148
23	0	1	24	0	136	2	21	19	203
24	0	3	13	6	147	6	29	20	224
25	0	0	29	0	193	2	40	42	306
26	0	1	19	1	211	2	56	46	336
27	0	0	32	1	274	2	79	56	444
28	0	2	19	3	283	6	87	62	462
29	0	1	42	0	363	2	131	95	634
30	0	4	22	4	380	2	162	91	665
31	0	1	39	0	480	2	216	131	869
32	0	1	27	2	485	8	238	135	896
Total	4	17	399	39	3,353	49	1,090	746	5,697

Let us call a 6-tuple *admissible*, if it represents a Heegaard diagram. Theorem 7.6.1 summarizes the results of enumerating admissible 6-tuples up to Heegaard complexity 32 and recognizing the corresponding 3-manifolds. Of course, before starting the recognition process, computer tried to simplify 6-tuples by different moves, which modify the tuples, but preserve the corresponding Heegaard decompositions.

Theorem 7.6.1. *The numbers of closed orientable irreducible genus 2 3-manifolds of Heegaard complexity ≤ 32 are given in Table 7.4.*

In Fig. 7.35 the growth of the number of manifolds of diverse types is shown graphically. We see that the growth within the given range is polynomial (at most cubic).

Let us list a few examples of 6-tuples, which represent manifolds possessing different geometries. For each geometry we show one of the manifolds having the minimal Heegaard complexity. We also show two additional manifold. The first manifold is the minimal composite manifold, which can be obtained from

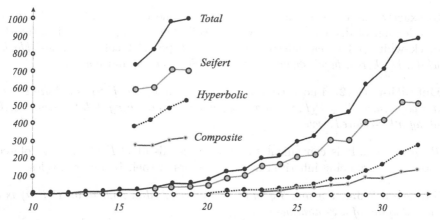

Fig. 7.35. Growth of the number of genus 2 manifolds

Table 7.5. Examples of genus 2-manifolds represented by 6-tuples

6-Tuple	Manifold	Homology	Geometry
2 1 1 1 2 3	$(S^2, (2,-1), (2,1), (2,1))$	$Z_2 \oplus Z_2$	S^3
2 2 2 2 1 1	$(S^2, (2,-1), (4,1), (4,1))$	$Z_2 \oplus Z$	E^3
2 3 2 2 1 1	$(S^2, (2,-1), (4,1), (5,1))$	Z_2	$\widetilde{SL_2R}$
2 3 2 2 2 1	$(S^3, (3,1), (3,1), (3,1)))$	$Z_3 \oplus Z_3$	Nil
4 2 1 3 8 2	$T^2 \times I / \begin{pmatrix} -2 & -1 \\ -1 & -1 \end{pmatrix}$	$Z_5 \oplus Z$	Sol
3 5 3 2 3 4	$(RP^2, (3,1), (3,-1))$	$Z_6 \oplus Z_6$	$H^2 \times R$
4 3 3 2 3 3	$(Q_1)_{(2,-3)}, V = 0.9427$	$Z_5 \oplus Z_5$	H^3
4 5 1 3 6 1	$(S^1 \times I, (2,1))/ \begin{pmatrix} 0 & 1 \\ 1 & 0 \end{pmatrix}$	$Z \oplus Z_3$	Composite
6 1 1 5 10 11	$(S^2, (2,-1), (2,-1), (2,1), (3,1))$	$Z_2 \oplus Z_2$	$\widetilde{SL_2R}$

$(S^1 \times I, (2,1))$ (the Seifert manifold fibered over an annulus with one exceptional fiber $(2,1)$) by identification of boundary tori via the given matrix. The second one is the first genus 2-manifold fibered over S^2 with four exceptional fibers. See Table 7.5

7.6.2 3-Manifolds Represented by Crystallizations with ≤ 32 Vertices

Let Γ be a regular graph of degree four whose edges are colored with four colors such that at each vertex edges of all four colors come together. In this case the pair (Γ, γ), where γ is the coloring, is called a *gem*. As indicated in the name (gem=Graph-Encoded Manifold), there is a close relationship between gems and manifolds.

Let (Γ, γ) be a 3-gem. Select two colors (say a, b) and consider the union $U(a, b)$ of all edges colored with a and b. Since any vertex of Γ is incident to exactly one edge of color a and exactly one edge of color b, $U(a,b)$ is a collection of disjoint circles. We will call them (a,b)-*cycles*. Thus for the black–white–red–green palette we have six types of 2-colored cycles: *black–white, black–red, black–green, white–red, white–green,* and *red–green*.

Definition 7.6.2. *A polyhedral realization of a 3-gem* (Γ, γ) *is a 2-dimensional polyhedron* $P = P(\Gamma, \gamma)$ *obtained from* Γ *by attaching 2-dimensional cells along all 2-colored cycles.*

Remark 7.6.3. It is convenient to think of 2-cells of $P(\Gamma, \gamma)$ to be 2-colored too, the colors being inherited from the colors of their boundary cycles.

Theorem 7.6.4. *For any 3-gem* (Γ, γ) *the polyhedral realization* $P(\Gamma, \gamma)$ *is a special spine of a 3-manifold.*

Proof. First we show that $P = P(\Gamma, \gamma)$ has singularities of the required types, see Definition 1.1.8. Since any color participates in three pairs of distinct colors, each edge e of G belongs to three 2-colored cycles. It follows that e is a triple line of P. Let v be a vertex of Γ. Since v is incident to four edges of distinct colors, exactly six 2-colored cycles pass through it such that every pair of edges belongs to exactly one cycle. It follows that v has a neighborhood of the required type. We conclude that P is a special polyhedron.

Let us show that P is a spine of a 3-manifold $M = M(\Gamma, \gamma)$ with boundary. By Theorem 1.1.20, it suffices to prove that the boundary curve δ of every 2-cell of P has the trivial normal bundle. Denote by a, b the colors of δ and choose a third color c from the given palette. Then a trivialization of the normal bundle of δ is given by the following rule: The normal vector must be directed inward the (a,c)- or (b,c)-colored cell provided that the initial point of the vector belongs to an a-colored or a b-colored edge of δ, respectively.   □

The boundary of $M = M(\Gamma, \gamma)$ always consists of four nonempty surfaces, possibly disconnected. If ∂M consists of four spheres, then the pair (Γ, γ) is called a *crystallization* of the closed manifold $\bar{M}$ obtained from M by filling these spheres by balls. It is known that every closed orientable 3-manifold admits a crystallization. See the paper [28], which is a good survey of the crystallization theory, including crystallizations of higher-dimensional manifolds. See also [19,24] for relation between the complexity of a 3-manifold M and the number of vertices of a minimal crystallization of M.

Theorem 7.6.5 summarizes the results of computer classification of closed orientable irreducible 3-manifolds which can be represented by crystallizations with at most 32 vertices.

Theorem 7.6.5. *The numbers of closed orientable irreducible 3-manifolds represented by crystallizations with* v *vertices are given in Table 7.6:*

The first row of the table was obtained by Lins [70]. However, Lins identified 17 of 67 manifolds only hypothetically, by means of their fundamental

Table 7.6. Manifolds represented by crystallizations with ≤ 32 vertices

v	E^3	$H^2 \times R$	S^3	Nil	SL_2R	Sol	H^3	Composite	Total
$v \leq 26$	6	0	44	8	7	2	0	0	67
$v = 30$	0	0	10	4	13	4	3	7	41
$v = 32$	0	2	38	4	34	10	5	17	110
Total	6	2	92	16	54	16	8	24	218

groups. The next step was made by Casali. She confirmed the results of Lins and classified all manifolds admitting crystallizations with ≤ 30 vertices [25]. At last, Tarkaev and Fominykh extended this classification to manifolds represented by crystallizations with ≤ 32 vertices. The computer program for enumeration of crystallizations was based on the same principles; the decisive progress was obtained thanks to a more refined algorithm and using 3-Manifold Recognizer. For $v \leq 30$ their results coincide with the results of Casali.

7.6.3 Classification of Crystallizations of Genus 2

Definition 7.6.6. *We say that a crystallization (Γ, γ) has genus 2 if the set of colors can be decomposed into two pairs a, b and c, d such that there are exactly three cycles colored by a, b and exactly three cycles colored by c, d.*

In order to explain the terminology, we note that the union F of all 2-cells of P glued along the remaining ac-, ad-, bc-, and bd-cycles form a closed surface of genus 2. This surface decomposes M into two handlebodies. The ab- and cd-colored cells are meridional discs of these handlebodies and decompose each of them into two balls. Therefore, it is not surprising that crystallizations of genus 2 are very close to Heegaard diagrams of the same genus.

Recall that a genus 2 Heegaard diagram is a collection $(F; u_1, u_2; v_1, v_2)$, where F is a closed orientable surface of genus 2 and u_1, u_2 and v_1, v_2 are two pairs of disjoint simple closed curves in F such that the complement to each system is connected (and homeomorphic to a 2-sphere with four holes). We modify that notion by increasing the number of curves in each system.

Definition 7.6.7. *A system u_1, u_2, u_3 of simple closed curves in a closed orientable surface F of genus 2 is called* admissible *if the curves cut F into two connected parts such that each curve separates one part from the other.*

Definition 7.6.8. *An extended Heegaard diagram of genus 2 is a collection $(F; u_1, u_2, u_3; v_1, v_2, v_3)$, where F is a closed orientable surface of genus 2 and (u_1, u_2, u_3), (v_1, v_2, v_3) are admissible systems of disjoint simple closed curves in F.*

Remark 7.6.9. Just as classic Heegaard diagrams, any extended Heegaard diagram $(F; u_1, u_2, u_3; v_1, v_2, v_3)$ determines a genus 2 Heegaard decomposition

$H_1 \cup H_2$ of the corresponding 3-manifold M such that H_1, H_2 are handlebodies and $\partial H_1 = \partial H_2 = F$. The curves $u_i, v_j, 1 \leq i, j \leq 3$ bound meridional discs of H_1, H_2, which split each handlebody $H_k, k = 1, 2$ into two balls B_1^k, B_2^k.

It follows from Definition 7.6.6 that any genus 2 crystallization of a 3-manifold M determines a genus 2 extended Heegaard diagram of the same manifold. Vice versa, any extended Heegaard diagram $(F; u_1, u_2, u_3; v_1, v_2, v_3)$ of $M = H_1 \cup H_2$ determines a genus 2 crystallization (Γ, γ) of M as follows. Let us assign four different colors to the four balls $B_1^k, B_2^k, k = 1, 2$ and take $\Gamma = \cup_i u_i \cup_j v_j$. Then each edge e of Γ is in the boundary of exactly three distinct balls of three different colors. We color e with the remaining fourth color.

It is known that all closed orientable manifolds of genus 2 can be represented by the so-called *symmetric* crystallizations that are encoded by numerical 6-tuples of the form $(n_1, n_2, n_3, k_1, k_2, k_3)$, (see [8]). We prefer to describe them in the language of extended Heegaard diagrams.

Let B be a 3-ball. By a *triangular pattern* on the sphere ∂B we mean a family D_1, D_2, D_3 of coherently oriented disjoint discs in ∂B together with a family L of arcs joining the discs. The arcs must be disjoint, and the ends of each arc must belong to the boundaries of distinct discs.

We denote by n_i the numbers of arcs joining ∂D_j to ∂D_k, where (i, j, k) is a cyclic permutation of the numbers $(1, 2, 3)$. Then the total number of arcs in L is equal to $n_1 + n_2 + n_3$. By a *primitive topological rotation* of a disc D_i we mean an orientation-preserving homeomorphism $R_i : D_i \to D_i$ taking each end of an arc in $L \cap \partial D_i$ to an end of the next arc (with respect to the positive orientation of the circle ∂D_i).

We construct a 6-parameter family of extended diagrams of genus 2 as follows. Let $(n_1, n_2, n_3, k_1, k_2, k_3)$ be an arbitrary 6-tuple of non-negative integers. We take two balls B and B' with the same triangular (n_1, n_2, n_3)-patterns located symmetrically with respect to some plane α, (see Fig. 7.36).

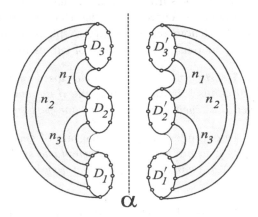

Fig. 7.36. Symmetric triangular (n_1, n_2, n_3)-patterns for $n_1 = 3$, $n_2 = 5$, $n_3 = 4$

We denote by s the symmetry with respect to this plane. Let us glue together the balls B and B' by identifying each disc D_i with the symmetric disc D_i', $1 \leq i \leq 3$. The gluing homeomorphism $h_i \colon D_i \to D_i'$ is given by rule $h_i = sR_i^{k_i}$, where s is the symmetry and $R_i^{k_i}$ is the k_ith power of the corresponding primitive topological rotation. The gluing results in a genus 2 handlebody H with a system $\lambda_1 \cup \lambda_2 \cup \ldots \cup \lambda_m$ of several simple closed curves on ∂H glued together from arcs in L and L'. We denote by $\mu_1 \cup \mu_2 \cup \mu_3$ the circles in ∂H obtained by identifying the boundaries of the discs D_i and D_i'.

Definition 7.6.10. *A numerical 6-tuple* $(n_1, n_2, n_3, k_1, k_2, k_3)$ *is said to be regular, if*

1. *$m = 3$, that is, we have three curves $\lambda_1, \lambda_2, \lambda_3$, and they are admissible.*
2. *After cutting ∂H along $\lambda_1 \cup \lambda_2 \cup \lambda_3$, the arcs into which the meridians μ_1, μ_2, μ_3 are cut form a triangular pattern.*

It follows from this definition that if a 6-tuple $(n_1, n_2, n_3, k_1, k_2, k_3)$ is regular, then the triple $(F; \mu_1, \mu_2, \mu_3; \lambda_1, \lambda_2, \lambda_3)$ is a genus 2 extended Heegaard diagram. We denote this diagram and the corresponding 3-manifold by $D(n_1, n_2, n_3, k_1, k_2, k_3)$ and $M(n_1, n_2, n_3, k_1, k_2, k_3)$, respectively. It is convenient to measure the complexity of the diagram $D(n_1, n_2, n_3, k_1, k_2, k_3)$ by the total number $N = 2(n_1 + n_2 + n_3)$ of crossing points of the meridians $\mu_i, \lambda_j, 1 \leq i, j \leq 3$.

Theorem 7.6.11. *Each genus 2 Heegaard decomposition $M = H_1 \cup H_2$ of a closed orientable 3-manifold M can be represented by a diagram of the form $D(n_1, n_2, n_3, k_1, k_2, k_3)$.*

Proof. Let us chose admissible meridians μ_1, μ_2, μ_3 for H_1 and admissible meridians $\lambda_1, \lambda_2, \lambda_3$ for H_2 such that they decompose $F = \partial H_1 = \partial H_2$ into regions without biangles. Let us cut F along $\mu_1 \cup \mu_2 \cup \mu_3$. Suppose that the pattern formed from resulting arcs of the curves in $\lambda_1, \lambda_2, \lambda_3$ is not triangular. Then there is an arc $\ell \subset \lambda_1 \cup \lambda_2 \cup \lambda_3$ such that both endpoints of ℓ lie in a meridian μ_i and ℓ has no other crossing point with $\mu_1 \cup \mu_2 \cup \mu_3$. Then the extended diagram $(F; \mu_1, \mu_2, \mu_3; \lambda_1, \lambda_2, \lambda_3)$ can be simplified by replacing μ_i with the connected sum along ℓ of the other two meridians μ_j, μ_k. Analogous move simplifies the diagram in the dual case in which the meridians μ_1, μ_2, μ_3 and $\lambda_1, \lambda_2, \lambda_3$ exchange their roles. Making these simplifications as long as possible, we achieve the situation in which both patterns are triangular. $\square$

We present the result of a computer enumeration of regular 6-tuples and the recognition of the corresponding manifolds $M(n_1, n_2, n_3, k_1, k_2, k_3)$ with $N = 2(n_1 + n_2 + n_3) \leq 64$.

Theorem 7.6.12. *The numbers of closed orientable irreducible 3-manifolds represented by genus 2 extended Heegaard diagrams with $N \leq 64$ crossing points are given in Table 7.7:*

Table 7.7. Manifolds of Heegaard genus 2

$N/2$	E^3	$H^2 \times R$	S^3	Nil	SL_2R	Sol	H^3	Composite	Total
9	0	0	1	0	0	0	0	0	1
10	0	0	0	0	0	0	0	0	0
11	0	0	1	0	0	0	0	0	1
12	0	0	3	0	0	0	0	0	3
13	0	0	1	0	0	0	0	0	1
14	0	0	4	0	0	0	0	0	4
15	1	0	2	0	0	0	0	0	3
16	1	0	5	1	0	0	0	0	7
17	2	0	3	1	0	0	0	0	6
18	0	0	8	4	3	0	0	0	15
19	0	0	1	0	3	0	0	0	4
20	0	0	7	4	10	2	0	0	23
21	0	0	3	0	4	0	1	2	10
22	0	0	8	1	20	2	0	0	31
23	0	1	3	0	4	0	3	4	15
24	0	0	8	3	37	3	1	0	52
25	0	0	4	2	8	0	5	6	25
26	0	0	9	1	47	3	0	6	66
27	0	0	1	0	11	0	10	8	30
28	0	1	8	1	70	2	4	5	91
29	0	0	5	0	15	0	23	11	54
30	0	0	15	1	85	2	8	4	115
31	0	1	4	0	14	0	33	18	70
32	0	1	10	1	120	3	15	13	163
Total	4	4	114	20	451	17	103	77	790

This result is stronger than the Bandieri–Gagliardi–Ricci Theorem [8] on existence of exactly 26 closed irreducible manifolds of genus 2 which admit a crystallization with $N \leq 34$ vertices. See also [86] for $N \leq 48$.

7.6.4 Recognition of Knots and Unknots

Let K be a knot in S^3. Then K is trivial if and only if its complementary space $S^3 \setminus \mathrm{Int}\, N(K)$ is a solid torus. Therefore, the 3-Manifold Recognizer can be used for recognition of the unknot. The result of the corresponding computer experiment was positive: The Recognizer successfully recognized that the diagrams shown in Figs. 7.37–7.39 represent the unknot.

Fig. 7.37. The Unknot of Gordon

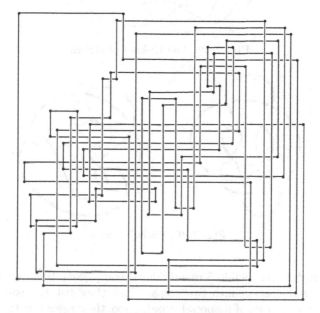

Fig. 7.38. The Unknot of Dynnikov

Figure 7.40 shows the famous Perko pair of knots, listed as distinct knots in many knot tables since the nineteenth century, until Kenneth Perko showed in 1974 that they were in fact the same knot. Less than in a minute Recognizer showed that their complements are homeomorphic (this implies that the knots are equivalent [37]).

7.7 Two-Step Enumeration of 3-Manifolds

We present here results of Martelli and Petronio [74] on enumeration of 3-manifolds. The main idea which enabled them to extend the tabulation of

Fig. 7.39. The Unknot of Haken

Fig. 7.40. Perko's knots

closed orientable irreducible 3-manifolds to complexity 9 consists in decomposing the enumeration into two steps. First they list the so-called prime bricks (building blocks of a special type). Next they assemble the bricks into 3-manifolds. A remarkable experimental discovery is that the number of different bricks needed for assembling all closed orientable irreducible 3-manifolds of complexity ≤ 9 is surprisingly small. Some of those bricks can be seen already at the level of complexity 6, where not only manifolds, but also minimal special spines had been enumerated [83, 84]. To describe the bricks, we need a relative version of the complexity theory. This version possess all important properties of the absolute complexity theory, and the upcoming Sect. 7.7.1 is devoted to establishing them.

7.7.1 Relative Spines and Relative Complexity

Recall that a graph $\Gamma \subset F$ is a spine of a closed connected surface F if $F \setminus \Gamma$ is an open disc. If F is closed but not connected, then a spine of F is the union of spines of its connected components. A spine of F is special, if it can be presented as a graph such that all its vertices have valence 3. Note that any special spine of any surface has a nonzero even number of vertices and that S^2 has no special spines.

Definition 7.7.1. *Let (M, Γ) be a 3-manifold with boundary pattern such that Γ is a spine of ∂M. Then a subpolyhedron $P \subset M$ is called a* relative spine *of (M, Γ), if the following holds:*

1. *P is a spine of $M \setminus Int\, D^3$, where D^3 is a 3-ball in M*
2. *$P \supset \partial M$*
3. *$\partial M \cap Cl(P \setminus \partial M) \subset \Gamma$*

A relative spine is almost simple, simple, *or* special *if it is a polyhedron of the corresponding type.*

Obviously, if M is closed, then any relative spine of $(M, \emptyset)$ is a spine of M in the sense of Definition 1.1.4. Theorem 7.7.2 can be easily proved by the same method as Theorem 1.1.13.

Theorem 7.7.2. *Let (M, Γ) be a 3-manifold with boundary pattern such that Γ is a special spine of ∂M. Then (M, Γ) has a special relative spine.*

Just as in the absolute case, any 3-manifold (M, Γ) can be reconstructed from any of its special relative spines in a unique way. Let us show this.

Theorem 7.7.3. *Let (M_i, Γ_i) be 3-manifolds with boundary pattern such that Γ_i is a special spine of $\partial M_i, i = 1, 2$. Suppose that (M_1, Γ_1) and (M_2, Γ_2) have homeomorphic special spines. Then (M_1, Γ_1) and (M_2, Γ_2) are homeomorphic (as pairs).*

Proof. Let P_1 and P_2 be homeomorphic special relative spines of (M_1, Γ_1) and (M_2, Γ_2). By Theorem 1.1.17 and Definition 7.7.1, $M_1 \setminus Int\, D^3$ is homeomorphic to $M_2 \setminus Int\, D^3$. Since P_1, P_2 are special and S^2 has no special spine, the boundaries of the removed 3-balls are the only spherical components of $\partial M_i \setminus Int\, D^3$. It follows that M_1 is homeomorphic to M_2. It remains to note that for $i = 1, 2$ the pattern Γ_i coincides with $\partial M_i \cap Cl(P_i \setminus \partial M_i)$ and thus is also determined by P_i. $\square$

We measure the complexity of an almost simple relative spine P of (M, Γ) by the number $c_{int}(P)$ of its *interior* true vertices, which lie strictly inside the manifold. In other words, counting $c_{int}(P)$, we forget about true vertices on ∂M. Each such vertex must be a vertex of Γ, so $c_{int}(P) \leq c(P) - 2n$, where n is the number of boundary components of M (each of them contains at least two vertices of Γ).

Definition 7.7.4. *Let (M, Γ) be a 3-manifold with boundary pattern such that Γ is a spine of ∂M. Then the* complexity $c(M, \Gamma)$ *of (M, Γ) is equal to k if*

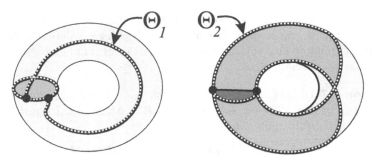

Fig. 7.41. Two relative spines B_1, B_2 of solid tori with different boundary patterns

(M, Γ) *possesses an almost simple relative spine with k interior true vertices and has no almost simple relative spines with a smaller number of interior true vertices. In other words, $c(M, \Gamma) = \min_P c_{int}(P)$, where the minimum is taken over all almost simple relative spines of (M, Γ).*

If M is closed, then this definition is equivalent to Definition 2.1.4. For nonclosed manifolds with one boundary component the absolute complexity $c(M)$ is, as a rule, smaller than $c(M, \Gamma)$. Indeed, puncturing the boundary component and collapsing the rest, we convert any almost simple relative spine P of (M, Γ) into an almost simple spine of M. This operation never creates new interior true vertices and quite often destroys some of them.

Example 7.7.5. Let V be a solid torus with a meridian μ. Consider two theta-curves Θ_1, Θ_2 on ∂V. Θ_1 consists of μ and an embedded arc that joins two different points of μ and approaches μ from different sides, Θ_2 is obtained by adding a subarc of μ to an embedded circle that intersects μ twice in the same direction. Then $c(V, \Theta_i) = 0$ for both theta-curves. Almost simple spines B_i of (V, Θ_i) are shown in Fig. 7.41. B_1 consists of ∂M and a meridional disc bounded by μ. It contains no true vertices at all. B_2 is the union of ∂V and a Möbius triplet inside V. It has two true vertices, but they are on the boundary of V and thus make no contribution to the complexity.

Definition 7.7.6. *An almost simple relative spine P of a 3-manifold (M, Γ) is called* minimal, *if it has exactly $c(M, \Gamma)$ interior true vertices.*

Denote by $\mathcal{T}$ the class of all manifolds (M, Γ) such that each component T of ∂M is a torus and $T \cap \Gamma$ is a nonseparating theta-curve. Closed manifolds are included.

Theorem 7.7.7. *Suppose $(M, \Gamma) \in \mathcal{T}$ is an irreducible 3-manifold such that (M, Γ) is neither one of the manifolds $(S^3, \emptyset), (RP^3, \emptyset), (L_{3,1}, \emptyset)$, nor the manifold (V, Θ_1) described in Example 7.7.5. Then (M, Γ) has a special relative spine which is minimal, i.e., has $c(M, \Gamma)$ interior true vertices.*

Proof. Let P be an almost simple relative spine of (M, Γ) which has $c(M, \Gamma)$ interior true vertices and thus is minimal. We can assume that P cannot be collapsed onto a smaller subpolyhedron. Let us prove that then P is special. If M is closed, then the conclusion follows from Theorem 2.2.4. Suppose that $\partial M \neq \emptyset$. If P has a 1-dimensional part or a noncell 2-component inside M, then one can simplify P by the same tricks as in the proof of Theorem 2.2.4. This contradicts the minimality of P.

Suppose that P has a 2-component α which is not a cell and is contained in ∂M. Since P is a spine of $M \setminus \text{Int } D^3$, then M is boundary reducible and hence is a solid torus. It follows that $\partial M \cap \text{Cl}(P \setminus \partial M)$ is a circle in Θ and thus $\text{Cl}(P \setminus \partial M)$ is a meridional disc. Therefore, $(M, \Gamma) = (V, \Theta_1)$ and we get a contradiction again.   □

Let us establish further properties of the relative complexity $c(M, \Gamma)$. The additivity property $c(M_1 \# M_2, \Gamma_1 \cup \Gamma_2) = c(M_1, \Gamma_1) + c(M_1, \Gamma_1)$ with respect to connected sums is true (since cutting M along a normal sphere creates no new interior vertices, a similar proof as for the absolute case works). For manifolds from the class $\mathcal{T}$ boundary connected sums are not defined, since we can get boundary components of genus ≥ 2. To prove the finiteness property we need the following lemma.

Lemma 7.7.8. *Let P be a special relative spine of an irreducible 3-manifold* $(M, \Gamma) \in \mathcal{T}$ *with nonempty boundary. Suppose that an edge e of P is not contained in ∂M, but joins two true vertices $v_1, v_2 \subset \partial M$ of P. Then either ∂M is a torus or M is $T^2 \times I$.*

Proof. CASE 1. Suppose that v_1, v_2 lie on the same boundary torus. Then they are vertices of the same theta-curve $\Theta \subset \partial M$. Note that $G = \Theta \cup e$ is a subgraph of the singular graph SP of P, G contains exactly two true vertices v_1, v_2 of P, and these vertices have valence 4 in G. It follows that $G = SP$. We can conclude that SP contains only one theta-curve and that ∂M is a torus.

CASE 2. Suppose that v_1, v_2 lie in different boundary tori T_1, T_2. Denote by C a unique 2-component of P contained in T_1. Since P is a spine of once punctured M, then $P \setminus C$ is a spine of M. Let us collapse $P \setminus C$ for as long as possible. Since the collapse preserves T_2 and destroys all 2-cells adjacent to e, we get an almost simple spine P' of M so that:

(i) P' contains T_2
(ii) P' has no triple points on T_2

Suppose that P' contains a 1-dimensional part. Then there exists a proper disc $D \subset M$ intersecting ∂M transversally at exactly one point. Since M is irreducible and not a solid torus, D cuts out a 3-ball from M. This means that cutting P' at the point $D \cap P'$ and collapsing, we get another spine of M which contains T_2 and has a smaller 1-dimensional part. Doing so for as long as possible, we get a spine of M which is simple and still possesses

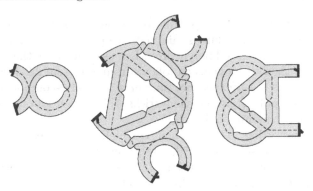

Fig. 7.42. Three special relative spines presented by regular neighborhoods of their interior singular graphs

properties (i), (ii) above. It follows that this spine coincides with T_2 and hence $M = T^2 \times I$. □

Proposition 7.7.9. *For any integer* k, *the class* $\mathcal{T}$ *contains only a finite number of distinct compact irreducible 3-manifolds* (M, Γ) *of relative complexity* k.

Proof. Let (M, Γ) be an irreducible 3-manifold of complexity k having $n > 0$ boundary tori. Suppose (M, Γ) is other than the exceptional manifold (V, Θ_1) considered in Example 7.7.5. Then (M, Γ) has a special relative spine P with $k + 2n$ true vertices: k vertices inside M and $2n$ vertices on ∂M (they correspond to the vertices of Γ). We claim that $n \leq 2k + 2$. The conclusion of the proposition follows from the claim, since then M has a relative special spine with $\leq 5k + 4$ true vertices, and we can apply Theorem 2.2.5.

Let us prove the claim. Suppose that P has an edge that joins two true vertices $v_1, v_2 \subset \partial M$ of P and is not contained in ∂M. Then the claim follows from Lemma 7.7.8. If P has no such edges, then any interior edge of P with an endpoint on ∂M has the other endpoint at an interior true vertex of P. Since any boundary torus of M contains two true vertices and every interior vertex has valence 4, we get $n \leq 2k$. □

For a graphical representation of spines in the absolute case, we used regular neighborhoods of singular graphs. The same method works for the relative case. Let P be a special relative spine of a 3-manifold (M, Γ) such that the boundary curve of every 2-component $\alpha \not\subset \partial M$ of P meets ∂M not more than once. Denote by SP_{int} its *interior singular graph* which consists of all true vertices and triples edges of P not contained in ∂M. Then P can be presented by a regular neighborhood $N = N(SP_{int})$ of SP_{int} in $\mathrm{Cl}(P \setminus \partial M)$. See Fig. 7.42, where we show a few relative spines. Free ends of N (each of them is marked by a fat triode, a wedge of three segments) correspond to the vertices of Γ. It is very easy to determine, which triodes lie in the same theta-curve: they must be joined by three boundary arcs.

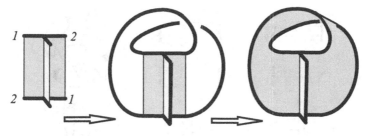

Fig. 7.43. Reconstruction of B_2

To reconstruct P, it suffices to do the following:

1. Locate all closed curves $C_1, \ldots, C_m$ and all theta-curves $\Theta_1, \ldots, \Theta_n$ in ∂N.
2. Attach m disjoint discs $D_1, \ldots, D_m$ to N such that $\partial D_i = C_i, 1 \le i \le m$.
3. Attach n other disjoint discs $D_1', \ldots, D_n'$ to N such that the boundary curve of each D_i' is contained in Θ_i and $D_i' \cup \Theta_i$ is a torus.

Suppose now that the above condition on P does not hold, i.e., that the boundary curves of some interior 2-components of P meet ∂M more than once. Then $N(SP_{int})$ does not determine P anymore, since we lack information on which pairs of endpoints of the fat triodes belong to the same edges of Γ. The endpoints are joined by arcs of ∂N. Some of these arcs show us right connections and thus can be considered as edges of Γ. In all other cases we enumerate the remaining edges of Γ and mark the remaining endpoints of the triodes by the corresponding numbers. Then the marked neighborhood N does determine P. Indeed, to get P, we reconstruct Γ as follows. First, we extend the union of triodes by adding those arcs of ∂N which show right connections and attaching new arcs joining endpoints with the same numbers. Next, we attach to $N \cup \Gamma$ disjoint discs $D_1, \ldots, D_m$ so that their boundaries run along $\partial N \cup \Gamma$ without passing through the vertices of Γ. Finally, we attach n disjoint discs $D_1', \ldots, D_n'$ to recover the boundary tori. See Fig. 7.43, where we illustrate the reconstruction process for the simplest relative special spine B_2 described in Example 7.7.5.

Graphical presentations of pines B_0, B_3, B_4 are shown in Fig. 7.44. First two spines B_0 and B_3 are special relative spines of $M_0 = M_3 = T^2 \times I$, the last one is a special relative spine of $M_4 = N^2 \times S^1$. These spines are important for the sequel, so we also present them by showing the fragments $\mathrm{Cl}(B_i \setminus \partial M_i)$ of $B_i, i = 0, 3, 4$. To get actual spines, we have to identify the arrows on the top with the corresponding arrows on the bottom, and add the boundary tori.

7.7.2 Assembling

Let $(M_i, \Gamma_i), i = 1, 2$, be two manifolds from $\mathcal{T}$ with nonempty boundaries and relative spines P_i. Choose two tori $T_1 \subset \partial M_1, T_2 \subset \partial M_2$ and a

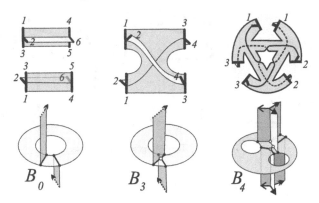

Fig. 7.44. Two relative spines of $T^2 \times I$ and the simplest relative spine of $N^2 \times S^1$, where T^2 is a torus and N^2 is a twice punctured disc

homeomorphism $\varphi \colon T_1 \to T_2$ taking theta-curve $\Theta_1 = \Gamma_1 \cap T_1$ to theta-curve $\Theta_2 = \Gamma_2 \cap T_2$. We can then construct a new manifold $(W, \Delta) \subset \mathcal{T}$, where $W = M_1 \cup_\varphi M_2$, and $\Delta = (\Gamma_1 \setminus \Theta_1) \cup (\Gamma_2 \setminus \Theta_2)$. Its relative spine P can be obtained by gluing P_1 and P_2 along φ and removing the 2-component of $P_1 \cup_\varphi P_2$ which is obtained by identifying $T_1 \setminus \Theta_1$ with $T_2 \setminus \Theta_2$.

Definition 7.7.10. *We say that the manifold $(W, \Delta) \in \mathcal{T}$ described above is obtained by* assembling *of (M_1, Γ_1) and (M_2, Γ_2). The same terminology is used for spines: P is obtained by assembling of P_1 and P_2. The assembling is* nontrivial, *if the spines are different from B_0.*

It follows from the definition that two relative manifolds as well as two relative spines can be assembled in a finite number of inequivalent ways. We point out that the relative complexity of spines is additive with respect to assembling: if P is obtained by assembling of P_1 and P_2, then $c(P, \Delta) = c(P_1, \Gamma_1) + c(P_2, \Gamma_2)$.

It is possible to assemble not only one relative manifold with another, but also a manifold with itself. Let (M, Γ) be a manifold from $\mathcal{T}$ with at least two boundary tori and a special relative spine P. Choose two tori $T_1, T_2 \subset \partial M$ and a homeomorphism $\varphi \colon T_1 \to T_2$ such that Θ_2 and the image $\Theta_1' = \varphi(\Theta_1)$ of Θ_1 intersect each other transversally in two points. This is the minimal number of intersection points of two transversal nonseparating theta-curves. We can then construct a new manifold $(W, \Delta) \subset \mathcal{T}$ with the boundary pattern $\Delta = \Gamma \setminus (\Theta_1 \cup \Theta_2)$ by identifying T_1 with T_2 via φ. Its relative spine Q is also obtained from P by identifying T_1 with T_2 via φ. In contrast to assembling of two relative manifolds, we do not remove the 2-component obtained by identifying $T_1 \setminus \Theta_1$ with $T_2 \setminus \Theta_2$. It is easy to see that Q has six new interior vertices: four of them correspond to the vertices of Θ_1', Θ_2, and two vertices are the intersection points of Θ_1' with Θ_2. It follows that $c(Q, \Delta) = c(P, \Gamma) + 6$.

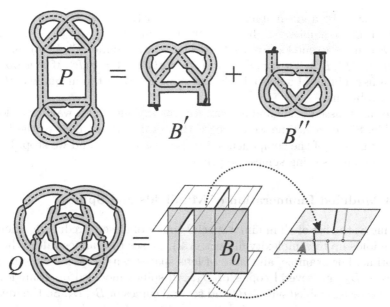

Fig. 7.45. Two relative spines of $T^2 \times I/(-E) = K^2 \tilde{\times} I$ which can be assembled, respectively, self-assembled from simpler special relative spines

As mentioned above, only a finite number of self-assemblings of a given manifold $(M, \Gamma) \in \mathcal{T}$ (or of its special relative spine) are possible. In the sequel we will refer to a combination of assemblings and self-assemblings simply as an assembling.

Example 7.7.11. Looking at the special spine P of the Stallings manifold M with fiber T^2 and the monodromy matrix $-E = \begin{pmatrix} -1 & 0 \\ 0 & -1 \end{pmatrix}$ (Fig. 7.45), one can easily see that it is assembled from two copies B', B'' of a special relative spine B of the thick Klein bottle $K^2 \tilde{\times} I$. It is not surprising, since M is homeomorphic to the double of $K^2 \tilde{\times} I$. Another special spine Q of M is self-assembled from the relative special spine B_0 of $T^2 \times I$. This is also not surprising, since M is a Stallings manifold with fiber T^2.

Definition 7.7.12. *A special relative spine P of a manifold $(M, \Gamma) \in \mathcal{T}$ is called prime, if it cannot be obtained by self-assembling or nontrivial assembling of some other spines of manifolds from $\mathcal{T}$.*

For example, blocks B_0–B_4 are prime while the ones shown in Fig. 7.42 are not.

Theorem 7.7.13. *Any special relative spine can be assembled from prime special relative spines.*

Proof. Let P be a given spine of a 3-manifold $(M, \Gamma) \in \mathcal{T}$. Denote by $\sigma(P)$ the first Betti number of the interior singular graph SP_{int} of P. Suppose that P can be assembled from two special relative spines P', P'' of manifolds $(M', \Gamma'), (M'', \Gamma'')$. Then $\sigma(P') + \sigma(P'') = \sigma(P) - 1 < \sigma(P)$. If P is obtained by self-assembling of a special relative spine P', then the inequality $\sigma(P') < \sigma(P)$ is obvious.

Let us disassemble P and all components arising in this way for as long as possible. Since each time we decrease the total Betti number of the interior singular graphs of the components, the process is finite and we stop. By construction, all resulting spines are prime.   □

7.7.3 Modified Enumeration of Manifolds and Spines

Looking at the list of all minimal special spines of closed irreducible orientable 3-manifolds up to complexity 6 (Sect. A.2), we see that almost all of them are not prime. For example, all spines of lens spaces can be assembled out of two copies of B_2 and several copies of B_3. All spines modeled on triangles with tails can be assembled out of B_4 and several copies of B_2, B_3. So the idea first to list prime spines up to a given complexity and only then assemble out of them other spines looks promising.

But how can we get a list of prime spines? At first glance, this problem seems to be even more difficult than enumeration of all spines: We should not only enumerate spines, but also select among them prime ones. Clearly, we are only interested in minimal spines. The answer is simple. There are several powerful criteria which allow us to recognize that a relative spine P of a 3-manifold (M, Γ) is either not minimal or not prime, and reject it at a very early step of its construction. We present two of them (see [74] for other criteria).

Proposition 7.7.14. *Let P be a special relative spine of $(M, \Gamma) \in \mathcal{T}$. Suppose that the boundary curve l of a 2-cell C of P passes through an interior edge e of P twice. Then P is either not minimal or not prime (or neither minimal nor prime).*

Proof. Suppose that l passes along e in two different directions, i.e., has a counterpass. Then P is not minimal, since we can simplify it by move U_2. Let l pass along e in only one direction. Then there is a circle $m \subset P$ that crosses SP transversally at a point of e. A regular neighborhood $N(m)$ of m in P is a Möbius triplet consisting of a Möbius strip and a disc, see Fig. 7.41, to the right. This means that P can be assembled from B_2 and another special relative spine.   □

Proposition 7.7.15. *Let P be a special relative spine of $(M, \Gamma) \in \mathcal{T}$. Suppose that a pair e_1, e_2 of interior edges of SP disconnects SP. Then P is either not minimal or not prime.*

Proof. Denote by C_1, C_2, C_3 the 2-cells of P adjacent to e_1. By Proposition 7.7.14, we can assume that they are distinct. Since e_1, e_2 disconnect SP, the cells C_1, C_2, C_3 are also adjacent to e_2. Choose points $X_1 \in e_1, X_2 \in e_2$ and join them by three arcs so that their union Θ (which is a theta-curve) is contained in the union of all 3-cells C_1, C_2, C_3, but not contained in the union of any two of them. One can easily construct a proper surface $F \subset M$ such that ∂F is contained in the spherical component of $M \setminus \text{Int } D^3$, $F \cap P = \Theta$, and Θ is a spine of F.

Since M is orientable and F decomposes it, F is also orientable. Thus we have two possibilities: F is a twice punctured disc and F is a punctured torus. In the first case P is not minimal, since one can perform simplifying move U_3 (see Sect. 7.2.5). In the second case we fill up ∂F by a disc $D \subset D^3 \subset M$ and get a torus T that decomposes (M, Γ) into two relative manifolds (M', Γ') and (M'', Γ''). Their special relative spines P', P'' can be obtained by cutting P along Θ and adding two copies of T. By construction, P is assembled from P', P''. Therefore, it is not prime. $\qquad\square$

These criteria significantly facilitate enumeration of prime minimal relative spines.

Let k be an integer. Suppose that we have constructed the set $\mathcal{S}_k$ of all prime minimal relative spines of complexity $\leq k$. Assembling them in all possible ways, we get a list of special spines containing all minimal special spines of closed irreducible orientable 3-manifolds of complexity $\leq k$. This two-step procedure is much faster than a straightforward enumeration of spines.

Moreover, if we are only interested in 3-manifolds, then we can replace $\mathcal{S}_k$ by a much smaller set $\mathcal{S}_k'$. Indeed, any 3-manifold (M, Γ) can have many different minimal relative spines. If all these spines are prime, then we include into $\mathcal{S}_k'$ only one of them. If (M, Γ) has at least one nonprime minimal relative spine, then no spine of (M, Γ) is included into $\mathcal{S}_k'$. Replacing spines from $\mathcal{S}_k'$ by corresponding relative manifolds (called *prime bricks*) and assembling them appropriately, we get a finite list containing all closed irreducible orientable 3-manifolds up to complexity k.

This idea was invented and carried out by Martelli and Petronio for $k = 9$, see [74]. Altogether there are 28 prime bricks of complexity ≤ 9. Exactly 9 of them are nonclosed. Assembling the bricks to closed 3-manifolds and casting off reducible manifolds and duplicates, Martelli and Petronio proved that there exist exactly 436 closed irreducible orientable 3-manifolds of complexity 8 and 1,156 such manifolds of complexity 9. Turaev–Viro invariants were used for proving that all these manifolds are distinct. See Sect. 2.3.1 and [74] for details.

The Turaev–Viro Invariants

8.1 The Turaev–Viro Invariants

These invariants were first described by Turaev and Viro [126]. They possess two important properties. First, just like homology groups, they are easy to calculate. Only the limitations of the computer at hand may cause some difficulties. Second, they are very powerful, especially if used together with the first homology group.

8.1.1 The Construction

We divide the construction of the Turaev–Viro invariants into six steps.

Step 1. Fix an integer $N \geq 1$.

Step 2. Consider the set $\mathcal{C} = \{0, 1, \dots, N - 1\}$ of integers. We will think of them as representing colors. To each integer $i = 0, 1, \dots, N - 1$ assign a complex number w_i called the *weight* of i.

Step 3. Let E be a butterfly, see Fig. 1.4. Recall that it has six wings. We will color the wings by colors from the palette $\mathcal{C}$ in order to get different *colored butterflies*. The butterfly admits exactly N^6 different colorings.

Definition 8.1.1. *Two colored butterflies are called* equivalent *if there exists a color preserving homeomorphism between them.*

The number of different colored butterflies up to equivalence is significantly less than N^6. It is because the butterfly is very symmetric: it inherits all the 24 symmetries of the regular tetrahedron, see Fig. 1.5. It is convenient also to present a colored butterfly by coloring the edges of a regular tetrahedron Δ. The body of the butterfly is the cone over the vertices of Δ while its wings are the cones over corresponding edges and have the same colors.

Step 4. To each colored butterfly, assign a complex number called the *weight* of the butterfly. There arises a problem: how to denote colored butterflies and their weights? Let us call two wings of a butterfly *opposite* if their

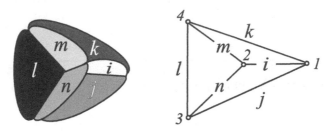

Fig. 8.1. The butterfly and its boundary graph

intersection is the vertex (not an edge). Note that any colored butterfly is determined (up to equivalence) by:

(a) Three pairs $(i, l), (j, m), (k, n)$ of colors that correspond to three pairs of opposite wings.
(b) A triple (i, j, k) of representatives of each pair that correspond to wings having a common edge.

An example of a colored butterfly and its boundary graph are shown in Fig. 8.1.

For typographic convenience, and following some earlier conventions, such a butterfly will be denoted by the (2×3)-matrix

$$\begin{pmatrix} i & j & k \\ l & m & n \end{pmatrix},$$

where the top row gives the colors of three adjacent wings and each column gives colors of opposite pairs of wings. The weight associated with the above butterfly is denoted by

$$\begin{vmatrix} i & j & k \\ l & m & n \end{vmatrix}$$

and called a $(q-6j)$-symbol, for reasons we will not go into here. An interested reader is referred to [125, 126] and references therein.

The $(q - 6j)$-symbol has many symmetries, corresponding to the symmetries manifest in the butterfly. The symmetry group of a butterfly presented as the cone over one-dimensional skeleton $\Delta^{(1)}$ of a regular tetrahedron is isomorphic to the symmetric group S_4 on four elements 1, 2, 3, 4 that correspond to the vertices of $\Delta^{(1)}$. Assume that the edges (1,2), (1,3), (1,4), (3,4), (2,4), and (2,3) have colors i, j, k, l, m, n, respectively. Then the following equalities correspond to generators (the transposition (2,3) and the cyclic permutation (1,2,3,4)) of S_4:

$$\begin{vmatrix} i & j & k \\ l & m & n \end{vmatrix} = \begin{vmatrix} j & i & k \\ m & l & n \end{vmatrix}, \quad \begin{vmatrix} i & j & k \\ l & m & n \end{vmatrix} = \begin{vmatrix} n & m & i \\ k & j & l \end{vmatrix}.$$

Step 5. Let P be a special polyhedron, $V(P)$ the set of its vertices, and $C(P)$ the set of its 2-cells.

Definition 8.1.2. *A coloring of P is a map $\xi : C(P) \to \mathcal{C}$.*

Denote by $Col(P)$ the set of all possible colorings of P. It consists of $N^{\#C(P)}$ elements, where N is the number of colors and $\#C(P)$ is the number of 2-cells in P. To each coloring $\xi \in Col(P)$ assign a *weight* $w(\xi)$ by the rule

$$w(\xi) = \prod_{v \in V(P)} \begin{vmatrix} i & j & k \\ l & m & n \end{vmatrix}_v \prod_{c \in C(P)} w_{\xi(c)}. \tag{8.1}$$

Note that any coloring ξ determines a coloring of a neighborhood of every vertex $v \in V(P)$. It means that in a neighborhood of v we see a colored butterfly

$$\begin{pmatrix} i & j & k \\ l & m & n \end{pmatrix}_v,$$

with the $(q-6j)$-symbol

$$\begin{vmatrix} i & j & k \\ l & m & n \end{vmatrix}_v.$$

Every 2-cell c of P is painted in the color $\xi(c)$ having the weight $w_{\xi(c)}$. Thus the right-hand part of the formula (8.1) is the product of all the symbols and the weights of all used colors (with multiplicity).

Definition 8.1.3. *Let P be a special polyhedron. Then the* weight *of P is given by the formula*

$$w(P) = \sum_{\xi \in Col(P)} w(\xi).$$

Step 6. Certainly, the weight of a special polyhedron P depends heavily on the weights w_i of colors and the values of $(q-6j)$-symbols. We will think of them as being *variables*; thus we have finitely many variables. If we fix their values, we get a well defined invariant of the topological type of P. Now let us try to subject the variables to constraints so that the weight of a special polyhedron will be invariant with respect to T-moves. In order to do that, let us write down the following system of equations:

$$\begin{vmatrix} i & j & k \\ l & m & n \end{vmatrix} \begin{vmatrix} i & j & k \\ l' & m' & n' \end{vmatrix} = \sum_z w_z \begin{vmatrix} i & m & n \\ z & n' & m' \end{vmatrix} \begin{vmatrix} j & l & n \\ z & n' & l' \end{vmatrix} \begin{vmatrix} k & l & m \\ z & m' & l' \end{vmatrix}, \tag{8.2}$$

where $i, j, k, l, m, n, l', m', n'$ run over all elements of the palette $\mathcal{C}$.

The geometrical meaning of the equations is indicated in Fig. 8.2 and explained in the proof of Theorem 8.1.4. We emphasize that the system is universal, i.e., it depends neither on manifolds nor on their spines.

In order to get a feeling of the system, let us estimate the number of variables and the number of equations. If we ignore the symmetries of symbols, then the number of variables is $N + N^6$: there are N weights of colors and N^6 symbols. The equations are parameterized by 9-tuples $(i, j, k, l, m, n, l', m', n')$

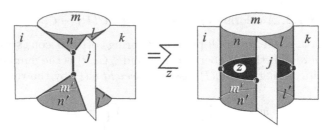

Fig. 8.2. Geometric presentation of equations

of colors. Therefore, we have N^9 equations (if we ignore symmetries of equations). In general, the system appears over-determined, but, as we shall see, solutions exist.

Let us show that every solution determines a 3-manifold invariant. Let M be a 3-manifold. Construct a special spine P of M having ≥ 2 vertices, and define an invariant $TV(M)$ by the formula $TV(M) = w(P)$, where $w(P)$ is the weight of P, see Definition 8.1.3.

Theorem 8.1.4. *If the $(q-6j)$-symbols and weights w_i are solutions of the system (8.2), then $w(P)$ does not depend on the choice of P. Therefore, $TV(M)$ is a well defined 3-manifold invariant.*

Proof. According to Theorem 1.2.5, it is sufficient to show that $w(P)$ is invariant with respect to T-moves. Let a special polyhedron P_2 be obtained from a special polyhedron P_1 by exactly one T-move, i.e., by removing a fragment E_T and inserting a fragment E'_T, see Definition 1.2.3. For any coloring ξ of P_1, let $Col_\xi(P_2)$ be the set of colorings of P_2 that coincide with ξ on $P_1 \backslash E_T = P_2 \backslash E'_T$. Since only one 2-cell of the fragment E'_T (the middle disc) has no common points with $\partial E'_T$, the set $Col_\xi(P_2)$ can be parameterized by the color z of this 2-cell. It follows that the set $Col_\xi(P_2)$ consists of N colorings $\zeta_z, 0 \leq z \leq N-1$.

Because of distributivity, the equation of the system (8.2) that corresponds to the 9-tuple (i,j,k,l,m,n,l',m',n') implies the equality $w(\xi) = \sum_z w(\zeta_z)$, see Fig. 8.2. To see this, multiply both sides of the equation by the constant factor that corresponds to the contribution made to the weights by the exteriors of the fragments. Summing up the equalities $w(\xi) = \sum_z w(\zeta_z)$ over all colorings of P_1, we get $w(P_1) = w(P_2)$.

Definition 8.1.5. *Any 3-manifold invariant obtained by the above construction will be called an* invariant of Turaev–Viro type. *The number $r = N+1$, where N is the number of colors in the palette $\mathcal{C}$, will be called the* order *of the invariant.*

8.1.2 Turaev–Viro Type Invariants of Order $r \leq 3$

There are no Turaev–Viro type invariants of order 1, since $r = N+1$ and $N \geq 1$. If $r = 2$, then $N = 1$. Hence we have a very poor palette consisting of only one color 0, and there is only one colored butterfly

$$\begin{pmatrix} 0 & 0 & 0 \\ 0 & 0 & 0 \end{pmatrix}.$$

Denote by w_0 and x the weight of the unique color and the symbol of the butterfly, respectively. In this case the system (8.2) consists of one equation $x^2 = w_0 x^3$. If $x = 0$, we get solutions that produce the trivial 3-manifold invariant $TV(M) \equiv 0$. Otherwise, we get a set of solutions $\{x = z^{-1}, w_0 = z\}$ parameterized by nonzero numbers z. Each of the solutions produces a 3-manifold invariant $TV(M) = z^{\chi(P)}$ (one should point out here that $\chi(P) = \chi(M)$ if $\partial M \neq \emptyset$, and $\chi(P) = 1$ if M is closed). Indeed, by Definition 8.1.3 we have $TV(M) = w(P) = z^{-V(P)} z^{C(P)} = z^{C(P)-V(P)}$, where $V(P)$ is the number of vertices of a special spine $P \subset M$ and $C(P)$ is the number of its 2-cells. Since every vertex of P is incident to exactly four edges, the number of edges of P is equal to $2V(P)$. It follows that $C(P) - V(P) = \chi(P)$.

Let us investigate the case $N = 2$, when there are two colors: 0 and 1. We will call them *white* and *black*, respectively. There are 11 different colored butterflies. Their symbols together with the weights w_0, w_1 of the colors form a set of 13 variables. Note that the transposition $\{0,1\} \leftrightarrow \{1,0\}$ of the colors induces an involution i on the set of variables. See Fig. 8.3, where the butterflies are presented by their boundary graphs. The lower indices of the corresponding variables show the number of black-colored wings. The weights of the black and white colors are also indicated.

It turns out that there are 74 equations. They correspond to different colorings of the boundary graph of the fragment E_T. See Fig. 8.4 for an example of a graphically expressed equation that corresponds to the equation

$$\begin{vmatrix} i & j & k \\ l & m & n \end{vmatrix} \begin{vmatrix} i & j & k \\ l' & m' & n' \end{vmatrix} = \sum_z w_z \begin{vmatrix} i & m & n \\ z & n' & m' \end{vmatrix} \begin{vmatrix} j & l & n \\ z & n' & l' \end{vmatrix} \begin{vmatrix} k & l & m \\ z & m' & l' \end{vmatrix}$$

of system (8.2) for $i = k = 1, j = l = m = n = l' = m' = n' = 0$.

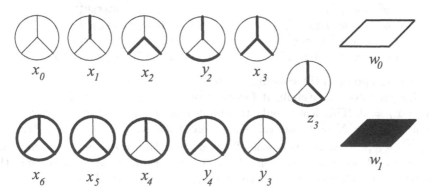

Fig. 8.3. Thirteen variables for $N = 2$

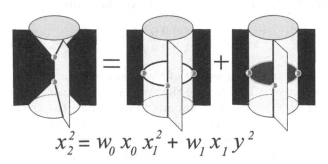

$$x_2^2 = w_0\, x_0\, x_1^2 + w_1\, x_1\, y^2$$

Fig. 8.4. An example of an equation

There are too many equations to list them all here, so we will take a shortcut. To simplify the calculations, we subject the symbols to the following constraints: if a butterfly contains one black and two white wings adjacent to the same edge, then the symbol must be zero. In other words, we assume that the symbols of the type

$$\begin{vmatrix} 0\;0\;1 \\ *\;*\;* \end{vmatrix},$$

that is, the variables x_1, x_2, y_2, x_3, z_3, and x_4 are zeros. The motivation for this restriction was *triangle inequality conditions* of Turaev–Viro, see Sect. 8.1.4. My former student Maxim Sokolov had verified that, in case $N = 2$, only restricted solutions of system (8.2) were interesting (unpublished). In other words, unrestricted solutions do not give any additional invariants. So we are left with seven variables x_0, y_3, y_4, x_5, x_6, and w_0, w_1. It is easy to see that this leaves only 14 equations:

(1) $x_0^2 = w_0 x_0^3$;

(2) $x_0 y_3 = w_1 y_3^3$;

(3) $y_3^2 = w_0 x_0 y_3^2$;

(4) $y_4^2 = w_0 y_3^2 y_4$;

(5) $y_3 y_4 = w_1 y_3 y_4^2$;

(6) $y_3^2 = w_0 y_4^3 + w_1 x_5^3$;

(7) $0 = w_0 y_4^2 x_5 + w_1 x_5^2 x_6$;

(8) $x_5^2 = w_1 y_4 x_5^2$;

(9) $y_3 x_5 = w_1 y_3 x_5^2$;

(10) $y_4 x_5 = w_1 x_5^3$;

(11) $x_5^2 = w_0 x_5 y_3^2$;

(12) $x_5^2 = w_0 y_4 x_5^2 + w_1 x_5 x_6^2$;

(13) $x_5 x_6 = w_1 x_5^2 x_6$;

(14) $x_6^2 = w_0 x_5^3 + w_1 x_6^3$.

They correspond to the colorings of ∂E_T shown in Fig. 8.5.

Let us solve the system. It follows from (4) and (8) that if $y_3 = 0$, then $y_4 = x_5 = 0$. This leaves only two equations $x_0^2 = w_0 x_0^3, x_6^2 = w_1 x_6^3$. Just as in the case $N = 1$, one can show that then we get the sum $TV(M) = w_0^{\chi(M)} + w_1^{\chi(M)}$ of the two-order 2 invariants. Hence one may assume that $y_3 \neq 0$ and, as it follows from the third equation, $x_0 \neq 0$. Note that the system is quasihomogeneous in the following sense: if we divide all x_i and y_j

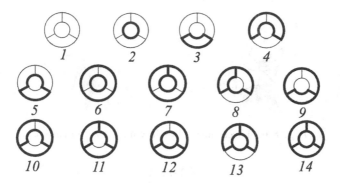

Fig. 8.5. Fourteen equations for $N = 2$

by x_0 and multiply w_0, w_1 by the same factor, we get an equivalent system. Hence we may assume that $x_0 = 1$ and, by the first equation, $w_0 = 1$.

The further events depend on whether or not $x_5 = 0$. Let $x_5 = 0$. Recall that $x_0 = w_0 = 1$. Set $w_1 = u$. It is easy to see that (2), (4), and (6) imply $y_3 = u^{-1/2}, y_4 = u^{-1}$, and $u^2 = 1$. All the other equations become identities except the last equation $x_6^2 = u x_6^3$. We get two solutions:

$$x_0 = w_0 = 1, w_1 = u, y_3 = u^{-1/2}, y_4 = u^{-1}, x_5 = x_6 = 0; \qquad (8.3)$$

$$x_0 = w_0 = 1, w_1 = u, y_3 = u^{-1/2}, y_4 = u^{-1}, x_5 = 0, x_6 = u^{-1}, \qquad (8.4)$$

where $u = \pm 1$ and for y_3 one may take any square root of u^{-1}.

Let $x_5 \neq 0$. Set $w_1 = \varepsilon$. Just as before, we get $y_3 = \varepsilon^{-1/2}$ and $y_4 = \varepsilon^{-1}$. From (9) and (7) one gets $x_5 = \varepsilon^{-1}$ and $x_6 = -\varepsilon^{-2}$. All other equations become identities except (6), (12), (14), that are equivalent to $\varepsilon^2 = 1 + \varepsilon$. We get a new solution:

$$x_0 = w_0 = 1, w_1 = \varepsilon, y_3 = \varepsilon^{-1/2}, y_4 = \varepsilon^{-1}, x_5 = \varepsilon^{-1}, x_6 = -\varepsilon^{-2}, \qquad (8.5)$$

where $\varepsilon = (1 \pm \sqrt{5})/2$.

Denote by $TV_{\pm}(M)$ the invariants corresponding to solution (8.3) for $u = \pm 1$. Let us describe a geometric interpretation of them. Any special polyhedron contains only finitely many different closed surfaces. Denote by $n_e(P)$ and $n_o(P)$ the total number of surfaces in P having *even* and, respectively, *odd* Euler characteristics.

Lemma 8.1.6. *For any special spine P of M we have $TV_{\pm}(M) = n_e(P) \pm n_o(P)$.*

Proof. There is a natural bijection between closed surfaces in P and black–white colorings of P with nonzero weights. Indeed, if we paint a surface $F \subset P$ in black, and the complement $P \backslash F$ in white, we get a coloring ξ of P such that it admits only three types of butterflies: the totally white butterfly

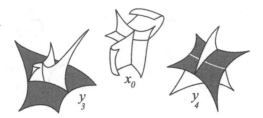

Fig. 8.6. Three butterflies having nonzero symbols

$$\begin{pmatrix} 0\ 0\ 0 \\ 0\ 0\ 0 \end{pmatrix},$$

and butterflies

$$\begin{pmatrix} 0\ 0\ 0 \\ 1\ 1\ 1 \end{pmatrix},$$

$$\begin{pmatrix} 0\ 1\ 1 \\ 0\ 1\ 1 \end{pmatrix},$$

see Fig. 8.6.

Since their symbols

$$x_0 = \begin{vmatrix} 0\ 0\ 0 \\ 0\ 0\ 0 \end{vmatrix}, y_3 = \begin{vmatrix} 0\ 0\ 0 \\ 1\ 1\ 1 \end{vmatrix},$$

and

$$y_4 = \begin{vmatrix} 0\ 1\ 1 \\ 0\ 1\ 1 \end{vmatrix}$$

are nonzero, the weight $w(\xi)$ is also nonzero.

Conversely, let ξ be a black–white coloring of P with a nonzero weight

$$w(\xi) = \prod_{v \in V(P)} \begin{vmatrix} i\ j\ k \\ l\ m\ n \end{vmatrix}_v \prod_{c \in C(P)} w_{\xi(c)}.$$

Denote by $F(\xi)$ the union of all black cells in P. Then $F(\xi)$ inherits the local structure of black parts of butterflies. Since $w(\xi) \neq 0$, the butterflies have nonzero symbols. In the case of solution (8.3), only the butterflies shown in Fig. 8.6 come into consideration. Hence $F(\xi)$ is a closed surface.

It turns out that the weight $w(\xi) \neq 0$ of a coloring ξ is closely related to the Euler characteristic $\chi(F)$ of the corresponding surface $F = F(\xi)$. Let us show that $w(\xi) = u^{\chi(F)}$. Denote by k_3 and k_4 the numbers of butterflies in P having the symbols y_3 and y_4, respectively. Then F inherits from P the cell structure with $k_3 + k_4$ vertices, $(3k_3 + 4k_4)/2$ edges, and some number of 2-cells, which we denote by $c_2(F)$. It follows that $\chi(F) = -k_3/2 - k_4 + c_2(F)$. Taking into account that $w_1 = u$, $y_3 = u^{-1/2}$, and $y_4 = u^{-1}$, we get that $w(\xi) = w_1^{c_2(F)} y_3^{k_3} y_4^{k_4} = u^{-k_3/2 - k_4 + c_2(F)} = u^{\chi(F)}$.

To conclude the proof, denote by $\xi_1, \ldots, \xi_n$ the colorings of P with nonzero weights. Then $w(P) = \sum_{i=1}^{n} w(\xi_i) = \sum_{i=1}^{n} u^{\chi(F(\xi_i))} = n_e \pm n_o$ for $u = \pm 1$.

It turns out that the invariants $TV_\pm(M)$ admit a very nice homological interpretation, see [126]. Note that for any compact 3-manifold M the homology group $H_2(M; Z_2)$ is finite, and any homology class $\alpha \in H_2(M; Z_2)$ can be presented by an embedded closed surface. We say that α is *even* or *odd* if it can be presented by an embedded surface in M having an even or odd Euler characteristic, respectively. Denote by $n_e(M)$ and $n_o(M)$ the number of even and, respectively, odd homology classes.

Proposition 8.1.7. *For any 3-manifold M we have:*

(a) $TV_+(M) = n_e(M) + n_o(M)$ (= the order of $H_2(M; Z_2)$);
(b) $TV_-(M) = n_e(M) - n_o(M)$;
(c)   *In case M is orientable either $TV_-(M) = TV_+(M)$ or $TV_-(M) = 0$ depending on whether or not M contains an odd surface.*

Proof. Choose a special spine P of M. Denote by $\mathcal{F}(P)$ the set of all closed surfaces in P. Each surface $F \in \mathcal{F}(P)$ represents an element of $H_2(P; Z_2)$. Thus we have a map $\varphi : \mathcal{F}(P) \to H_2(P; Z_2) = H_2(M; Z_2)$. It is easy to see that, due to the nice local structure of simple polyhedra, φ is a bijection. We may apply Lemma 8.1.6 and get $TV_\pm(M) = n_e(P) \pm n_o(P)$. Since P is a deformation retract of M, we have $n_e(P) = n_e(M), n_o(P) = n_o(M)$, that implies (a) and (b). To get (c), consider the map $H_2(M; Z_2) \to Z_2$ that takes even classes to 0 and odd classes to 1. If M is orientable, the map is a homomorphism. Hence even elements of $H_2(M; Z_2)$ form a subgroup that either coincides with $H_2(M; Z_2)$ or has index 2 (if there is at least one odd element). It follows that either $n_o = 0$ (and we get $TV_-(M) = TV_+(M)$) or $n_e = n_o$ (and we get $TV_-(M) = 0$).

Examples. The values of $TV_\pm$-invariants for some 3-manifolds are given in the following table. The list contains all closed orientable prime manifolds of complexity ≤ 2 (see Chap. 2), and two nonorientable manifolds: $S^1 \times RP^2$ and $K^2 \times S^1$, where K^2 is the Klein bottle.

M	TV_+	TV_-	M	TV_+	TV_-
S^3	1	1	$L_{8,3}$	2	2
$S^2 \times S^1$	2	2	$L_{5,2}$	1	1
RP^3	2	0	$L_{5,1}$	1	1
$L_{3,1}$	1	1	$L_{7,2}$	1	1
$L_{4,1}$	2	2	S^3/Q_8	4	4
$S^1 \times RP^2$	4	2	$K^2 \times S^1$	8	4

Note that TV_--invariant distinguishes $S^2 \times S^1$ and RP^3 even though the manifolds have isomorphic homology groups with coefficients in Z_2. Nevertheless, it does not distinguish between $L_{3,1}$ and S^3, or some other pairs from among the manifolds listed above.

Remark 8.1.8. In view of Proposition 8.1.7, the first five lines of the table are evident, since the corresponding 3-manifolds are orientable and only RP^3 contains an odd surface. Let us explain the last statement. The manifold $S^1 \times RP^2$ contains four homologically distinct surfaces that realize elements of $H_2(M; Z_2)$: the empty surface, projective plane $\{*\} \times RP^2$, the torus $S^1 \times RP^1$, and the Klein bottle $S^1 \tilde{\times} RP^1$. The best way to imagine $S^1 \tilde{\times} RP^1 \subset S^1 \times RP^2$ is to let a point $x \in S^1$ move around S^1, rotating simultaneously $\{x\} \times RP^1$ inside $\{x\} \times RP^2$ such that the total rotation angle would be 180°. Since only one of the surfaces (projective plane) has an odd Euler characteristic, $n_o = 1$ and $n_e = 3$. It follows from Proposition 8.1.7 that $TV_+(S^1 \times RP^2) = 4$ and $TV_-(S^1 \times RP^2) = 2$. The following conjecture was stated by Kauffman and Lins:

Conjecture [60]. *Consider an arbitrary closed 3-manifold M, and let X be a special spine for M. Let n_e be the number of closed surfaces contained in X that have even Euler characteristic and n_o the number of closed surfaces in X that have odd Euler characteristic. Then either $n_e = n_o$ or $n_o = 0$.*

Moreover, $n_e = n_o$ if and only if the same is true for all special spines of M, and $n_o = 0$ if and only if the values of Turaev–Viro invariants for $\theta = (2 \pm 1)\pi/4$ are integers and equal.

As we have seen above, for orientable manifolds the first part of the conjecture is true while the manifold $S^1 \times RP^2$ disproves it for the nonorientable case. The second part of the conjecture is also wrong, see [113] and Sect. 8.1.5.

Let us turn now our attention to solution (8.4). One can easily see that it gives nothing new, since we get the sum of the $TV_\pm$-invariant and of an order two invariant $u^{\chi(P)}$. The reason is that if the weight $w(\xi)$ of a black–white coloring ξ of a special polyhedron P is nonzero, then the black part of P is either a closed surface or coincides with P. Thus solution (8.4) produces invariants $TV_\pm(M) + (\pm 1)^{\chi(P)}$. On the contrary, the invariants corresponding to the solution (8.5) are very interesting since they are actually the simplest nontrivial invariants of Turaev–Viro type. We consider them in Sect. 8.1.3.

8.1.3 Construction and Properties of the ε-Invariant

We start with an alternative description of the new invariant. Let P be a simple polyhedron. Denote by $\mathcal{F}(P)$ the set of all simple subpolyhedra of P including P and the empty set.

Lemma 8.1.9. $\mathcal{F}(P)$ *is finite.*

Proof. It is easy to see that if a simple subpolyhedron $F \subset P$ contains at least one point of a 2-component α of P, then $\alpha \subset F$. It follows that for describing F it is sufficient to specify which 2-components of P are contained in F. Thus the total number of simple subpolyhedra of P is no greater than 2^n, where n is the number of 2-components in P.

Let us associate to each simple polyhedron F its ε-*weight*

$$w_\varepsilon(F) = (-1)^{V(F)} \varepsilon^{\chi(F)-V(F)},$$

where $V(F)$ is the number of vertices of F, $\chi(F)$ is its Euler characteristic, and ε is a solution of the equation $\varepsilon^2 = \varepsilon + 1$. One may take $\varepsilon = (1+\sqrt{5})/2$ as well as $\varepsilon = (1-\sqrt{5})/2$.

Definition 8.1.10. *The ε-invariant $t(P)$ of a simple polyhedron P is given by the formula $t(P) = \sum_{F \in \mathcal{F}(\mathcal{P})} w_\varepsilon(F)$.*

Below we will prove that the ε-invariant of a special polyhedron P coincides with the weight $w(P)$ that corresponds to solution (8.5), see Sect. 8.1.2. Hence it is invariant under T-moves. Nevertheless, we prefer to give an independent proof since it reveals better the geometric nature of the invariance.

Theorem 8.1.11. $t(P)$ *is invariant under T-moves.*

Proof. Let a simple polyhedron P_2 be obtained from a simple polyhedron P_1 by the move T. Denote by E_T the fragment of P_1 which is cut out and replaced by a fragment E'_T of P_2. It is convenient to assume that the complement $P_1 \backslash E_T$ of E_T and the complement $P_2 \backslash E'_T$ of the fragment E'_T do coincide. Let us analyze the structures of E_T and E'_T.

The fragment E_T consists of two cones and three sheets called *wings*. Each cone consists of three-curved triangles. E'_T consists of six-curved rectangles, the middle disc, and three wings. Let us divide the set $\mathcal{F}(\mathcal{P}_\infty)$ of all simple subpolyhedra of P_1 into two subsets. A simple subpolyhedron $F \in \mathcal{F}(P_1)$ is called *rich* (with respect to E_T), if $F \cap E_T$ contains all six triangles of E_T, and *poor* otherwise.

We wish to arrange a finite-to-(one or zero) correspondence between simple subpolyhedra of P_2 and those of P_1 such that the correspondence respects ε-weights.

(a) Let F_1 be a poor subpolyhedron of P_1. Since E_T without a triangle is homeomorphic to E'_T without the corresponding rectangle, there exists exactly one simple subpolyhedron F_2 of P_2 such that $F_1 \cap (P_1 - E_T) = F_2 \cap (P_2 - E'_T)$. Moreover, F_2 is homeomorphic to F_1 and hence has the same ε-weight.

(b) Let F_1 be a rich subpolyhedron of P_1. Then there exist exactly two simple subpolyhedra F'_2 and F''_2 of P_2 such that $F_1 \cap (P_1 \backslash E_T) = F'_2 \cap (P_2 \backslash E'_T) = F''_2 \cap (P_2 \backslash E'_T)$, namely, the one that contains the middle disc, and the other that does not. The intersection $F_1 \cap E_T$ can contain 0, 2, or 3 wings. It is easy to verify that in all three cases the equation $w_\varepsilon(F_1) = w_\varepsilon(F'_2) + w_\varepsilon(F''_2)$ is equivalent to $\varepsilon^2 = \varepsilon + 1$. See Fig. 8.7 for the case of 0 wings: C denotes the product of weights and symbols that correspond to 2-cells and vertices outside E_T and E'_T.

(c) We have not considered simple subpolyhedra of P_2 the intersections of which with E'_T contain six rectangles and exactly one wing. The set of such polyhedra can be decomposed onto pairs F'_2, F''_2 such that

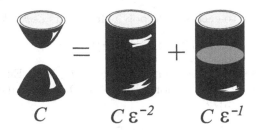

$$C \qquad\qquad C\,\varepsilon^{-2} \qquad\qquad C\,\varepsilon^{-1}$$

Fig. 8.7. The behavior of a rich surface that does not contain wings

$F_2' \cap (P_2 \setminus E_T') = F_2'' \cap (P_2 \setminus E_T')$, and exactly one of the subpolyhedra F_2', F_2'' contains the middle disc. For each such pair we have $w_\varepsilon(F_2') + w_\varepsilon(F_2'') = 0$. We can conclude that $t(P_1) = t(P_2)$.

Theorems 8.1.11 and 1.2.5 show that the following definition makes sense.

Definition 8.1.12. *Let M be a compact 3-manifold. Then the ε-invariant $t(M)$ of M is given by the formula $t(M) = t(P)$, where P is a special spine of M.*

Now we relate the ε-invariant to the TV-invariant that corresponds to solution (8.5) in Sect. 8.1.11.

Proposition 8.1.13. *The ε-invariant coincides with the TV-invariant corresponding to solution (8.5).*

Proof. We use the same ideas as in the proof of Lemma 8.1.6. Assign to any black–white coloring ξ of a special spine P the union $F(\xi)$ of all black cells of P. Note that the black part of any butterfly has a singularity allowed for simple polyhedra if and only if the corresponding symbol (with respect to solution (8.5)) is nonzero. Hence the assignment $\xi \to F(\xi)$ induces a bijection between colorings with nonzero weights and simple subpolyhedra of P.

Now, let us verify that the ε-weight $w_\varepsilon(F)$ of a simple subpolyhedron $F = F(\xi) \subset P$ coincides with the weight $w(\xi)$ of the coloring ξ. Denote by k_3, k_4, k_5, k_6 the numbers of butterflies in P with symbols y_3, y_4, x_5, x_6, respectively. Clearly, k_6 is the number of totally black butterflies in P and thus coincides with the number $V(F)$ of true vertices of F. Note that the first butterfly has three black edges while the other three have four black edges each. $F(\xi)$ inherits from P the cell structure with $k_3 + k_4 + k_5 + k_6$ vertices, $(3k_3 + 4k_4 + 4k_5 + 4k_6)/2$ edges, and some number of 2-cells that we denote by $c_2(F)$. It follows that $\chi(F) = -k_3/2 - k_4 - k_5 - k_6 + c_2(F)$. Taking into account that $w_1 = \varepsilon$, $y_3 = \varepsilon^{-1/2}$, $y_4 = x_5 = \varepsilon^{-1}$, and $x_6 = -\varepsilon^{-2}$, we have $w(\xi) = w_1^{c_2(F)} y_3^{k_3/2} y_4^{k_4} x_5^{k_5} x_6^{k_6} = (-1)^{V(P)} \varepsilon^{\chi(F) - V(P)}$. Taking sums of the weights, we get the conclusion.

Example. Let us compute "by hand" the ε-invariant of S^3. One should take a special spine of S^3, no matter which one. Let us take the Abalone A, see Fig. 1.1.4. It contains two 2-cells; one of them is the meridional disc D of the tube. There are three simple subpolyhedra of A:

1. The empty subpolyhedron with the ε-weight 1.
2. The whole Abalone with the ε-weight $(-1)^V \varepsilon^{\chi(A)-V(A)} = -1$, since $V(A) = 1$ and $\chi(A) = 1$.
3. The subpolyhedron $A\backslash\mathrm{Int}\, D$, i.e., the subpolyhedron covered by the remaining 2-cell. It contains no vertices and has zero Euler characteristic. Hence it has ε-weight 1.

Summing up, we get $t(S^3) = 1$.

One can see from this example that the calculation of the ε-invariant is theoretically simple, but may be cumbersome in practice, especially when the manifold is complicated. Sokolov wrote a computer program that, given a special spine, calculates the ε-invariant of the corresponding 3-manifold. The results are presented in Table 8.1.

8.1.4 Turaev–Viro Invariants of Order $r \geq 3$

As we have seen in Sect. 8.1.1, the number of variables of the system (8.2) grows as N^6, where $N = r - 1$ is the number of colors in the palette $C = \{0, 1, \ldots, N - 1\}$. One may decrease the number of variables by imposing some constraints on butterflies with nonzero symbols.

Definition 8.1.14. *An unordered triple i, j, k of colors taken from the palette C is called* admissible *if*

1. $i + j \geq k, j + k \geq i, k + i \geq j$ (triangle inequalities).
2. $i + j + k$ is even.
3. $i + j + k \leq 2r - 4$.

Remark 8.1.15. In the original paper [126], where this definition is taken from, Turaev and Viro used the half-integer palette $\{0, 1/2, \ldots, (r - 2)/2\}$. There are some deep reasons behind this choice but we prefer to consider the integer palette C. In any case, this is only a problem of notation.

Let us give a geometric interpretation of admissibility. Consider a disc D with three adjacent strips that contain i, j, and k strings, respectively. Then the triple (i, j, k) satisfies conditions 1, 2 if and only if the strings can be joined together in a nonsingular way as shown in Fig. 8.8.

To be more precise, the united strings should be disjoint and no string should return to the strip it is coming out of. The third condition $i + j + k \leq 2r - 4$ is of technical nature and can be avoided.

Definition 8.1.16. *A coloring ξ of a special polyhedron P is called* admissible *if the colors of any three wings adjacent to the same edge form an admissible triple. The set of all admissible colorings will be denoted by* $\mathrm{Adm}(P)$.

Table 8.1. ε-Invariants of closed irreducible orientable 3-manifolds up to complexity 5

c_i	M	$t(M)$	c_i	M	$t(M)$
0_1	S^3	1	5_2	$L_{13,2}$	$\varepsilon+1$
0_2	RP^3	$\varepsilon+1$	5_3	$L_{16,3}$	1
0_3	$L_{3,1}$	$\varepsilon+1$	5_4	$L_{17,3}$	$\varepsilon+1$
1_1	$L_{4,1}$	1	5_5	$L_{17,4}$	$\varepsilon+1$
1_2	$L_{5,2}$	0	5_6	$L_{19,4}$	1
2_1	$L_{5,1}$	$\varepsilon+2$	5_7	$L_{20,9}$	$\varepsilon+2$
2_2	$L_{7,2}$	$\varepsilon+1$	5_8	$L_{22,5}$	$\varepsilon+1$
2_3	$L_{8,3}$	$\varepsilon+1$	5_9	$L_{23,5}$	$\varepsilon+1$
2_4	S^3/Q_8	$\varepsilon+3$	5_{10}	$L_{23,7}$	$\varepsilon+1$
3_1	$L_{6,1}$	1	5_{11}	$L_{24,7}$	1
3_2	$L_{9,2}$	1	5_{12}	$L_{25,7}$	0
3_3	$L_{10,3}$	0	5_{13}	$L_{25,9}$	$\varepsilon+2$
3_4	$L_{11,3}$	1	5_{14}	$L_{26,7}$	1
3_5	$L_{12,5}$	$\varepsilon+1$	5_{15}	$L_{27,8}$	$\varepsilon+1$
3_6	$L_{13,5}$	$\varepsilon+1$	5_{16}	$L_{29,8}$	1
3_7	S^3/Q_{12}	$\varepsilon+3$	5_{17}	$L_{29,12}$	1
4_1	$L_{7,1}$	$\varepsilon+1$	5_{18}	$L_{30,11}$	$\varepsilon+2$
4_2	$L_{11,2}$	1	5_{19}	$L_{31,12}$	1
4_3	$L_{13,3}$	$\varepsilon+1$	5_{20}	$L_{34,13}$	1
4_4	$L_{14,3}$	$\varepsilon+1$	5_{21}	$S^3/Q_8 \times Z_5$	$\varepsilon+2$
4_5	$L_{15,4}$	$\varepsilon+2$	5_{22}	$S^3/Q_{12} \times Z_5$	$\varepsilon+2$
4_6	$L_{16,7}$	1	5_{23}	$S^3/Q_{16} \times Z_3$	$\varepsilon+1$
4_7	$L_{17,5}$	$\varepsilon+1$	5_{24}	S^3/Q_{20}	$3\varepsilon+2$
4_8	$L_{18,5}$	$\varepsilon+1$	5_{25}	$S^3/Q_{20} \times Z_3$	$-\varepsilon+2$
4_9	$L_{19,7}$	1	5_{26}	S^3/D_{40}	$-\varepsilon+2$
4_{10}	$L_{21,8}$	1	5_{27}	S^3/D_{48}	$\varepsilon+3$
4_{11}	$S^3/Q_8 \times Z_3$	$2\varepsilon+3$	5_{28}	$S^3/P_{24} \times Z_5$	$\varepsilon+2$
4_{12}	S^3/Q_{16}	1	5_{29}	S^3/P_{48}	$\varepsilon+1$
4_{13}	S^3/D_{24}	$2\varepsilon+3$	5_{30}	S^3/P'_{72}	$\varepsilon+3$
4_{14}	S^3/P_{24}	$2\varepsilon+3$	5_{31}	S^3/P_{120}	$3\varepsilon+2$
5_1	$L_{8,1}$	$\varepsilon+1$			

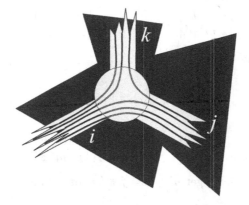

Fig. 8.8. Disjoint strings without returns

Similarly, one may speak about admissible butterflies: a colored butterfly

$$\begin{pmatrix} i & j & k \\ l & m & n \end{pmatrix}$$

is *admissible* if all the triples (i,j,k), (k,l,m), (m,n,i), (j,l,n) are admissible (they represent the wings that meet together along the four edges of the butterfly).

The constraint on butterflies we have mentioned above is the following:

   The symbols of nonadmissible butterflies must be zeros.

Another way of saying this is that we define the weight (see Definition 8.1.3) of a special polyhedron P by taking the sum over admissible colorings only:

$$w(P) = \sum_{\xi \in Adm(P)} w(\xi)$$

The following nonrigorous considerations show that the system (8.2) (subjected to the admissibility restrictions from Definition 8.1.14) should not have too many solutions (if any). For simplicity assume that the symbols of all admissible butterflies are nonzero. Since all equations are quasihomogeneous, we may assume $w_0 = 1$ (see Sect. 8.1.2). Denote by s_k the symbol

$$\begin{vmatrix} 0 & 0 & 0 \\ k & k & k \end{vmatrix} = \begin{vmatrix} k & 0 & k \\ 0 & k & 0 \end{vmatrix}.$$

(a) Write down an equation of system (8.2) for the case $l = 0$. If $j \neq n$ or $k \neq m$, all terms in both sides of the equation contain symbols of nonadmissible butterflies (this is because a triple of the type $(0, x, y)$ is admissible if and only if $x = y$). This annihilates the equation. We may assume therefore that $n = j$ and $m = k$. Similarly, $l' = z$, and we get the equation

$$\begin{vmatrix} i & j & k \\ 0 & k & j \end{vmatrix} \begin{vmatrix} i & j & k \\ z & m' & n' \end{vmatrix} = w_z \begin{vmatrix} i & k & j \\ z & n' & m' \end{vmatrix} \begin{vmatrix} j & 0 & j \\ z & n' & z \end{vmatrix} \begin{vmatrix} k & 0 & k \\ z & m' & z \end{vmatrix},$$

that after dividing both sides by

$$\begin{vmatrix} i & j & k \\ z & m' & n' \end{vmatrix} = \begin{vmatrix} i & k & j \\ z & n' & m' \end{vmatrix}$$

gives

$$\begin{vmatrix} i & j & k \\ 0 & k & j \end{vmatrix} = w_z \begin{vmatrix} j & 0 & j \\ z & n' & z \end{vmatrix} \begin{vmatrix} k & 0 & k \\ z & m' & z \end{vmatrix}.$$

(b) Taking $z = 0$, we get $n' = n = j, m' = m = k$, and

$$\begin{vmatrix} i & j & k \\ 0 & k & j \end{vmatrix} = s_j s_k.$$

This converts the preceding equation to $s_j s_k = w_z s_j s_z s_k s_z$ or, equivalently, to $w_z = s_z^{-2}$.

(c) Next, let us write down the equation of system (8.2) for $i = j = k = 0$. The admissibility implies that $l' = m' = n'$ and $l = m = n$. Taking into account that

$$\begin{vmatrix} 0 & l & l \\ z & l' & l' \end{vmatrix} = \begin{vmatrix} z & l' & l \\ 0 & l & l' \end{vmatrix} = s_l s_{l'},$$

we get the equation $s_l s_{l'} = \sum_z w_z (s_l s_{l'})^3$, which is equivalent to $w_l w_{l'} = \sum_z w_z$ (both sums are taken over all $z \le r - 2$ such that the triple (l, l', z) is admissible). In particular, for $l' = 1$ and $1 \le l \le r - 2$ we get the system

$$w_1 w_1 = w_0 + w_2$$
$$w_2 w_1 = w_1 + w_3$$
$$\cdots$$
$$w_{r-3} w_1 = w_{r-4} + w_{r-2}$$
$$w_{r-2} w_1 = w_{r-3}$$

To solve it, present w_1 in the form

$$w_1 = -(q + q^{-1}) = -\frac{q^2 - q^{-2}}{q - q^{-1}},$$

where q is a new variable. Since

$$\frac{q^{i+1} - q^{-i-1}}{q - q^{-1}} = \frac{q^i - q^{-i}}{q - q^{-1}}(q + q^{-1}) - \frac{q^{i-1} - q^{-i+1}}{q - q^{-1}},$$

we get inductively

$$w_i = (-1)^i \frac{q^{i+1} - q^{-i-1}}{q - q^{-1}} \quad \text{for } 1 \le i \le r - 2 \text{ and } \frac{q^r - q^{-r}}{q - q^{-1}} = 0.$$

We can conclude that the solutions of the system have the form

$$w_i = (-1)^i \frac{q^{i+1} - q^{-i-1}}{q - q^{-1}}, 1 \le i \le r - 2,$$

where q runs over all roots of unity of degree $2r$.

It follows from the above considerations that system (8.2) is very restrictive because a very small part of it allows us to find the weights w_i of all colors and some symbols. It is surprising that any solutions exist! Below we present solutions found by Turaev and Viro [126]. To adjust our notation to the original one (see Remark 8.1.15), we adopt the following notational convention:

$$\hat{k} = k/2$$

for any integer k. Let q be a $2r$-th root of unity such that q^2 is a primitive root of unity of degree r. Note that q itself may not be primitive.

For an integer n set

$$[n] = \frac{q^n - q^{-n}}{q - q^{-1}}. \tag{8.6}$$

Note that all $[n]$ are real numbers and $[n] = 0$ if and only if $n = 0 \bmod r$.

Define the quantum factorial $[n]!$ by setting

$$[n]! = [n][n-1]\ldots[2][1].$$

In particular, $[1]! = [1] = 1$. Just as for the usual factorial, set by definition, $[0]! = 1$.

For an admissible triple (i, j, k) put

$$\Delta(i, j, k) = \left(\frac{[\hat{i} + \hat{j} - \hat{k}]! \, [\hat{j} + \hat{k} - \hat{i}]! \, [\hat{k} + \hat{i} - \hat{j}]!}{[\hat{i} + \hat{j} + \hat{k} + 1]!} \right)^{1/2}.$$

Remark 8.1.17. As we will see later, it does not matter, which square root of the expression in the round brackets is taken for $\Delta(i, j, k)$. The resulting 3-manifold invariant will be the same.

Now we are ready to present the solution. The weights of colors from the palette $\mathcal{C} = \{0, 1, \ldots, r - 2\}$ are given by

$$w_i = (-1)^i [i + 1]. \tag{8.7}$$

The symbol

$$\begin{vmatrix} i & j & k \\ l & m & n \end{vmatrix}$$

of the butterfly

$$\begin{pmatrix} i & j & k \\ l & m & n \end{pmatrix}$$

is given by

$$\begin{vmatrix} i & j & k \\ l & m & n \end{vmatrix} = \sum_z \frac{(-1)^z [z+1]! \, A(i,j,k,l,m,n)}{B\left(z, \begin{pmatrix} i & j & k \\ l & m & n \end{pmatrix}\right) C\left(z, \begin{pmatrix} i & j & k \\ l & m & n \end{pmatrix}\right)}, \tag{8.8}$$

where

$$A(i,j,k,l,m,n) = \boldsymbol{i}^{(i+j+k+l+m+n)} \Delta(i,j,k)\Delta(i,m,n)\Delta(j,l,n)\Delta(k,l,m),$$

$$B\left(z, \begin{pmatrix} i & j & k \\ l & m & n \end{pmatrix}\right) = [\hat{z} - \hat{i} - \hat{j} - \hat{k}]! \, [z - \hat{i} - \hat{m} - \hat{n}]! \, [z - \hat{j} - \hat{l} - \hat{n}]! \, [z - \hat{k} - \hat{l} - \hat{m}]!,$$

$$C\left(z, \begin{pmatrix} i & j & k \\ l & m & n \end{pmatrix}\right) = [\hat{i} + \hat{j} + \hat{l} + \hat{m} - z]! \, [\hat{i} + \hat{k} + \hat{l} + \hat{n} - z]! \, [\hat{j} + \hat{k} + \hat{m} + \hat{n} - z]!,$$

and the sum is taken over all integer z such that all expressions in the square brackets are nonnegative. In other words, one should have $\alpha \leq z \leq \beta$, where

$$\alpha = \max(\hat{i} + \hat{j} + \hat{k}, \hat{i} + \hat{m} + \hat{n}, \hat{j} + \hat{l} + \hat{n}, \hat{k} + \hat{l} + \hat{m}),$$

$$\beta = \min(\hat{i} + \hat{j} + \hat{l} + \hat{m}, \hat{i} + \hat{k} + \hat{l} + \hat{n}, \hat{j} + \hat{k} + \hat{m} + \hat{n})$$

(it follows from the triangle inequalities that $\alpha \leq \beta$).

Remark 8.1.18. The bold letter $\boldsymbol{i}$ in the above expression for $A(i,j,k,l,m,n)$ is the imaginary unit (do not confuse with the symbol i denoting a color from the palette $\mathcal{C}$). If we replace $\boldsymbol{i}$ by $-\boldsymbol{i}$, we get a different solution producing the same 3-manifold invariant. It is because for any coloring with a nonzero weight the number of butterflies

$$\begin{pmatrix} i & j & k \\ l & m & n \end{pmatrix}$$

with an odd number $(i + j + k + l + m + n)$ is even.

Why do the presented values of variables form a solution to the system (8.2)? Turaev and Viro proved this by a reference to a paper of Kirillov and Reshetikhin [64], who had used the so-called Biederharn-Elliot identity [10,27] to obtain a solution to a similar system. Meanwhile there appeared many different ways to prove the existence of the invariants. Probably, one of the simplest approaches is based on remarkable results of Kauffman, Lickorish and others, and belongs to Roberts, see [108] and references therein. An exhaustive exposition of the subject along with deep connections to quantum groups, motivating ideas in physics and to other areas of mathematics can be found in the fundamental monograph [125].

Definition 8.1.19. *The 3-manifold invariant corresponding to the above solution will be called the* order r Turaev–Viro invariant *and denoted by* $TV_q(M)$.

Remark 8.1.20 (On the Terminology). We distinguish between *invariants of Turaev–Viro type* and *Turaev–Viro invariants* $TV_q(M)$. The former correspond to arbitrary solutions (that potentially would be found in future), the latter are related to the particular solutions given by (8.7) and (8.8). For example, the ε-invariant is of Turaev–Viro type but it is not a Turaev–Viro invariant.

Remark 8.1.21. One should point out that our exposition of results in [126] differs from the original approach. In the first place, to simplify the construction, we do not pay any attention to the relative case, which is very important from the point of view of category theory. In particular, we do not reveal the functorial nature of the invariants, nor how they fit into the conception of Topological Quantum Field Theory (TQFT) [7]. On the other hand, it is sometimes convenient to consider (as we do) absolute invariants of not necessarily closed 3-manifolds

Secondly, our version of the invariants is S^3-normalized, i.e., $TV_q(S^3) = 1$ for all q. The invariant $|\, M \,|$ presented in [126] for a degree $2r$ root of unity $q = q_0$ is related to $TV_q(M)$ by the formula

$$| \, M \,| = -\frac{(q - q^{-1})^2}{2r} TV_q(M).$$

Thirdly, the solution given by (8.7) and (8.8) satisfies additional equations of the type

$$\sum_z w_z \begin{vmatrix} i & l & m \\ z & m' & l' \end{vmatrix} \begin{vmatrix} j & l & m \\ z & m' & l' \end{vmatrix} = \delta^i_j,$$

where i, j, l, m, l', m' run over all elements of the palette $\mathcal{C}$ and δ^i_j is the Kronecker symbol. These equations guarantee that the weight of a simple polyhedron is invariant under lune moves, see Fig. 1.16. Moreover, one can calculate the invariants starting from any simple (not necessarily special) spine of a manifold. The only difference is that one should take into account the Euler characteristics of 2-components by defining the weight of a coloring ξ by

$$w(\xi) = \prod_{v \in V(P)} \begin{vmatrix} i & j & k \\ l & m & n \end{vmatrix}_v \prod_{c \in C(P)} w^{\chi(c)}_{\xi(c)},$$

instead of corresponding formula (8.1) for the case of disc 2-components.

Finally, for the solution given by (8.7) and (8.8), the following holds: there exists a number w (it is equal to $-2r/(q - q^{-1})^2$) such that for all j

$$w_j = w^{-1} \sum_{(k,l)} w_k w_l,$$

where the sum is taken over all k, l such that the triple (j, k, l) is admissible. This condition guarantees that performing a bubble move on a simple polyhedron is equivalent to multiplying its weight by w.

8.1.5 Computing Turaev–Viro Invariants

Just after discovering the invariants, Turaev and Viro calculated all of them for the sphere S^3, real projective space RP^3, for $S^2 \times S^1$, and the lens space $L_{3,1}$. The calculation was facilitated by the fact that these manifolds have simple spines without vertices. For example, $L_{3,1}$ admits a spine consisting of one 2-cell and having one triple circle. It can be presented as the identification space of a disc by a free action of the group Z_3 on the boundary. Sometimes it is called *triple hat*. Note that the three wings adjacent to any segment of the triple circle belong to the same 2-cell, and hence have the same color $i \in C$ (for any coloring). The admissibility conditions (see Definition 8.1.14) imply that i must be even and no greater than $(2r - 4)/3$. It follows that

$$TV_q(L_{3,1}) = \sum_i w_i, \tag{8.9}$$

where the sum is taken over all even i such that $0 \le i \le (2r - 4)/3$.

If every simple spine of a given 3-manifold contains vertices, obtaining explicit expressions for order r Turaev–Viro invariants (as functions on q) for all r is difficult, see [137] for the case of lens spaces. On the other hand, if r is fixed, the problem has a purely combinatorial nature and can be solved by means of a computer program. One should construct a special spine of the manifold and, using formulas (8.6)–(8.8), calculate the weights and symbols. Then one can enumerate all colorings and find out the value of the invariant by taking the sum of their weights. Extensive numerical tables of that kind can be found in [60, 113].

Let us make a digression. Soon after the discovery of the invariants, many mathematicians (and certainly Turaev and Viro) noticed that the invariants of M were actually sums of invariants of pairs (M, h), where M is a 3-manifold and $h \in H_2(M; Z_2)$. We describe this observation in detail. Let ξ be an admissible coloring of a special spine P of a 3-manifold M by colors taken from the palette $C = \{0, 1, \ldots, N - 1\}$. Reducing all colors mod 2, we get a black–white coloring ξ mod 2 that happens to be also admissible owing to condition 2 of Definition 8.1.14. As we know from the proof of Lemma 8.1.6, admissible black–white colorings correspond to black (i.e., colored by the color 1) surfaces in P or, what is just the same, to elements of $H_2(M; Z_2)$. This decomposes the set of all admissible colorings of P into classes corresponding to elements of $H_2(M; Z_2)$: two colorings belong to the same class $Adm(P, h)$ if their mod 2 reductions determine the same homology class $h \in H_2(M; Z_2)$ (and hence the same surface in P).

Assume now that the pair (M, h) and a special spine P of M are given. Define an invariant $TV_q(M, h)$ by setting $TV_q(M, h) = w(P, h)$, where the

h-weight $w(P, h)$ is given by

$$w(P,h) = \sum_{\xi \in A\mathrm{dm}(P,h)} w(\xi).$$

The same proof as given for Theorem 8.1.4 shows that this definition is correct, i.e., $TV_q(M, h)$ does not depend on the choice of P. We need only one additional observation. Let a special spine P_2 of M be obtained from a special spine P_1 by exactly one T-move, i.e., by removing a fragment E_T and inserting a fragment E'_T, see Definition 1.2.3. For any admissible coloring ξ of P_1, let $Col_\xi(P_2)$ be the set of admissible colorings of P_2 that coincide with ξ on $P_1 \backslash E_T = P_2 \backslash E'_T$. Then all the colorings in $Col_\xi(P_2)$ determine the same homology class $h \in H_2(M; Z_2)$ as the coloring ξ.

It follows from the definition of $TV_q(M, h)$ that the Turaev–Viro invariant $TV_q(M)$ is the sum of $TV_q(M, h)$ taken over all $h \in H_2(M; Z_2)$. Especially important is the *homologically trivial part* $TV_q(M)_0$ of the Turaev–Viro invariant that corresponds to the zero element of $H_2(M; Z_2)$. Recall that $h \in H_2(M; Z_2)$ is even or odd, if it can be realized by a closed surface in M having the Euler characteristic of the same parity. We follow [113] and denote by $TV_q(M)_1$ the *odd* part of $TV_q(M)$, that is equal to the sum of $TV_q(M, h)$ over all odd elements $h \in H_2(M; Z_2)$. Similarly, by $TV_q(M)_2$ we denote the sum of $TV_q(M, h)$ taken over all even elements $h \in H_2(M; Z_2)$ different from 0. Clearly, $TV_q(M) = TV_q(M)_0 + TV_q(M)_1 + TV_q(M)_2$.

Remark 8.1.22. Note that since any special spine $P \subset M$ is two-dimensional, the 2-cycle group $C_2(P, Z_2)$ coincides with $H_2(P; Z_2)$. Therefore, the mod 2 reduction of an admissible coloring $\xi \in Adm(P)$ determines the trivial element of $H_2(P; Z_2)$ if and only if just even colors $0, 2, \ldots$ have been used. Thus, the only difference between $TV_q(M)$ and its homologically trivial part $TV_q(M)_0$ is that we consider all admissible colorings in the first case and only even ones in the second.

At the end of the book we reproduce from [116] (with notational modifications) tables of Turaev–Viro invariants of order ≤ 7 and their summands for all closed orientable irreducible 3-manifolds up to complexity 6 (Table A.1; see Chap. 2 for the definition of complexity). We subject q to the following constraint: q must be a primitive root of unity of degree $2r$. This constraint is slightly stronger than the one in the definition of Turaev–Viro invariants, see Sect. 8.1.4. Nevertheless, we do not lose any information because of the following relation proved in [116]: $TV_q(M)_\nu = (-1)^\nu TV_{-q}(M)_\nu$, where $\nu \in \{0, 1, 2\}$.

The invariants are presented by polynomials of q. This presentation is much better than the numerical form since we simultaneously encode the invariants evaluated at all degree $2r$ primitive roots of unity, and avoid problems with the precision of calculations. For the sake of compactness of notation, we write σ_k instead of $q^k + q^{-k}$. For instance, we set $\sigma_1 = q + q^{-1}, \sigma_2 = q^2 + q^{-2}$,

Table 8.2. Turaev–Viro invariants of order $r \leq 7$ and their summands for closed orientable irreducible 3-manifolds of complexity ≤ 2

M	$\nu\backslash r$	3	4	5	6	7
S^3	0	1	1	1	1	1
	1	0	0	0	0	0
	2	0	0	0	0	0
	$\sum$	1	1	1	1	1
RP^3	0	1	2	$\sigma_2 + 2$	4	$-\sigma_3 + 2\sigma_2 + 3$
	1	-1	$-\sigma_1$	$-\sigma_2 - 2$	$-2\sigma_1$	$\sigma_3 - 2\sigma_2 - 3$
	2	0	0	0	0	0
	$\sum$	0	$-\sigma_1 + 2$	0	$-2\sigma_1 + 4$	0
$L_{3,1}$	0	1	1	$\sigma_2 + 2$	3	$\sigma_2 + 2$
	1	0	0	0	0	0
	2	0	0	0	0	0
	$\sum$	1	1	$\sigma_2 + 2$	3	$\sigma_2 + 2$
$L_{4,1}$	0	1	2	1	4	$\sigma_2 + 2$
	1	0	0	0	0	0
	2	1	0	1	0	$\sigma_2 + 2$
	$\sum$	2	2	2	4	$2\sigma_2 + 4$
$L_{5,2}$	0	1	1	0	1	$-\sigma_3 + 2\sigma_2 + 3$
	1	0	0	0	0	0
	2	0	0	0	0	0
	$\sum$	1	1	0	1	$-\sigma_3 + 2\sigma_2 + 3$
$L_{5,1}$	0	1	1	$\sigma_2 + 3$	1	$-\sigma_3 + 2\sigma_2 + 3$
	1	0	0	0	0	0
	2	0	0	0	0	0
	$\sum$	1	1	$\sigma_2 + 3$	1	$-\sigma_3 + 2\sigma_2 + 3$
$L_{7,2}$	0	1	1	$\sigma_2 + 2$	1	0
	1	0	0	0	0	0
	2	0	0	0	0	0
	$\sum$	1	1	$\sigma_2 + 2$	1	0
$L_{8,3}$	0	1	2	$\sigma_2 + 2$	4	1
	1	0	0	0	0	0
	2	1	2	$\sigma_2 + 2$	0	1
	$\sum$	2	4	$2\sigma_2 + 4$	4	2
S^3/Q_8	0	1	4	$\sigma_2 + 4$	10	$2\sigma_2 + 7$
	1	0	0	0	0	0
	2	3	6	$3\sigma_2 + 12$	18	$6\sigma_2 + 21$
	$\sum$	4	10	$4\sigma_2 + 16$	28	$8\sigma_2 + 28$

For each M the first three lines present $TV_q(M)_\nu$; the fourth line contains the values of $TV_q(M)$. For brevity, we write σ_k instead of $q^k + q^{-k}$

and so on. For your convenience, a small part of the table (for manifolds of complexity ≤ 2) is given.

Items (1–4) below are devoted to analysis of the table and commentaries.

(1) Selected testing has shown that the table agrees with the ones presented in $[60, 61, 63]$, as well as with the above-mentioned calculations made by the authors of the invariants.

(2) The manifold $4_{12} = S^3/Q_{16}$ disproves the second part of the Kauffman–Lins Conjecture (see Remark 8.1.8). We see from Table A.9 that TV_q (S^3/Q_{16}) is equal to 6 for every primitive root of unity q of degree 8, including $q = exp((2 \pm 1)\pi/4)$. Nevertheless, since $TV_q(S^3/Q_{16})_1 \neq 0$, there is at least one surface with odd Euler characteristic. Therefore, $n_o \neq 0$.

(3) Let us call 3-manifolds *twins* if their Turaev–Viro invariants of order ≤ 7 have the same triples of summands. The distribution of twins is shown in Fig. 8.9. Each line of the table consists of twin manifolds. Cells painted in gray contain *genuine twins*, i.e., manifolds having the same TV-invariants of all orders. They cannot be distinguished by Turaev–Viro invariants.

Let us comment on the table. There are no twins up to complexity 3. First two pairs of twins appear on the level of complexity ≤ 4: manifold $3_4 (= L_{11,3})$ is a twin of $4_2 (= L_{11,2})$, and $3_6 (= L_{13,5})$ is a twin of $4_3 (= L_{13,3})$. At the level of complexity ≤ 5 there appear new twin pairs and twin triples, and at the level ≤ 6 we can find even a 7-tuple of twins.

Note that $TV_q(M)_1$ and $TV_q(M)_2$ are *not* invariants of Turaev–Viro type (see Definition 8.1.5) since the constraints on mod 2 reduction of colorings

0	1	2	3	4	5		6			
0_1					5_{16}	5_{17} 6_{11}	6_{26}	6_{27}	6_{28}	
			3_2			6_1				
			3_4	4_2		5_{19} 6_{13}	6_{14}			
			3_6	4_3	5_2	6_{29}				
				4_7	5_4 5_5					
				4_9	5_6	6_3				
					5_{15} 6_9					
					5_{20} 6_{18}					
					5_9 5_{10} 6_6	6_{21}	6_{22}	6_{33}	
						6_{16}	6_{17}			
						6_{23}	6_{24}			
						6_{65}	6_{67}			
						6_{68}	6_{69}			
						6_{70}	6_{71}			

Fig. 8.9. Each line of the table contains twin manifolds. Cells painted in *gray* contain *genuine twins* (manifolds having the same TV-invariants of all orders)

have global nature. On the other hand, one can extract from Table A.9 that $TV_q(M)_1$ and $TV_q(M)_2$ add actually nothing to information given by $TV_q(M)_0$ and $TV_q(M)$. For that reason, for manifolds of complexity 6 we include only values of $TV_q(M)$ and $TV_q(M)_0$.

(4) Analyzing the tables, we can notice that $TV_3(M)TV_q(M)_0 = TV_q(M)$ for odd r the following holds: In fact, this equality is always true and follows from the similar formula of Kirby and Melvin for Reshetikhin–Turaev invariants [63], and the Turaev–Walker theorem saying that each Turaev–Viro invariant is equal to the square of the absolute value of the corresponding Reshetikhin–Turaev invariant (see, for instance, [108, 125]).

(5) It is not difficult to observe that all coefficients in the polynomials presenting Turaev–Viro invariants in our tables are integers. This is not an accident. Robers and Masbaum gave in [75] an elegant proof that values of Turaev–Viro invariants are algebraic integers.

(6) As we have mentioned above, there exist explicit expressions for Turaev–Viro invariants for lens spaces. Using Yamada's formulas [137], Sokolov found a simple solution to the following interesting problem: *Which lens spaces can be distinguished by Turaev–Viro invariants?*

For any integer v define a *characteristic* function $h_v : Z \to Z_2$ by setting $h_v(k) = 1$ if $k = \pm 1$ mod v, and $h_v(k) = 0$, otherwise.

Theorem 8.1.23. *[117] Lens spaces L_{p_1,q_1} and L_{p_2,q_2} have the same Turaev–Viro invariants of all orders if and only if $p_1 = p_2$ and for any divisor $v > 2$ of p_1 we have $h_v(q_1) = h_v(q_2)$.*

Sokolov noticed also that if $p_1 \neq p_2$, then L_{p_1,q_1} and L_{p_2,q_2} can be distinguished by Turaev–Viro invariant TV_q of some order $r \leq 2R$, where R is the minimal natural number such that R is coprime with p_1, p_2, and $p_1 \neq p_2$ mod R. In case $p_1 = p_2$ it is sufficient to consider only invariants of order $r \leq p_1$. If p is prime, then the criterion is especially simple.

Corollary 8.1.24. *If p is prime, then L_{p,q_1} and L_{p,q_2} have the same Turaev–Viro invariants if and only if for $i = 1, 2$ either $q_i = \pm 1$ mod p or $q_i \neq \pm 1$ mod p.*

Note that all lines of the table in Fig. 8.9 except the last three contain only lens spaces. Thus Theorem 8.1.23 and the above corollary are sufficient for selecting genuine twins among them. See Remark 8.2.15 in Sect. 8.2 for an explanation why the last three lines contain genuine twin pairs.

Remark 8.1.25. It is interesting to note that if p is prime, then Reshetikhin–Turaev invariants can distinguish any two nonhomeomorphic lens spaces L_{p,q_1} and L_{p,q_2}, see [51].

8.1.6 More on ε-Invariant

Comparing Tables 8.1 and A.9, one can see that ε-invariant coincides with the homologically trivial part $TV_q(M)_0$ of order 5 Turaev–Viro invariant. Let us prove that.

Theorem 8.1.26. *[89] Let M be a closed 3-manifold. Then $\varepsilon(M) = TV_q(M)_0$, where $\varepsilon = (1+\sqrt{5})/2$ for $q = exp(\pm\frac{\pi}{5}i)$ and $\varepsilon = (1-\sqrt{5})/2$ for $q = exp(\pm\frac{3\pi}{5}i)$.*

Proof. The values of q and ε presented above are related by the equality $\varepsilon = q^2 + 1 + q^{-2}$. Indeed, $q^{10} = 1$ implies

$$\frac{q^5 - q^{-5}}{q + q^{-1}} = q^4 + q^2 + 1 + q^{-2} + q^{-4} = 0,$$

which is equivalent to $(q^2 + 1 + q^{-2})^2 = (q^2 + 1 + q^{-2}) + 1$. Recall that for calculating ε-invariant we use two colors: 0 and 1. For calculating $TV_q(M)_0$ in the case $r = 5$ we use *even* colors from the palette $\{0, 1, 2, 3\}$, i.e., two colors 0 and 2 (see Remark 8.1.22). It remains to verify that the correspondence $(0, 1) \rightarrow (0, 2)$ transforms weights of colors and symbols for ε-invariant (see solution (8.5) on page (389)) to those for $TV_q(M)_0$ (see formulas (8.7) and (8.8) on page 399). For instance, the weight of the color 1, in the case of ε-invariant, equals ε while the weight of the color 2, in the case of Turaev–Viro invariant, equals $q^2 + 1 + q^{-2}$. $\quad\blacksquare$

Let us discuss briefly the relation between ε-invariant and TQFT. A three-dimensional TQFT is a functor $\mathcal{F}$ from the three-dimensional cobordism category to the category of vector spaces. The functor should satisfy some axioms, see [7]. In particular, "quantum" means that $\mathcal{F}$ takes the disjoint union of surfaces to the tensor product of vector spaces. In our case the base field corresponding to the empty surface is the field R of real numbers. It follows that to every closed 3-manifold there corresponds a linear map from R to R, that is, the multiplication by a number. This number is an invariant of the manifold.

As explained in [126], Turaev–Viro invariants fit into conception of TQFT. The only difference is that instead of the cobordism category one should consider a category whose objects have the form (F, Γ), where F is a surface and Γ is a fixed one-dimensional special spine of F. Here a one-dimensional special polyhedron is a regular graph of valence 3. Morphisms between objects $(F_-, \Gamma_-), (F_+, \Gamma_+)$ have the form (M, i_-, i_+), where M is a 3-manifold with boundary ∂M presented as the union of two disjoint surfaces $\partial_- M, \partial_+ M+$, and $i_\pm : F_\pm \rightarrow \partial_\pm M$ are homeomorphisms.

The ε-invariant, as any other Turaev–Viro type invariant, admits a similar interpretation. From general categorical considerations (see [126, Sect. 2.4]) it follows that there arises a homomorphism Φ from the mapping class group of the two-dimensional torus $T^2 = S^1 \times S^1$ to the matrix group $GL_5(R)$. Given a homomorphism $h : T^2 \rightarrow T^2$, we construct the cobordism $(T^2 \times I, i_-, i_+(h))$,

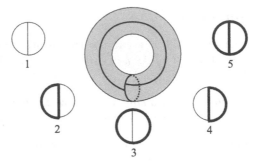

Fig. 8.10. Five *black–white* colorings for a spine of a torus

where $i_- : T^2 \to T^2 \times \{0\} \subset T^2 \times I$ is the standard inclusion and $i_+(h) : T^2 \to T^2 \times \{1\} \subset T^2 \times I$ is the inclusion induced by h. We assume that T^2 is equipped with a fixed special spine Θ. By definition, put $\Phi(h) = \mathcal{F}(T^2 \times I, i_-, i_+(h))$, where $\mathcal{F}$ is the functor corresponding to the ε-invariant. Then $\Phi(h)$ is a linear map $R^5 \to R^5$. The dimension is 5, since Θ admits exactly five black–white colorings (see Fig. 8.10) that are admissible in the following sense: one black and two white edges never meet at the same vertex.

The mapping class group of the torus is generated by twists τ_m and τ_l along a meridian and a longitude, respectively. The twists satisfy the relations $\tau_m \tau_l \tau_m = \tau_l \tau_m \tau_l$ and $(\tau_m \tau_l \tau_m)^4 = 1$. Denote by a and b the matrices of the corresponding linear maps $R^5 \to R^5$. One can verify that $a^5 = 1$, $(aba)^2 = 1$ and that the group $\langle a, b \mid aba = bab, (aba)^2 = 1, a^5 = 1 \rangle$ is finite. Actually, the presentation coincides with the standard presentation $\langle a, b \mid aba = bab, (a^2 b)^2 = a^5 = 1 \rangle$ of the alternating group A_5.

The following theorem is a direct consequence of this observation [89]:

Theorem 8.1.27. *Let F be a closed surface and n a nonnegative integer. Denote by $\mathcal{M}(F, n)$ the set of all Seifert manifolds over F with n exceptional fibers. Then the set $\{t(M), M \in \mathcal{M}(F, n)\}$ of the values of the ε-invariant is finite.*

The number 60^n (60 is the order of the alternating symmetric group A_5) serves as an upper estimate for the number of values of $t(M)$. Certainly, the estimate is very rough. More detailed considerations show that for lens spaces the number of values of ε-invariant is equal to 4. We give without proof an exact expression for the ε-invariant of the lens space $L(p, q)$.

Theorem 8.1.28.

$$t(L_{p,q}) = \begin{cases} 1, & \text{if } p \equiv \pm 1 \bmod 5; \\ \varepsilon + 1, & \text{if } p \equiv \pm 2 \bmod 5; \\ \varepsilon + 2, & \text{if } p \equiv 0 \bmod 5 \text{ and } q \equiv \pm 1 \bmod 5; \\ 0, & \text{if } p \equiv 0 \bmod 5 \text{ and } q \equiv \pm 2 \bmod 5. \end{cases}$$

8.2 3-Manifolds Having the Same Invariants of Turaev–Viro Type

This section is based on the following observation of Lickorish [69]: if two 3-manifolds M_1, M_2 have special spines with the same incidence relation between 2-cells and vertices (in a certain strong sense), then their Turaev–Viro invariants of all orders coincide. Manifolds having spines as above are called *similar*. We construct a simple example of similar 3-manifolds with different homology groups, and present a result of Nowik and the author [88] stating that under certain conditions similar manifolds are homeomorphic.

Let P be a special spine of a 3-manifold M and $V = V(P)$ the set of its vertices. Denote by $N(V, P)$ a regular neighborhood of V in P. It consists of some number of disjoint copies of the butterfly E. The intersection of the union of all open 2-cells in P with each butterfly consists of exactly six wings.

Definition 8.2.1. *Two special polyhedra P_1 and P_2 are called* similar *if there exists a homeomorphism $\varphi : N(V(P_1), P_1) \to N(V(P_2), P_2)$ such that for any two wings w_1 and w_2 of P_1 the following condition holds: w_1 and w_2 belong to the same 2-cell of P_1 if and only if $\varphi(w_1)$ and $\varphi(w_2)$ belong to the same 2-cell of P_2. The homeomorphism φ is called a* similarity homeomorphism.

A good way to think of it is the following: let us paint the 2-cells of P_1 in different colors and the corresponding 2-cells of P_2 in the same colors. Then the similarity homeomorphism φ is required to preserve the colors of wings. In other words, P_2 must contain exactly the same colored butterflies as P_2.

Definition 8.2.2. *Two 3-manifolds M_1 and M_2 are said to be* similar *if a special spine of M_1 is similar to a special spine of M_2.*

Examples of similar but nonhomeomorphic 3-manifolds will be presented later. The following proposition is based on an idea of Lickorish [69]. It is related to all invariants of Turaev–Viro type, not only to Turaev–Viro ones (see Definition 8.1.5 and Remark 8.1.20).

Proposition 8.2.3. *Similar manifolds have the same invariants of Turaev–Viro type.*

Proof. Let us look carefully through the construction of Turaev–Viro type invariants (Sect. 8.1.1). We come to the conclusion that all what we need to know to calculate the invariants is just the number of vertices and 2-cells, and the incidence relation between vertices and 2-cells, see Definition 8.2.1. For similar spines these data coincide and hence produce the same invariants.

Below we describe moves on special polyhedra and moves on manifolds that transform them into similar ones. We start with moves on manifolds.

Let M be a (not necessarily orientable) 3-manifold and $F \subset \text{Int } M$ a closed connected surface such that F is two-sided in M and $\chi(F) \geq 0$. The last condition means that F is homeomorphic to S^2, RP^2, $T^2 = S^1 \times S^1$, or to the Klein bottle K^2. Choose a homeomorphism $r : F \to F$ such that:

(1) If $F = S^2$, then r reverses the orientation.

(2) If $F = RP^2$, then r is the identity.

(3) If $F = T^2$ or $F = K^2$, then r induces multiplication by -1 in $H_1(F; Z)$.

It is clear that r is unique up to isotopy. In case (3) one can explicitly describe it as follows: present the torus or the Klein bottle as a square with identified opposite edges. Then r is induced by the symmetry of the square with respect to the center.

Now cut M along F and repaste the two copies of F thus obtained according to the homeomorphism r. We get a new 3-manifold M_1.

Definition 8.2.4. *We say the new 3-manifold M_1 arising in such a way is obtained from M by the* manifold move *along F.*

Remark 8.2.5. The manifold move along RP^2 does not change the manifold, and neither does the move along any trivial (i.e., bounding a ball) 2-sphere. Suppose $F = T^2$ and F bounds a solid torus in M. Since $r : T^2 \to T^2$ can be extended to the interior of the solid torus, we have $M_1 = M$. The same is true for any Klein bottle that bounds in M a solid Klein bottle $S^1 \tilde{\times} D^2$.

Now let us turn our attention to moves on special polyhedra. Let G be a connected graph with two vertices of valence 3. There exist two such graphs: a theta-curve (a circle with a diameter) and an *eyeglass curve* (two circles joined by a segment). Choose a homeomorphism $\varrho : G \to G$ such that:

(1) If G is a theta-curve, then $\varrho = \varrho_1$, where $\varrho_1 : G \to G$ permutes the vertices and takes each of the three edges into itself.

(2) If G is an eyeglass curve, then $\varrho = \varrho_2$, where $\varrho_2 : G \to G$ leaves the joining segment fixed and inverses both loops, see Fig. 8.11.

Definition 8.2.6. *An one-dimensional subpolyhedron G of a special polyhedron P is called* proper *if a regular neighborhood $N(G, P)$ of G in P is a twisted or untwisted I-bundle over G. If $N(G, P) \approx G \times I$, then G is called* two-sided.

Let $G \subset P$ be a two-sided theta-curve or an eyeglass curve in a special polyhedron P. Cut P along G and repaste the two copies of G thus obtained according to the homeomorphism ϱ. We get a new special polyhedron P_1.

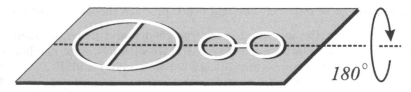

Fig. 8.11. Involution ϱ on the theta-curve and eyeglass curve

Definition 8.2.7. *We say the new special polyhedron P_1 is obtained from P by a spine move σ_i along G, where $i = 1$ if $\varrho = \varrho_1$ and $i = 2$ if $\varrho = \varrho_2$.*

Proposition 8.2.8. *Spine moves transform special polyhedra to similar ones.*

Proof. Let $G \subset P$ be a two-sided theta-curve or a two-sided eyeglass curve in a special polyhedron P. The edges of G decompose some 2-cells of P into smaller parts that are glued together to new 2-cells. Since ϱ takes each edge of G to itself, the boundary curves of the new 2-cells run along the same edges as before, although may pass along them in a different order. Nothing happens near vertices. It follows that the new special polyhedron is similar to P.

Recall that we have two types of spine moves: theta-move σ_1 and glasses-move σ_2. It is convenient to introduce the third move σ_3.

Let $G \subset P$ be a proper theta-curve with edges l_1, l_2, l_3 such that

(1) G separates P.
(2) l_1 and l_2 belong to the same 2-cell C of P.

Choose a homeomorphism $\varrho_3 : G \to G$ such that ϱ_3 leaves l_3 fixed and permutes l_1 and l_2. Cut P along G and repaste the two copies of G thus obtained according to ϱ_3. We say that the new special polyhedron P_1 arising in such a way is obtained from P by the move σ_3.

Lemma 8.2.9. *σ_3 can be expressed through σ_1 and σ_2.*

Proof. Let l_1, l_2 be the two edges of G which are contained in the same 2-cell C such that σ_3 transposes them. Then there exists a simple arc $l \subset C$ such that $l \cap G = \partial l$ and l connects l_1 with l_2. Consider a regular neighborhood $N = N(G \cup l)$ of $G \cup l$ in P. Since G separates P, it has a neighborhood homeomorphic to $G \times [0, 1]$. Hence N can be presented as $G \times [0, 1]$ with a twisted or an untwisted band B attached to $G \times \{1\}$, see Fig. 8.12.

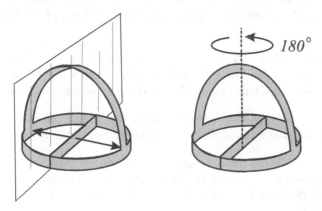

Fig. 8.12. Two types of $N = (G \times [0, 1]) \cup B$; the rotation by $180°$ determines a homeomorphism of N

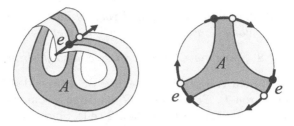

Fig. 8.13. Spine move across e: we cut out the region A and paste it back by a homeomorphism that permutes the *white* and *black* vertices and is invariant on edges

If the band is untwisted, then N is bounded by G_1 and G_2, where G_1 is a theta-curve isotopic to G and G_2 is an eyeglass curve. There exists a homeomorphism $h : N \to N$ such that $h|_{G_1} = \varrho_3$ and $h|_{G_2} = \varrho_2$ (h can be visualized as the symmetry in the vertical plane shown on Fig. 8.12). It follows that the move σ_3 along G_1 (and along G) is equivalent to the move σ_2 along G_2.

Let the band be twisted. Then N is bounded by two theta-curves G_1 and G_2, where G_1 is isotopic to G. There exists a homeomorphism $h : N \to N$ (this time the rotation by 180° around the vertical axis) such that $h|_{G_1} = \varrho_1\varrho_3$ and $h|_{G_2} = \varrho_1$. Hence, the superposition of the moves σ_1 and σ_3 along G_1 is equivalent to the move σ_1 along G_2. Taking into account that $\varrho_1^2 = 1$, we can conclude that the move σ_3 along G is equivalent to the superposition of the move σ_1 along G_1 and the move σ_1 along G_2.

Suppose the boundary curve of a 2-cell C of a special spine P passes along an edge e of P three times. Choose two points on e and join them by three arcs in C as it is shown on Fig. 8.13. The union G of the arcs is a proper two-sided theta or an eyeglass curve in P. One can consider the spine move along G. To distinguish this type of spine move we supply it with a special name.

Definition 8.2.10. *Let G be a proper two-sided theta-curve or an eyeglass curve in a special polyhedron P such that both vertices of G lie in the same edge. Then the spine move along G is called a* spine move across e.

Our next goal is to prove that spine moves induce moves on manifolds, and vice versa, manifold moves can be realized by spine moves.

Lemma 8.2.11. *Let G be a proper theta or an eyeglass curve in a special spine P of a closed 3-manifold M. Then there exists a closed connected surface $F \subset M$ such that $\chi(F) \geq 0$, $F \cap P = G$, and F is transversal to the singular graph SP of P.*

Proof. Let $N = N(P, M)$ be a regular neighborhood of P in M. Present N as the mapping cylinder $C_f = P \cup (\partial N \times [0, 1])/ \sim$ of an appropriate locally homeomorphic map $f : \partial N \to P$. Then $F_1 = (f^{-1}(G) \times [0, 1])/ \sim$ is a surface in N such that

(1) $F_1 \cap \partial N = \partial F_1$, $F_1 \cap P = G$, and F_1 is transversal to SP.
(2) G is a spine of F_1.

To obtain F, attach disjoint 2-cells contained in the 3-cell $M \setminus N$ to the boundary components of F_1. Since $\chi(F_1) = \chi(G) = -1$, we have $\chi(F) \geq 0$.

Proposition 8.2.12. *Let P be a special spine of a closed 3-manifold M, and let $G \subset P$ be a two-sided theta-curve or a two-sided eyeglass curve. Denote by P_1 the special polyhedron obtained from P by the spine move along G. Then*

(1) P_1 is a spine of a closed 3-manifold M_1.
(2) M_1 can be obtained from M by a manifold move.

Proof. Let $F \subset M$ be the surface constructed in Lemma 8.2.11. Since G is two-sided, F is also two-sided. The homeomorphism $\varrho : G \to G$ can be extended to a homeomorphism $r : F \to F$. It is clear that r satisfies conditions (1)–(3) preceding Definition 8.2.4 of a manifold move. Denote by M_1 the 3-manifold obtained from M by the manifold move along F. Since $r\big|_G = \varrho$, P_1 is a spine of M_1.

Proposition 8.2.13. *Let a closed 3-manifold M_1 be obtained from a closed 3-manifold M by a manifold move along a surface $F \subset M$. Then M and M_1 are similar.*

Proof. Let us construct a special spine P of M such that $G = P \cap F$ is a proper two-sided theta-curve. To do it, remove an open ball D^3 from M such that $D = D^3 \cap F$ consists of one open disc if $F = T^2, K^2$, and of three open discs if $F = S^2$. We do not take $F = RP^2$ since in this case the manifold move is trivial. Denote by F_1 the surface $F \setminus D$. Starting from $F_1 \times \partial I$, collapse a regular neighborhood $N = F_1 \times I$ in $M \setminus D^3$ onto $G \times I$, where G is a theta-graph in F_1. The collapsing can be easily extended to a collapsing of $M \setminus D^3$ onto a special spine $P \supset G \times I$.

Apply to P the spine move along G. It follows from Proposition 8.2.12 that the special polyhedron P_1 thus obtained is a spine of M_1. Since P and P_1 are similar, the same is true for M and M_1.

Example 8.2.14. We are ready now to construct two similar manifolds with different homology groups. Take $M_1 = S^1 \times S^1 \times S^1$ and consider the torus $T^2 = S^1 \times S^1 \times \{*\} \subset M$. To construct M_2, perform the manifold move on M_1 along T^2. By Proposition 8.2.13, M_2 is similar to M_1. A simple calculation shows that $H_1(M_1; Z) = Z \oplus Z \oplus Z$, and $H_1(M_2; Z) = Z_2 \oplus Z_2 \oplus Z$.

Remark 8.2.15. According to Preposition 8.2.3, manifolds M_1 and M_2 above have the same Turaev–Viro invariants. For instance, if $q = exp(\frac{\pi}{7}i)$, then $TV_q(M_1) = TV_q(M_2) = -63q^3 + 189q^2 + 378 + 189q^{-2} - 63q^{-3}$, see Table A.1, where M_1, M_2 have names 6_{70}, 6_{71}, and Fig. 8.9, where they are shown as genuine twins. Manifolds 6_{65}, 6_{67} as well as 6_{68}, 6_{69} occupying the neighboring lines, also form genuine twin pairs since they are related by the same manifold move.

It is interesting to recall here that Turaev–Viro invariants of order 2 determine the order of the second homology group with coefficients Z_2, see Sect. 8.1.7. This agrees with the observation that $H_2(M_1; Z_2) = H_2(M_2; Z_2) = Z_2 \oplus Z_2 \oplus Z_2$.

As we have claimed at the beginning of this section, under certain conditions similarity of 3-manifolds implies homeomorphism. The idea of the proof is to transform a special spine P_1 of the first manifold into a similar special spine P_2 of the second one step by step. Our first goal is to define *graph moves* for transforming the singular graph of P_1 to the one of P_2.

Let Γ be a finite (multi)graph. Fix a finite set A. By a *coloring* of Γ by A we mean a map $c : E(\Gamma) \to A$, where $E(\Gamma)$ is the set of all open edges of Γ. Denote by $V(\Gamma)$ the set of vertices of Γ and by $N(V, \Gamma)$ a regular neighborhood of V in Γ. The intersection of open edges with $N(V, \Gamma)$ consists of half-open 1-cells, which are called *thorns*.

Definition 8.2.16. *Two colored graphs Γ_1 and Γ_2 are called* similar, *if there exists a homeomorphism $\varphi : N(V(\Gamma_1), \Gamma_1) \to N(V(\Gamma_2), \Gamma_2)$ preserving the colors of thorns. The homeomorphism φ is called a* similarity homeomorphism.

Let Γ be a colored graph. Choose two edges e_1 and e_2 of the same color and cut each of them in the middle. Repaste the four "half edges" thus obtained into two new edges which do not coincide with the initial ones.

Definition 8.2.17. *We say the new colored graph Γ_1 arising in such a way is obtained from Γ by a* graph move *along e_1 and e_2. The graph move is called* admissible, *if Γ and Γ_1 are connected.*

Remark 8.2.18. For any given e_1 and e_2 there exist two different graph moves along e_1 and e_2. Suppose Γ is connected and $\Gamma \setminus \text{Int}\ (e_1 \cup e_2)$ consists of two connected components such that each of them contains one vertex of each edge. Then precisely one of the moves is admissible, see Fig. 8.14. If $\Gamma \setminus \text{Int}\ (e_1 \cup e_2)$ is connected, then both moves are admissible.

Lemma 8.2.19. *Let Γ_1 and Γ_2 be similar colored graphs. If they are connected, then one can pass from Γ_1 to Γ_2 by a sequence of admissible graph moves.*

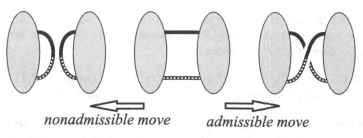

nonadmissible move   admissible move

Fig. 8.14. Admissible and nonadmissible graph moves

Proof. It follows from Definition 8.2.16 that there exists a homeomorphism $\varphi : N(V(\Gamma_1), \Gamma_1) \to N(V(\Gamma_2), \Gamma_2)$ preserving the colors of thorns. We call an edge e in Γ_1 *correct* if φ maps the two thorns t_1 and t_2 contained in it into the same edge f in Γ_2. The thorns t_1 and t_2, the edge f and the thorns $\varphi(t_1), \varphi(t_2) \subset f$ are also called *correct*. The homeomorphism φ can be extended to an edge e if and only if e is correct, so to prove Lemma 8.2.19 it is sufficient to show that the number of correct edges can be increased by admissible graph moves on Γ_1 and Γ_2.

Let t_1 be an incorrect thorn in Γ_1 and let $t_2, t_3, \ldots, t_{2n}$ be all other incorrect thorns of the same color (say, red). We shall say that a thorn t_i, $2 \le i \le 2n$, is *good* (with respect to t_1), if t_1 and t_i belong to the same edge or if they can be transferred to the same edge by an admissible graph move on Γ_1. Denote by T the set $\{\tau_i = \varphi(t_i), \ 1 \le i \le 2n\}$ of all red incorrect thorns in Γ_2. We shall say that a thorn τ_i, $2 \le i \le 2n$, is *good*, if τ_1 and τ_i belong to the same edge or if they can be transferred to the same edge by an admissible graph move on Γ_2.

Consider two subsets A_1 and A_2 of the set T. The subset $A_1 \subset T$ consists of the images of good thorns in Γ_1, the subset $A_2 \subset T$ is the set of all good thorns in Γ_2. Let $\#X$ denote the number of elements in X. Since any red incorrect edge in Γ_1 and Γ_2 contains at least one good thorn, we have $\#A_1 \ge n$ and $\#A_2 \ge n$. Note that $\#T = 2n$ and, because t_1 and $\tau_1 = \varphi(t_1)$ are not good, τ_1 does not belong to $A_1 \cup A_2$. Hence, $\#(A_1 \cup A_2) < 2n$, and $A_1 \cap A_2 \ne \emptyset$. We can conclude that there exist i and j, $2 \le i, \ j \le 2n$, such that t_i and τ_j are good. By definition of good edges, we can perform admissible graph moves such that after these moves t_1 and t_2 belong to the same edge and τ_1 and τ_2 also belong to the same edge. The moves are performed along incorrect edges. Hence, all correct edges are preserved, but now a new correct edge has appeared (just the one containing t_1 and t_2).

Our next step is to prove Proposition 8.2.22 below stating that under certain conditions any similarity homeomorphism between neighborhoods of vertices of special spines can be extended to the union of edges. We need two lemmas.

Lemma 8.2.20. *Let P be a special spine of a closed 3-manifold M. Suppose that every surface $F \subset M$ with $\chi(F) \geq 0$ separates M. Then each proper theta or eyeglass curve $G \subset P$ separates P.*

Proof. Let $F \subset M$ be the surfaces constructed in Lemma 8.2.11. Since F separates M and P is a spine of M, $\Gamma = F \cap P$ separates P.

Suppose P is a special spine of a closed 3-manifold M. Let us color the 2-cells of P in different colors. At each edge of P three 2-cells meet, and so to each edge there corresponds some unordered triplet of colors (possibly with multiplicity). We call this triplet the *tricolor* of the edge. Thus, we may consider SP as a colored graph. Note that each spine move on P induces an admissible graph move on SP. It turns out that under certain conditions all admissible graph moves can be obtained in this way.

Lemma 8.2.21. *Let P be a special spine of a 3-manifold M. If every surface $F \subset M$ with $\chi(F) \geq 0$ is separating, then each admissible graph move γ on SP is induced by a spine move on P.*

Proof. Let γ be performed along edges e_1 and e_2. Then e_1 and e_2 have the same tricolor. Connect the middle points of e_1 and e_2 by three disjoint arcs $l_j \subset P$ ($j = 1, 2, 3$) in such a way that $G = l_1 \cup l_2 \cup l_3$ is a proper theta-curve. If the tricolor has multiplicity, this is also possible. By Lemma 8.2.20, G separates P into two parts such that each part contains one vertex of e_1 and one vertex of e_2. Denote by σ_1 the spine move along G. Then σ_1 induces an admissible graph move along e_1 and e_2. Since such a move is unique (see Remark 8.2.18), it coincides with γ.

Recall that if a closed 3-manifold M is irreducible, then any compressible torus or Klein bottle in M bounds a solid torus or Klein bottle, respectively. There exist no compressible projective planes at all. It follows that if an irreducible M contains no closed incompressible surfaces with nonnegative Euler characteristic, then the following holds:

(1) Every surface $F \subset M$ with $\chi(F) \geq 0$ separates M.
(2) Every manifold move on M produces a homeomorphic manifold, see Remark 8.2.5.

Proposition 8.2.22. *Let M_1 and M_2 be similar closed 3-manifolds. Suppose M_1 is irreducible and does not contain closed incompressible surfaces with nonnegative Euler characteristic. Then there exist special spines P_i of M_i ($i = 1, 2$) and a homeomorphism $\psi : N_1 \cup SP_1 \rightarrow N_2 \cup SP_2$ such that $\psi|_{N_1} : N_1 \rightarrow N_2$ is a similarity homeomorphism, where $N_i = N(V(P_i), P_i)$.*

Proof. Let $\varphi : N_1 \rightarrow N_2$ be a similarity homeomorphism, where P_1 and P_2 are special spines of M_1 and M_2, respectively. We imagine the 2-cells of P_1 and P_2 as being painted in different colors such that φ preserves the colors of wings. As above, we paint also each edge in the corresponding tricolor. Then φ induces a similarity homeomorphism between SP_1 and SP_2. If all edges of SP_1 are correct, then φ can be extended to a homeomorphism ψ satisfying the conclusion of the proposition. If not, we use Lemma 8.2.19 to correct them by a sequence of graph moves. By Lemma 8.2.21 this sequence can be realized by a sequence of spine moves. It remains to note that each move on a spine of M_1 produces a spine of the same manifold, so we do not violate the assumption on M.

Let P_1 and P_2 be special spines of M_1 and M_2, and let a homeomorphism $\psi : N_1 \cup SP_1 \to N_2 \cup SP_2$ induce a similarity homeomorphism ψ' between $N_1 = N(V(P_1), P_1)$ and $N_2 = N(V(P_2), P_2)$. Identify $N_1 \cup SP_1$ and $N_2 \cup SP_2$ via ψ. We obtain two special spines P_1 and P_2 such that their singular graphs and wings coincide.

Let e be an edge of P_1. It contains two thorns t_1, t_2. Let $\omega_1^{(i)}, \omega_2^{(i)}, \omega_3^{(i)}$ be the wings adjacent to t_i, $i = 1, 2$. A regular neighborhood $N(e \backslash \mathrm{Int}\, (t_1 \cup t_2), P_1)$ of a middle part of e in P_1 is homeomorphic to $Y \times I$, where Y is a wedge of three segments. Hence, we have a natural bijection $a_{1e} : \{\omega_1^{(1)}, \omega_2^{(1)}, \omega_3^{(1)}\} \to \{\omega_1^{(2)}, \omega_2^{(2)}, \omega_3^{(2)}\}$. In the same way a direct product structure on $N(e \backslash \mathrm{Int}\, (t_1 \cup t_2), P_2)$ determines a natural bijection $a_{2e} : \{\omega_1^{(1)}, \omega_2^{(1)}, \omega_3^{(1)}\} \to \{\omega_1^{(2)}, \omega_2^{(2)}, \omega_3^{(2)}\}$. Denote by β_e the permutation $a_{2e}^{-1} a_{1e}$.

Definition 8.2.23. *An edge e is called* even *(odd) if β_e is an even (odd) permutation.*

Let C be a 2-cell of P_1. Denote by E_C the collection of edges incident to C. We allow multiplicity, so if the boundary curve of C passes along an edge e two (three) times, then e is included in E_C two (three) times. Note that E_C coincides with the set of edges incident to the 2-cell of P_2 having the same color.

Lemma 8.2.24. *For any 2-cell C of P_1 the collection E_C contains an even number of odd edges.*

Proof. Regular neighborhoods $N(V(P_i), M_i)$ $(i = 1, 2)$ consist of 3-balls. Choose orientations of the 3-balls such that the similarity homeomorphism $\psi' : N(V(P_1), P_1) \to N(V(P_2), P_2)$ is extendible to an orientation preserving homeomorphism between $N(V(P_1), M_1)$ and $N(V(P_2), M_2)$. The orientations induce a cyclic order on the set $\{\omega_1^{(j)}, \omega_2^{(j)}, \omega_3^{(j)}\}$ of wings adjacent to each thorn of P_1 or P_2. We shall say that an edge e is *orientation reversing* with respect to P_i, if the corresponding bijection $a_{ie} : \{\omega_1^{(1)}, \omega_2^{(1)}, \omega_3^{(1)}\} \to \{\omega_1^{(2)}, \omega_2^{(2)}, \omega_3^{(2)}\}$ preserves the cyclic order, $i = 1, 2$. Since the boundary curve of each 2-cell in a 3-manifold is orientation preserving, E_C contains an even number of orientation reversing edges with respect to P_1 and an even number of orientation reversing edges with respect to P_2. It remains to note that e is odd if and only if e is orientation reversing with respect to one of spines P_1, P_2, and orientation preserving with respect to the other.

Theorem 8.2.25. *Let M_1 and M_2 be similar closed 3-manifolds. Suppose M_1 is irreducible and does not contain closed incompressible surfaces with non-negative Euler characteristics. Then M_1 and M_2 are homeomorphic.*

Proof. According to Proposition 8.2.22, there exist special spines P_i of M_i $(i = 1, 2)$ and a homeomorphism $\psi : N_1 \cup SP_1 \to N_2 \cup SP_2$ such that $\psi\big|_{N_1} :$ $N_1 \to N_2$ is a similarity homeomorphism, where $N_i = N(V(P_i), P_i)$. As above, identify $N_1 \cup SP_1$ with $N_2 \cup SP_2$ via ψ. We define an edge e of P_1 to be *strongly correct* (SC) if the corresponding permutation β_e is trivial. In other words, e is SC if and only if the identification ψ can be extended to a neighborhood of e in P_1. Note that if all edges are SC, then ψ can be extended to a homeomorphism between P_1 and P_2 and to a homeomorphism between M_1 and M_2. We claim that one can perform spine moves on P_1 until all edges become SC. This will prove Theorem 8.2.25, because each spine move can be extended to a manifold move on M_1 that does not change its homeomorphism type.

As above, we paint the 2-cells of P_1 and P_2 in different colors and the edges in tricolors. Note that if the tricolor of an edge e consists of three different colors, then e is obviously SC. Assume that the tricolor of e is bichromatic (i.e., it has the form (x, y, y), $x \neq y$), and that e is not SC. Then e is odd. It follows from Lemma 8.2.24 that there is another non-SC edge e' of tricolor (x, z, z) (possibly $z = x$ or $z = y$). Assuming first that $z \neq y$, we construct a proper eyeglass curve G with the vertices on e and e' (this is also possible when $z = x$). By Lemma 8.2.20, G is two-sided, and the spine move σ_2 along G can be performed. The edge e will now be SC. If $z = y$, there are two possibilities for the relative displacement of e and e' along the boundary curve of y-colored 2-cell: the displacement (e, e, e', e') and the displacement (e, e', e, e'). In the first case we can still construct an eyeglass curve with vertices on e and e' and perform σ_2. In the second case we construct a proper theta-curve G with the vertices on e and e'. The move σ_3 along G makes e strongly correct.

Assume now e is a monochromatic non-SC edge of tricolor (x, x, x), and assume that there is another edge e' with the same tricolor. Denote by C_x the x-colored 2-cell of P_1. We shall say that e and e' are *linked* if the boundary curve of C_x cannot be decomposed into two arcs d and d', such that d passes three times along l and d' passes three times along l'. Suppose that l and l' are linked. In order to make l strongly correct, we use spine moves σ_3 along theta-curves with vertices on l and l'. Each such move changes β_e by some permutation. It is sufficient to show that each transposition τ of wings can be achieved. In essence, there are two possibilities for the relative displacement of e and e' on the boundary curve of C_x. It is clear that in both cases τ can be realized by a move σ_3 along the theta-curve $G = l_1 \cup l_2 \cup l_3$, see Fig. 8.15.

Suppose now that each two non-SC edges of tricolor (x, x, x) are unlinked. If e is an odd edge with tricolor (x, x, x), then there is another odd edge e'

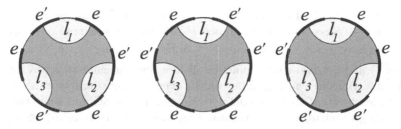

Fig. 8.15. Two linked and one unlinked positions of edges e, e' in the boundary of a 2-cell

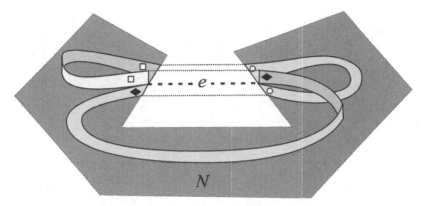

Fig. 8.16. Decomposition of wings into pairs

with the same tricolor. We use the manifold move along $G = l_1 \cup l_2 \cup l_3$ (see Fig. 8.15) to make e and e' even.

It remains to consider the following situation: all non-SC edges are monochromatic and even, and there are no linked edges among them. Let e be a non-SC edge with tricolor (x, x, x). Denote by P_3 the spine obtained from P_1 by the spine move across e, see Definition 8.2.10. Let t_1 and t_2 be the thorns in e and let $w_1^{(i)}, w_2^{(i)}, w_3^{(i)}$ be the wings adjacent to t_i, $i = 1, 2$. The direct product structures on regular neighborhoods of $e \setminus \operatorname{Int} (t_1 \cup t_2)$ in P_i determine natural bijections $a_{ie} : \{w_1^{(1)}, w_2^{(1)}, w_3^{(1)}\} \to \{w_1^{(2)}, w_2^{(2)}, w_3^{(2)}\}$, $i = 1, 2, 3$. It is sufficient to prove that a_{2e} coincides with a_{3e}, because this means that the spine move across e makes e strongly correct.

Consider a regular neighborhood N of $SP_1 \setminus e$ in P_1. The difference $N \setminus SP_1$ consists of some number of half-open annuli and precisely three x-colored half-open discs. Each of the discs contains two wings from the set $W = \{w_j^{(i)}, 1 \le j \le 3, i = 1, 2\}$. Thus, we have a decomposition of the set W into three pairs. In Fig. 8.16 the wings forming each pair are marked with similar signs. Taking P_2 or P_3 instead of P_1, we obtain two other decompositions. A very important observation: since all non-monochromatic edges are SC and e is not linked with any other edge, these three decompositions coincide.

At least one pair of the decomposition contains a wing adjacent to the pair, $1 \leq j,\ k \leq 3$. Since each of the spines P_1, P_2, P_3 contains only one x-colored 2-cell, we have $a_{ie}(w_j^{(1)}) \neq w_k^{(2)}$, $1 \leq i \leq 3$. Hence, among $a_{1e}(w_j^{(1)})$, $a_{2e}(w_j^{(1)})$, $a_{3e}(w_j^{(1)})$ at least two wings coincide. Taking into account that any two different bijections a_{1e}, a_{2e}, a_{3e} differ on an even permutation, we can conclude that at least two of them do coincide. Since e is not SC and since the spine move across e changes the corresponding bijection, we have $a_{1e} \neq a_{2e}$ and $a_{1e} \neq a_{3e}$. It follows that $a_{2e} \equiv a_{3e}$.

A

Appendix

A.1 Manifolds of Complexity ≤ 6

By means of an arbitrary ordering we order closed orientable irreducible 3-manifolds of each complexity $k \leq 6$ (Tables A.1–A.7) and write k_i for the manifold number i among those of complexity k. This method of notation is borrowed from knot theory (for example, see [109]).

We present 3-manifolds as follows.

I. $L_{p,q}$ is the lens space with parameters p, q.
II. S^3/G is the quotient space of S^3 by a free linear action of a nonabelian finite group G. All such groups are known, see [94]. They are:
   (a) Finite cyclic groups
   (b) Groups $Q_{4n}, n \geq 2$
   (c) Groups $D_{2^k(2n+1)}, k \geq 3, n \geq 1$
   (d) Groups P_{24}, P_{48}, P_{120}, and $P'_{8(3^k)}, k \geq 2$
   (e) Direct products of any of these groups with a cyclic group of coprime order

The subscripts show the orders of the groups. Presentations by generators and relations, and abelian quotients of the groups (coinciding with the first homology groups of the corresponding 3-manifolds) are the following:

1. $Q_{4n} = \langle x, y | x^2 = (xy)^2 = y^n \rangle; H_1 = Z_2 \oplus Z_2$ if n is even, and Z_4 if n is odd.
2. $D_{2^k(2n+1)} = \langle x, y | x^{2^k} = 1, y^{2n+1} = 1, xyx^{-1} = y^{-1} \rangle; H_1 = Z_{2^k}$.
3. $P_{24} = \langle x, y | x^2 = (xy)^3 = y^3, x^4 = 1 \rangle; H_1 = Z_3$.
4. $P_{48} = \langle x, y | x^2 = (xy)^3 = y^4, x^4 = 1 \rangle; H_1 = Z_2$.
5. $P_{120} = \langle x, y | x^2 = (xy)^3 = y^5, x^4 = 1 \rangle; H_1 = 0$.
6. $P'_{8(3^k)} = \langle x, y, z | x^2 = (xy)^2 = y^2, zxz^{-1} = y, zyz^{-1} = xy, z^{3^k} = 1 \rangle; H_1 = Z_{3^k}$.

III. Let $h_A : T \to T$ be a self-homeomorphism of the torus $T = S^1 \times S^1$ corresponding to an unimodular integer matrix $A = (a_{ij})$ of order 2. This means that h_A takes any curve of the type (m, n) to a curve of the type $(a_{11}m + a_{12}n, a_{21}m + a_{22}n$. Then $T \times I/A$ is the Stalling manifold with fiber T and monodromy map h_A. In other words, $T \times I/A$ is obtained from $T \times I$ by identifying its boundary tori via h_A.

IV. Recall that the boundary of the orientable I-bundle $K \tilde\times I$ over the Klein bottle K is a torus. Choose on it a coordinate system (m, l) such that m projects onto a meridian (i.e., a nonseparating orientation-preserving circle) of K and l double covers a longitude (an orientation-reversing circle) of K. Let A be an integer matrix of order 2 with determinant (-1). Then $K \tilde\times I \cup K \tilde\times I/A$ denotes the manifold obtained by pasting together two copies of $K \tilde\times I$ via h_A.

Finally, $(F, (p_1, q_1), \ldots, (p_k, q_k))$ is the orientable Seifert manifold with the base surface F and k fibers with non-normalized parameters $(p_i, q_i), 1 \leq i \leq k$. We do not write Seifert structures and homology groups of lens spaces, since they are well known: $L_{p,q} = (S^2, (q, p))$ and $H_1(L_{p,q}; Z) = Z_p$.

Remark A.1.1. For reader's convenience, we recall the definition of the Seifert manifold $M = (F, (p_1, q_1), \ldots, (p_k, q_k)), k \geq 1$, where F is a compact surface and (p_i, q_i) are pairs of coprime integers. Consider a surface F_1 obtained from F by removing the interiors of k disjoint discs. The boundary circles of these discs are denoted by $c_1, \ldots, c_k$. Let $c_{k+1}, \ldots, c_n$ be all the remaining circles of ∂F. If F is closed, then this set is empty.

Consider an orientable S^1-bundle M_1 over F_1. In other words, $M_1 = F_1 \times S^1$ or $M_1 = F_1 \tilde\times S^1$, depending on whether or not F_1 is orientable. We choose an orientation of M_1 and a section $s : F_1 \to M_1$ of the projection map $p : M_1 \to F_1$. On each torus $T_i = p^{-1}(c_i), 1 \leq i \leq n$, we choose a coordinate system (μ_i, λ_i) taking $s(c_i)$ as μ_i and a fiber $p^{-1}(\{*\})$ as λ_i. The orientations of the coordinate curves must satisfy the following conditions:

1. In case $M_1 = F_1 \times S^1$ the orientations of λ_i must be induced by a fixed orientation of S^1. If $M_1 = F_1 \tilde\times S^1$, then the orientations of λ_i can be chosen arbitrarily.
2. The intersection number of μ_i with λ_i must be 1.

Now, let us attach solid tori $V_i = D_i^2 \times S^1, 1 \leq i \leq k$, to M_1 via homeomorphisms $h_i : \partial V_i \to T_i$ such that each h_i takes the meridian $\partial D_i^2 \times \{*\}$ of V_i into a curve of the type (p_i, q_i). The resulting manifold is M. We emphasize that the remaining boundary tori $T_i, k + 1 \leq i \leq n$, of M still possess coordinate systems (μ_i, λ_i).

Table A.1. Complexity 0

c_i	M
0_1	S^3
0_2	RP^3
0_3	$L_{3,1}$

Table A.2. Complexity 1

c_i	M
1_1	$L_{4,1}$
1_2	$L_{5,2}$

Table A.3. Complexity 2

c_i	M
2_1	$L_{5,1}$
2_2	$L_{7,2}$
2_3	$L_{8,3}$

c_i	M	Seifert structure	$H_1(M; Z)$
2_4	S^3/Q_8	$(S^2, (2,1), (2,1)(2,-1))$	$Z_2 \oplus Z_2$

Table A.4. Complexity 3

c_i	M		c_i	M
3_1	$L_{6,1}$		3_4	$L_{11,3}$
3_2	$L_{9,2}$		3_5	$L_{12,5}$
3_3	$L_{10,3}$		3_6	$L_{13,5}$

c_i	M	Seifert structure	$H_1(M; Z)$
3_7	S^3/Q_{12}	$(S^2, (2,1), (2,1), (3,-2))$	Z_4

Table A.5. Complexity 4

c_i	M		c_i	M
4_1	$L_{7,1}$		4_6	$L_{16,7}$
4_2	$L_{11,2}$		4_7	$L_{17,5}$
4_3	$L_{13,3}$		4_8	$L_{18,5}$
4_4	$L_{14,3}$		4_9	$L_{19,7}$
4_5	$L_{15,4}$		4_{10}	$L_{21,8}$

c_i	M	Seifert structure	$H_1(M; Z)$
4_{11}	$S^3/Q_8 \times Z_3$	$(S^2, (2,1), (2,1), (2,1))$	$Z_2 \oplus Z_6$
4_{12}	S^3/Q_{16}	$(S^2, (2,1), (2,1), (4,-3))$	$Z_2 \oplus Z_2$
4_{13}	S^3/D_{24}	$(S^2, (2,1), (2,1), (3,-1))$	Z_8
4_{14}	S^3/P_{24}	$(S^2, (2,1), (3,1), (3,-2))$	Z_3

Table A.6. Complexity 5

c_i	M
5_1	$L_{8,1}$
5_2	$L_{13,2}$
5_3	$L_{16,3}$
5_4	$L_{17,3}$
5_5	$L_{17,4}$
5_6	$L_{19,4}$
5_7	$L_{20,9}$
5_8	$L_{22,5}$
5_9	$L_{23,5}$
5_{10}	$L_{23,7}$

c_i	M
5_{11}	$L_{24,7}$
5_{12}	$L_{25,7}$
5_{13}	$L_{25,9}$
5_{14}	$L_{26,7}$
5_{15}	$L_{27,8}$
5_{16}	$L_{29,8}$
5_{17}	$L_{29,12}$
5_{18}	$L_{30,11}$
5_{19}	$L_{31,12}$
5_{20}	$L_{34,13}$

c_i	M	Seifert structure	$H_1(M;Z)$
5_{21}	$S^3/Q_8 \times Z_5$	$(S^2,(2,1),(2,1),(2,3))$	$Z_2 \oplus Z_{10}$
5_{22}	$S^3/Q_{12} \times Z_5$	$(S^2,(2,1),(2,1),(3,2))$	Z_{20}
5_{23}	$S^3/Q_{16} \times Z_3$	$(S^2,(2,1),(2,1),(4,-1))$	$Z_2 \oplus Z_6$
5_{24}	S^3/Q_{20}	$(S^2,(2,1),(2,1),(5,-4))$	Z_4
5_{25}	$S^3/Q_{20} \times Z_3$	$(S^2,(2,1),(2,1),(5,-2))$	Z_{12}
5_{26}	S^3/D_{40}	$(S^2,(2,1),(2,1),(5,-3))$	Z_8
5_{27}	S^3/D_{48}	$(S^2,(2,1),(2,1),(3,1))$	Z_{16}
5_{28}	$S^3/P_{24} \times Z_5$	$(S^2,(2,1),(3,2),(3,-1))$	Z_{15}
5_{29}	S^3/P_{48}	$(S^2,(2,1),(3,1),(4,-3))$	Z_2
5_{30}	S^3/P'_{72}	$(S^2,(2,1),(3,2),(3,-2))$	Z_9
5_{31}	S^3/P_{120}	$(S^2,(2,1),(3,1),(5,-4))$	0

Table A.7. Complexity 6

c_i	M	c_i	M	c_i	M	c_i	M
6_1	$L_{9,1}$	6_{10}	$L_{28,5}$	6_{19}	$L_{35,8}$	6_{28}	$L_{41,16}$
6_2	$L_{15,2}$	6_{11}	$L_{29,9}$	6_{20}	$L_{36,11}$	6_{29}	$L_{43,12}$
6_3	$L_{19,3}$	6_{12}	$L_{30,7}$	6_{21}	$L_{37,8}$	6_{30}	$L_{44,13}$
6_4	$L_{20,3}$	6_{13}	$L_{31,7}$	6_{22}	$L_{37,10}$	6_{31}	$L_{45,19}$
6_5	$L_{21,4}$	6_{14}	$L_{31,11}$	6_{23}	$L_{39,14}$	6_{32}	$L_{46,17}$
6_6	$L_{23,4}$	6_{15}	$L_{32,7}$	6_{24}	$L_{39,16}$	6_{33}	$L_{47,13}$
6_7	$L_{24,5}$	6_{16}	$L_{33,7}$	6_{25}	$L_{40,11}$	6_{34}	$L_{49,18}$
6_8	$L_{24,11}$	6_{17}	$L_{33,10}$	6_{26}	$L_{41,11}$	6_{35}	$L_{50,19}$
6_9	$L_{27,5}$	6_{18}	$L_{34,9}$	6_{27}	$L_{41,12}$	6_{36}	$L_{55,21}$

c_i	M	Seifert structure	$H_1(M;Z)$
6_{37}	$S^3/Q_8 \times Z_7$	$(S^2,(2,1),(2,1),(2,5))$	$Z_2 \oplus Z_{14}$
6_{38}	$S^3/Q_{12} \times Z_7$	$(S^2,(2,1),(2,1),(3,4))$	Z_{28}
6_{39}	$S^3/Q_{16} \times Z_5$	$(S^2,(2,1),(2,1),(4,1))$	$Z_2 \oplus Z_{10}$
6_{40}	$S^3/Q_{16} \times Z_7$	$(S^2,(2,1),(2,1),(4,3))$	$Z_2 \oplus Z_{14}$
6_{41}	$S^3/Q_{20} \times Z_7$	$(S^2,(2,1),(2,1),(5,2))$	Z_{28}
6_{42}	S^3/Q_{24}	$(S^2,(2,1),(2,1),(6,-5))$	$Z_2 \oplus Z_2$
6_{43}	$S^3/Q_{28} \times Z_3$	$(S^2,(2,1),(2,1),(7,-4))$	Z_{12}

Table A.7. (continued)

6_{44}	$S^3/Q_{28} \times Z_5$	$(S^2, (2,1), (2,1), (7,-2))$	Z_{20}
6_{45}	$S^3/Q_{32} \times Z_3$	$(S^2, (2,1), (2,1), (8,-5))$	$Z_2 \oplus Z_6$
6_{46}	$S^3/Q_{32} \times Z_5$	$(S^2, (2,1), (2,1), (8,-3))$	$Z_2 \oplus Z_{10}$
6_{47}	S^3/D_{56}	$(S^2, (2,1), (2,1), (7,-5))$	Z_8
6_{48}	S^3/D_{80}	$(S^2, (2,1), (2,1), (5,-1))$	Z_{16}
6_{49}	S^3/D_{96}	$(S^2, (2,1), (2,1), (3,5))$	Z_{32}
6_{50}	S^3/D_{112}	$(S^2, (2,1), (2,1), (7,-3))$	Z_{16}
6_{51}	S^3/D_{160}	$(S^2, (2,1), (2,1), (5,3))$	Z_{32}
6_{52}	$S^3/P_{24} \times Z_7$	$(S^2, (2,1), (3,1), (3,1))$	Z_{21}
6_{53}	$S^3/P_{24} \times Z_{11}$	$(S^2, (2,1), (3,2), (3,2))$	Z_{33}
6_{54}	$S^3/P_{48} \times Z_5$	$(S^2, (2,1), (3,2), (4,-3))$	Z_{10}
6_{55}	$S^3/P_{48} \times Z_7$	$(S^2, (2,1), (3,1), (4,-1))$	Z_{14}
6_{56}	$S^3/P_{48} \times Z_{11}$	$(S^2, (2,1), (3,2), (4,-1))$	Z_{22}
6_{57}	$S^3/P_{120} \times Z_7$	$(S^2, (2,1), (3,1), (5,-3))$	Z_7
6_{58}	$S^3/P_{120} \times Z_{13}$	$(S^2, (2,1), (3,1), (5,-2))$	Z_{13}
6_{59}	$S^3/P_{120} \times Z_{17}$	$(S^2, (2,1), (3,2), (5,-3))$	Z_{17}
6_{60}	$S^3/P_{120} \times Z_{23}$	$(S^2, (2,1), (3,2), (5,-2))$	Z_{23}
6_{61}	S^3/P'_{216}	$(S^2, (2,1), (3,1), (3,2))$	Z_{27}
6_{62}		$(S^2, (3,1), (3,1), (3,-1))$	$Z_3 \oplus Z_3$
6_{63}		$(S^2, (3,2), (3,2), (3,-2))$	$Z_3 \oplus Z_6$
6_{64}		$(S^2, (3,2), (3,2), (3,-1))$	$Z_3 \oplus Z_9$

c_i	M	Seifert structure	$H_1(M; Z)$
6_{65}	$T \times I/ \begin{pmatrix} 1 & -1 \\ 1 & 0 \end{pmatrix}$	$(S^2, (2,-1), (3,1), (6,1))$	Z
6_{66}	$T \times I/ \begin{pmatrix} 0 & 1 \\ -1 & 0 \end{pmatrix}$	$(S^2, (2,-1), (4,1), (4,1))$	$Z_2 \oplus Z$
6_{67}	$T \times I/ \begin{pmatrix} 0 & 1 \\ -1 & -1 \end{pmatrix}$	$(S^2, (3,1), (3,1), (3,-2))$	$Z_3 \oplus Z$
6_{68}	$T \times I/ \begin{pmatrix} -1 & 0 \\ -1 & -1 \end{pmatrix}$	$(K, (1,1))$	$Z_4 \oplus Z$
6_{69}	$T \times I/ \begin{pmatrix} 1 & 0 \\ 1 & 1 \end{pmatrix}$	$(T, (1,1))$	$Z \oplus Z$
6_{70}	$T \times I/ \begin{pmatrix} -1 & 0 \\ 0 & -1 \end{pmatrix}$	$(S^2, (2,1), (2,-1), (2,1), (2,-1)) = K \tilde{\times} S^1$	$Z_2 \oplus Z_2 \oplus Z$
6_{71}	$T \times I/ \begin{pmatrix} 1 & 0 \\ 0 & 1 \end{pmatrix}$	$T \times S^1$	$Z \oplus Z \oplus Z$
6_{72}	$K \tilde{\times} I \cup K \tilde{\times} I/ \begin{pmatrix} -1 & 0 \\ -1 & 1 \end{pmatrix}$	$(S^2, (2,1), (2,1), (2,1), (2,-1))$	$Z_2 \oplus Z_2 \oplus Z_4$
6_{73}	$K \tilde{\times} I \cup K \tilde{\times} I/ \begin{pmatrix} 0 & 1 \\ 1 & 0 \end{pmatrix}$	$(RP^2, (2,1), (2,-1))$	$Z_4 \oplus Z_4$
6_{74}	$K \tilde{\times} I \cup K \tilde{\times} I/ \begin{pmatrix} 1 & 1 \\ 1 & 0 \end{pmatrix}$	$(RP^2, (2,1), (2,1))$	$Z_4 \oplus Z_4$

A.2 Minimal Spines of Manifolds up to Complexity 6

For any manifold $k_i, k \leq k \leq 6$, we present all its minimal special spines. Manifolds 0_1–0_3 of complexity 0 are presented by their minimal almost simple spines. The spines are given by regular neighborhoods of their singular graphs.

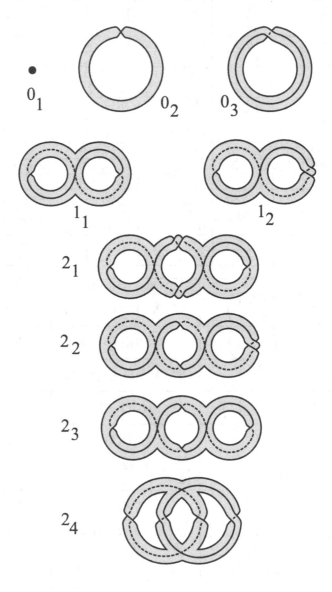

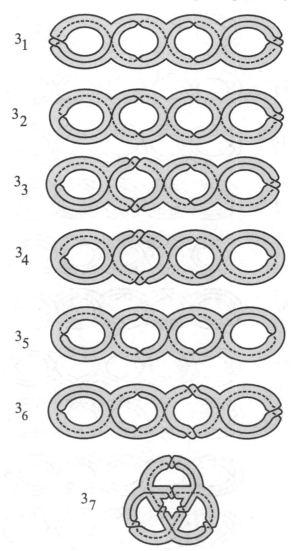

3_1

3_2

3_3

3_4

3_5

3_6

3_7

4_1

4_2

4_3

4_4

4_5

4_6

4_7

4_8

4_9

4_{10}

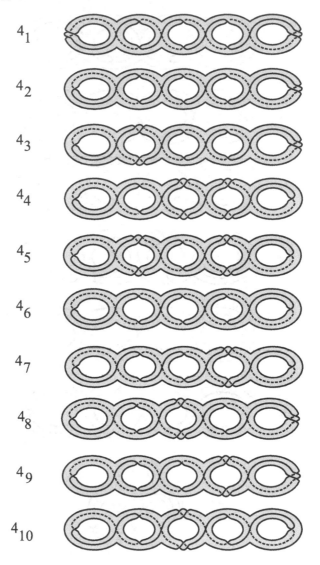

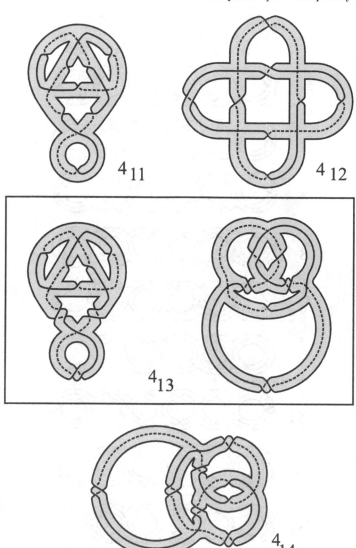

4_{11}

4_{12}

4_{13}

4_{14}

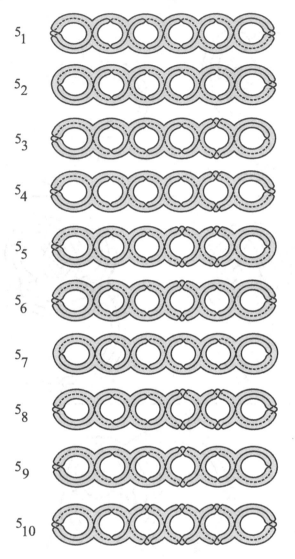

5_{11}

5_{12}

5_{13}

5_{14}

5_{15}

5_{16}

5_{17}

5_{18}

5_{19}

5_{20}

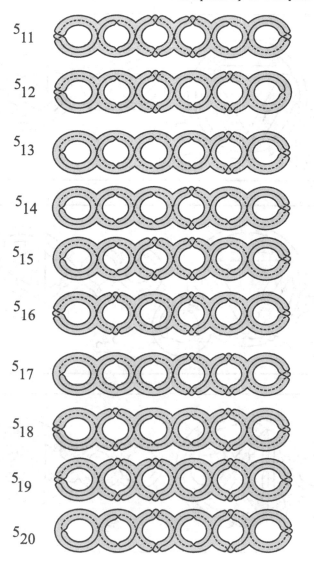

5_{21}

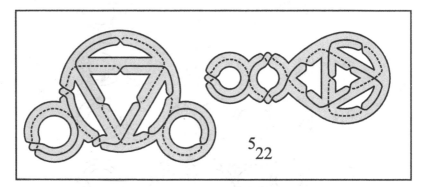

5_{22}

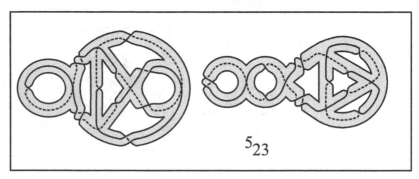

5_{23}

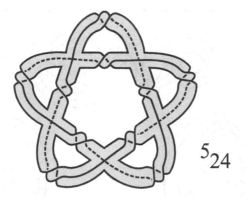

5_{24}

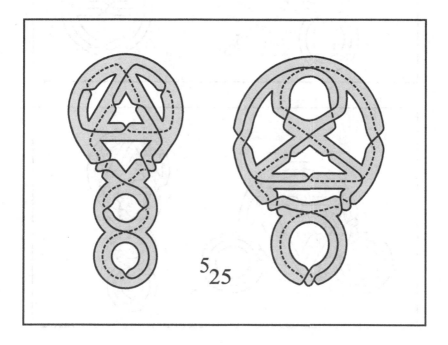

5_{25}

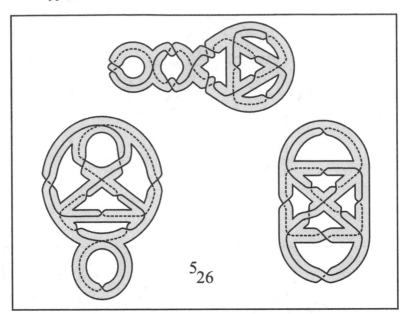

5_{26}

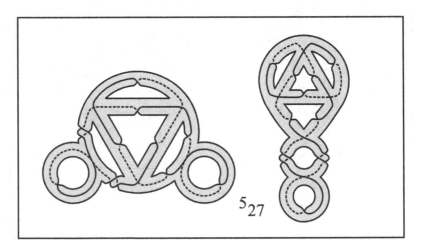

5_{27}

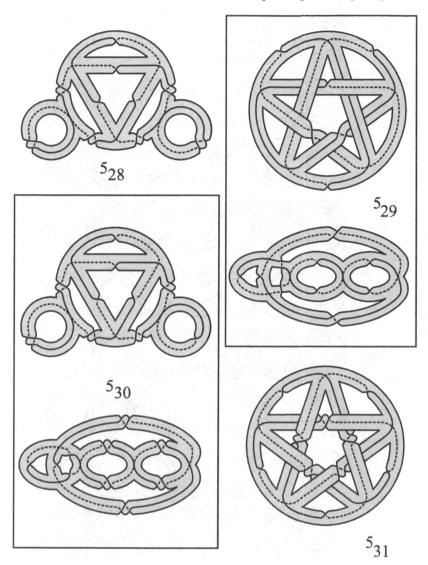

5_{28}

5_{29}

5_{30}

5_{31}

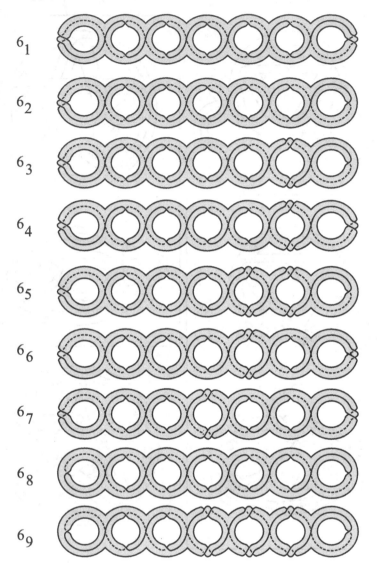

6_{10}

6_{11}

6_{12}

6_{13}

6_{14}

6_{15}

6_{16}

6_{17}

6_{18}

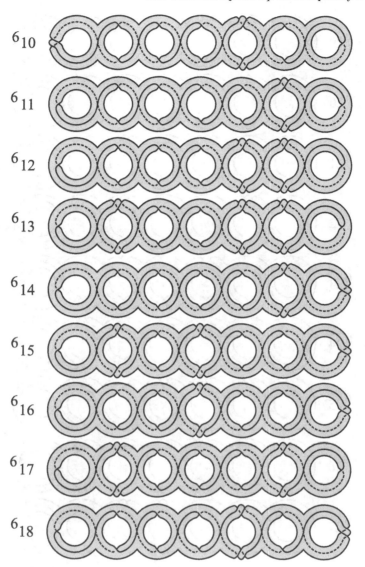

6_{19}

6_{20}

6_{21}

6_{22}

6_{23}

6_{24}

6_{25}

6_{26}

6_{27}

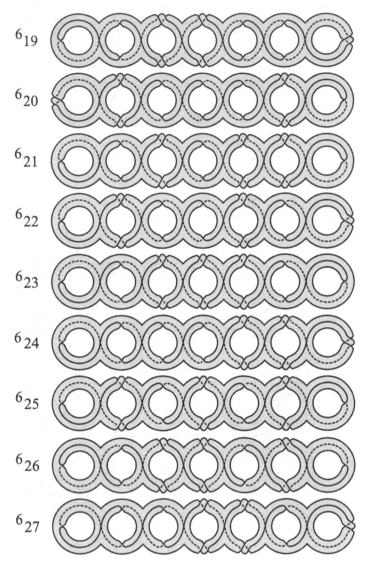

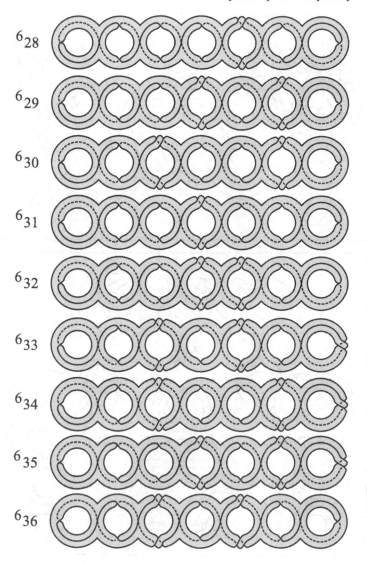

6_{28}

6_{29}

6_{30}

6_{31}

6_{32}

6_{33}

6_{34}

6_{35}

6_{36}

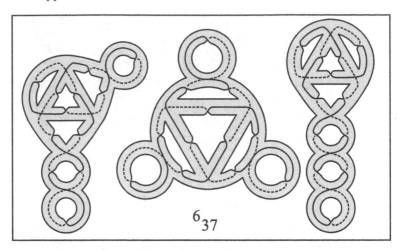

6_{37}

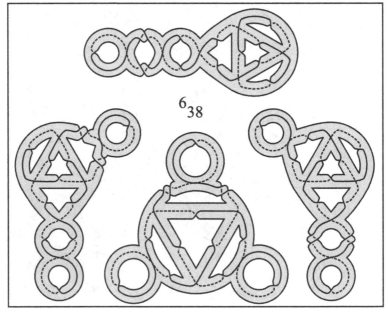

6_{38}

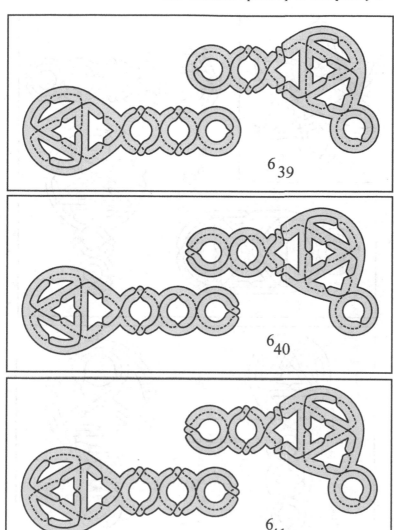

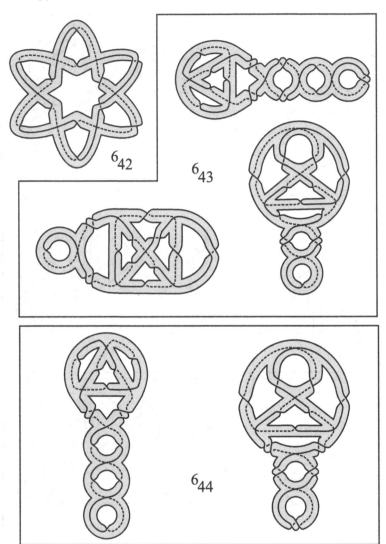

6_{42}

6_{43}

6_{44}

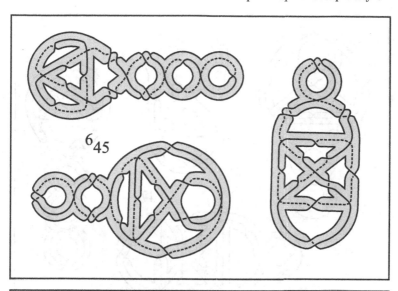

6_{45}

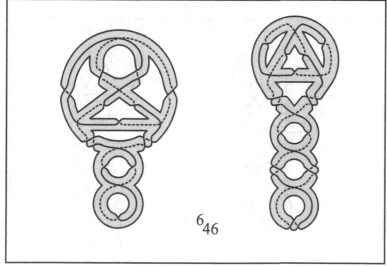

6_{46}

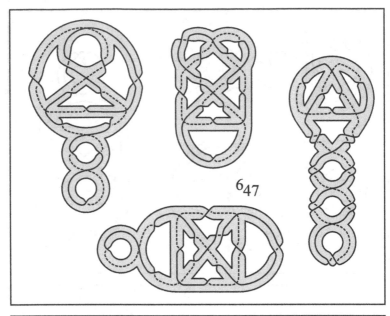

6_{47}

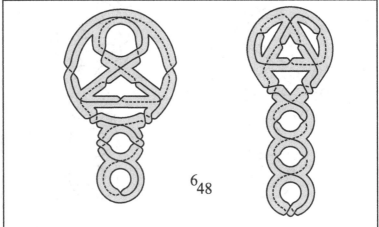

6_{48}

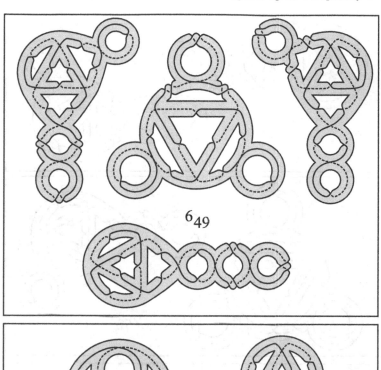

6_{49}

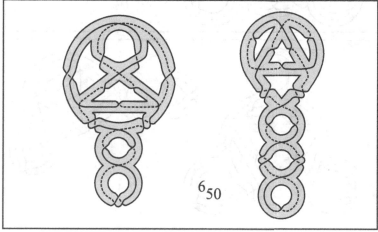

6_{50}

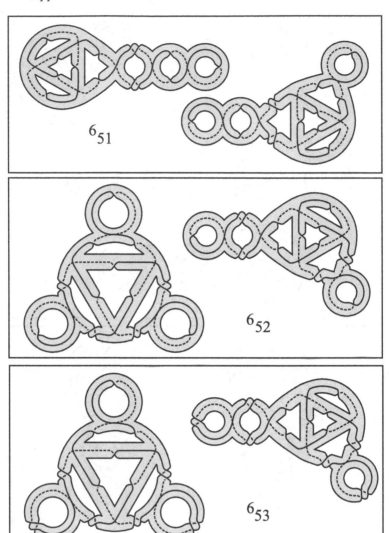

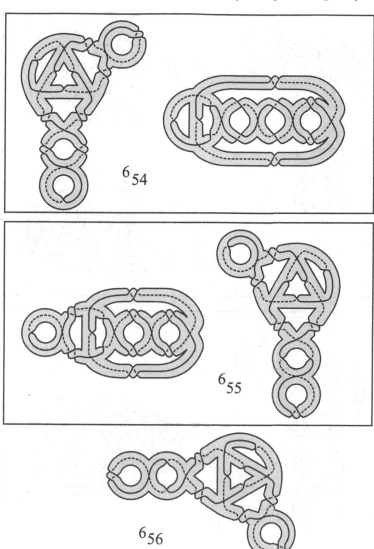

6_{54}

6_{55}

6_{56}

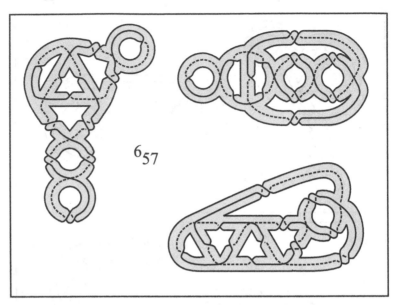

6_{57}

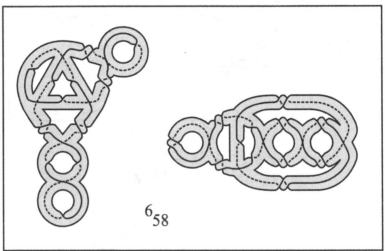

6_{58}

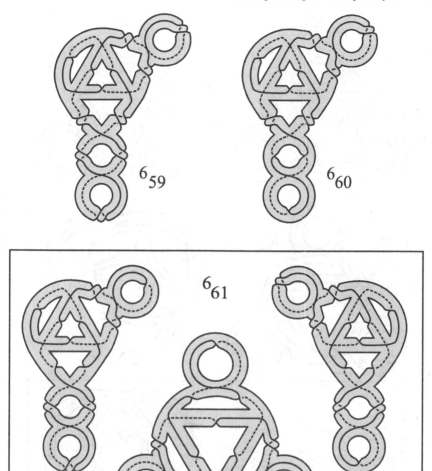

6_{59}

6_{60}

6_{61}

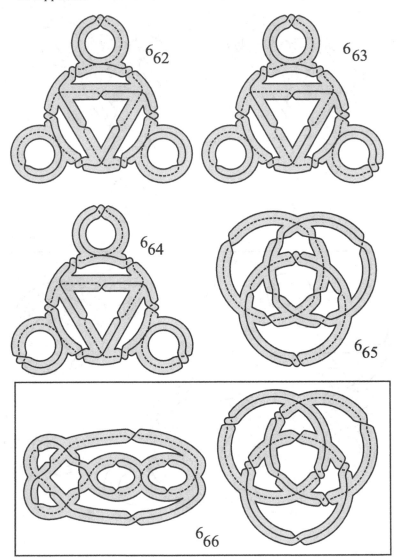

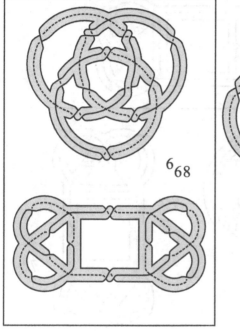

6_{67}

6_{68}

6_{69}

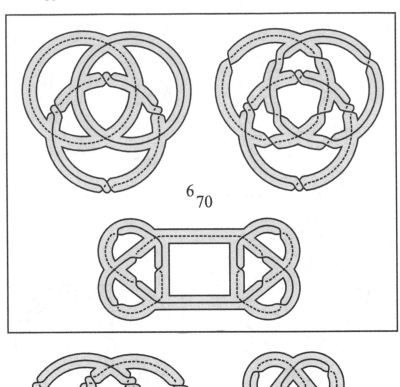

6_{70}

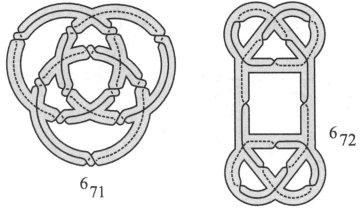

6_{71}

6_{72}

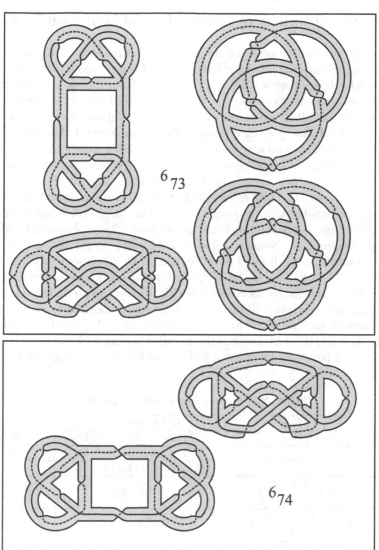

A.3 Minimal Spines of Some Manifolds of Complexity 7

Looking at the earlier table of manifolds of complexity ≤ 6, one can see that the great majority of them have minimal spines modeled on nonclosed chains and triangles with tails. The same tendency holds for manifolds of complexity 7. In order to save space, we reproduce only a part of Ovchinnikov's table [102, 103] of all closed orientable irreducible 3-manifolds of complexity 7. Namely, we list below only those manifolds of complexity 7 that do not admit minimal spines of the two aforementioned types. Each manifold is represented by only one minimal spine chosen arbitrarily. We number the manifold in the same order as they appear in [102, 103]. The presence of an asterisk as in 7_k^* tells us that the numbering differs from the original one (for example, manifold 7_1^* coincides with the manifold 7_{88} from [102,103]). A list of all closed orientable irreducible 3-manifolds up to complexity 9 was obtained by Martelli and Petronio, see [74]. Nonorientable manifolds are considered in [2, 3].

Let us introduce additional notation needed to describe manifolds of complexity 7 (Table A.8). Let $D_m = (D^2, (2, -1), (3, m))$, where m is not divisible by 3, be the Seifert manifold fibered over the disc with two exceptional fibers. Recall that the boundary torus of D_m possesses a coordinate system (μ, λ), see Remark A.1.1. Let A be an integer matrix of order 2 with determinant (-1). Then $D_m \cup K\tilde{\times}I/A$ denotes the manifold obtained by attaching D_m to $K\tilde{\times}I/A$ via a homeomorphism $h_A: \partial D_m \to \partial(K\tilde{\times}I)$ corresponding to A.

Table A.8. Some manifolds of complexity 7

c_i	M		Seifert structure	$H_1(M; Z)$
7_1^*	S^3/Q_{28}		$(S^2, (2, -1), (2, 1), (7, 1))$	Z_4
7_2^*			$(S^2, (2, -1), (3, 1), (7, 1))$	0
7_3^*			$(S^2, (2, -1), (4, 1), (5, 1))$	Z_2
7_4^*	$T \times I/\begin{pmatrix} 1 & 0 \\ 2 & 1 \end{pmatrix}$		$(T^2, (1, 2))$	$Z \oplus Z \oplus Z_2$
7_5^*	$T \times I/\begin{pmatrix} -1 & 0 \\ -2 & -1 \end{pmatrix}$		$(K^2, (1, 2))$	$Z_2 \oplus Z_2 \oplus Z$
7_6^*	$T \times I/\begin{pmatrix} -2 & -1 \\ -1 & -1 \end{pmatrix}$		$-$	$Z \oplus Z_5$
7_7^*	$T \times I/\begin{pmatrix} 2 & 1 \\ 1 & 1 \end{pmatrix}$		$-$	Z
7_8^*	$K\tilde{\times}I \cup K\tilde{\times}I/\begin{pmatrix} 1 & 0 \\ -2 & -1 \end{pmatrix}$		$(S^2, (2, 1), (2, 1), (2, 1), (2, 1))$	$Z_2 \oplus Z_2 \oplus Z_8$
7_9^*	$K\tilde{\times}I \cup K\tilde{\times}I/\begin{pmatrix} -1 & -1 \\ 1 & 2 \end{pmatrix}$		$-$	$Z_4 \oplus Z_4$
7_{10}^*	$K\tilde{\times}I \cup K\tilde{\times}I/\begin{pmatrix} 1 & -1 \\ -2 & 1 \end{pmatrix}$		$-$	$Z_4 \oplus Z_8$

Table A.8. (continued)

7^*_{11}	$K\tilde{\times}I \cup K\tilde{\times}I/\begin{pmatrix} 0 & 1 \\ 1 & -2 \end{pmatrix}$	$(RP^2,(2,1),(2,3))$	$Z_4 \oplus Z_4$
7^*_{12}	$K\tilde{\times}I \cup K\tilde{\times}I/\begin{pmatrix} 1 & -2 \\ -1 & 1 \end{pmatrix}$	$-$	$Z_2 \oplus Z_2 \oplus Z_4$
7^*_{13}	$D_2 \cup K\tilde{\times}I/\begin{pmatrix} 0 & 1 \\ 1 & 0 \end{pmatrix}$	$(RP^2,(2,1),(3,-1))$	Z_{24}
7^*_{14}	$D_2 \cup K\tilde{\times}I/\begin{pmatrix} 1 & 0 \\ -1 & -1 \end{pmatrix}$	$(S^2,(2,1),(2,1),(2,1)(3,-1))$	$Z_2 \oplus Z_{14}$
7^*_{15}	$D_1 \cup K\tilde{\times}I/\begin{pmatrix} 1 & 0 \\ -1 & -1 \end{pmatrix}$	$(S^2,(2,1),(2,1),(2,1)(3,-2))$	$Z_2 \oplus Z_{10}$
7^*_{16}	$D_2 \cup K\tilde{\times}I/\begin{pmatrix} -1 & -1 \\ 0 & 1 \end{pmatrix}$	$-$	Z_4
7^*_{17}	$D_1 \cup K\tilde{\times}I/\begin{pmatrix} -1 & -1 \\ 0 & 1 \end{pmatrix}$	$-$	Z_4
7^*_{18}	$D_2 \cup K\tilde{\times}I/\begin{pmatrix} -1 & 0 \\ 0 & 1 \end{pmatrix}$	$(S^2,(2,1),(2,1),(2,1)(3,-4))$	$Z_2 \oplus Z_2$
7^*_{19}	$D_2 \cup K\tilde{\times}I/\begin{pmatrix} 0 & 1 \\ 1 & -1 \end{pmatrix}$	$(RP^2,(2,1),(3,2))$	Z_{24}
7^*_{20}	$D_1 \cup K\tilde{\times}I/\begin{pmatrix} 0 & 1 \\ 1 & -1 \end{pmatrix}$	$(RP^2,(2,1),(3,1))$	Z_{24}
7^*_{21}	$D_2 \cup K\tilde{\times}I/\begin{pmatrix} 1 & -1 \\ -1 & 0 \end{pmatrix}$	$-$	Z_{28}
7^*_{22}	$D_1 \cup K\tilde{\times}I/\begin{pmatrix} 1 & -1 \\ -1 & 0 \end{pmatrix}$	$-$	Z_{20}

7^*_1

7^*_2

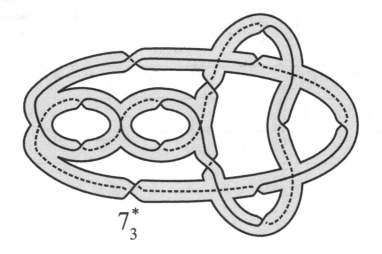

7^*_3

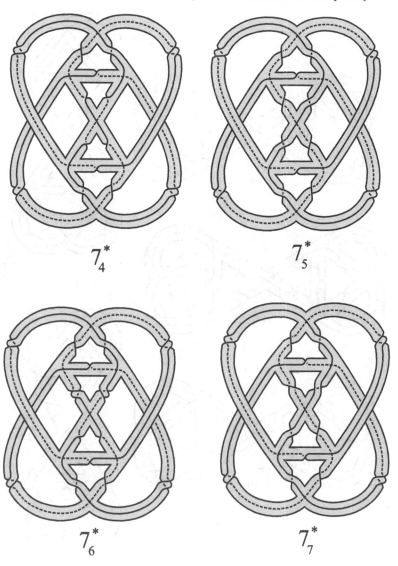

7^*_4

7^*_5

7^*_6

7^*_7

7_8^*

7_9^*

7_{10}^*

7_{11}^*

7_{12}^*

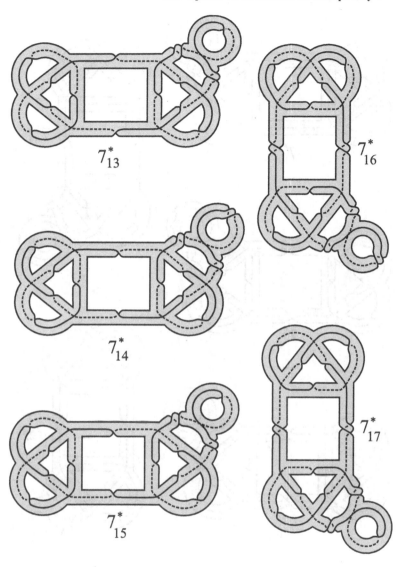

7^{*}_{13}

7^{*}_{16}

7^{*}_{14}

7^{*}_{15}

7^{*}_{17}

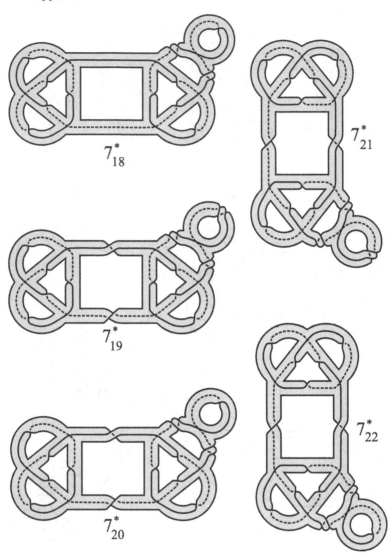

7^*_{18}

7^*_{19}

7^*_{20}

7^*_{21}

7^*_{22}

A.4 Tables of Turaev–Viro Invariants

Tables A.9–A.13 were composed by my former student Sokolov. They contain the values of Turaev–Viro invariants of order ≤ 7 and their summands for all closed orientable irreducible 3-manifolds up to complexity 6. The invariants are presented by polynomials of q, where q is a primitive root of unity of degree $2r$. For the sake of compactness of notation, we write σ_k instead of $q^k + q^{-k}$.

Table A.9. TV-invariants for manifolds of complexity 0–2

c_i	M	ν	3	4	5	6	7
0_1	S^3	0	1	1	1	1	1
		1	0	0	0	0	0
		2	0	0	0	0	0
		Σ	1	1	1	1	1
0_2	RP^3	0	1	2	σ_2+2	4	$-\sigma_3+2\sigma_2+3$
		1	-1	$-\sigma_1$	$-\sigma_2-2$	$-2\sigma_1$	$\sigma_3-2\sigma_2-3$
		2	0	0	0	0	0
		Σ	0	$-\sigma_1+2$	0	$-2\sigma_1+4$	0
0_3	$L_{3,1}$	0	1	1	σ_2+2	3	σ_2+2
		1	0	0	0	0	0
		2	0	0	0	0	0
		Σ	1	1	σ_2+2	3	σ_2+2
1_1	$L_{4,1}$	0	1	2	1	4	σ_2+2
		1	0	0	0	0	0
		2	1	0	1	0	σ_2+2
		Σ	2	2	2	4	$2\sigma_2+4$
1_2	$L_{5,2}$	0	1	1	0	1	$-\sigma_3+2\sigma_2+3$
		1	0	0	0	0	0
		2	0	0	0	0	0
		Σ	1	1	0	1	$-\sigma_3+2\sigma_2+3$
2_1	$L_{5,1}$	0	1	1	σ_2+3	1	$-\sigma_3+2\sigma_2+3$
		1	0	0	0	0	0
		2	0	0	0	0	0
		Σ	1	1	σ_2+3	1	$-\sigma_3+2\sigma_2+3$
2_2	$L_{7,2}$	0	1	1	σ_2+2	1	0
		1	0	0	0	0	0
		2	0	0	0	0	0
		Σ	1	1	σ_2+2	1	0
2_3	$L_{8,3}$	0	1	2	σ_2+2	4	1
		1	0	0	0	0	0
		2	1	2	σ_2+2	0	1
		Σ	2	4	$2\sigma_2+4$	4	2
2_4	Q_8	0	1	4	σ_2+4	10	$2\sigma_2+7$
		1	0	0	0	0	0
		2	3	6	$3\sigma_2+12$	18	$6\sigma_2+21$
		Σ	4	10	$4\sigma_2+16$	28	$8\sigma_2+28$

Table A.10. TV-invariants for manifolds of complexity 3

c_i	M	ν	3	4	5	6	7
3_1	$L_{6,1}$	0	1	2	1	6	1
		1	-1	σ_1	-1	0	-1
		2	0	0	0	0	0
		Σ	0	σ_1+2	0	6	0
3_2	$L_{9,2}$	0	1	1	1	3	$-\sigma_3+2\sigma_2+3$
		1	0	0	0	0	0
		2	0	0	0	0	0
		Σ	1	1	1	3	$-\sigma_3+2\sigma_2+3$
3_3	$L_{10,3}$	0	1	2	0	4	σ_2+2
		1	-1	σ_1	0	$2\sigma_1$	$-\sigma_2-2$
		2	0	0	0	0	0
		Σ	0	σ_1+2	0	$2\sigma_1+4$	0
3_4	$L_{11,3}$	0	1	1	1	1	σ_2+2
		1	0	0	0	0	0
		2	0	0	0	0	0
		Σ	1	1	1	1	σ_2+2
3_5	$L_{12,5}$	0	1	2	σ_2+2	6	$-\sigma_3+2\sigma_2+3$
		1	0	0	0	0	0
		2	1	0	σ_2+2	6	$-\sigma_3+2\sigma_2+3$
		Σ	2	2	$2\sigma_2+4$	12	$-2\sigma_3+4\sigma_2+6$
3_6	$L_{13,5}$	0	1	1	σ_2+2	1	1
		1	0	0	0	0	0
		2	0	0	0	0	0
		Σ	1	1	σ_2+2	1	1
3_7	Q_{12}	0	1	2	σ_2+4	10	$-\sigma_3+2\sigma_2+5$
		1	0	0	0	0	0
		2	1	0	σ_2+4	6	$-\sigma_3+2\sigma_2+5$
		Σ	2	2	$2\sigma_2+8$	16	$-2\sigma_3+4\sigma_2+10$
4_1	$L_{7,1}$	0	1	1	σ_2+2	1	$-\sigma_3+3\sigma_2+6$
		1	0	0	0	0	0
		2	0	0	0	0	0
		Σ	1	1	σ_2+2	1	$-\sigma_3+3\sigma_2+6$
4_2	$L_{11,2}$	0	1	1	1	1	σ_2+2
		1	0	0	0	0	0
		2	0	0	0	0	0
		Σ	1	1	1	1	σ_2+2

Table A.11. TV-invariants for manifolds of complexity 4

c_i	M	ν	3	4	5	6	7
4_3	$L_{13,3}$	0	1	1	σ_2+2	1	1
		1	0	0	0	0	0
		2	0	0	0	0	0
		Σ	1	1	σ_2+2	1	1
4_4	$L_{14,3}$	0	1	2	1	4	0
		1	-1	$-\sigma_1$	-1	$2\sigma_1$	0
		2	0	0	0	0	0
		Σ	0	$-\sigma_1+2$	0	$2\sigma_1+4$	0
4_5	$L_{15,4}$	0	1	1	σ_2+3	3	1
		1	0	0	0	0	0
		2	0	0	0	0	0
		Σ	1	1	σ_2+3	3	1
4_6	$L_{16,7}$	0	1	2	1	4	$-\sigma_3+2\sigma_2+3$
		1	0	0	0	0	0
		2	1	2	1	0	$-\sigma_3+2\sigma_2+3$
		Σ	2	4	2	4	$-2\sigma_3+4\sigma_2+6$
4_7	$L_{17,5}$	0	1	1	σ_2+2	1	σ_2+2
		1	0	0	0	0	0
		2	0	0	0	0	0
		Σ	1	1	σ_2+2	1	σ_2+2
4_8	$L_{18,5}$	0	1	2	σ_2+2	6	σ_2+2
		1	-1	$-\sigma_1$	$-\sigma_2-2$	0	$-\sigma_2-2$
		2	0	0	0	0	0
		Σ	0	$-\sigma_1+2$	0	6	0
4_9	$L_{19,7}$	0	1	1	1	1	$-\sigma_3+2\sigma_2+3$
		1	0	0	0	0	0
		2	0	0	0	0	0
		Σ	1	1	1	1	$-\sigma_3+2\sigma_2+3$
4_{10}	$L_{21,8}$	0	1	1	1	3	$-\sigma_3+3\sigma_2+6$
		1	0	0	0	0	0
		2	0	0	0	0	0
		Σ	1	1	1	3	$-\sigma_3+3\sigma_2+6$
4_{11}	$Q_8 \times Z_3$	0	1	4	$2\sigma_2+5$	12	$-2\sigma_3+7\sigma_2+12$
		1	0	0	0	0	0
		2	3	6	$6\sigma_2+15$	36	$-6\sigma_3+21\sigma_2+36$
		Σ	4	10	$8\sigma_2+20$	48	$-8\sigma_3+28\sigma_2+48$
4_{12}	Q_{16}	0	1	4	1	10	$-\sigma_3+2\sigma_2+5$
		1	-2	0	-2	$4\sigma_1$	$2\sigma_3-4\sigma_2-10$
		2	1	2	1	6	$-\sigma_3+2\sigma_2+5$
		Σ	0	6	0	$4\sigma_1+16$	0

Table A.12. TV-invariants for manifolds of complexity 4, 5

c_i	M	ν	3	4	5	6	7
4_{13}	D_{24}	0	1	2	$2\sigma_2+5$	10	$-2\sigma_3+5\sigma_2+8$
		1	0	0	0	0	0
		2	1	2	$2\sigma_2+5$	6	$-2\sigma_3+5\sigma_2+8$
		Σ	2	4	$4\sigma_2+10$	16	$-4\sigma_3+10\sigma_2+16$
4_{14}	P_{24}	0	1	1	$2\sigma_2+5$	3	$-2\sigma_3+4\sigma_2+6$
		1	0	0	0	0	0
		2	0	0	0	0	0
		Σ	1	1	$2\sigma_2+5$	3	$-2\sigma_3+4\sigma_2+6$
5_1	$L_{8,1}$	0	1	2	σ_2+2	4	1
		1	0	0	0	0	0
		2	1	-2	σ_2+2	0	1
		Σ	2	0	$2\sigma_2+4$	4	2
5_2	$L_{13,2}$	0	1	1	σ_2+2	1	1
		1	0	0	0	0	0
		2	0	0	0	0	0
		Σ	1	1	σ_2+2	1	1
5_3	$L_{16,3}$	0	1	2	1	4	$-\sigma_3+2\sigma_2+3$
		1	0	0	0	0	0
		2	1	-2	1	0	$-\sigma_3+2\sigma_2+3$
		Σ	2	0	2	4	$-2\sigma_3+4\sigma_2+6$
5_4	$L_{17,3}$	0	1	1	σ_2+2	1	σ_2+2
		1	0	0	0	0	0
		2	0	0	0	0	0
		Σ	1	1	σ_2+2	1	σ_2+2
5_5	$L_{17,4}$	0	1	1	σ_2+2	1	σ_2+2
		1	0	0	0	0	0
		2	0	0	0	0	0
		Σ	1	1	σ_2+2	1	σ_2+2
5_6	$L_{19,4}$	0	1	1	1	1	$-\sigma_3+2\sigma_2+3$
		1	0	0	0	0	0
		2	0	0	0	0	0
		Σ	1	1	1	1	$-\sigma_3+2\sigma_2+3$
5_7	$L_{20,9}$	0	1	2	σ_2+3	4	1
		1	0	0	0	0	0
		2	1	0	σ_2+3	0	1
		Σ	2	2	$2\sigma_2+6$	4	2

Table A.12. (continued)

c_i	M	ν	3	4	5	6	7
5_8	$L_{22,5}$	0	1	2	$\sigma_2 + 2$	4	1
		1	-1	σ_1	$-\sigma_2 - 2$	$-2\sigma_1$	-1
		2	0	0	0	0	0
		Σ	0	$\sigma_1 + 2$	0	$-2\sigma_2 + 4$	0
5_9	$L_{23,5}$	0	1	1	$\sigma_2 + 2$	1	$-\sigma_3 + 2\sigma_2 + 3$
		1	0	0	0	0	0
		2	0	0	0	0	0
		Σ	1	1	$\sigma_2 + 2$	1	$-\sigma_3 + 2\sigma_2 + 3$
5_{10}	$L_{23,7}$	0	1	1	$\sigma_2 + 2$	1	$-\sigma_3 + 2\sigma_2 + 3$
		1	0	0	0	0	0
		2	0	0	0	0	0
		Σ	1	1	$\sigma_2 + 2$	1	$-\sigma_3 + 2\sigma_2 + 3$
5_{11}	$L_{24,7}$	0	1	2	1	6	$\sigma_2 + 2$
		1	0	0	0	0	0
		2	1	-2	1	-6	$\sigma_2 + 2$
		Σ	2	0	2	0	$2\sigma_2 + 4$
5_{12}	$L_{25,7}$	0	1	1	0	1	$\sigma_2 + 2$
		1	0	0	0	0	0
		2	0	0	0	0	0
		Σ	1	1	0	1	$\sigma_2 + 2$
5_{13}	$L_{25,9}$	0	1	1	$\sigma_2 + 3$	1	$\sigma_2 + 2$
		1	0	0	0	0	0
		2	0	0	0	0	0
		Σ	1	1	$\sigma_2 + 3$	1	$\sigma_2 + 2$
5_{14}	$L_{26,7}$	0	1	2	1	4	$-\sigma_3 + 2\sigma_2 + 3$
		1	-1	σ_1	-1	$-2\sigma_1$	$\sigma_3 - 2\sigma_2 - 3$
		2	0	0	0	0	0
		Σ	0	$\sigma_1 + 2$	0	$-2\sigma_1 + 4$	0
5_{15}	$L_{27,8}$	0	1	1	$\sigma_2 + 2$	3	1
		1	0	0	0	0	0
		2	0	0	0	0	0
		Σ	1	1	$\sigma_2 + 2$	3	1
5_{16}	$L_{29,8}$	0	1	1	1	1	1
		1	0	0	0	0	0
		2	0	0	0	0	0
		Σ	1	1	1	1	1

Table A.12. TV-invariants for manifolds of complexity 5 (continued)

c_i	M	ν	3	4	5	6	7
5_{17}	$L_{29,12}$	0	1	1	1	1	1
		1	0	0	0	0	0
		2	0	0	0	0	0
		Σ	1	1	1	1	1
5_{18}	$L_{30,11}$	0	1	2	$\sigma_2 + 3$	6	$-\sigma_3 + 2\sigma_2 + 3$
		1	-1	$-\sigma_1$	$-\sigma_2 - 3$	0	$\sigma_3 - 2\sigma_2 - 3$
		2	0	0	0	0	0
		Σ	0	$-\sigma_1 + 2$	0	6	0
5_{19}	$L_{31,12}$	0	1	1	1	1	$\sigma_2 + 2$
		1	0	0	0	0	0
		2	0	0	0	0	0
		Σ	1	1	1	1	$\sigma_2 + 2$
5_{20}	$L_{34,13}$	0	1	2	1	4	1
		1	-1	$-\sigma_1$	-1	$2\sigma_1$	-1
		2	0	0	0	0	0
		Σ	0	$-\sigma_1 + 2$	0	$2\sigma_1 + 4$	0
5_{21}	$Q_8 \times Z_5$						
		0	1	4	$\sigma_2 + 3$	10	$-3\sigma_3 + 6\sigma_2 + 11$
		1	0	0	0	0	0
		2	3	6	$3\sigma_2 + 9$	18	$-9\sigma_3 + 18\sigma_2 + 33$
		Σ	4	10	$4\sigma_2 + 12$	28	$-12\sigma_3 + 24\sigma_2 + 44$
5_{22}	$Q_{12} \times Z_5$						
		0	1	2	$\sigma_2 + 3$	10	$-2\sigma_3 + 5\sigma_2 + 8$
		1	0	0	0	0	0
		2	1	0	$\sigma_2 + 3$	6	$-2\sigma_3 + 5\sigma_2 + 8$
		Σ	2	2	$2\sigma_2 + 6$	16	$-4\sigma_3 + 10\sigma_2 + 16$
5_{23}	$Q_{16} \times Z_3$						
		0	1	4	$\sigma_2 + 2$	12	$2\sigma_2 + 5$
		1	-2	0	$-2\sigma_2 - 4$	0	$-4\sigma_2 - 10$
		2	1	2	$\sigma_2 + 2$	12	$2\sigma_2 + 5$
		Σ	0	6	0	24	0

Table A.12. (continued)

c_i	M	ν	3	4	5	6	7
5_{24}	Q_{20}	0	1	2	$3\sigma_2+5$	4	$2\sigma_2+7$
		1	0	0	0	0	0
		2	1	0	$3\sigma_2+5$	0	$2\sigma_2+7$
		Σ	2	2	$6\sigma_2+10$	4	$4\sigma_2+14$
5_{25}	$Q_{20}\times Z_3$	0	1	2	$-\sigma_2+1$	6	$-2\sigma_3+7\sigma_2+12$
		1	0	0	0	0	0
		2	1	0	$-\sigma_2+1$	6	$-2\sigma_3+7\sigma_2+12$
		Σ	2	2	$-2\sigma_2+2$	12	$-4\sigma_3+14\sigma_2+24$
5_{26}	D_{40}	0	1	2	$-\sigma_2+1$	4	$-3\sigma_3+6\sigma_2+11$
		1	0	0	0	0	0
		2	1	2	$-\sigma_2+1$	0	$-3\sigma_3+6\sigma_2+11$
		Σ	2	4	$-2\sigma_2+2$	4	$-6\sigma_3+12\sigma_2+22$
5_{27}	D_{48}	0	1	2	σ_2+4	10	$2\sigma_2+5$
		1	0	0	0	0	0
		2	1	2	σ_2+4	6	$2\sigma_2+5$
		Σ	2	4	$2\sigma_2+8$	16	$4\sigma_2+10$
5_{28}	$P_{24}\times Z_5$	0	1	1	σ_2+3	3	$2\sigma_2+4$
		1	0	0	0	0	0
		2	0	0	0	0	0
		Σ	1	1	σ_2+3	3	$2\sigma_2+4$
5_{29}	P_{48}	0	1	2	σ_2+2	10	2
		1	-1	σ_1	$-\sigma_2-2$	0	-2
		2	0	0	0	0	0
		Σ	0	σ_1+2	0	10	0
5_{30}	P'_{72}	0	1	1	σ_2+4	3	2
		1	0	0	0	0	0
		2	0	0	0	0	0
		Σ	1	1	σ_2+4	3	2
5_{31}	P_{120}	0	1	1	$3\sigma_2+5$	1	$-2\sigma_3+5\sigma_2+8$
		1	0	0	0	0	0
		2	0	0	0	0	0
		Σ	1	1	$3\sigma_2+5$	1	$-2\sigma_3+5\sigma_2+8$

Table A.13. TV-invariants for manifolds of complexity 6

c_i	M	ν	3	4	5	6	7
6_1	$L_{9,1}$	0	1	1	1	3	$-\sigma_3 + 2\sigma_2 + 3$
		1	0	0	0	0	0
		2	0	0	0	0	0
		Σ	1	1	1	3	$-\sigma_3 + 2\sigma_2 + 3$
6_2	$L_{15,2}$	0	1	1	0	3	1
		1	0	0	0	0	0
		2	0	0	0	0	0
		Σ	1	1	0	3	1
6_3	$L_{19,3}$	0	1	1	1	1	$-\sigma_3 + 2\sigma_2 + 3$
		1	0	0	0	0	0
		2	0	0	0	0	0
		Σ	1	1	1	1	$-\sigma_3 + 2\sigma_2 + 3$
6_4	$L_{20,3}$	0	1	2	0	4	1
		1	0	0	0	0	0
		2	1	0	0	0	1
		Σ	2	2	0	4	2
6_5	$L_{21,4}$	0	1	1	1	3	0
		1	0	0	0	0	0
		2	0	0	0	0	0
		Σ	1	1	1	3	0
6_6	$L_{23,4}$	0	1	1	$\sigma_2 + 2$	1	$-\sigma_3 + 2\sigma_2 + 3$
		1	0	0	0	0	0
		2	0	0	0	0	0
		Σ	1	1	$\sigma_2 + 2$	1	$-\sigma_3 + 2\sigma_2 + 3$
6_7	$L_{24,5}$	0	1	2	1	6	$\sigma_2 + 2$
		1	0	0	0	0	0
		2	1	2	1	-6	$\sigma_2 + 2$
		Σ	2	4	2	0	$2\sigma_2 + 4$
6_8	$L_{24,11}$	0	1	2	1	6	$\sigma_2 + 2$
		1	0	0	0	0	0
		2	1	2	1	6	$\sigma_2 + 2$
		Σ	2	4	2	12	$2\sigma_2 + 4$
6_9	$L_{27,5}$	0	1	1	$\sigma_2 + 2$	3	1
		2	0	0	0	0	0
		2	0	0	0	0	0
		Σ	1	1	$\sigma_2 + 2$	3	1

Table A.13. (continued)

c_i	M	ν	3	4	5	6	7
6_{10}	$L_{28,5}$	0	1	2	$\sigma_2 + 2$	4	0
		1	0	0	0	0	0
		2	1	0	$\sigma_2 + 2$	0	0
		Σ	2	2	$2\sigma_2 + 4$	4	0
6_{11}	$L_{29,9}$	0	1	1	1	1	1
		1	0	0	0	0	0
		2	0	0	0	0	0
		Σ	1	1	1	1	1
6_{12}	$L_{30,7}$	0	1	2	0	6	$-\sigma_3 + 2\sigma_2 + 3$
		1	-1	$-\sigma_1$	0	0	$\sigma_3 - 2\sigma_2 - 3$
		2	0	0	0	0	0
		Σ	0	$-\sigma_1 + 2$	0	6	0
6_{13}	$L_{31,7}$	0	1	1	1	1	$\sigma_2 + 2$
		1	0	0	0	0	0
		2	0	0	0	0	0
		Σ	1	1	1	1	$\sigma_2 + 2$
6_{14}	$L_{31,11}$	0	1	1	1	1	$\sigma_2 + 2$
		1	0	0	0	0	0
		2	0	0	0	0	0
		Σ	1	1	1	1	$\sigma_2 + 2$
6_{15}	$L_{32,7}$	0	1	2	$\sigma_2 + 2$	4	$\sigma_2 + 2$
		1	0	0	0	0	0
		2	1	2	$\sigma_2 + 2$	0	$\sigma_2 + 2$
		Σ	2	4	$2\sigma_2 + 4$	4	$2\sigma_2 + 4$
6_{16}	$L_{33,7}$	0	1	1	$\sigma_2 + 2$	3	$-\sigma_3 + 2\sigma_2 + 3$
		1	0	0	0	0	0
		2	0	0	0	0	0
		Σ	1	1	$\sigma_2 + 2$	3	$-\sigma_3 + 2\sigma_2 + 3$
6_{17}	$L_{33,10}$	0	1	1	$\sigma_2 + 2$	3	$-\sigma_3 + 2\sigma_2 + 3$
		1	0	0	0	0	0
		2	0	0	0	0	0
		Σ	1	1	$\sigma_2 + 2$	3	$-\sigma_3 + 2\sigma_2 + 3$

Table A.13. TV-invariants for manifolds of complexity 6 (continued)

c_i	M	ν	3	4	5	6	7
6_{18}	$L_{34,9}$	0	1	2	1	4	1
		1	-1	$-\sigma_1$	-1	$2\sigma_2$	-1
		2	0	0	0	0	0
		Σ	0	$-\sigma_1+2$	0	$2\sigma_1+4$	0
6_{19}	$L_{35,8}$	0	1	1	0	1	$-\sigma_3+3\sigma_2+6$
		1	0	0	0	0	0
		2	0	0	0	0	0
		Σ	1	1	0	1	$-\sigma_3+3\sigma_2+6$
6_{20}	$L_{36,11}$	0	1	2	1	6	1
		1	0	0	0	0	0
		2	1	0	1	-6	1
		Σ	2	2	2	0	2
6_{21}	$L_{37,8}$	0	1	1	σ_2+2	1	$-\sigma_3+2\sigma_2+3$
		1	0	0	0	0	0
		2	0	0	0	0	0
		Σ	1	1	σ_2+2	1	$-\sigma_3+2\sigma_2+3$
6_{22}	$L_{37,10}$	0	1	1	σ_2+2	1	$-\sigma_3+2\sigma_2+3$
		1	0	0	0	0	0
		2	0	0	0	0	0
		Σ	1	1	σ_2+2	1	$-\sigma_3+2\sigma_2+3$
6_{23}	$L_{39,14}$	0	1	1	1	3	σ_2+2
		1	0	0	0	0	0
		2	0	0	0	0	0
		Σ	1	1	1	3	σ_2+2
6_{24}	$L_{39,16}$	0	1	1	1	3	σ_2+2
		1	0	0	0	0	0
		2	0	0	0	0	0
		Σ	1	1	1	3	σ_2+2
6_{25}	$L_{40,11}$	0	1	2	σ_2+3	4	$-\sigma_3+2\sigma_2+3$
		1	0	0	0	0	0
		2	1	2	σ_2+3	0	$-\sigma_3+2\sigma_2+3$
		Σ	2	4	$2\sigma_2+6$	4	$-2\sigma_3+4\sigma_2+6$
6_{26}	$L_{41,11}$	0	1	1	1	1	1
		1	0	0	0	0	0
		2	0	0	0	0	0
		Σ	1	1	1	1	1

Table A.13. (continued)

c_i	M	ν	3	4	5	6	7
6_{27}	$L_{41,12}$	0	1	1	1	1	1
		1	0	0	0	0	0
		2	0	0	0	0	0
		Σ	1	1	1	1	1
6_{28}	$L_{41,16}$	0	1	1	1	1	1
		1	0	0	0	0	0
		2	0	0	0	0	0
		Σ	1	1	1	1	1
6_{29}	$L_{43,12}$	0	1	1	$\sigma_2 + 2$	1	1
		1	0	0	0	0	0
		2	0	0	0	0	0
		Σ	1	1	$\sigma_2 + 2$	1	1
6_{30}	$L_{44,13}$	0	1	2	1	4	$-\sigma_3 + 2\sigma_2 + 3$
		1	0	0	0	0	0
		2	1	0	1	0	$-\sigma_3 + 2\sigma_2 + 3$
		Σ	2	2	2	4	$-2\sigma_3 + 4\sigma_2 + 6$
6_{31}	$L_{45,19}$	0	1	1	$\sigma_2 + 3$	3	$\sigma_2 + 2$
		1	0	0	0	0	0
		2	0	0	0	0	0
		Σ	1	1	$\sigma_2 + 3$	3	$\sigma_2 + 2$
6_{32}	$L_{46,17}$	0	1	2	1	4	$\sigma_2 + 2$
		1	-1	$-\sigma_1$	-1	$-2\sigma_1$	$-\sigma_2 - 2$
		2	0	0	0	0	0
		Σ	0	$-\sigma_1 + 2$	0	$-2\sigma_1 + 4$	0
6_{33}	$L_{47,13}$	0	1	1	$\sigma_2 + 2$	1	$-\sigma_3 + 2\sigma_2 + 3$
		1	0	0	0	0	0
		2	0	0	0	0	0
		Σ	1	1	$\sigma_2 + 2$	1	$-\sigma_3 + 2\sigma_2 + 3$
6_{34}	$L_{49,18}$	0	1	1	1	1	0
		1	0	0	0	0	0
		2	0	0	0	0	0
		Σ	1	1	1	1	0
6_{35}	$L_{50,19}$	0	1	2	$\sigma_2 + 3$	4	1
		1	-1	$-\sigma_1$	$-\sigma_2 - 3$	$-2\sigma_1$	-1
		2	0	0	0	0	0
		Σ	0	$-\sigma_1 + 2$	0	$-2\sigma_1 + 4$	0

Table A.13. TV-invariants for manifolds of complexity 6 (continued)

c_i	M	ν	3	4	5	6	7
6_{36}	$L_{55,21}$	0	1	1	$\sigma_2 + 3$	1	1
		1	0	0	0	0	0
		2	0	0	0	0	0
		Σ	1	1	$\sigma_2 + 3$	1	1
6_{37}	$Q_8 \times Z_7$						
		0	1	4	$2\sigma_2 + 5$	10	$-\sigma_3 + 3\sigma_2 + 6$
		1	0	0	0	0	0
		2	3	6	$6\sigma_2 + 15$	18	$-3\sigma_3 + 9\sigma_2 + 18$
		Σ	4	10	$8\sigma_2 + 20$	28	$-4\sigma_3 + 12\sigma_2 + 24$
6_{38}	$Q_{12} \times Z_7$						
		0	1	2	$2\sigma_2 + 5$	10	$-\sigma_3 + 3\sigma_2 + 6$
		1	0	0	0	0	0
		2	1	0	$2\sigma_2 + 5$	6	$-\sigma_3 + 3\sigma_2 + 6$
		Σ	2	2	$4\sigma_2 + 10$	16	$-2\sigma_3 + 6\sigma_2 + 12$
6_{39}	$Q_{16} \times Z_5$						
		0	1	4	$\sigma_2 + 3$	10	$-2\sigma_3 + 5\sigma_2 + 8$
		1	-2	0	$-2\sigma_2 - 6$	$-4\sigma_1$	$4\sigma_3 - 10\sigma_2 - 16$
		2	1	2	$\sigma_2 + 3$	6	$-2\sigma_3 + 5\sigma_2 + 8$
		Σ	0	6	0	$-4\sigma_1 + 16$	0
6_{40}	$Q_{16} \times Z_7$						
		0	1	4	$\sigma_2 + 2$	10	$-\sigma_3 + 3\sigma_2 + 6$
		1	-2	0	$-2\sigma_2 - 4$	$-4\sigma_1$	$2\sigma_3 - 6\sigma_2 - 12$
		2	1	2	$\sigma_2 + 2$	6	$-\sigma_3 + 3\sigma_2 + 6$
		Σ	0	6	0	$-4\sigma_1 + 16$	0
6_{41}	$Q_{20} \times Z_7$						
		0	1	2	$-\sigma_2 + 1$	4	$-\sigma_3 + 3\sigma_2 + 6$
		1	0	0	0	0	0
		2	1	0	$-\sigma_2 + 1$	0	$-\sigma_3 + 3\sigma_2 + 6$
		Σ	2	2	$-2\sigma_2 + 2$	4	$-2\sigma_3 + 6\sigma_2 + 12$

Table A.13. (continued)

c_i	M	ν	3	4	5	6	7
6_{42}	Q_{24}	0	1	4	1	16	σ_2+2
		1	0	0	0	0	0
		2	3	-2	3	12	$3\sigma_++6$
		Σ	4	2	4	28	$4\sigma_2+8$
6_{43}	$Q_{28}\times Z_3$						
		0	1	2	$2\sigma_2+5$	6	$-\sigma_3-\sigma_2+2$
		1	0	0	0	0	0
		2	1	0	$2\sigma_2+5$	6	$-\sigma_3-\sigma_2+2$
		Σ	2	2	$4\sigma_2+10$	12	$-2\sigma_3-2\sigma_2+4$
6_{44}	$Q_{28}\times Z_5$						
		0	1	2	σ_2+3	4	$3\sigma_3-\sigma_2+2$
		1	0	0	0	0	0
		2	1	0	σ_2+3	0	$3\sigma_3-\sigma_2+2$
		Σ	2	2	$2\sigma_2+6$	4	$6\sigma_3-2\sigma_2+4$
6_{45}	$Q_{32}\times Z_3$						
		0	1	4	$2\sigma_2+5$	12	$-\sigma_3+2\sigma_2+3$
		1	-2	$-4\sigma_1$	$-4\sigma_2-10$	0	$2\sigma_3-4\sigma_2-6$
		2	1	2	$2\sigma_2+5$	12	$-\sigma_3+2\sigma_2+3$
		Σ	0	$-4\sigma_1+6$	0	24	0
6_{46}	$Q_{32}\times Z_5$						
		0	1	4	σ_2+3	10	1
		1	-2	$-4\sigma_1$	$-2\sigma_2-6$	$4\sigma_1$	-2
		2	1	2	σ_2+3	6	1
		Σ	0	$-4\sigma_1+6$	0	$4\sigma_1+16$	0
6_{47}	D_{56}	0	1	2	$2\sigma_2+5$	4	$3\sigma_3-\sigma_2+2$
		1	0	0	0	0	0
		2	1	2	$2\sigma_2+5$	0	$3\sigma_3-\sigma_2+2$
		Σ	2	4	$4\sigma_2+10$	4	$6\sigma_3-2\sigma_2+4$

Table A.13. TV-invariants for manifolds of complexity 6 (continued)

c_i	M	ν	3	4	5	6	7
6_{48}	D_{80}	0	1	2	$3\sigma_2 + 5$	4	$-2\sigma_3 + 7\sigma_2 + 12$
		1	0	0	0	0	0
		2	1	2	$3\sigma_2 + 5$	0	$-2\sigma_3 + 7\sigma_2 + 12$
		Σ	2	4	$6\sigma_2 + 10$	4	$-4\sigma_3 + 14\sigma_2 + 24$
6_{49}	D_{96}	0	1	2	$2\sigma_2 + 5$	10	$-\sigma_3 + 2\sigma_2 + 5$
		1	0	0	0	0	0
		2	1	2	$2\sigma_2 + 5$	6	$-\sigma_3 + 2\sigma_2 + 5$
		Σ	2	4	$4\sigma_2 + 10$	16	$-2\sigma_3 + 4\sigma_2 + 10$
6_{50}	D_{112}	0	1	2	$\sigma_2 + 4$	4	$-\sigma_3 - \sigma_2 + 2$
		1	0	0	0	0	0
		2	1	2	$\sigma_2 + 4$	0	$-\sigma_3 - \sigma_2 + 2$
		Σ	2	4	$2\sigma_2 + 8$	4	$-2\sigma_3 - 2\sigma_2 + 4$
6_{51}	D_{160}	0	1	2	$-\sigma_2 + 1$	4	$2\sigma_2 + 7$
		1	0	0	0	0	0
		2	1	2	$-\sigma_2 + 1$	0	$2\sigma_2 + 7$
		Σ	2	4	$-2\sigma_1 + 2$	4	$4\sigma_2 + 14$
6_{52}	$P_{24} \times Z_7$						
		0	1	1	$\sigma_2 + 4$	3	$-\sigma_3 + 3\sigma_2 + 6$
		1	0	0	0	0	0
		2	0	0	0	0	0
		Σ	1	1	$\sigma_2 + 4$	3	$-\sigma_3 + 3\sigma_2 + 6$
6_{53}	$P_{24} \times Z_{11}$						
		0	1	1	$2\sigma_2 + 5$	3	2
		1	0	0	0	0	0
		2	0	0	0	0	0
		Σ	1	1	$2\sigma_2 + 5$	3	2
6_{54}	$P_{48} \times Z_5$						
		0	1	2	$\sigma_2 + 3$	10	$-2\sigma_3 + 4\sigma_2 + 6$
		1	-1	$-\sigma_1$	$-\sigma_2 - 3$	0	$2\sigma_3 - 4\sigma_2 - 6$
		2	0	0	0	0	0
		Σ	0	$-\sigma_1 + 2$	0	10	0

Table A.13. (continued)

c_i	M	ν	3	4	5	6	7
6_{55}	$P_{48} \times Z_7$						
		0	1	2	1	10	$-\sigma_3 + 3\sigma_2 + 6$
		1	-1	σ_1	-1	0	$\sigma_3 - 3\sigma_2 - 6$
		2	0	0	0	0	0
		Σ	0	$\sigma_1 + 2$	0	10	0
6_{56}	$P_{48} \times Z_{11}$						
		0	1	2	$\sigma_2 + 2$	10	$2\sigma_2 + 4$
		1	-1	$-\sigma_1$	$-\sigma_2 - 2$	0	$-2\sigma_2 - 4$
		2	0	0	0	0	0
		Σ	0	$-\sigma_1 + 2$	0	10	0
6_{57}	$P_{120} \times Z_7$						
		0	1	1	$-\sigma_2 + 1$	1	$-\sigma_3 + 3\sigma_2 + 6$
		1	0	0	0	0	0
		2	0	0	0	0	0
		Σ	1	1	$-\sigma_2 + 1$	1	$-\sigma_3 + 3\sigma_2 + 6$
6_{58}	$P_{120} \times Z_{13}$						
		0	1	1	$-\sigma_2 + 1$	1	$-2\sigma_3 + 5\sigma_2 + 8$
		1	0	0	0	0	0
		2	0	0	0	0	0
		Σ	1	1	$-\sigma_2 + 1$	1	$-2\sigma_3 + 5\sigma_2 + 8$
6_{59}	$P_{120} \times Z_{17}$						
		0	1	1	$-\sigma_2 + 1$	1	$-\sigma_3 + 2\sigma_2 + 5$
		1	0	0	0	0	0
		2	0	0	0	0	0
		Σ	1	1	$-\sigma_2 + 1$	1	$-\sigma_3 + 2\sigma_2 + 5$
6_{60}	$P_{120} \times Z_{23}$						
		0	1	1	$-\sigma_2 + 1$	1	$2\sigma_2 + 5$
		1	0	0	0	0	0
		2	0	0	0	0	0
		Σ	1	1	$-\sigma_2 + 1$	1	$2\sigma_2 + 5$

Table A.13. TV-invariants for manifolds of complexity 6 (continued)

c_i	M	ν	3	4	5	6	7
6_{61}	P'_{216}	0	1	1	$2\sigma_2 + 5$	3	$2\sigma_2 + 4$
		1	0	0	0	0	0
		2	0	0	0	0	0
		Σ	1	1	$2\sigma_2 + 5$	3	$2\sigma_2 + 4$
6_{62}	$M(S^2,(3,2),(3,1),(3,-2))$						
		0	1	2	$2\sigma_2 + 5$	12	$-2\sigma_3 + 6\sigma_2 + 9$
		1	-1	$-\sigma_1$	$-2\sigma_2 - 5$	$-6\sigma_1$	$2\sigma_3 - 6\sigma_2 - 9$
		2	0	0	0	0	0
		Σ	0	$-\sigma_1 + 2$	0	$-6\sigma_1 + 12$	0
6_{63}	$M(S^2,(3,2),(3,2),(3,-2))$						
		0	1	1	$\sigma_2 + 4$	9	$-2\sigma_3 + 3\sigma_2 + 6$
		1	0	0	0	0	0
		2	0	0	0	0	0
		Σ	1	1	$\sigma_2 + 4$	9	$-2\sigma_3 + 3\sigma_2 + 6$
6_{64}	$M(S^2,(3,2),(3,2),(3,-1))$						
		0	1	1	$2\sigma_2 + 5$	9	$\sigma_3 + 3$
		1	0	0	0	0	0
		2	0	0	0	0	0
		Σ	1	1	$2\sigma_2 + 5$	9	$\sigma_3 + 3$
6_{65}	$T \times I / \begin{pmatrix} 1 & -1 \\ 1 & 0 \end{pmatrix}$						
		0	1	2	$\sigma_2 + 3$	12	0
		1	0	0	0	0	0
		2	1	-2	$\sigma_2 + 3$	0	0
		Σ	2	0	$2\sigma_2 + 6$	12	0

Table A.13. (continued)

c_i	M	ν	3	4	5	6	7
6_{66}	$T \times I / \begin{pmatrix} 0 & 1 \\ -1 & 0 \end{pmatrix}$						
		0	1	4	0	12	$-\sigma_3 + 3\sigma_2 + 6$
		1	-2	0	0	0	$2\sigma_3 - 6\sigma_2 - 12$
		2	1	0	0	0	$-\sigma_3 + 3\sigma_2 + 6$
		Σ	0	4	0	12	0
6_{67}	$T \times I / \begin{pmatrix} 0 & 1 \\ -1 & -1 \end{pmatrix}$						
		0	1	2	$\sigma_2 + 3$	12	0
		1	0	0	0	0	0
		2	1	-2	$\sigma_2 + 3$	0	0
		Σ	2	0	$2\sigma_2 + 6$	12	0
6_{68}	$T \times I / \begin{pmatrix} -1 & 0 \\ -1 & -1 \end{pmatrix}$						
		0	1	4	$-\sigma_2 + 2$	18	$-2\sigma_3 + 6\sigma_2 + 12$
		1	0	0	0	0	0
		2	3	0	$-3\sigma_2 + 6$	18	$-6\sigma_3 + 18\sigma_2 + 36$
		Σ	4	4	$-4\sigma_2 + 8$	36	$-8\sigma_3 + 24\sigma_2 + 48$
6_{69}	$T \times I / \begin{pmatrix} 1 & 0 \\ 1 & 1 \end{pmatrix}$						
		0	1	4	$-\sigma_2 + 2$	18	$-2\sigma_3 + 6\sigma_2 + 12$
		1	0	0	0	0	0
		2	3	0	$-3\sigma_2 + 6$	18	$-6\sigma_3 + 18\sigma_2 + 36$
		Σ	4	4	$-4\sigma_2 + 8$	36	$-8\sigma_3 + 24\sigma_2 + 48$

Table A.13. TV-invariants for manifolds of complexity 6 (continued)

c_i	M	ν	3	4	5	6	7
6_{70}	$T \times I / \begin{pmatrix} -1 & 0 \\ 0 & -1 \end{pmatrix}$						
		0	1	8	$4\sigma_2 + 12$	48	$-9\sigma_3 + 27\sigma_2 + 54$
		1	0	0	0	0	0
		2	7	28	$28\sigma_2 + 84$	252	$-63\sigma_3 + 189\sigma_2 + 378$
		Σ	8	36	$32\sigma_2 + 96$	300	$-72\sigma_3 + 216\sigma_2 + 432$
6_{71}	$T \times I / \begin{pmatrix} 1 & 0 \\ 0 & 1 \end{pmatrix}$						
		0	1	8	$4\sigma_2 + 12$	48	$-9\sigma_3 + 27\sigma_2 + 54$
		1	0	0	0	0	0
		2	7	28	$28\sigma_2 + 84$	252	$-63\sigma_3 + 189\sigma_2 + 378$
		Σ	8	36	$32\sigma_2 + 96$	300	$-72\sigma_3 + 216\sigma_2 + 432$
6_{72}	$K\tilde{\times}I \cup K\tilde{\times}I / \begin{pmatrix} -1 & 0 \\ -1 & 1 \end{pmatrix}$						
		0	1	4	$4\sigma_2 + 8$	28	$-9\sigma_3 + 19\sigma_2 + 30$
		1	0	0	0	0	0
		2	3	12	$12\sigma_2 + 24$	72	$-27\sigma_3 + 57\sigma_2 + 90$
		Σ	4	16	$16\sigma_2 + 32$	100	$-36\sigma_3 + 76\sigma_2 + 120$

Table A.13. (continued)

c_i	M	ν	3	4	5	6	7
6_{73}	$K\tilde{\times}I \cup K\tilde{\times}I/\begin{pmatrix} 0 & 1 \\ 1 & 0 \end{pmatrix}$						
		0	1	8	$3\sigma_2 + 10$	46	$-6\sigma_3 + 22\sigma_2 + 44$
		1	0	0	0	0	0
		2	7	24	$21\sigma_2 + 70$	198	$-42\sigma_3 + 154\sigma_2 + 308$
		Σ	8	32	$24\sigma_2 + 80$	244	$-48\sigma_3 + 176\sigma_2 + 352$
6_{74}	$K\tilde{\times}I \cup K\tilde{\times}I/\begin{pmatrix} 1 & 1 \\ 1 & 0 \end{pmatrix}$						
		0	1	4	$\sigma_2 + 4$	22	$-4\sigma_3 + 10\sigma_2 + 16$
		1	0	0	0	0	0
		2	3	4	$3\sigma_2 + 12$	30	$-12\sigma_3 + 30\sigma_2 + 48$
		Σ	4	8	$4\sigma_2 + 16$	52	$-16\sigma_3 + 40\sigma_2 + 64$

References

1. Alexander, J. W.: The combinatorial theory of complexes. Ann. of Math. **31**, 29–320 (1930)
2. Amendola, G., Martelli, B.: Non-orientable 3-manifolds of small complexity. Topology Appl. **133**, 157–178 (2003)
3. Amendola, G., Martelli B.: Non-orientable 3-manifolds of complexity up to 7. Topology Appl. **150**, 179–195 (2005)
4. Andrews, J. J., Curtis, M. L.: Free groups and handlebodies. Proc. Amer. Math. Soc. **16**, 192–195 (1965)
5. Anisov, S.: Toward lower bounds for complexity of 3-manifolds: a program. arXiv:math.GT/0103169, 1–43 (2001)
6. Anisov, S.: Exact values of complexity for an infinite number of 3-manifolds. Mosc. Math. J. **5**, no. 2, 305–310, 493 (2005)
7. Atiyah, M.: Topological quantum field theories. Publ. Math. IHES, **68**, 175–186 (1989)
8. Bandieri, P., Gagliardi, C., Ricci, L., Classifying genus two 3-manifolds up to 34 tetrahedra. Acta Appl. Math. **86**, no. 3, 267–283 (2005)
9. Bestvina, M., Handel, M.: Train-tracks for surface homeomorphisms. Topology, **34**, no. 1, 109–140 (1995)
10. Biederharn, L. C.: An identity satisfied by Racah coefficients. J. Math. Phys., **31**, 287–293 (1953)
11. Bolsinov, A. V., Matveev, S. V., Fomenko, A. T.: Topological classification of integrable Hamiltonian systems with two degrees of freedom. A list of systems of small complexity. (Russian) Uspekhi Mat. Nauk, **45**, no. 2(272), 49–77, 240 (1990); English translation in Russian Math. Surveys, **45**, no. 2, 59–94 (1990)
12. Bolsinov, A. V., Fomenko, A. T.: Introduction to the topology of integrable Hamiltonian systems. (Russian) "Nauka", Moscow (1997)
13. Burton, B. A., Introducing Regina, the 3-manifold topology software. Experiment. Math. **13**, no. 3, 267–272 (2004)
14. Burton, B. A., Structures of small closed non-orientable 3-manifold triangulations. arXiv:math.GT/0311113 v2, Nov. 2003 and 15 Sep. 2005 (corrected version)
15. Burton, B. A., Efficient enumeration of 3-manifold triangulations. Austral. Math. Soc. Gaz. **31**, no. 2, 108–114 (2004)

16. Burton, B. A., Observation from the 8-tetrahedron non-orientable census. arXiv:math.GT/0509345 v1, 15 Sep. 2005

17. Casali, M. R., Classification of nonorientable 3-manifolds admitting decompositions into ≤ 26 coloured tetrahedra. Acta Appl. Math. **54**, no. 1, 75–97 (1998)

18. Casali, M. R, Representing and recognizing torus bundles over S^1. Bol. Soc. Mat. Mexicana **(3) 10**, Special Issue, 89–105 (2004)

19. Casali, M. R., Computing Matveev's complexity of non-orientable 3-manifolds via crystallization theory. Topology Appl. **144**, no. 1-3, 201–209 (2004)

20. Casler, B. G.: An imbedding theorem for connected 3-manifolds with boundary. Proc. Amer. Math. Soc., **16**, 559–566 (1965)

21. Casson, A. J., Bleiler, S. A.: Automorphisms of surfaces after Nielsen and Thurston. London Mathematical Society Student Texts, **9**. Cambridge University Press, Cambridge (1988)

22. Chinburg, T.: A small arithmetic hyperbolic three-manifold. Proc. Amer. Math. Soc., **100**, no. 1, 140–144 (1987)

23. Chinburg, T., Friedman, E., Jones, K. J., Reid, A. W.: The arithmetic hyperbolic 3-manifold of smallest volume. Ann. Scuola Norm. Sup. Pisa Cl. Sci (4), Vol. **XXX**, 1–40 (2001)

24. Cristofori, P., Casali, M. R., Computing Matveev's complexity via crystallization theory: the orientable case. Acta Appl. Math. **92**, no. 2, 113–123 (2006)

25. Cristofori, P., Casali, M. R., A catalogue of orientable 3-manifolds triangulated by 30 colored tetrahedra. J. Knot Theory Ramifications. (to appear) (2007)

26. Dunwoody, M. J.: The homotopy type of a two-dimensional complex. Bull. London Math. Soc., **8**, no. 3, 282–285 (1976)

27. Elliot, J. P.: Theoretical studies in nuclear structure V: The matrix elements of non-central forces with application to the 2p-shell. Proc. Roy. Soc., **A218**, 345–370 (1953)

28. Ferri, M., Gagliardi, C., Grasselli, L., A graph-theoretical representation of PL-manifolds—a survey on crystallizations. Aequationes Math. 31 (1986), no. 2–3, 121–141.

29. Fomenko, A. T., Kuznecov, V. E., Volodin, I. A.: The problem of discriminating algorithmically the standard three-dimensional sphere. Appendix by S. P. Novikov. Russiam Mathematical Surveys, 1974, **29**, no. 5(179), 71–172.

30. Fomenko, A. T.: The topology of surfaces of constant energy of integrable Hamiltonian systems and obstructions to integrability. (Russian) Izv. Akad. Nauk SSSR Ser. Mat., **50**, no. 6, 1276–1307, 1344 (1986)

31. Fomenko, A. T., Matveev, S. V.: Algorithmic and computer methods for three-manifolds. Mathematics and its Applications, 425. Kluwer Academic Publishers, Dordrecht (1997)

32. Fomenko, A. T., Matveev, S. V.: Isoenergetic surfaces of Hamiltonian systems, the enumeration of three-dimensional manifolds in order of growth of their complexity, and the calculation of the volumes of closed hyperbolic manifolds. (Russian) Uspekhi Mat. Nauk, **43**, no. 1(259), 5–22, 247 (1988); translation in Russian Math. Surveys, **43**, no. 1, 3–24 (1988)

33. Frigerio, R., Martelli, B., Petronio, C.: Complexity and Heeagaard genus of an infinite class of compact 3-manifolds. Pacific J. Math. **210**, 283-297 (2003)

34. Gabai, D.: Foliations and the topology of 3-manifolds. III. J. Differential Geom., **26**, no. 3, 479–536 (1987)

35. Gillman, D., Laszlo, P.: A computer search for contractible 3-manifolds. Topology Appl., **16**, no. 1, 33–41 (1983)
36. Gillman, D., Rolfsen, D.: The Zeeman conjecture for standard spines is equivalent to the Poincaré conjecture. Topology **22**, no. 3, 315–323 (1983)
37. Gordon, C. McA., Luecke, J.: Knots are determined by their complements. J. Amer. Math. Soc., **2**, no. 2, 371–415 (1989)
38. Haken, W.: Theorie der Normalflächen. (German) Acta Math., **105**, 245–375 (1961)
39. Haken, W.: Über das Homöomorphieproblem der 3-Mannigfaltigkeiten. I. (German) Math. Z., **80**, 89–120 (1962)
40. Health, D. J.: On classification of Heegaard splittings, Osaka J. Math., **34**, 497–523 (1997)
41. Hemion, G.: On the classification of homeomorphisms of 2-manifolds and the classification of 3-manifolds. Acta Math., **142**, no. 1–2, 123–155 (1979)
42. Hemion, G.: The classification of knots and 3-dimensional spaces. Oxford Science Publications. The Clarendon Press, Oxford University Press, New York (1992)
43. Hempel, J.: 3-Manifolds. Ann. of Math. Studies, No. **86**. Princeton University Press, Princeton, N. J.; University of Tokyo Press, Tokyo (1976)
44. Hoste, J., Thistlethwaite, M., Weeks, J.: The first 1,701,936 knots. Math. Intelligencer, **20**, no. 4, 33–48 (1998)
45. Hu, Sze-Tsen: Homotopy theory. Pure and Applied Mathematics. A series of Monographs and Textbooks. New York, Academic Press (1959)
46. Hildebrand, M., Weeks, J.: A computer generated census of cusped hyperbolic 3-manifolds. Computers and mathematics (Cambridge, MA, 1989), 53–59, Springer, New York (1989)
47. Hog-Angeloni, C. Metzler, W., Sieradski, A. J.: Two-dimensional homotopy and combinatorial group theory. London Mathematical Society Lecture Note Series, **197**. Cambridge University Press, Cambridge (1993)
48. Homma, T., Ochiai, M., Takahashi, M.: An Algorithm for Recognizing S^3 in 3-Manifolds with Heegaard Splittings of Genus Two, Osaka J. Math., **17**, 625–648 (1980)
49. Ikeda, H.: Contractible 3-manifolds admitting normal spines with maximal elements. Kobe J. Math., **1**, no. 1, 57–66 (1984)
50. Ikeda, H., Inoue, Y.: Invitation to DS-diagrams. Kobe J. Math., **2**, no. 2, 169–186 (1985)
51. Jeffrey, L. C.: Chern-Simons-Witten invariants of lens spaces and torus bundles, and the semiclassical approximation. Comm. Math. Phys., **147**, no. 3, 563–604 (1992)
52. Jaco, W.: Lectures on three-manifold topology. CBMS Regional Conference Series in Mathematics, 43. American Mathematical Society, Providence, R.I., (1980)
53. Jaco, W., Shalen, P. B.: Surface homeomorphisms and periodicity. Topology, **16**, no. 4, 347–367 (1977)
54. Jaco, W., Shalen, P. B.: A new decomposition theorem for irreducible sufficiently large 3-manifolds. Algebraic and geometric topology (Proc. Sympos. Pure Math., Stanford Univ., Stanford, Calif., 1976), Part 2, pp. 71–84, Proc. Sympos. Pure Math., **XXXII**, Amer. Math. Soc., Providence, R.I. (1978)
55. Jaco, W. H., Shalen, P. B.: Seifert fibered spaces in 3-manifolds. Mem. Amer. Math. Soc., **21**, no. 220 (1979)

484   References

56. Jaco, W., Oertel, U.: An algorithm to decide if a 3-manifold is a Haken manifold. Topology, **23**, no. 2, 195–209 (1984)

57. Johannson, K.: Homotopy equivalences of 3-manifolds with boundaries. Lecture Notes in Mathematics, **761**. Springer, Berlin (1979)

58. Johannson, K.: Topologie und Geometrie von 3-Mannigfaltigkeiten. (German) [Topology and geometry of 3-manifolds] Jahresber. Deutsch. Math.-Verein., **86**, no. 2, 37–68 (1984)

59. Johannson, K.: Classification problems in low-dimensional topology. Geometric and algebraic topology, 37–59, Banach Center Publ., **18**, PWN, Warsaw (1986)

60. Kauffman, L. H., Lins, S.: Computing Turaev–Viro invariants for 3-manifolds. Manuscripta Math., **72**, 81–94 (1991)

61. Kauffman, L. H., Lins, S. L.: Temperley-Lieb recoupling theory and invariants of 3-manifolds. Annals of Mathematics Studies, **134**. Princeton University Press, Princeton, NJ (1994)

62. Kapovich, M.: Hyperbolic manifolds and discrete groups. Lectures on Thurston's Hyperbolization. (2002)

63. Kirby, R., Melvin, P.: The 3-manifold invariants of Witten and Reshetikhin-Turaev for sl$(2, C)$. Invent. Math., **105**, no. 3, 473–545 (1991)

64. Kirillov, A. N., Reshetikhin, N. Yu.: Representations of the algebra $U_q(\mathrm{sl}(2))$, q-orthogonal polynomials and invariants of links. Infinite-dimensional Lie algebras and groups (Luminy-Marseille, 1988), 285–339, Adv. Ser. Math. Phys., **7**, World Sci. Publishing, Teaneck, NJ (1989)

65. Klette, R.: Cell complexes through time. Proc. Vision Geometry IX, San Diego, **SPIE-4117**, 134–145 (2000)

66. Kneser, H.: Geschlossene Flächen in dreidimensionalen Mannigfaltigkeiten. Jahresber. Deut. Math. Ver., **38**, 248–260 (1929)

67. Kovalevsky, V. A.: Finite topology as applied to image analysis. Computer Vision, Graphics, and Image Processing, **46**, 141–161 (1989)

68. Lickorish, W. B. R.: A finite set of generators for the homeotopy group of a 2-manifold. Proc. Cambridge Philos. Soc., **60**, 769–778 (1964)

69. Lickorish, W. B. R.: Distinct 3-manifolds with all SU$(2)_q$ invariants the same. Proc. Amer. Math. Soc. **117**, no. 1, 285–292 (1993)

70. Lins, S.: Gems, Computers and Attractors for 3-Manifolds, Knots and Everything 5, World Scientific, (1995)

71. Lustig, M.: Nielsen equivalence and simple-homotopy type. Proc. London Math. Soc. (3), **62**, no. 3, 537–562 (1991)

72. Mandelbaum, R.: Four-dimensional topology: an introduction. Bull. Amer. Math. Soc. (N.S.), **2**, no. 1, 1–159 (1980)

73. Manning, J.: Algorithmic detection and description of hyperbolic structures on closed 3-manifolds with solvable word problem. Geometry and Topology, **6**, 1–26 (2002)

74. Martelli, B., Petronio, C.: Three-manifolds having complexity at most 9. Experimental Math., **10**, 207–236 (2001)

75. Masbaum, G., Roberts, J. D.: A simple proof of integrality of quantum invariants at prime roots of unity. Math. Proc. Cambridge Philos. Soc., **121**, no. 3, 443–454 (1997)

76. Matveev, S. V.: Universal 3-deformations of special polyhedra. (Russian) Uspekhi Mat. Nauk **42**, no. 3(255), 193–194 (1987)

77. Matveev, S. V.: The Zeeman conjecture for nonthickenable special polyhedra is equivalent to the Andrews-Curtis conjecture. (Russian) Sibirsk. Mat. Zh., **28**, no. 6, 66–80 (1987)

78. Matveev, S. V.: Transformations of special spines, and the Zeeman conjecture. (Russian) Izv. Akad. Nauk SSSR Ser. Mat., **51**, no. 5, 1104–1116 (1987); translation in Math. USSR-Izv. 31, no. 2, 423–434 (1988)

79. Matveev, S. V.: The theory of the complexity of three-dimensional manifolds. (Russian) Akad. Nauk Ukrain. SSR Inst. Mat. Preprint, no. **13**, 32 pp. (1988)

80. Matveev, S. V.: Complexity theory of three-dimensional manifolds. Acta Appl. Math., **19**, no. 2, 101–130 (1990)

81. Matveev, S. V. : Algorithms for the recognition of the three-dimensional sphere (after A. Thompson). (Russian) Mat. Sb., **186**, no. 5, 69–84 (1995); English translation in Sb. Math., **186**, no. 5, 695–710 (1995)

82. Matveev, S. V.: Classification of sufficiently large 3-manifolds. (Russian) Uspekhi Mat. Nauk, **52**, no. 5(317), 147–174 (1997); translation in Russian Math. Surveys **52**, no. 5, 1029–1055 (1997)

83. Matveev, S. V.: Tables of 3-manifolds up to complexity 6. Max-Planck Institute Preprint MPI 1998-67, 1–50 (1998)

84. Matveev, S. V.: Tables of spines and 3-manifolds up to complexity 7. Max-Planck Institute Preprint MPI 2002-71, 1–65 (2002)

85. Matveev, S. V.: Generalized graph manifolds and their efficient recognition. (Russian) Mat. Sb. **189**, no. 10, 89–104 (1998); translation in Sb. Math., **189**, no. 9–10, 1517–1531 (1998)

86. S. Matveev, Recognition and Tabulation of 3-manifolds, Dokl. RAS, **400**, No. 1, 26–28 (2005) (Russian; English transl. in Doklady Mathematics, **71**, 20–22 (2005))

87. S. Matveev, Tabulation of 3-manifolds, Uspekhi Mat. Nauk, **60:4**, 97-122 (2005)(Russian; English translation in Russian Math. Surveys **60**, no. 4, 673–698 (2005))

88. Matveev, S. V., Nowik, T.: On 3-manifolds having the same Turaev–Viro invariants. Russian J. Math. Phys., **2**, no. 3, 317–324 (1994)

89. Matveev, S. V., Ovchinnikov, M. A., Sokolov, M. V.: On a simple invariant of Turaev–Viro type. Zap. Nauchn. Sem. S.-Peterburg. Otdel. Mat. Inst. Steklov. (POMI), **234**, Differ. Geom. Gruppy Li i Mekh. 15-1, 137–142, 263 (1996); translation in J. Math. Sci. (New York), **94**, no. 2, 1226–1229 (1999)

90. Matveev, S. V., Pervova, E. L.: Lower bounds for the complexity of three-dimensional manifolds. (Russian) Dokl. Akad. Nauk, **378**, no. 2, 151–152 (2001)

91. Matveev, S. V., Savvateev, V. V.: Three-dimensional manifolds having simple special spines. (Russian) Colloq. Math., **32**, 83–97 (1974)

92. Metzler, W.: Über den Homotopietyp zweidimensionaler CW-Komplexe und Elementartransformationen bei Darstellungen von Gruppen durch Erzeugende und definierende Relationen. J. Reine Angew. Math., **285**, 7–23 (1976)

93. Metzler, W.: Die Unterscheidung von Homotopietyp und einfachem Homotopietyp bei zweidimensionalen Komplexen. (German) [Distinguishing between homotopy type and simple homotopy type in two-dimensional complexes] J. Reine Angew. Math., **403**, 201–219 (1990)

94. Milnor, J.: Groups which act on S^n without fixed points. Amer. J. Math., **79**, 623–630 (1957)

95. Milnor, J.: Whitehead torsion. Bull. Amer. Math. Soc. **72**, 358–426 (1966)

96. Neumann, W. D.: Seifert manifolds, plumbing, μ-invariant and orientation reversing maps. Lecture Notes in Mathematics, **664**. Springer, Berlin (1978)

97. Neumann, W. D., Swarup, G. A.: Canonical decompositions of 3-manifolds. Geom. Topol., **1**, 21–40 (1997)

98. Nielsen, J.: Abbildungsklassen endlicher Ordnung. Acta math., **75**, 23–115 (1943)

99. Orlik, P.: Seifert manifolds. Lecture Notes in Mathematics, **291**. Springer-Verlag, Berlin-New York (1972)

100. Otal, J.-P.: Thurston's hyperbolization of Haken manifolds. Surveys in differential geometry, Vol. **III** (Cambridge, MA, 1996), 77–194, Int. Press, Boston, MA (1998)

101. Otal, J.-P.: The hyperbolization theorem for fibered 3-manifolds. Translated from the 1996 French original by Leslie D. Kay. SMF/AMS Texts and Monographs, **7**. American Mathematical Society, Providence, RI; Société Mathématique de France, Paris (2001)

102. Ovchinnikov, M. A.: The table of 3-manifolds of complexity 7. Preprint, Chelyabinsk State University (1997)

103. Ovchinnikov, M. A: Construction of special spines for 3-manifolds of Waldhausen. PhD Thesis, Chelyabinsk State University, Chelyabinsk (2000)

104. Pachner, U.: Bistellare Equivalenz kombinatorischer Mannigfaltigkeiten. (German) Arch. Math. (Basel), **30**, no. 1, 89–98 (1978)

105. Pachner, U.: P. L. homeomorphic manifolds are equivalent by elementary shellings. European J. Combin., **12**, no. 2, 129–145 (1991)

106. Papakyriakopoulos, C. D.: On Dehn's lemma and the asphericity of knots. Ann. of Math. (2), **66**, 1–26 (1957)

107. Piergallini, R.: Standard moves for standard polyhedra and spines. Third National Conference on Topology (Italian) (Trieste, 1986). Rend. Circ. Mat. Palermo (2) Suppl. No. **18**, 391–414 (1988)

108. Roberts, J.: Skein theory and Turaev–Viro invariants. Topology, **34**, no. 4, 771–787 (1995)

109. Rolfsen, D.: Knots and links. Mathematics Lecture Series, No. **7**. Publish or Perish, Inc., Berkeley, Calif. (1976)

110. Rourke, C. P., Sanderson, B. J.: Introduction to piecewise-linear topology. Ergebnisse der Mathematik und ihrer Grenzgebiete, Band **69**. Springer-Verlag, New York-Heidelberg (1972)

111. Scott, P.: The geometries of 3-manifolds. Bull. London Math. Soc., **15**, no. 5, 401–487 (1983)

112. Seifert, H.: Topologie dreidimensionaler gefaserter Räume. Acts math., **60**, 147–238 (1933)

113. Sokolov, M. V.: The Turaev–Viro invariant for 3-manifolds is a sum of three invariants. Canadian Math. Bull., **39**, (4), 468–475 (1996)

114. Sela, Z.: Topics in 3-manifolds. Ph. D. thesis, Hebrew University (1991)

115. Sela, Z.: The isomorphism problem for hyperbolic groups I. Ann. of Math. **141**, 217–283 (1995)

116. Sokolov, M. V.: Turaev–Viro invariants of manifolds up to complexity 6. Preprint, Chelyabinsk State University (1995)

117. Sokolov, M. V.: Which lens spaces are distinguished by Turaev–Viro invariants? (Russian) Mat. Zametki, **61**, no. 3, 468–470 (1997); translation in Math. Notes, **61**, no. 3–4, 384–387 (1997)

118. Stallings, J.: On fibering certain 3-manifolds. Topology of 3-manifolds and related topics (Proc. The Univ. of Georgia Institute, 1961) pp. 95–100, Prentice-Hall, Englewood Cliffs, N.J. (1962)

119. Thompson, A.: Thin position and the recognition problem for S^3. Math. Res. Lett., **1**, no. 5, 613–630 (1994)

120. Thurston, W.: The geometry and topology of 3-manifolds, mimeographed notes, Math. Dept., Princeton Univ., Princeton, NJ (1979)

121. Thurston, W. P.: Three-dimensional manifolds, Kleinian groups and hyperbolic geometry. Bull. Amer. Math. Soc. (N.S.), **6**, no. 3, 357–381 (1982)

122. Thurston, W. P.: Hyperbolic structures on 3-manifolds. I. Deformation of acylindrical manifolds. Ann. of Math. (2), **124**, no. 2, 203–246 (1986)

123. Thurston, W. P.: Hyperbolic structures on 3-manifolds. II. Surface groups and 3-manifolds which fibre over the circle, Preprint, Princeton University. (1986)

124. Thurston, W. P.: On the geometry and dynamics of diffeomorphisms of surfaces. Bull. Amer. Math. Soc. (N.S.), **19**, no. 2, 417–431 (1988)

125. Turaev, V. G.: Quantum invariants of knots and 3-manifolds. de Gruyter Studies in Mathematics, **18**. Walter de Gruyter & Co., Berlin (1994)

126. Turaev, V. G., Viro, O. Ya.: State sum invariants of 3-manifolds and quantum $6j$-symbols. Topology, **31**, no. 4, 865–902 (1992)

127. Viro, O. Ja.: Links, two-sheeted branching coverings and braids. (Russian) Mat. Sb. (N.S.), **87(129)**, 216–228 (1972)

128. Volodin, I. A., Kuznecov, V. E., Fomenko, A. T.: The problem of the algorithmic discrimination of the standard three-dimensional sphere. (Russian) Appendix by S. P. Novikov. Uspehi Mat. Nauk, **29** , no. 5(179), 71–168 (1974)

129. Waldhausen, F.: Eine Klasse von 3-dimensionalen Mannigfaltigkeiten. I, II. (German) Invent. Math., **3**, 308–333; ibid. **4**, (1967), 87–117 (1967)

130. Waldhausen, F.: On irreducible 3-manifolds which are sufficiently large. Ann. of Math. (2), **87**, 56–88 (1968)

131. Waldhausen, F.: Recent results on sufficiently large 3-manifolds. Algebraic and geometric topology (Proc. Sympos. Pure Math., Stanford Univ., Stanford, Calif., 1976), Part 2, pp. 21–38, Proc. Sympos. Pure Math., **XXXII**, Amer. Math. Soc., Providence, R.I. (1978)

132. Weeks, J.: SnapPea (hyperbolic 3-manifold software), available from http://www.geomtrygames.org/SnapPea

133. Weeks, J.: Hyperbolic Structures on 3-manifolds, Ph. D. Thesis, Princeton University (1985)

134. Whitehead, J. H. C.: Simple homotopy types. Amer. J. Math., **72**, 1–57 (1950)

135. Wall, C. T. C.: Formal deformations. Proc. London Math. Soc. (3), **16**, 342–352 (1966)

136. Wolf, J. A.: Spaces of constant curvature. McGraw-Hill Book Co., New York-London-Sydney (1967)

137. Yamada, Shuji: The absolute value of the Chern-Simons-Witten invariants of lens spaces. J. Knot Theory Ramifications, **4**, no. 2, 319–327 (1995)

138. Zeeman, E. C.: On the dunce hat. Topology, **2**, 341–358 (1964)

Index